This book is due

Learning Radiology

Learning Radiology

RECOGNIZING THE BASICS 3rd EDITION

William Herring, MD, FACR

Vice Chairman and Residency Program Director
Albert Einstein Medical Center
Philadelphia, Pennsylvania

ELSEVIER

ELSEVIER

1600 John F. Kennedy Blvd.
Ste 1800
Philadelphia, PA 19103-2899

LEARNING RADIOLOGY: RECOGNIZING THE BASICS, 3rd EDITION ISBN: 978-0-323-32807-4
Copyright © 2016, 2012, 2007 by Saunders, an imprint of Elsevier Inc.

Notices

Library of Congress Cataloging-in-Publication Data
Herring, William, author.
  Learning radiology : recognizing the basics / William Herring.—3rd edition.
    p. ; cm.
Includes bibliographical references and index.
ISBN 978-0-323-32807-4 (paperback : alk. paper)
I. Title.
[DNLM: 1. Radiography—methods. 2. Diagnosis, Differential. WN 200]
R899
616.07′572—dc23
                                        2015006990

Senior Content Strategist: James Merritt
Content Development Specialist: Katy Meert
Publishing Services Manager: Anne Altepeter
Senior Project Manager: Doug Turner
Designer: Xiaopei Chen

Printed in the United States of America

Last digit is the print number: 9 8 7 6 5 4 3 2 1

Working together
to grow libraries in
developing countries

www.elsevier.com • www.bookaid.org

To my wife, Patricia,
And our family

Contributor

Daniel J. Kowal, MD
Computed Tomography Division Director
Radiology Elective Director
Department of Radiology
Saint Vincent Hospital
Worcester, Massachusetts
Chapter 22: Magnetic Resonance Imaging: Understanding the Principles and Recognizing the Basics

Preface

I've checked, and most prefaces to a third edition or later start out with something like, "It's hard to believe that this is the third edition of…" Not this text. I know how much work it's taken, so I definitely **can** believe it. But thank you if you have contributed in any way, including reading this preface, to the success of this book.

In the first edition, I asked you to suppose for a moment that your natural curiosity drove you to wonder what kind of bird with a red beak just landed on your window sill. You could get a book on birds that listed all of them alphabetically from *albatross* to *woodpecker* and spend time looking through hundreds of bird pictures. Or you could get a book that lists birds by the colors of their beaks and thumb through a much shorter list to find that your feathered visitor is a cardinal.

This book is a red beak book. Where possible, groups of diseases are first described by the way they **look** rather than by what they're **called**. Imaging diagnoses frequently, but not always, rest on a recognition of a reproducible visual picture of that abnormality. That is called the ***pattern recognition approach*** to identifying abnormalities, and the more experience you have looking at imaging studies, the more comfortable and confident you'll be with that approach.

Before diagnostic images can help you decide what disease the patient may have, you must first be able to differentiate between what is normal in appearance and what is not. That isn't as easy as it may sound. Recognizing the difference between normal and abnormal probably takes as much practice, if not more, than deciding what disease a person has.

Radiologists spend their entire lives performing just such differentiations. You won't be a radiologist after you've completed this book, but you should be able to recognize abnormalities and interpret images better and, by so doing, perhaps participate in the care of patients with more assurance and confidence.

When pattern recognition doesn't work, this text will try wherever possible to give you a logical **approach** to reaching a diagnosis. By learning an approach, you'll have a method you can apply to similar problems again and again. An analytic approach will enable you to apply a rational solution to diagnostic imaging problems.

This text was written to make complimentary use of the platform on which radiologic images are now almost universally viewed: the digital display. Although digital displays may be ideal for looking at images, some people do not want to read large volumes of text from their digital devices. So we've joined the text in the printed book with photos, videos, quizzes, and tutorials—many of them interactive—and made them available online at StudentConsult/Inkling.com in a series of **web enhancements** that accompany the book. I think you'll really enjoy them.

This text is not intended to be encyclopedic. Many wonderful radiology reference texts are available, some of which contain thousands of pages and weigh slightly less than a Mini Cooper. This text is oriented more toward students, interns, residents, residents-to-be, and other health care professionals who are just starting out.

This book emphasizes conventional radiography because that is the type of study most patients undergo first and because the same imaging principles that apply to reaching the diagnosis on conventional radiographs can frequently be applied to making the diagnosis on more complex modalities.

Let's get started. Or, if you're the kind of person (like I am) who reads the preface **after** you've read the book, I hope you enjoyed it.

William Herring, MD, FACR

Acknowledgments

I am again grateful to the many thousands of you whom I have never met but who found a website called Learning Radiology helpful, making it so popular that it played a role launching the first edition of this book, which itself was so popular that it led to this third edition.

For their help and suggestions, I thank David Saul, MD, one of our radiology residents, who made invaluable suggestions about how this edition could be changed. Daniel Kowal, MD, a radiologist who graduated from our program, did an absolutely wonderful job in simplifying the complexities of MRI again in the chapter he wrote. Jeffrey Cruz, MD, one of our residents, helped out with the online Radiation Safety and Dose module, and Sherif Saad, MD, contributed an illustration.

I thank Chris Kim, MD; Susan Summerton, MD; Mindy Horrow, MD; Peter Wang, MD; and Huyen Tran, MD, for supplying additional images for this edition. And thanks to Mindy Horrow, MD; Eric Faerber, MD; and Brooke Devenney-Cakir, MD, for reviewing chapters from this text.

I certainly want to recognize and again thank Jim Merritt and Katy Meert from Elsevier for their support and assistance.

I also acknowledge the hundreds of radiology residents and medical students who, over the years, have provided me with an audience of motivated learners, without whom a teacher would have no one to teach.

Finally, I want to thank my wonderful wife, Pat, who has encouraged me throughout the project, and my family.

William Herring, MD, FACR

Contents

Video Contents

CHAPTER 1
Recognizing Anything

AN INTRODUCTION TO IMAGING MODALITIES

It's always exciting when a class starts out with a surprise quiz. No pencils are necessary. Here are six images with brief histories presented as unknowns. Each is diagnostic. If you don't know the answers, that is perfectly fine because that's what you are here to learn. The answers are at the very end of this book (Figs. 1-1 to 1-6).

- You are about to learn about each of the imaging modalities, about how to approach imaging studies, about the six diseases represented in the figures, and much more as you complete this text.

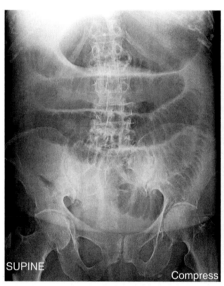

FIGURE 1-1 A 56-year-old patient with abdominal pain.

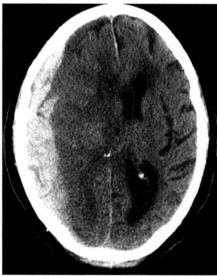

FIGURE 1-2 A 49-year-old who fell off a ladder.

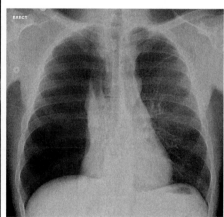

FIGURE 1-3 A 22-year-old with sudden chest pain.

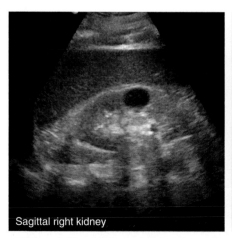

FIGURE 1-4 Incidental finding on abdominal ultrasound.

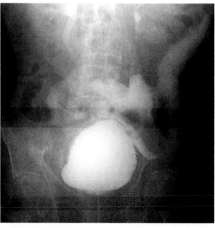

FIGURE 1-5 Cystogram of a 56-year-old who was in an automobile accident.

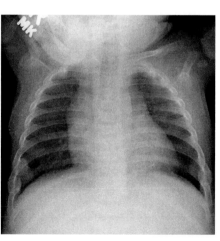

FIGURE 1-6 A 4-month-old with irritability.

FROM DARKNESS ... LIGHT

- In 1895, Wilhelm Röntgen (or Roentgen), working in a darkened laboratory in Würzburg, Germany, noticed that a screen painted with a fluorescent material in the same room, but a few feet from a cathode ray tube he had energized and made lightproof, started to glow (fluoresce). Sensing something important had happened, he recognized that the screen was responding to the nearby production of an unknown ray transmitted invisibly through the room. He named the new rays "x-rays," using the mathematical symbol "x" for something unknown. It didn't take long before almost everyone was taking x-rays of almost everything imaginable.
- For about 100 years after that, radiographic images survived their brief birth as a burst of ionizing radiation nestled comfortably on a piece of film. In some places, film is the medium still used, but it's much less common.
- Today, like in 1895, conventional radiographic images (usually shortened to **x-rays**) are produced by a combination of ionizing radiation and light striking a **photosensitive surface,** which, in turn, produces a **latent image** that is subsequently **processed.** At first, the processing of film was carried out in a darkroom containing trays with various chemicals; the films were then, literally, hung out and then up to dry.
    - ♦ When an immediate reading was requested, the films were interpreted while still dripping with chemicals, and thus the term *wet reading* for a "stat" interpretation was born.
    - ♦ Films were then viewed on lighted view boxes (almost always backward or upside-down if the film placement was being done as part of a movie or television show).
- This workflow continued for many decades, but it had **two major drawbacks:**
    - ♦ It required a great deal of **physical storage space** for the ever-growing number of films. Even though each film is very thin, many films in thousands of patients' folders take up a great deal of space (eFig. 1-1).
    - ♦ The other drawback was that the radiographic films could physically be in only **one place at a time,** which was not necessarily where they might be needed to help in the care of a patient.
- So, eventually, **digital radiography** came into being, in which the photographic film was replaced by a **photosensitive cassette** or **plate** that could be processed by an **electronic reader** and the resulting image could be stored in a **digital format.** This electronic processing no longer required a darkroom to develop the film or a large room to store the films. Countless images could be stored in the space of one spinning hard disk on a computer server. Even more important, the images could be viewed by anyone with the right to do so, anywhere in the world, at any time.
- The images were maintained on computer servers, where they could be **stored** and **archived** for posterity and from which they could be **communicated** to others. This system is referred to as *PACS,* which stands for **picture archiving, communications, and storage.**
- Using PACS systems, images created using all modalities can be stored and retrieved. Conventional radiography, computed tomography (CT), ultrasonography, magnetic resonance imaging (MRI), fluoroscopy, and nuclear medicine are examples of images that can be stored in this way.
- We will look briefly at each of these modalities in the sections that follow.

CONVENTIONAL RADIOGRAPHY

- Images produced through the use of ionizing radiation (i.e., the production of x-rays, but without added contrast material such as barium or iodine) are called *conventional radiographs* or, more often, **plain films.**
- The major advantage of conventional radiographs is that the images are relatively **inexpensive** to produce, can be obtained almost **anywhere** by using portable or mobile machines, and are still the most widely obtained imaging studies.
- They require a **source** to produce the x-rays (the "x-ray machine"), a method to **record** the image (a film, cassette, or photosensitive plate), and a way to **process** the recorded image (using either chemicals or a digital reader).
- Common uses for conventional radiography include the ubiquitous chest x-ray, plain films of the abdomen, and virtually every initial image of the skeletal system to evaluate for fractures or arthritis.
- The major disadvantages of conventional radiography are the **limited range of densities** it can demonstrate and that it **uses ionizing radiation.**

THE FIVE BASIC DENSITIES

- **Conventional radiography is limited to demonstrating five basic densities,** arranged here from **least** to **most** dense (Table 1-1):
    - ♦ **Air,** which appears the blackest on a radiograph
    - ♦ **Fat,** which is shown in a lighter shade of gray than air
    - ♦ **Soft tissue or fluid** (because both soft tissue and fluid appear the same on conventional radiographs, it's impossible to differentiate the heart muscle from the blood inside of the heart on a chest radiograph)
    - ♦ **Calcium** (usually contained within bones)
    - ♦ **Metal,** which appears the whitest on a radiograph
        - • Objects of metal density are not normally present in the body. Radiologic **contrast media** and **prosthetic knees** or **hips** are **examples of metal densities** artificially placed in the body (Fig. 1-7).
- Although conventional radiographs are produced by ionizing radiation in relatively low doses, radiation has the potential

TABLE 1-1 FIVE BASIC DENSITIES SEEN ON CONVENTIONAL RADIOGRAPHY

Density	Appearance
Air	Absorbs the least x-ray and appears "blackest" on conventional radiographs
Fat	Gray, somewhat darker (blacker) than soft tissue
Fluid or soft tissue	Both fluid (e.g., blood) and soft tissue (e.g., muscle) have the same densities on conventional radiographs
Calcium	The most dense, naturally occurring material (e.g., bones); absorbs most x-rays
Metal	Usually absorbs all x-rays and appears the "whitest" (e.g., bullets, barium)

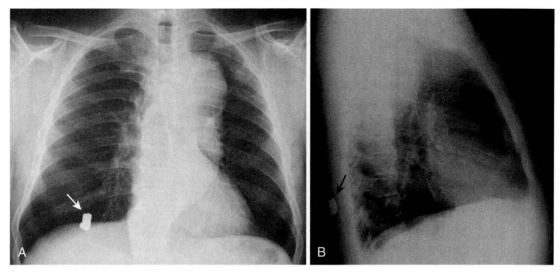

FIGURE 1-7 **Bullet in the chest. A,** The dense (white) metallic foreign body overlying the right lower lung field (*white arrow*) is a bullet. It is much denser (whiter) than the bones (calcium density), represented by the ribs, clavicles, and spine. Fluid (such as the blood in the heart) and soft-tissue density (such as the muscle of the heart) have the same density, which is why we cannot differentiate the two using conventional radiography. The air in the lungs is the least dense (blackest). **B,** Two views at 90° angles to each other, such as these frontal and lateral chest radiographs, are called *orthogonal* views. With only one view, it would be impossible to know the location of the bullet. On the lateral view, the bullet can be seen lying in the soft tissues of the back (*black arrow*). Orthogonal views are used throughout conventional radiography to localize structures in all parts of the body.

to produce cell mutations, which could lead to many forms of cancer or anomalies. Public health data on lower levels of radiation vary with regard to assessment of risk, but it is generally held that only medically necessary diagnostic examinations should be performed and that imaging using x-rays should be avoided during potentially teratogenic times, such as pregnancy. (More information about radiation dose and safety is available at StudentConsult.com.)

COMPUTED TOMOGRAPHY

- CT (or "CAT") scanners, first introduced in the 1970s, brought a quantum leap to medical imaging.
- Using a gantry with a rotating x-ray beam and multiple detectors in various arrays (which themselves rotate continuously around the patient), along with sophisticated computer algorithms to process the data, a large number of two-dimensional, slicelike images (each of which is millimeters in size) can be formatted in multiple imaging planes (Video 1-1).
- **A CT scanner** is connected to a **computer** that processes the data though various **algorithms** to produce **images** of diagnostic quality.
- A CT image is composed of a matrix of thousands of tiny squares called *pixels,* each of which is computer-assigned a **CT number** from −1000 to +1000 measured in **Hounsfield units (HUs),** named after Sir Godfrey Hounsfield, the man credited with developing the first CT scanner (for which he won the Nobel Prize in Medicine in 1979 with Allan Cormack).
    - ◆ The CT number will **vary according to the density of the tissue** scanned and is a **measure of how much of the x-ray beam is absorbed** by the tissues at each point in the scan. By convention, **air is assigned a Hounsfield number of −1000 HU** and **bone about**

400 HU to 600 HU. Fat is −40 to −100 HU, water is 0, and **soft tissue is 20 HU to 100 HU.**

- CT images are displayed or viewed using a range of Hounsfield numbers preselected to best demonstrate the tissues being studied (e.g., from −100 to +300), and anything within that range of CT numbers is displayed over the levels of density in the available gray scale. This range is called the *window.*

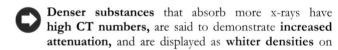

Denser substances that absorb more x-rays have **high CT numbers,** are said to demonstrate **increased attenuation,** and are displayed as **whiter densities** on CT scans.

- ◆ On conventional radiographs, these substances (e.g., **metal** and **calcium**) would also appear whiter and would be said to have **increased density** or to be **more opaque.**
- **Less dense substances** that absorb fewer x-rays have **low CT numbers,** are said to demonstrate **decreased attenuation,** and are displayed as **blacker densities** on CT scans.
    - ◆ On conventional radiographs, these substances (e.g., **air** and **fat**) would also appear as blacker densities and would be said to have **decreased density** (or **increased lucency**).
- CT scans can also be windowed in a way that optimizes the visibility of different types of pathology **after** they are obtained, a benefit called *postprocessing,* which digital imaging, in general, markedly advanced. Postprocessing allows for additional manipulation of the raw data to best demonstrate the abnormality **without repeating a study** and without reexposing the patient to radiation (Fig. 1-8).
- Traditionally, CT images were viewed mostly in the axial plane. Now, because of volumetric acquisition of data, CT

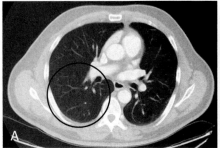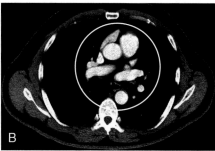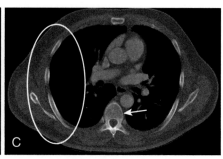

FIGURE 1-8 Windowing the thorax. Chest computed tomography scans are usually "windowed" and displayed in several formats to optimize anatomical definition. **A,** Lung windows are chosen to maximize the ability to image abnormalities of the lung parenchyma and to identify normal and abnormal bronchial anatomy *(black circle).* **B,** Mediastinal windows are chosen to display the mediastinal, hilar, and pleural structures to best advantage *(white circle).* **C,** Bone windows are utilized as a third way of displaying the data to visualize the bony structures to their best advantage *(white oval and arrow).* It is important to recognize that the displays of these different windows are manipulations of the data obtained during the original scan and do not require rescanning the patient.

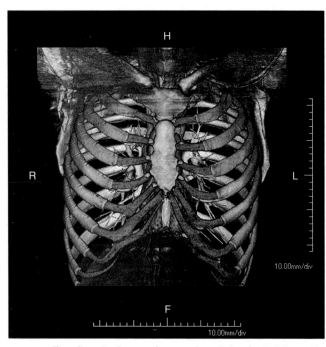

FIGURE 1-9 Three-dimensional computed tomography rendering of normal rib cage. This grayscale version (color online) of a three-dimensional surface rendering of the rib cage is made possible by the acquisition of multiple, thin computed tomographic sections through the body. These sections can then be reconstructed to demonstrate surface anatomy, as in this illustration. The same data set could have been manipulated to show the heart or lungs (which are digitally removed here) and not the rib cage. Such renderings are especially helpful in demonstrating the exact anatomic relationships of structures, especially for surgical planning. *F,* Foot; *L,* left; *H,* head; *R,* right.

scans can be shown in any plane: axial, sagittal, or coronal. Volumetric data consist of a series of thin sections that can be **reassembled** for a three-dimensional reconstruction. Surface and volume rendering in three dimensions can produce CT images of amazing, realistic quality (Fig. 1-9).

- One of the major benefits of CT scanning over conventional radiography is its ability to **expand the gray scale,** which enables differentiation of many more than the five basic densities available on conventional radiographs.

- Because of increasingly sophisticated arrays of detectors and acquisition of hundreds of slices simultaneously, **multislice CT scanners permit very fast imaging** (from head to toe in less than 10 seconds), which has allowed the development of new applications for CT, such as **virtual colonoscopy** and **virtual bronchoscopy, cardiac calcium scoring,** and **CT coronary angiography** (Video 1-2).

- CT examinations can contain 1000 or more images; therefore the older convention of filming each image for study on a viewbox is impractical, and such scans are almost always viewed on computer workstations.

- **CT scans are the cornerstone of cross-sectional imaging** and are widely available, although not as yet truly portable. Production of CT images requires an expensive scanner, a space dedicated to its installation, and sophisticated computer processing power. Like conventional x-ray machines, CT scanners utilize ionizing radiation (x-rays) to produce their images.

ULTRASONOGRAPHY

- Ultrasound probes utilize acoustic energy above the audible frequency of humans to produce images, instead of using x-rays as both conventional radiography and CT scans do (see Chapter 21).

- An ultrasound **probe** or **transducer** both produces the ultrasonic signal and records it. The signal is processed for its characteristics by an onboard computer. Ultrasound images are recorded digitally and are easily stored in a PACS system. Images are displayed either as static images or in the form of a movie (or "cine") (Video 1-3).

- Ultrasound scanners are relatively **inexpensive** compared with CT and MRI scanners. They are **widely available** and can be made **portable** to the point of being handheld.

- Because ultrasonography utilizes no ionizing radiation, it is particularly useful in obtaining images of children and women of **childbearing age** and **during pregnancy.**

- **Ultrasonography is widely used in medical imaging.** It is usually **the study of first choice** in imaging the female pelvis and in pediatric patients, in differentiating cystic versus solid lesions in patients of all ages, in noninvasive vascular imaging, in imaging of the fetus and placenta during pregnancy, and in real-time, image-guided fluid aspiration and biopsy.

■ Other common uses are evaluation of cystic versus solid breast masses, thyroid nodules, and tendons and in assessment of the brain, hips, and spine in newborns. Ultrasonography is used in settings ranging from intraoperative scanning in the surgical suite to the medical tent in the battlefield and in locations as remote as Antarctica.

■ Ultrasonography is generally considered to be **a very safe imaging modality** that has no known major side effects when used at medically diagnostic levels.

MAGNETIC RESONANCE IMAGING

■ MRI utilizes the potential energy stored in the body's **hydrogen atoms.** The atoms are manipulated by very strong magnetic fields and radiofrequency pulses to produce enough localizing and tissue-specific energy to allow highly sophisticated computer programs to generate two- and three-dimensional images (see Chapter 22).

■ MRI scanners are **not as widely available** as CT scanners. They are **expensive** to acquire and require careful site construction to operate properly. In general, they also have a relatively **high ongoing operating cost.**

■ However, they utilize **no ionizing radiation** and produce much higher contrast between different types of soft tissues than is possible with CT.

■ MRI is widely used in neurologic imaging and is particularly sensitive in imaging soft tissues such as the muscles, tendons, and ligaments.

■ There are **safety issues** associated with the extremely strong magnetic fields of an MRI scanner, both for objects within the body (e.g., cardiac pacemakers) and for ferromagnetic projectiles in the MRI scanner environment (e.g., metal oxygen tanks in the room). There are also known side effects from the radiofrequency waves that such scanners produce and possible adverse effects due to some MRI contrast agents.

FLUOROSCOPY

■ Fluoroscopy (or "fluoro") is a modality in which ionizing radiation (x-rays) is used in performing **real-time visualization** of the body in a way that allows for evaluation of the motion of body parts, real-time positioning changes of bones and joints, and the location and path of externally administered barium or iodine contrast agents through the gastrointestinal and genitourinary tracts and blood vessels. Images can be viewed as they are acquired on video screens and captured as either a series of static images or moving (video) images (Video 1-4).

■ Fluoroscopy requires an x-ray unit specially fitted to allow for controlled motion of the x-ray tube, as well as the imaging sensor and the patient, to find the best projection to demonstrate the body part being studied. To do this, fluoroscopic tables are made to tilt and the fluoroscopic tube can be moved freely back and forth above the patient (Fig. 1-10).

■ Instantaneous "snapshots" obtained during the procedure are called *spot films.* They are combined with other images obtained by using an overhead x-ray machine in multiple

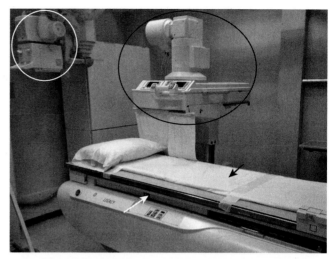

FIGURE 1-10 **A standard radiology room equipped for both conventional radiography and fluoroscopy.** The patient lies on the table *(black arrow),* which has the capacity to tilt up or down. Images can be obtained using the tube on the fluoroscopic carriage *(black oval),* which can be moved over the patient by the operator and then manipulated more or less freely to follow the barium column. Static images can be obtained using the overhead x-ray tube *(white circle).* The x-ray tube can be moved into place over an x-ray cassette, which would be located under the patient *(white arrow).*

projections (usually by both the radiologist and the radiologic technologist) during the performance of barium studies for whatever part of the gastrointestinal tract is being studied, depending on the nature of the abnormality and the mobility of the patient.

■ In **interventional radiology,** iodinated contrast is selectively injected into blood vessels or other ducts that can be imaged fluoroscopically to demonstrate normal anatomy, pathology, or the position of catheters or other devices (Video 1-5).

■ Fluoroscopy units can be made **mobile,** although they are still relatively large and heavy. They carry the same warnings regarding exposure to radiation as any other modality using ionizing radiation.

■ Radiation doses in fluoroscopy can be **substantially higher** than those used in conventional radiography because so many images are acquired for every minute of fluoroscopy time. Therefore the dose is reduced by using the **shortest possible fluoroscopy time** to obtain diagnostic images.

NUCLEAR MEDICINE

■ A **radioactive isotope (radioisotope)** is an unstable form of an element that emits radiation from its nucleus as it decays. Eventually, the end product is a stable, nonradioactive isotope of another element.

■ Radioisotopes can be produced **artificially** (most frequently by neutron enrichment in a nuclear reactor or in a cyclotron) or may occur **naturally.** Naturally occurring radioisotopes include **uranium** and **thorium.** The **vast majority of radioisotopes** used in medicine are produced **artificially.**

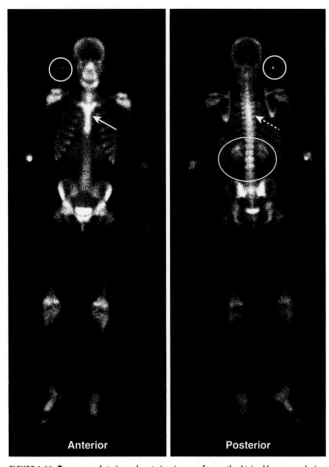

| Anterior | Posterior |

FIGURE 1-11 Bone scan. Anterior and posterior views are frequently obtained because each view brings different structures closer to the gamma camera for optimum imaging, such as the sternum on the anterior view *(solid white arrow)* and the spine on the posterior view *(dotted white arrow)*. Notice that the kidneys are normally visible on the posterior view *(white oval)*. Unlike the convention used in viewing other studies in radiology, the patient's right side is not always on your left in nuclear medicine scans. On posterior views, the patient's right side is on your right. This can be confusing, so make sure you look for the labels on the scan. In many cases, a white marker dot will be located on the patient's right side *(white circles)*.

- **Radiopharmaceuticals** are combinations of **radioisotopes** attached to a **pharmaceutical** that has binding properties that allow it to concentrate in certain body tissues, such as the lungs, thyroid, or bones. Radioisotopes used in clinical nuclear medicine are also referred to as *radionuclides, radiotracers,* or, sometimes, simply *tracers.*
- Various body organs have a specific affinity for, or absorption of, different biologically active chemicals. For example, the thyroid takes up **iodine;** the brain utilizes **glucose;** bones utilize **phosphates;** and **particles** of a **certain size** can be trapped in the lung capillaries (Fig. 1-11).
- After the radiopharmaceutical is carried to a tissue or organ in the body, usually via the bloodstream, its radioactive emissions allow it to be measured and imaged using a detection device called a *gamma camera.*
- **Single-photon emission computed tomography (SPECT)** is a nuclear medicine modality in which a gamma

camera is used to acquire several two-dimensional images from **multiple angles,** which are then reconstructed by computer into a **three-dimensional data set** that can be manipulated to produce thin slices in any projection. To acquire SPECT scans, the **gamma camera rotates around the patient.**
- **Positron emission tomography (PET)** is used to produce three-dimensional images that depict the body's biochemical and metabolic processes at a molecular level. It is performed using a **positron (positive electron)**-producing radioisotope attached to a targeting **pharmaceutical.**
- PET scanning is most often used in the **diagnosis and treatment follow-up of cancer.** It is frequently used **to locate hidden metastases** from a known tumor or to **detect recurrence.** Oncologic PET scans make up about 90% of the clinical use of PET. Some tumors take up more of the radiotracer than others and are referred to as *FDG-avid* tumors, with *FDG* referring to the contrast agent fluorodeoxyglucose (Video 1-6).
- Unlike other modalities that use ionizing radiation, the patient can briefly be the **source** of radiation exposure to others (e.g., technologists) in nuclear medicine studies. To limit exposure to others, the principles of **decreasing the time** in close proximity to the patient, **increasing the distance** from the source, and **appropriate shielding** are used (see online section on Radiation Dose and Safety).
- Compared with CT and fluoroscopy, nuclear medicine studies, in general, produce less patient exposure. The types of scans that deliver the highest dose relative to other nuclear scans are cardiac studies and PET examinations. (An additional online chapter on nuclear medicine is available to registered users at StudentConsult.com.)

CONVENTIONS USED IN THIS BOOK

- And now, a word from our sponsor. **Bold type** is used liberally throughout this text to **highlight important points,** and because this is a book filled with a large number of extraordinarily important points, **there is much bold type.**
- **Diagnostic pitfalls,** potential false-positive or false-negative traps on the sometimes perilous journey to the correct interpretation of an image, are signaled by this icon:
- **Important points** that are so important that not even **boldface** type does them justice are signaled by this icon:
- Online-only content is listed throughout the chapter (as eFigures, eTables, Videos, and so forth). Also, additional or complementary instructional material available on the *StudentConsult/Inkling.com* website for registered users is listed at the end of each chapter. Web-only extras include quizzes, imaging anatomy modules, expanded text and an additional chapter on nuclear medicine, color photos, and videos.
- **"Take-home" points** are listed at the end of each chapter in a Take-Home Points table.
  - ♦ You may use these points anywhere, not only in your home.

Recognizing Anything | **7**

TAKE-HOME POINTS

RECOGNIZING ANYTHING: AN INTRODUCTION TO IMAGING MODALITIES

Today, almost all images are stored electronically on a picture archiving, communications, and storage system called *PACS*.

Conventional radiographs (plain films) are produced using ionizing radiation generated by x-ray machines and viewed on a monitor or light box.

Such x-ray machines are relatively inexpensive and widely available, and they can be made portable. The images are limited as to the sensitivity of findings they are capable of displaying.

There are five basic radiographic densities, arranged in order from that which appears the whitest to that which appears the blackest: metal, calcium (bone), fluid (soft tissue), fat, and air.

Computed tomography (CT) utilizes rapidly spinning arrays of x-ray sources and detectors and sophisticated computer processing to increase the sensitivity of findings visible and display them in any geometric plane.

CT scanners have become the foundation of cross-sectional imaging. They are moderately expensive and also use ionizing radiation to produce their images.

Ultrasonography produces images using the acoustic properties of tissue and does not employ ionizing radiation. It is thus safe for use in children and in women of childbearing age and during pregnancy. It is particularly useful in analyzing soft tissues and blood flow.

Ultrasonography units are less expensive, are in widespread use, and have been produced as small as handheld devices.

Magnetic resonance imaging (MRI) produces images based on the energy derived from hydrogen atoms placed in a very strong magnetic field and subjected to radiofrequency pulsing. The data thus derived are analyzed by powerful computer algorithms to produce images in any plane.

MRI units are relatively expensive, require site construction for their placement, and are usually relatively high in cost to operate. They have become the cornerstone of neuroimaging and are of particular use in studying muscles, ligaments, and tendons.

Fluoroscopy utilizes ionizing radiation to produce real-time visualization of the body that allows for evaluation of motion, positioning, and the visualization of barium or iodine contrast agents moving through the gastrointestinal and genitourinary tracts and blood vessels.

Nuclear medicine utilizes radioisotopes that have been given the property to "target" different organs of the body to evaluate the physiology and anatomy of those organs. Unlike other modalities that use ionizing radiation, the patient can briefly be the source of radiation exposure in nuclear medicine studies.

 WEBLINK

Visit StudentConsult.Inkling.com for videos and more information.

For your convenience, the following QR code may be used to link to **StudentConsult.com**. You must register this title using the PIN code on the inside front cover of the text to access online content.

CHAPTER 2
Recognizing a Technically Adequate Chest Radiograph

■ You have to be able to quickly determine if a study is **technically adequate** so that you don't mistake technical deficiencies for abnormalities. We will focus on the chest radiograph in this chapter. This chapter will enable you to evaluate the technical adequacy of a chest x-ray by helping you become more familiar with the diagnostic pitfalls certain technical artifacts can introduce.

EVALUATING THE CHEST RADIOGRAPH FOR TECHNICAL ADEQUACY

■ Evaluating **five technical factors** will help you to determine if a chest radiograph is **adequate** for interpretation or whether certain artifacts may have been introduced that can lead you astray (Table 2-1):
   ♦ **Penetration**
   ♦ **Inspiration**
   ♦ **Rotation**
   ♦ **Magnification**
   ♦ **Angulation**

Penetration

■ Unless x-rays adequately pass through the body part being studied, you may not visualize everything necessary on the image produced.
   ♦ To determine if a frontal chest radiograph is adequately penetrated, **you should be able to see the thoracic spine through the heart shadow** (Fig. 2-1).

 Pitfalls of underpenetration (inadequate penetration)

♦ You can tell **if a frontal chest radiograph is underpenetrated** (too light) if you are **not able to see the thoracic spine through the heart.**
♦ Underpenetration can introduce at least **two errors** into your interpretation:
   • First, the left hemidiaphragm may not be visible on the frontal film because the left lung base may appear opaque. This technical artifact could either mimic or hide true disease in the left lower lung field (e.g., left lower lobe pneumonia or left pleural effusion) (Fig. 2-2).
      ▪ **Solution:** Look at the lateral chest radiograph to confirm the presence of disease at the left base (see "The Lateral Chest" in Chapter 3).
   • Second, the pulmonary markings, which are mostly the blood vessels in the lung, may appear more prominent than they really are. You may mistakenly think the patient is in congestive heart failure or has pulmonary fibrosis.
      ▪ **Solutions:** Look for other radiologic signs of congestive heart failure (see Chapter 13). Look at

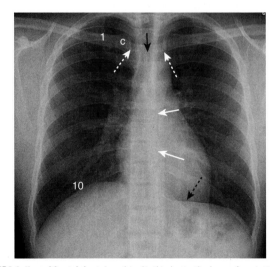

FIGURE 2-1 Normal frontal chest. As explained in this chapter, the degree of penetration shown here is adequate because we can see the spine through the heart *(solid white arrows)*. There is a good inspiration, with almost 10 posterior ribs visible. We can determine that the patient is not rotated, because the spinous process *(solid black arrow)* is midway between the heads of the clavicles *(dotted white arrows)*. There is little magnification because this is a posteroanterior chest image. The medial end of the clavicle (C) superimposes onto the anterior 1st rib (1), so there is no angulation. Note that the left hemidiaphragm is visible *(dotted black arrow)*, as it should be.

TABLE 2-1 WHAT DEFINES A TECHNICALLY ADEQUATE CHEST RADIOGRAPH?

Factor	What You Should See
Penetration	Should be able to see the spine through the heart
Inspiration	Should see at least eight to nine posterior ribs
Rotation	Spinous process should fall equidistant between the medial ends of the clavicles
Magnification	Anteroposterior films (mostly portable chest x-rays) magnify the heart slightly
Angulation	Clavicle normally has an S shape, and medial end superimposes onto the 3rd or 4th rib

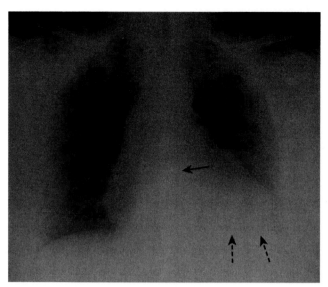

FIGURE 2-2 Underpenetrated frontal chest radiograph. The spine *(solid black arrow)* is not visible through the cardiac shadow. The left hemidiaphragm is also not visible *(dotted black arrows)*, and the degree of underpenetration makes it impossible to differentiate between actual disease at the left base versus nonvisualization of the left hemidiaphragm due to underpenetration. A lateral radiograph of the chest would help to differentiate between artifact of technique and true disease.

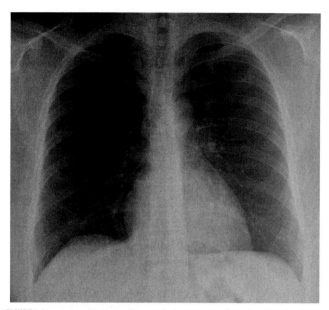

FIGURE 2-3 Overpenetrated frontal chest radiograph. The overpenetration makes lung markings difficult to see, mimicking some of the findings in emphysema or possibly suggesting a pneumothorax. How lucent (dark) the lungs appear on a radiograph is a poor way of evaluating for the presence of emphysema because of artifacts introduced by technique. In emphysema, the lungs are frequently hyperinflated and the diaphragm flattened (see Chapter 12). To diagnose a pneumothorax, you should see the pleural white line (see Chapter 10).

the lateral chest film to confirm the presence of increased markings, airspace disease, or effusion at the left base that you suspected on the basis of the frontal radiograph.

 Pitfall of overpenetration

- If the study is overpenetrated (too dark), the lung markings may seem decreased or absent (Fig. 2-3). You could mistakenly think the patient has emphysema or a pneumothorax, or if the degree of overpenetration is marked, it could render findings such as a pulmonary nodule almost invisible.
  - **Solutions:** Look for other radiographic signs of emphysema (see Chapter 12) or pneumothorax (see Chapter 10). Ask the radiologist if the film should be repeated.

Inspiration

- A full inspiration ensures a reproducible radiographic image for comparison from one time to the next and eliminates artifacts that may be confused for, or obscure, disease.
  - The **degree of inspiration can be assessed by counting the number of posterior ribs visible** above the diaphragm on the frontal chest radiograph.
    - To help in differentiating the **anterior** from the **posterior** ribs, see Box 2-1.
  - If **10 posterior ribs are visible, it is an excellent inspiration** (Fig. 2-4).
  - In many **hospitalized patients, visualization of eight to nine posterior ribs** is a degree of inspiration usually **adequate** for accurate interpretation of the image.

 Pitfall: Poor inspiration

  - A **poor inspiratory effort will compress and crowd the lung markings**, especially at the bases of the lungs

BOX 2-1 DIFFERENTIATING BETWEEN ANTERIOR AND POSTERIOR RIBS

Posterior ribs are immediately more apparent to the eye on frontal chest radiographs.

The posterior ribs are oriented more or less horizontally.

Each pair of posterior ribs attaches to a thoracic vertebral body.

Anterior ribs are visible, but more difficult to see, on the frontal chest radiograph.

Anterior ribs are oriented downward toward the feet.

Anterior ribs attach to the sternum or to each other with cartilage that usually is not visible until later in life, when the cartilage may calcify.

near the diaphragm (Fig. 2-5). This may lead you to mistakenly think the study shows lower lobe pneumonia.
- **Solution:** Look at the lateral chest radiograph to confirm the presence of pneumonia (see "The Lateral Chest Radiograph" in Chapter 3).

Rotation

- **Significant rotation** (the patient turns the body to one side or the other) **may alter the expected contours of the heart and great vessels, the hila, and hemidiaphragms.**
- The easiest way to assess whether the patient is rotated toward the left or right is by studying **the position of the medial ends of each clavicle relative to the spinous process** of the thoracic vertebral body between the clavicles (Fig. 2-6).
  - The **medial ends of the clavicles are anterior structures.**
  - The **spinous process is a posterior structure** (Fig. 2-7).

♦ If **the spinous process appears to lie equidistant from the medial end of each clavicle** on the frontal chest radiograph, **there is no rotation** (see Fig. 2-7, *A*).

♦ If the spinous process appears **closer to the medial end of the left clavicle, the patient is rotated toward his or her own right side** (see Fig. 2-7, *B*).

♦ If the spinous process appears **closer to the medial end of the right clavicle**, the patient is **rotated toward her or his own left side** (see Fig. 2-7, *C*).

♦ These relationships hold true regardless of whether the patient was facing the x-ray tube or facing the cassette at the time of exposure.

 Pitfalls of excessive rotation

♦ Even minor degrees of rotation can distort the normal anatomic appearance of the heart and great vessels, the hila, and hemidiaphragms.

♦ **Marked rotation can introduce errors in interpretation:**

• **The hilum may appear larger** on the side rotated farther away from the imaging cassette, because objects farther from the imaging cassette tend to be more magnified than objects closer to the cassette.

  ▪ **Solutions: Look at the hilum on the lateral chest view** to see if that view confirms hilar enlargement (see "The Hilar Region" in Chapter 3).

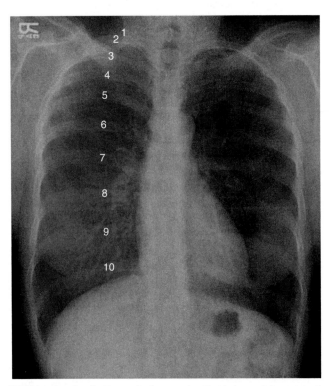

FIGURE 2-4 Counting ribs. The posterior ribs are numbered in this photograph. Ten posterior ribs are visible above the right hemidiaphragm, an excellent inspiration. In most hospitalized patients, the presence of eight or nine visible posterior ribs in the frontal projection indicates an inspiration that is adequate for accurate interpretation of the image. When counting ribs, make sure you do not miss counting the 2nd posterior rib, which frequently overlaps the 1st rib.

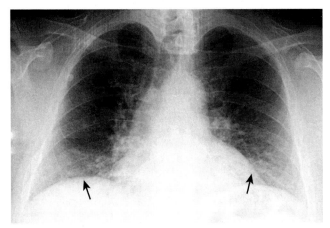

FIGURE 2-5 Suboptimal inspiration. Only eight posterior ribs are visible on this frontal chest radiograph. A poor inspiration may "crowd," and therefore accentuate, the lung markings at the bases *(black arrows)* and may make the heart seem larger than it actually is. The crowded lung markings may mimic the appearance of aspiration or pneumonia. A lateral chest radiograph should help eliminate the possibility, or confirm the presence, of basilar airspace disease suspected on the basis of the frontal radiograph.

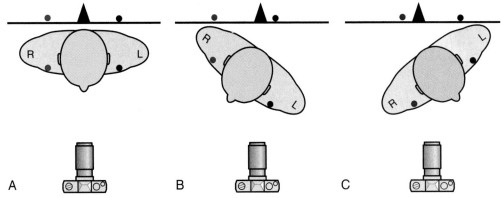

FIGURE 2-6 How to determine if the patient is rotated. A, The patient is not rotated, and the medial ends of the right *(orange dot)* and left *(black dot)* clavicles are projected on the radiograph *(black line)* equidistant from the spinous process *(black triangle).* **B,** The patient is rotated toward his own right side. Notice how the medial end of the left clavicle *(black dot)* is projected closer to the spinous process than is the medial end of the right clavicle *(orange dot).* **C,** This patient is rotated toward his own left side. The medial end of the right clavicle *(orange dot)* is projected closer to the spinous process than is the medial end of the left clavicle *(black dot).* The camera icon depicts this as an anteroposterior projection, but the same relationships would hold true for a posteroanterior projection as well. Figure 2-7 shows how this applies to radiographs.

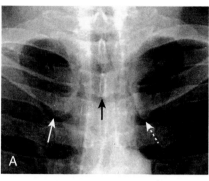
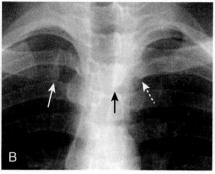
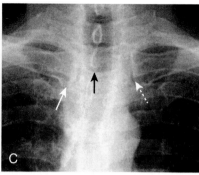

FIGURE 2-7 How to evaluate for rotation. A, Close-up view of the heads of the clavicles demonstrates that each *(white arrows)* is about equidistant from the spinous process of the vertebral body between them *(black arrow)*. This indicates the patient is not rotated. **B,** Close-up view of the heads of the clavicles in a patient rotated toward his own right side. (Remember that you are viewing the study as if the patient were facing you.) The spinous process *(black arrow)* projects much closer to the left clavicular head *(dotted white arrow)* than to the right clavicular head *(solid white arrow)*. **C,** Close-up view of the heads of the clavicles in a patient rotated toward his own left side. The spinous process *(black arrow)* is much closer to the right clavicular head *(solid white arrow)* than it is to the left *(dotted white arrow)*.

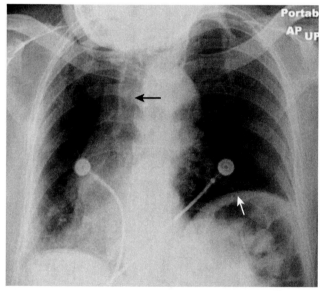

FIGURE 2-8 Distorted appearance due to severe rotation. Frontal chest radiograph of a patient markedly rotated toward her own right side. Notice how the left hemidiaphragm, being farther from the cassette than the right hemidiaphragm because of the rotation, appears higher than it normally would *(white arrow)*. The heart and the trachea *(black arrow)* appear displaced into the right hemithorax because of the rotation.

> Compare the current study with a previous study of the same patient to assess for change.

- Rotation may also distort the appearance of the normal contours of the heart and hila.
- The **hemidiaphragm may appear higher** on the side rotated away from the imaging cassette (Fig. 2-8).
  - **Solution:** Compare the current study with a previous study of the same patient.

Magnification

- Magnification usually is not an issue in assessing normal pulmonary anatomy, but depending on the position of the patient relative to the imaging cassette, magnification can play a role in assessment of the size of the heart.

> ⮞ **The closer any object is to the surface on which it is being imaged, the more true to its actual size the**

resultant image will be. As a corollary, **the farther any object is from the surface on which it is being imaged, the more magnified that object will appear.**

- In the standard **PA** chest radiograph (i.e., one obtained in the **posteroanterior projection**) **the heart,** being an anterior structure, is closer to the imaging surface and thus **truer to its actual size.** In a PA study, the x-ray beam enters at "P" (posterior) and exits at "A" (anterior). The standard frontal chest radiograph is usually a PA exposure.
- In an AP image (i.e., one obtained in the **anteroposterior projection**) **the heart** is **farther** from the imaging cassette and is therefore **slightly magnified.** In an AP study, the x-ray beam enters at "A" (anterior) and exits at "P" (posterior). Portable, bedside chest radiographs are almost always AP.
- Therefore **the heart will appear slightly larger on an AP image than will the same heart on a PA image** (Fig. 2-9).
- There is another reason the heart looks larger on a portable AP chest image than on a standard PA chest radiograph:
  - The **distance between the x-ray tube and the patient is shorter when a portable AP image is obtained** (about 40 inches) **than** when a **standard PA chest radiograph** is exposed (taken by convention at 72 inches). The greater the distance the x-ray source is from the patient, the less the degree of magnification.
- **To learn how to determine if the heart is really enlarged on an AP chest radiograph, see Chapter 4.**

Angulation

- Normally, the x-ray beam passes **horizontally (parallel to the floor) for an upright chest** study, and in that position, the **plane of the thorax** is **perpendicular** to the x-ray beam.
- **Hospitalized patients** in particular may not be able to sit completely upright in bed in order that **the x-ray beam may enter the thorax with the patient's head and thorax tilted backward.** This has the same effect as angling the x-ray beam toward the patient's head, and the image obtained thus is called an *apical lordotic* view of the chest.
  - On apical lordotic views, **anterior structures in the chest** (such as the clavicles) are projected **higher** on

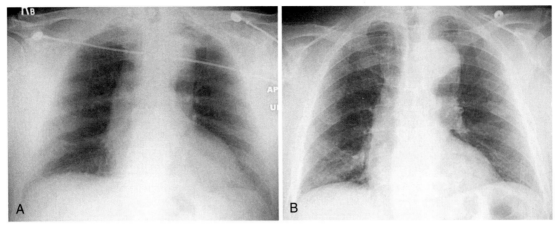

FIGURE 2-9 Effect of positioning on magnification of the heart. A, Frontal chest radiograph done in the anteroposterior projection shows the heart to be slightly larger than in **B,** which shows the same patient's chest exposed minutes later in the posteroanterior projection. Because the heart lies anteriorly in the chest, it is farther from the imaging surface in **(A)** and is therefore magnified more than in **(B),** in which the heart is closer to the imaging surface. In actual practice, there is very little difference in heart size between anteroposterior and posteroanterior exposures, as long as the patient's inspiration was equal in both.

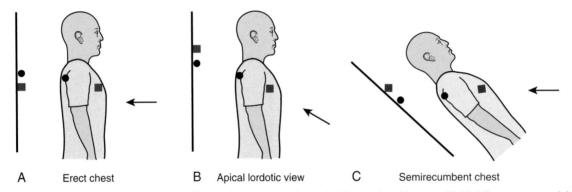

| A Erect chest | B Apical lordotic view | C Semirecumbent chest |

FIGURE 2-10 Diagram of apical lordotic effect. A, The x-ray beam *(black arrow)* is correctly oriented perpendicularly to the plane of the cassette *(black line).* The *orange square* symbolizes an anterior structure (such as the clavicles), and the *black circle* indicates a posterior structure (such as the spine). **B,** The x-ray beam is angled upward, which is the manner in which an apical lordotic view of the chest is obtained. The x-ray beam is no longer perpendicular to the cassette, which has the effect of projecting anterior structures higher than posterior structures on the radiograph. **C,** The positions of the x-ray beam and patient lead to exactly the same end result shown in **(B),** which is how semirecumbent bedside studies are frequently obtained in patients who are not able to sit or stand upright. The anterior structures shown in **(C)** are projected higher than posterior structures.

the resultant radiographic image than **posterior structures** in the chest, which are projected lower (Fig. 2-10).

 Pitfall of excessive angulation

- ♦ You can recognize an apical lordotic chest study when you see the clavicles project at or above the 1st ribs on the frontal image. An apical lordotic view distorts the appearance of the clavicles, straightening their normal S-shaped appearance (Fig. 2-11).
- ♦ Apical lordotic views may also distort the appearance of other structures in the thorax. The **heart may have an unusual shape**, which sometimes mimics cardiomegaly and distorts the normal appearance of the cardiac borders. The **sharp border of the left hemidiaphragm may be lost,** which could be mistaken as a sign of a left pleural effusion or left lower lobe pneumonia.
  - **Solutions:** Know how to recognize technical artifacts and understand how they can distort normal anatomy. Consult with a radiologist about confusing images.

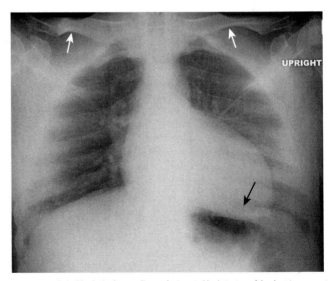

FIGURE 2-11 Apical lordotic chest radiograph. An apical lordotic view of the chest is now most frequently obtained inadvertently in patients who are semirecumbent at the time of the study. Notice how the clavicles are projected above the 1st ribs and that their usual S shape is now straight *(white arrows).* The lordotic view also distorts the shape of the heart and produces spurious obscuration of the left hemidiaphragm *(black arrow).* Unless the artifacts of technique are understood, these findings could be mistaken for disease that does not exist.

TAKE-HOME POINTS

RECOGNIZING A TECHNICALLY ADEQUATE CHEST RADIOGRAPH

There are five parameters that define an adequate chest examination, and recognition of them is important to accurately differentiate abnormalities from technically produced artifacts.

These parameters are penetration, inspiration, rotation, magnification, and angulation.

If the chest is adequately penetrated, you should be able to see the spine through the heart. Underpenetrated (too light) studies obscure the left lung base and tend to spuriously accentuate the lung markings; overpenetrated studies (too dark) may mimic emphysema or pneumothorax.

If the patient has taken an adequate inspiration, you should see at least eight or nine posterior ribs above the diaphragm. Poor

inspiratory efforts may mimic basilar lung disease and may make the heart appear larger.

The spinous process should fall equidistant between the medial ends of the clavicles to indicate the patient is not rotated. Rotation can introduce numerous artifactual anomalies affecting the contour of the heart and the appearance of the hila and diaphragm.

Anteroposterior (AP) films (mostly portable chest x-rays) will magnify the heart slightly compared with the standard posteroanterior (PA) chest radiograph (usually done in the radiology department).

Frontal views of the chest obtained with the patient semirecumbent in bed (tilted backward) may produce apical lordotic images that distort normal anatomy.

 WEBLINK

Visit StudentConsult.Inkling.com for more information and a quiz on the Correct Chest X-ray Technique.

For your convenience, the following QR Code may be used to link to **StudentConsult.com**. You must register this title using the PIN code on the inside front cover of the text to access online content.

CHAPTER 3
Recognizing Normal Pulmonary Anatomy

In this chapter, you will learn the normal anatomy of the lungs as depicted by conventional radiography and chest computed tomography (CT). To become more comfortable with interpreting images of the chest, you should first be able to recognize fundamental, normal anatomy to enable you to differentiate it from what is abnormal.

WHICH "SYSTEM" WORKS BEST

What is the best system to look at an imaging study such as a chest x-ray? I am glad you asked.

Some folks systematically look at imaging studies, such as chest radiographs, from the outside of the image to the inside of the image; others look at them from the inside out or from top to bottom. Some systems for reminding you to examine every part of an image have catchy acronyms and mnemonics.

The fact is: **It does not matter which system you use, as long as you look at everything** on the image. So, **use whichever system works** for you, but be sure to look at everything. "Looking at everything," by the way, includes looking at **all of the views available in a given study,** not just everything on *one* view. (Do not forget that lateral chest radiograph in a two-view study of the chest.)

Experienced radiologists usually have no system at all. Burned-in images are bad for computer monitors, but they are great for radiologists. "Burned" into the neurons of a radiologist's brain are mental images of what a normal frontal chest radiograph looks like, what thoracic sarcoidosis looks like, and so on. They frequently use a **"gestalt"** impression of a study that they see in their mind's eye within seconds of looking at an image. If the image does or does not correspond to the mental image that resides in their brains, **then** they systematically study the images. This is not magic; this ability comes only with experience, so at least for now, you are probably not quite ready to use the gestalt approach.

The most valuable system to use in interpreting images is the **system in which you routinely increase your knowledge.** If you do not know what you are looking for, you can stare at an image for hours or days, or in the case of the lateral chest radiograph, you can ignore an image entirely, and the end result will be the same: You will not see the findings. There is an axiom in radiology: **You only see what you look for, and you only look for what you know.** So if you don't know what to look for, you will never recognize the finding, no matter what system you use or how long you stare at the image.

By reading this book, you will gain the knowledge that will allow you to recognize what it is you are looking at—the best system of all.

THE NORMAL FRONTAL CHEST RADIOGRAPH

- Figure 3-1 displays some of the normal anatomic features visible on the frontal chest radiograph.

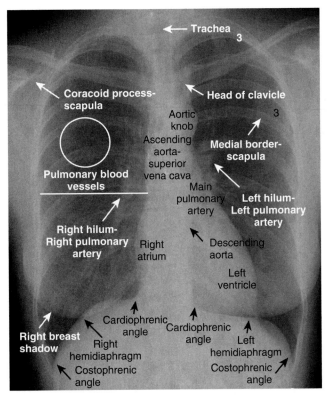

FIGURE 3-1 Well-exposed frontal view of a normal chest. Notice how the spine is just visible through the heart shadow. Both the right and left lateral costophrenic angles are sharply and acutely angled. The *white line* demarcates the approximate level of the minor or horizontal fissure, which is usually visible in the frontal view because it is seen on end. There is no minor fissure on the left side. The *white circle* contains lung markings that are blood vessels. Note that the left hilum is normally slightly higher than the right. The *white "3"* lies on the posterior 3rd rib, and the *black "3"* lies on the anterior 3rd rib.

- Vessels and bronchi—normal lung markings
  - Virtually all of the "white lines" you see in the lungs on a chest radiograph **are blood vessels.** Blood vessels characteristically branch and taper gradually from the hila centrally to the peripheral margins of the lungs. You cannot accurately differentiate between pulmonary arteries and pulmonary veins on a conventional radiograph.
  - **Bronchi are mostly invisible** on a normal chest radiograph because they are normally very **thin-walled,** they **contain air,** and they are **surrounded by air.**
- Pleura—normal anatomy
  - The **pleura is composed of two layers, the outer parietal and inner visceral layers, with the pleural space between them.** The visceral pleura is adherent

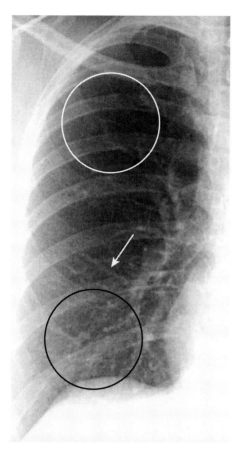

FIGURE 3-2 Normal pulmonary vasculature. The right lung is shown. In the upright position, the lower-lobe vessels *(black circle)* are larger in size than the upper-lobe vessels *(white circle)*, and all vessels taper gradually from central to peripheral *(white arrow)*. Alterations in pulmonary flow or pressure may change these relationships.

to the lung and enfolds to form the major and minor fissures.

♦ Normally, there are **several milliliters of fluid but no air in the pleural space.**

♦ **Neither the parietal pleura nor the visceral pleura is normally visible** on a conventional chest radiograph, except where the two layers of visceral pleura enfold to form the fissures. Even then, they are **usually no thicker than a line drawn with the point of a sharpened pencil.**

NORMAL PULMONARY VASCULATURE

In the upright position, the blood flow to the bases is normally greater than the flow to the apices because of the effect of gravity. Therefore the **size** of the vessels at the **base** is normally **larger** than the size of the vessels at the **apex** of the lung.

■ Normally, blood vessels branch and taper gradually from central (the hila) to peripheral (near the chest wall) (Fig. 3-2).

■ Changes in pressure or flow can alter the normal dynamics of the pulmonary vasculature, some of which are described in Chapter 13.

■ For more on recognizing normal pulmonary vasculature and an imaging approach to diagnosing heart disease in adults, registered users can view "The ABCs of Heart Disease:

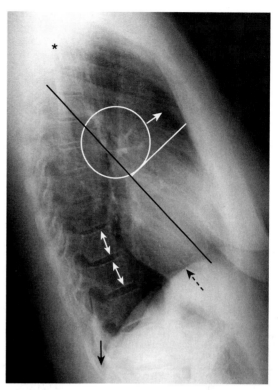

FIGURE 3-3 Normal left lateral chest radiograph. There is a clear space behind the sternum *(solid white arrow)*. The hila produce no discrete shadow *(white circle)*. The vertebral bodies are approximately of equal height, and their end plates are parallel to each other *(double white arrows)*. The posterior costophrenic angles *(solid black arrow)* are sharp. Notice how the thoracic spine appears to become blacker (darker) from the shoulder girdle *(black star)* to the diaphragm because there is less dense tissue for the x-ray beam to traverse at the level of the diaphragm. The heart normally touches the anterior aspect of the left hemidiaphragm and usually obscures (silhouettes) it. The superior surface of the right hemidiaphragm is frequently seen continuously from back to front *(dotted black arrow)* because it is not obscured by the heart. Notice the normal space posterior to the heart and anterior to the spine; this will be important in assessing cardiomegaly (see Chapter 13). The *black line* represents the approximate location of the major or oblique fissure; the *white line* is the approximate location of the minor or horizontal fissure. Both are frequently visible because they are seen on end on a lateral view radiograph.

Recognizing Adult Heart Disease from the Frontal Chest Radiograph" online.

THE NORMAL LATERAL CHEST RADIOGRAPH

■ As part of the standard two-view chest examination, an **upright, frontal** chest radiograph and an **upright, left lateral** view of the chest are obtained.

■ A **left lateral chest x-ray** (with the patient's left side against the detector) is of **great diagnostic value** but is **sometimes ignored by beginners** because of their lack of familiarity with the findings visible in that projection.

■ Figure 3-3 displays some of the normal anatomic features visible on the lateral chest radiograph.

■ **Why look at the lateral chest?**
   ♦ It can help you **determine the location** of disease you already identified as being present on the frontal image.
   ♦ It can **confirm the presence of disease** you may be unsure of on the basis of the frontal image alone, such as a mass or pneumonia.
   ♦ It can **demonstrate disease not visible on the frontal image** (Fig. 3-4).

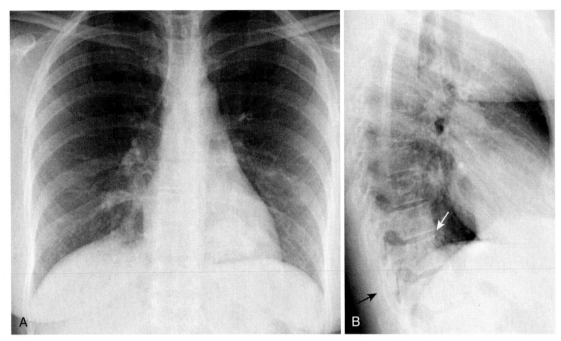

FIGURE 3-4 The spine sign. Frontal **(A)** and lateral **(B)** views of the chest demonstrate airspace disease on the lateral image **(B)** in the left lower lobe that may not be immediately apparent on the frontal image. (Look closely at **A** and you may see the pneumonia in the left lower lobe behind the heart.) Normally, the thoracic spine appears to get "blacker" as you view it from the neck to the diaphragm because there is less dense tissue for the x-ray beam to traverse just above the diaphragm than in the region of the shoulder girdle (see Fig. 3-3). In this case, a left lower lobe pneumonia superimposed on the lower spine in the lateral view *(white arrow)* makes the spine appear "whiter" (more dense) just above the diaphragm. This is called the **spine sign.** Note that on a well-positioned lateral projection image, the right and left posterior ribs almost superimpose on each other *(black arrow),* a sign of a true lateral view.

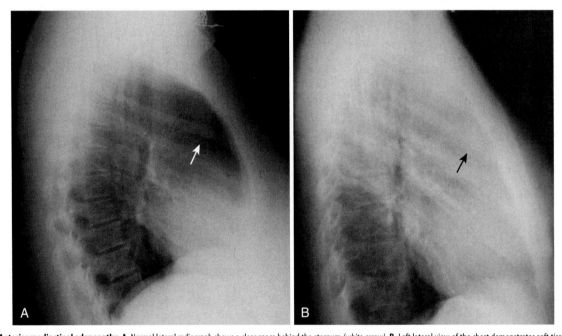

FIGURE 3-5 Anterior mediastinal adenopathy. A, Normal lateral radiograph shows a clear space behind the sternum *(white arrow).* **B,** Left lateral view of the chest demonstrates soft tissue that is "filling in" the normal clear space behind the sternum *(black arrow).* This represents anterior mediastinal lymphadenopathy in a patient with lymphoma. Adenopathy is probably the most frequent reason the retrosternal clear space is obscured. Thymoma, teratoma, and substernal thyroid enlargement also can produce anterior mediastinal masses, but they do not usually produce exactly this appearance.

Five Key Areas on the Lateral Chest X-Ray
(see Fig. 3-3 and Table 3-1)

- The retrosternal clear space
- The hilar region
- The fissures
- The thoracic spine
- The diaphragm and posterior costophrenic sulci

The retrosternal clear space

- Normally, there is a relatively lucent crescent just behind the sternum and anterior to the shadow of the ascending aorta. **Look for this clear space to "fill in"** with soft-tissue density when there is an **anterior mediastinal mass** present (Fig. 3-5).

TABLE 3-1   THE LATERAL CHEST: A QUICK GUIDE OF WHAT TO LOOK FOR

Region	What You Should See
Retrosternal clear space	Lucent crescent between sternum and ascending aorta
Hilar region	No discrete mass present
Fissures	Major and minor fissures should be pencil-point thin, if visible at all
Thoracic spine	Rectangular vertebral bodies with parallel end plates; disk spaces maintain height from top to bottom of thoracic spine
Diaphragm and posterior costophrenic sulci	Right hemidiaphragm slightly higher than left; sharp posterior costophrenic sulci

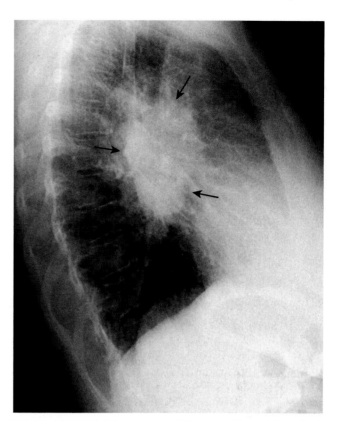

FIGURE 3-7 Hilar mass on lateral radiograph. Left lateral view of the chest shows a discrete, lobulated mass in the region of the hila *(black arrows)*. Normally, the hila do not cast a shadow that is easily detectable on the lateral projection. This patient had bilateral hilar adenopathy from sarcoidosis, but any cause of hilar adenopathy or a primary tumor in the hilum would have a similar appearance.

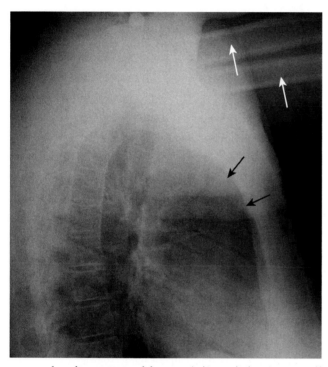

FIGURE 3-6 Arms obscure retrosternal clear space. In this example, the patient was not able to hold her arms over her head for the lateral chest examination, as patients are instructed to do to eliminate the shadows of the arms from overlapping the lateral chest. The humeri are clearly visible *(white arrows)*, so even though the soft tissue of the patient's arms appears to fill in the retrosternal clear space *(black arrows)*, this should not be mistaken for an abnormality such as anterior mediastinal adenopathy (see Fig. 3-5).

 Pitfall: Be careful not to mistake the soft tissue of the patient's superimposed arms for "filling-in" of the clear space. Although patients are asked to hold their arms over their head for a lateral chest exposure, many are too weak to raise their arms.

♦ **Solution:** You should be able to identify the patient's arm on the radiograph by spotting the humerus (Fig. 3-6).

The hilar region

■ The hila may be difficult to assess on the frontal view, especially if both hila are slightly enlarged, because comparison with the opposite normal side is impossible.

■ The lateral view may help. Most of the hilar densities are made up of the pulmonary arteries. Normally, there is no discrete mass visible in the hilar region on the lateral view radiograph.

■ When there is a hilar mass, such as that which might occur with enlargement of hilar lymph nodes, the hilum (or hila) will cast a distinct, lobulated, masslike shadow on the lateral radiograph (Fig. 3-7).

The fissures

■ On the **lateral film,** both the **major (oblique) and minor (horizontal) fissures may be visible** as fine, white lines (about as thick as a line made with the point of a sharpened pencil). The fissures demarcate the upper and lower lobes on the left and the upper, middle, and lower lobes on the right.

■ The **major fissures** course obliquely, roughly **from the level of the 5th thoracic vertebra to** a point on the diaphragmatic surface of the pleura **a few centimeters behind the sternum.**

■ The **minor fissure lies at the level of the 4th anterior rib** (on the right side only) and is **horizontally** oriented (see Fig. 3-3).

■ Both the major and minor fissures may be visible on the lateral view, but because of the oblique plane of the major fissure, **only the minor fissure is usually visible on the frontal view.**

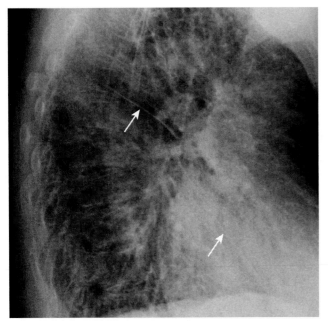

FIGURE 3-8 Fluid in the major fissures. Left lateral view of the chest shows thickening of both the right and left major fissures *(white arrows)*. This patient was in congestive heart failure, and this thickening represents fluid in the fissures. Normally, the fissures are invisible; if visible, they appear as fine, white lines of uniform thickness no larger than a line made with the point of a sharpened pencil. The major or oblique fissure runs from the level of the 5th thoracic vertebral body to a point on the anterior diaphragm about 2 cm behind the sternum. Notice the increased interstitial markings that are visible throughout the lungs, which are due to fluid in the interstitium of the lung.

■ **When a fissure contains fluid or develops fibrosis from a chronic process, it will become thickened** (Fig. 3-8). **Thickening of the fissure by fluid is almost always associated with other signs of fluid in the chest,** such as Kerley B lines and pleural effusions (see Chapter 13). **Thickening of the fissure by fibrosis** is the more likely cause if there are **no other signs of fluid in the chest.**

The thoracic spine

■ Normally, the **thoracic vertebral bodies are** roughly **rectangular in shape,** and **each** vertebral body's **end plate parallels the end plate of the vertebral body above and below it. Each intervertebral disk space either becomes slightly taller than or remains the same as the one above it** throughout the thoracic spine.

■ Degeneration of the disk can lead to narrowing of the disk space and the development of small, bony spurs *(osteophytes)* at the margins of the vertebral bodies.

■ When there is a compression fracture, most often from osteoporosis, the vertebral body loses height. Compression fractures very commonly first involve depression of the superior end plate of the vertebral body (Fig. 3-9).

■ **Do not forget to look at the thoracic spine when studying the lateral chest radiograph** for valuable clues about systemic disorders (see Chapter 26).

The diaphragm and posterior costophrenic sulci

■ Because the diaphragm is composed of soft tissue (muscle) and the abdomen below it contains soft-tissue structures such as the liver and spleen, only the upper border of the diaphragm, abutting an air-filled lung, is usually visible on conventional radiographs.

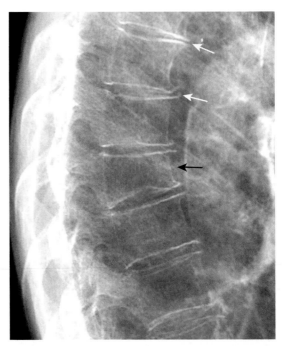

FIGURE 3-9 Osteoporotic compression fracture and degenerative disk disease. Do not forget to look at the thoracic spine when studying the lateral chest radiograph for valuable information about a host of systemic diseases. In this study, there is loss of stature of the 8th thoracic vertebral body due to osteoporosis *(black arrow)*. Compression fractures frequently involve the superior end plate first. There are small osteophytes present at multiple levels due to degenerative disk disease *(white arrows)*.

■ Even though we have one diaphragm that separates the thorax from the abdomen, we usually do not normally see the entire diaphragm from side to side on conventional radiographs because of the position of the heart in the center of the chest.

 ♦ Therefore, radiographically, we refer to the right half of the diaphragm as the *right hemidiaphragm* and the left half of the diaphragm as the *left hemidiaphragm.*

■ **How can the right hemidiaphragm be differentiated from the left hemidiaphragm on the lateral radiograph?**

 ♦ The **right hemidiaphragm is usually visible for its entire length from front to back.** Normally, the **right hemidiaphragm is slightly higher than the left,** a relationship that tends to hold true on the lateral as well as the frontal radiograph.

 ♦ The **left hemidiaphragm is seen sharply posteriorly but is silhouetted by the muscle of the heart anteriorly** (i.e., its edge disappears anteriorly) (Fig. 3-10).

 ♦ **Air in the stomach or splenic flexure of the colon appears immediately below the left hemidiaphragm.** The liver lies below the right hemidiaphragm, and bowel gas is usually not seen between the liver and the right hemidiaphragm.

The posterior costophrenic angles (posterior costophrenic sulci)

■ Each hemidiaphragm produces a rounded dome that indents the central portion of the base of each lung like the bottom of a wine bottle. This produces a depression, or **sulcus,** that surrounds the periphery of each lung and represents the lowest point of the pleural space when the patient is upright.

- On a frontal chest radiograph, this sulcus is most easily viewed in profile at the outer edge of the lung as the **lateral costophrenic sulcus** (also called the *lateral costophrenic angle*) and on the lateral radiograph as the **posterior costophrenic sulcus** (also known as the *posterior costophrenic angle*) (see Figs. 3-1 and 3-3).
- **Normally,** all of the **costophrenic sulci are sharply outlined and acutely angled.**
- Pleural effusions accumulate in the deep recesses of the costophrenic sulci with the patient upright, filling in their acute angles. This is called *blunting of the costophrenic angles* (see Chapter 8).
- It requires only about **75 mL** of fluid (or less) to **blunt the posterior costophrenic angle** on the lateral film, whereas

it takes about **250 to 300 mL** to **blunt the lateral costophrenic angles** on the frontal film (see Fig. 3-10 and Table 3-1).

NORMAL CT ANATOMY OF THE CHEST

- By convention, **CT scans of the chest,** like most other radiologic studies, **are viewed** with the **patient's right on your left and the patient's left on your right.** If the patient is scanned in the supine position, as most usually are, the **top** of each image is **anterior** and the **bottom** of each image is **posterior.**

➡️ Chest CT scans are usually "windowed" and **displayed in** at least **two formats** designed to be viewed as parts of the same study to optimize anatomic definition.

- ♦ **Lung windows** are chosen to maximize the ability to image abnormalities of the **lung parenchyma** and to identify **normal and abnormal bronchial anatomy.** The mediastinal structures usually appear as a homogeneous white density on lung windows.
- ♦ **Mediastinal windows** are chosen to display the **mediastinal, hilar,** and **pleural structures** to best advantage. The lungs usually appear completely black when viewed with mediastinal windows.
- ♦ **Bone windows** are also utilized quite often as a third way of displaying the data, demonstrating the bony structures to their best advantage.
- ♦ It is important to know that the displays of these different windows are **manipulations of the data** obtained during the **original scan** and **do not require rescanning the patient.**

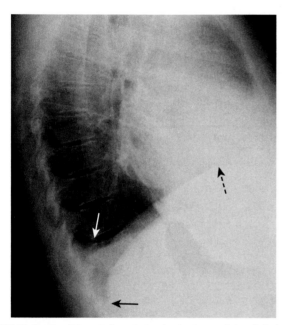

FIGURE 3-10 Blunting of the posterior costophrenic sulcus by a small pleural effusion. Left lateral view of the chest shows fluid blunting the posterior costophrenic sulcus *(solid white arrow).* The other posterior costophrenic angle *(solid black arrow)* is sharp. The pleural effusion is on the right side because the hemidiaphragm involved can be traced anteriorly farther forward *(dotted black arrow)* than the other hemidiaphragm (the left), which is normally silhouetted by the heart and not visible anteriorly.

NORMAL CT ANATOMY OF THE LUNGS

- All of the anatomy visible on conventional radiographs of the chest is visible on CT scans of the chest, but in greater detail. With reconstruction of thin-section CT images, the lungs can be visualized in any plane, although the three most common ones are the axial, sagittal, and coronal planes (Fig. 3-11).

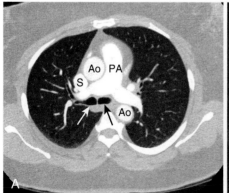
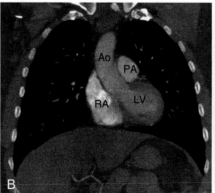
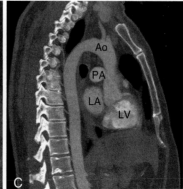

FIGURE. 3-11 Thorax. Axial **(A),** coronal **(B),** and sagittal **(C)** views of the thorax. The three standard planes for imaging the thorax are shown above. Remember that all of these data were acquired at the time of the same scanning session, but that thin-section image acquisition allows reformatting in any plane. The left main bronchus *(black arrow)* and the right main bronchus *(white arrow)* can be seen in **(A).** *Ao,* Aorta; *LA,* left atrium; *LV,* left ventricle; *PA,* pulmonary artery; *RA,* right atrium; *S,* superior vena cava.

- Blood vessels are visible for almost their entire course from the hilar to pleural surfaces. Pulmonary arteries can be differentiated from pulmonary veins (Fig. 3-12).
- Bronchi and bronchioles are also visible, and as a rule, bronchi are normally smaller than their accompanying pulmonary arteries (Fig. 3-13).
- The **trachea,** usually oval shaped, is about 2 cm in diameter.
- In most people, there is a space visible just underneath the arch of the aorta but above the pulmonary artery called the *aortopulmonary window.* The aortopulmonary window is an important landmark, because it is a common location for **enlarged lymph nodes to appear.** At or slightly below this level, the trachea bifurcates at the **carina** into the **right** and **left main bronchi** (Fig. 3-14).
- Slightly more inferior are the **right** and **left main bronchi** and the **bronchus intermedius.** The **right main bronchus**

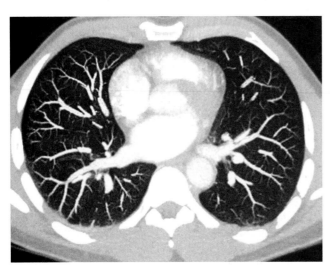

FIGURE 3-12 MIP of pulmonary vasculature. MIP stands for **maximum intensity projection** and is a way to display certain structures of a given density, preferentially making them stand out more easily. It is a computer postprocessing manipulation of the same data acquired at the time of the original scan. It produces an image that looks like an angiogram and is used particularly for computed tomography angiography (as here) and is also utilized for finding pulmonary nodules (see Video 3-1 for MIP of pulmonary vessels).

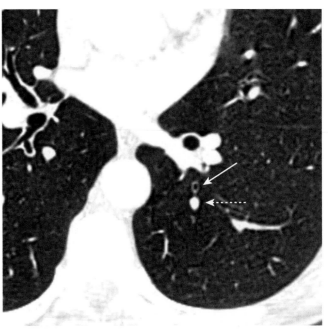

FIGURE 3-13 Bronchus-artery relationship. The normal relationship between the bronchus *(solid white arrow)* and its accompanying pulmonary artery *(dotted white arrow)* is that the artery is usually larger than the bronchus. In bronchiectasis, that relationship is reversed, with the bronchus becoming larger than the artery **(signet-ring sign)** (see Fig. 12-29).

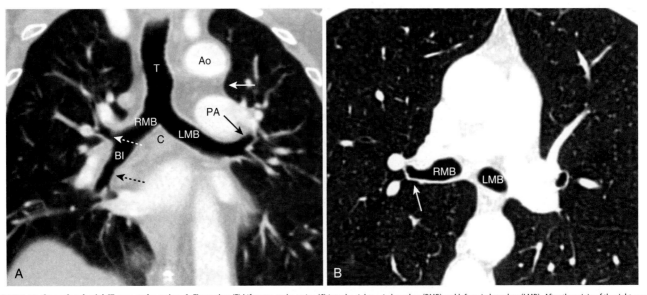

FIGURE 3-14 Coronal and axial CT scan at the carina. A, The trachea (T) bifurcates at the carina (C) into the right main bronchus (RMB) and left main bronchus (LMB). After the origin of the right upper lobe bronchus *(dotted white arrow),* the bronchus intermedius (BI) gives rise to the right lower lobe bronchus *(dotted black arrow)* and middle lobe bronchus (not shown). The left upper lobe bronchus is shown by the *solid black arrow.* The aortopulmonary "window" *(solid white arrow)* lies between the aorta (Ao) and the pulmonary artery (PA). **B,** Just distal to the carina, the right main bronchus gives rise to the upper lobe bronchus *(white arrow).* The LMB can also be seen at this level.

will appear as a circular, air-containing structure that will become tubular as the **right upper-lobe** bronchus comes into view. There should be nothing but lung tissue posterior to the bronchus intermedius (Fig. 3-15).

■ The **left main bronchus** will appear as an air-containing circular structure on the left.

THE FISSURES

■ Depending on slice thickness, the **fissures** will be visible either as **thin white lines or** by an **avascular band** about 2 cm thick as they travel obliquely through the lungs (Fig. 3-16).

■ The **minor fissure** travels in the same horizontal plane as an axial CT image so that it normally is not visible, except in the sagittal or coronal plane. Like the major fissures, though, the **location of the minor fissure can be inferred by an avascular zone** between the right upper and middle lobes (see Fig. 3-16, *A*).

■ The major fissure demarcates the upper lobe from the lower lobes. On the right, the minor fissure demarcates the middle lobe. Its analog on the left is the lingular segment of the left upper lobe (Fig. 3-17).

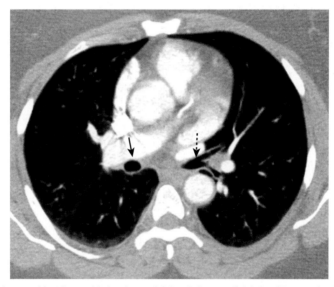

FIGURE 3-15 Bronchus intermedius. Distal to the origin of the right upper lobe bronchus is a short bronchial section called the bronchus intermedius *(solid black arrow)*. The bronchus intermedius divides into the middle and lower lobe bronchi more caudal to this image. There should normally be nothing but lung tissue posterior to the bronchus intermedius. The left main bronchus is shown by the *dotted black arrow.*

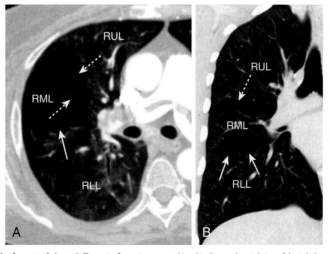

FIGURE 3-16 Fissures seen on axial and coronal reformatted views. A, The major fissure is seen as a thin white line on the axial view of the right lung *(solid white arrows)*, and the minor fissure, traveling obliquely in the plane of the image, is seen as an avascular zone *(dotted white arrow)*. **B,** The appearance of the fissures is reversed on the coronal reformatted image of the right lung. The minor fissure *(dotted white arrow)* is seen in tangent, and the major fissure *(solid white arrows)* travels obliquely at this level. *RLL,* Right lower lobe; *RML,* right middle lobe; *RUL,* right upper lobe.

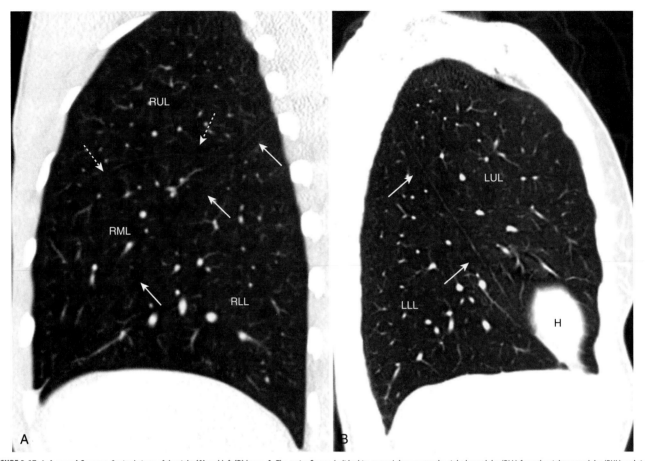

FIGURE 3-17 **Lobes and fissures.** Sagittal views of the right **(A)** and left **(B)** lungs. **A,** The major fissure *(solid white arrows)* demarcates the right lower lobe (RLL) from the right upper lobe (RUL) and right middle lobe (RML). On the right, the minor fissure *(dotted white arrows)* demarcates the middle lobe, separating the upper and lower lobes anteriorly. **B,** On the left, the analogs of the middle lobe are the lingular segments of the left upper lobe. A portion of the heart (H) is seen at the left.

TAKE-HOME POINTS

RECOGNIZING NORMAL PULMONARY ANATOMY

The best "system" to use for carefully looking at any imaging study is one grounded in a solid knowledge base of the appearance of normal anatomy and the most common deviations from normal.

Virtually all of the lung markings are composed of pulmonary blood vessels; most bronchi are too thin-walled to be visible on conventional radiographs.

Normal pulmonary vasculature tapers gradually from central to peripheral, and the vessels are normally larger at the base than at the apex on an upright chest radiograph.

The lateral chest radiograph can provide invaluable information and should always be studied when available.

Five key areas to inspect on the lateral projection radiograph are the retrosternal clear space, the hilar region, fissures, the thoracic spine, and the diaphragm/posterior costophrenic sulci.

There is normally a retrosternal "clear space" on a lateral radiograph that can "fill in" with a mediastinal mass, such as lymphoma.

Although the pulmonary arteries themselves can normally be seen in the hila on the lateral projection radiograph, a discrete mass in the hilum is abnormal and should alert one to the possibility of tumor or adenopathy.

The minor fissure, not the major fissure, will usually be visible on a frontal view radiograph. On the lateral view radiograph, both the

major and minor fissures can be seen normally. When visible, they are very thin lines of uniform size of about 1-2 mm in thickness.

The thoracic spine should appear to become blacker from the upper to the lower portion of the spine because of greater overlying soft tissue more superiorly. Increased density at the base, such as a pneumonia, can produce the reverse of this normal pattern, called the *spine sign*.

On the lateral view radiograph, the left hemidiaphragm will be obscured (silhouetted) anteriorly by the heart. The right hemidiaphragm is usually higher than the left and can be seen in its entirety from front to back.

The costophrenic angles are normally acute and sharply outlined. Pleural effusions cause blunting of the costophrenic angles.

CT scans of the chest display much more detail than conventional radiographs and, owing to rapid acquisition of very thin slices, can be displayed in any plane using the original data set.

The normal anatomy of the trachea and main bronchi is outlined.

Both the major and minor fissures are visible on CT scans either as thin white lines or as avascular bands, depending on the orientation of the fissure relative to the plane in which the scan is displayed.

WEBLINK

Visit StudentConsult.Inkling.com for videos and more information.

For your convenience, the following QR Code may be used to link to **StudentConsult.com**. You must register this title using the PIN code on the inside front cover of the text to access online content.

CHAPTER 4
Recognizing Normal Cardiac Anatomy

Initially, the emphasis will be on conventional radiography, beginning with an assessment of heart size, then a description of the normal and abnormal contours of the heart on the frontal radiograph, and finally a discussion about the normal anatomy of the heart as seen on computed tomography (CT) and magnetic resonance imaging (MRI).

EVALUATING THE HEART ON CHEST RADIOGRAPHS
Recognizing a Normal-Sized Heart

 You can estimate the size of the cardiac silhouette on the frontal chest radiograph using the *cardiothoracic ratio,* which is a measurement of the **widest transverse diameter of the heart** compared with the **widest internal diameter of the rib cage** (from inside of rib to inside of rib at the level of the diaphragm) (Fig. 4-1).

- In most normal adults at full inspiration, the cardiothoracic ratio is **<50%.** That is, the size of the heart is usually less than half of the internal diameter of the thoracic rib cage.

The Normal Cardiac Contours
- The **normal cardiac contours** comprise a series bumps and indentations visible on the frontal chest radiograph. They are demonstrated in Figure 4-2.

 Key points about the cardiac contours:

- The **ascending aorta** should normally not project farther to the right than the right heart border (i.e., right atrium).

- The **aortic knob** is normally <35 mm (measured from the edge of the air-filled trachea) and will normally push the trachea slightly to the right.
- The **main pulmonary artery** segment is usually concave or flat. In younger females, it may normally be convex outward.
- The **normal left atrium** does not contribute to the border of the heart on a nonrotated frontal chest radiograph.
- An **enlarged left atrium** "fills-in" and straightens the normal concavity just inferior to the main pulmonary artery segment and may sometimes be visible on the right side of the heart.

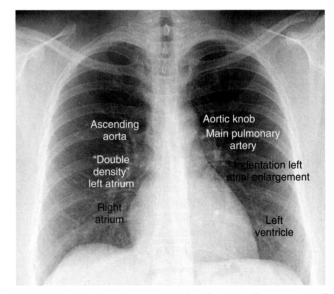

FIGURE 4-2 Normal cardiac contours seen in the frontal projection. There are seven identifiable cardiac contours on the frontal chest radiograph. On the right side of the heart, the first contour is a low-density, almost straight edge visible just lateral to the trachea reflecting the size of the **ascending aorta.** Where the contour of the ascending aorta meets the contour of the right atrium, there is usually a **slight indentation** where the left atrium may appear when it enlarges (called the *double-density sign*). The right heart border is formed by the **right atrium.** On the left, the first contour is the **aortic knob,** a radiographic structure formed by the foreshortened aortic arch superimposed on a portion of the proximal descending aorta. The next contour below the aortic knob is the **main pulmonary artery,** before it divides into a right and left pulmonary artery. Just below the main pulmonary artery segment there is normally a **slight indentation** where an enlarged left atrium may appear on the left side of the heart. The last contour of the heart on the left is formed by the **left ventricle.** The descending aorta almost disappears with the shadow of the spine.

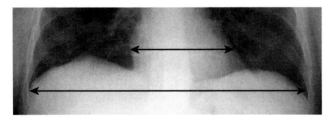

FIGURE 4-1 The cardiothoracic ratio. To estimate the cardiothoracic ratio, the widest diameter of the heart *(upper double arrow)* is compared with the widest internal diameter of the thoracic cage from the inside of the right rib to the inside of the left rib *(lower double arrow).* The widest internal diameter of the thorax is usually at the level of the diaphragm. The cardiothoracic ratio should be <50% in most normal adults on a standard postero-anterior frontal radiograph taken with an adequate inspiration (about nine posterior ribs showing).

- The lower portion of the left heart border is made up of the **left ventricle.** Remember that the left ventricle is really a posterior ventricle and the right ventricle is an anterior ventricle.
- Normally, the **descending aorta** parallels the spine and is barely visible on the frontal radiograph of the chest. When it becomes **tortuous** or **uncoiled,** it swings farther away from the thoracic spine towards the patient's left (Fig. 4-3).

GENERAL PRINCIPLES

- As you interpret cardiac abnormalities, no matter what modality is being used, the following principles hold true:
- The **ventricles respond to obstruction** to their outflow **by first undergoing hypertrophy** rather than dilatation. Therefore the heart may **not** appear enlarged at first with lesions like **aortic stenosis, coarctation of the aorta, pulmonic stenosis, or systemic hypertension.** When the ventricular wall becomes thicker, the lumen actually becomes smaller, and it is only when the muscle begins to fail and the heart decompensates that the heart visibly enlarges.
- **Cardiomegaly,** as recognized on chest radiographs, **is primarily produced by ventricular enlargement,** not by isolated enlargement of the atria. Therefore the heart usually appears normal in size in early **mitral stenosis,** even though the left atrium is enlarged.
- In general, **the most marked chamber enlargement will occur from volume overload** rather than **pressure overload,** so that the largest chambers are usually produced by **regurgitant valves** rather than stenotic valves. Therefore the heart will usually be larger with aortic regurgitation than aortic stenosis, and the left atrium will usually be larger in mitral regurgitation than mitral stenosis (Fig. 4-4).

EVALUATING THE HEART ON CARDIAC CT

- **CT scanning of the heart** is done using a fast, multislice CT scanner, usually with intravenous iodinated contrast and electrocardiographic (ECG)-gated acquisition to reduce motion artifact.
- Both cardiac CT and cardiac MRI utilize ECG-gating, which allows for a series of images to be obtained either prospectively or retrospectively over several cardiac cycles and parsed together with powerful computer algorithms.
- Cardiac CT can be used to evaluate the coronary arteries and valves and search for cardiac masses. By reconstructing multiple phases of the cardiac cycle, it is also possible to analyze wall motion and evaluate ejection fraction and myocardial perfusion.
- The three standard planes for viewing CT images of the heart are the axial (transverse), sagittal (lateral), and coronal (frontal). Figures 4-5 to 4-10 demonstrate the major normal CT anatomy of the heart and great vessels.

NORMAL CARDIAC CT ANATOMY

- We will cover only a few of the major anatomic landmarks demonstrable on chest CT, and all of the scans utilized will be *contrast-enhanced,* that is, the patient will have received intravenous contrast to opacify the blood vessels.

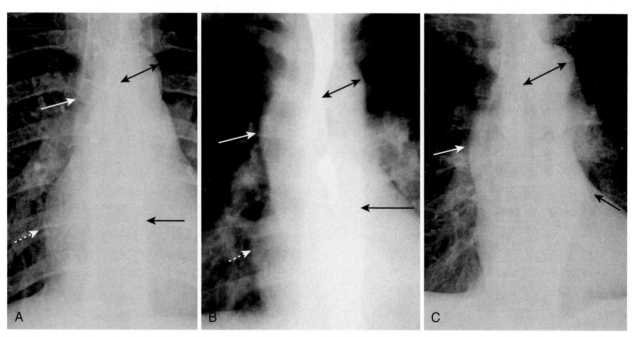

FIGURE 4-3 Appearances of the aorta. A, Normal. The ascending aorta is a low-density, almost straight edge *(solid white arrow)* and does not project beyond the right heart border *(dotted white arrow).* The aortic knob is not enlarged *(double arrow),* and the descending aorta *(solid black arrow)* almost disappears with the shadow of the thoracic spine. **B,** Aortic stenosis. The ascending aorta is abnormal as it projects convex outward *(solid white arrow)* almost as far as the right heart border *(dotted white arrow).* This is due to *poststenotic dilatation.* The aortic knob *(double arrow)* and descending aorta *(solid black arrow)* remain normal. **C,** Hypertension. Both the ascending *(solid white arrow)* and descending aorta *(solid black arrow)* project too far to the right and left, respectively. The aortic knob is enlarged *(double black arrows).*

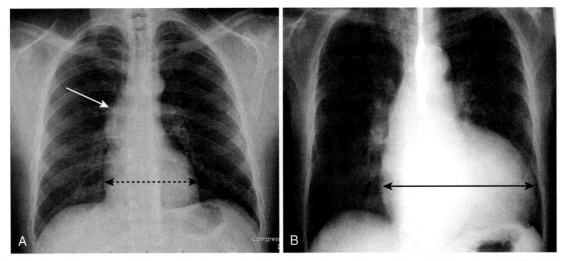

FIGURE 4-4 Heart size with stenotic versus regurgitant valve. A, There is poststenotic dilatation of the ascending aorta *(white arrow)* from turbulent flow in this patient with aortic stenosis. Notice that the heart is not enlarged even though this lesion produces left ventricular hypertrophy *(dotted black double arrows).* **B,** This patient has aortic regurgitation. Note the extremely large left ventricle *(solid black double arrows).* Volume overload will cause a greater increase in chamber size than will increased pressure.

- It is best to read the text in conjunction with its associated photograph. Any references to "right" or "left" mean the patient's right or left side, not yours.
- We will start at the top of the chest and progress inferiorly, highlighting the major structures visible at **six key levels.** This is a good way to systematically study every CT study of the chest.

Five-Vessel Level (Fig. 4-5)

- At this level, you should be able to identify the **lungs, the trachea, and the esophagus.** The **trachea** is black because it contains air; it is usually oval in shape and about 2 cm in diameter. The **esophagus** lies posterior and then to either the left or right of the trachea. It is usually collapsed but may contain swallowed air.
- Depending on the exact level of the image, several of the great vessels will be visible. The **venous structures** tend to be more **anterior than the arterial.**
- The **brachiocephalic veins** lie just posterior to the sternum. From the patient's **right** to the patient's **left,** the visible arteries may include the **innominate artery, left common carotid,** and **left subclavian arteries.**

Aortic Arch Level (Fig. 4-6)

- At this level, you should be able to identify the **aortic arch, superior vena cava, and azygous vein.**
- The *aortic arch* is an upside-down U-shaped tube. If the scan skims the very top of the arch, it will appear as a comma-shaped tubular structure with roughly the same diameter anteriorly as posteriorly.
- The *superior vena cava,* into which the *azygous vein* enters, will be located to the right of the trachea.

Aortopulmonary Window Level (Fig. 4-7)

- At this level you should be able to identify the **ascending and descending aorta, superior vena cava,** and **uppermost aspect of the left pulmonary artery** (maybe).

As we scan lower and scan through the opening of the upside-down U-shaped aortic arch, the **ascending aorta**

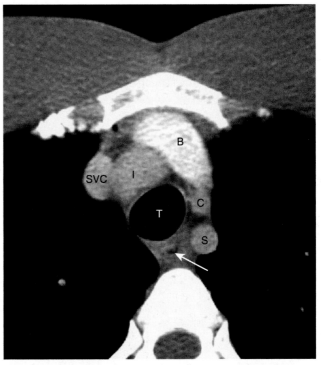

FIGURE 4-5 Five vessel level. At this level, you should be able to identify the lungs, the trachea (T), and the esophagus *(white arrow).* Depending on the exact level of the image, several of the great vessels will be visible. The superior vena cava (SVC) is the large vessel to the right of the trachea. The brachiocephalic vein (B) lies just posterior to the sternum. From the patient's right to the patient's left, the arteries you see may include the innominate artery (I), left common carotid (C), and left subclavian arteries (S).

will appear as a rounded density **anteriorly,** whereas the descending aorta will appear as a rounded density **posteriorly** and to the **left** of the spine. The **ascending aorta** usually measures **2.5 to 3.5 cm** in diameter and the **descending aorta** is slightly smaller at **2 to 3 cm.**

- In most people, there is a space visible just underneath the arch of the aorta but above the pulmonary artery called the

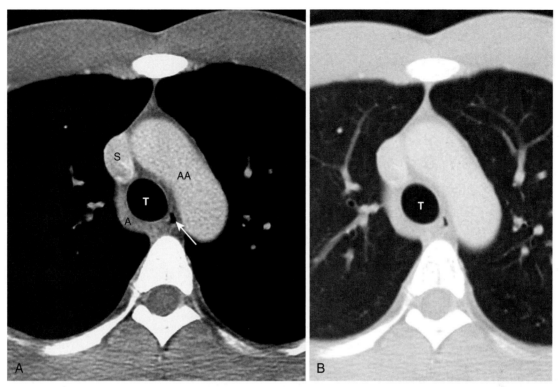

FIGURE 4-6 Aortic arch level (A) mediastinal window and (B) lung window. A, At this level, you should be able to identify the aortic arch (AA), superior vena cava (S), and azygous vein (A). The aortic arch is an upside-down U-shaped tube. It will appear as a comma-shaped tubular structure with about the same diameter anteriorly as posteriorly, as is seen here. The *white arrow* points to air in the esophagus. **B,** The same image as **(A)** but windowed to best visualize lung anatomy. Lung windows are chosen to maximize our ability to image abnormalities of the lung parenchyma and to identify normal and abnormal bronchial anatomy.

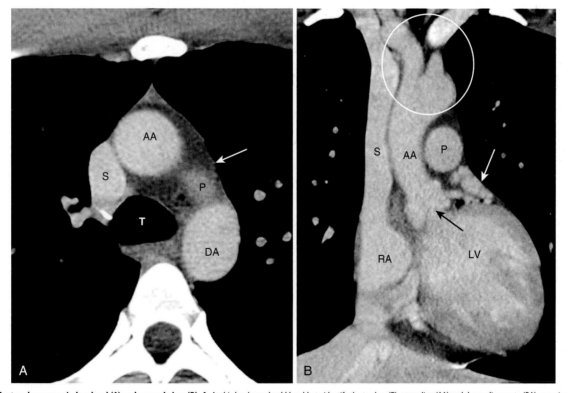

FIGURE 4-7 Aortopulmonary window level (A) and coronal view (B). A, At this level you should be able to identify the trachea (T), ascending (AA) and descending aorta (DA), superior vena cava (S), and possibly the uppermost aspect of the left pulmonary artery (P). In most people, there is a space visible just under the arch of the aorta and above the pulmonary artery called the *aortopulmonary window (white arrow)*. **B,** Coronal reformatted CT enables us to also see the right atrium (RA), superior vena cava (S), left ventricle (LV), aortic valve *(black arrow)*, ascending aorta (AA), main pulmonary artery (P), left atrial appendage *(white arrow)*, and origin of the great vessels *(white circle)*.

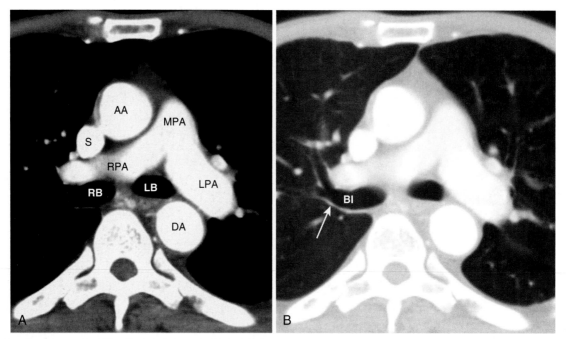

FIGURE 4-8 Main pulmonary artery level (A) mediastinal window and (B) lung window. A, At this level, you should be able to identify the main (MPA), right (RPA) and left pulmonary arteries (LPA), the right (RB) and left main bronchi (LB), and the superior vena cava (S). The left pulmonary artery passes anterior to the descending aorta (DA). The right pulmonary artery passes posterior to the ascending aorta (AA) and crosses to the right side. **B,** Distal to the takeoff of the right upper lobe bronchus is the bronchus intermedius (BI). The posterior wall of the right upper lobe bronchus is 2 to 3 mm in thickness with only aerated lung normally posterior to it *(white arrow)*.

aortopulmonary window. The aortopulmonary window is an important landmark because it is a favorite location for **enlarged lymph nodes** to appear.

- At or slightly below this level, the trachea bifurcates at the **carina** into the **right and left main bronchi.**

Main Pulmonary Artery Level (Fig. 4-8)

- At these levels (it may require more than one image to see all of these structures), you should be able to identify the **main, right, and left pulmonary arteries; the right and left main bronchi;** and the **bronchus intermedius.**
- The **left pulmonary artery** is higher than the right and appears as if it were a continuation of the main pulmonary artery. The **right pulmonary artery** originates at a 90° angle to the main pulmonary artery and crosses to the right side.
- The **right main bronchus** will appear as a circular, air-containing structure, which will then become tubular as the **right upper lobe** bronchus comes into view. There should be nothing but lung tissue posterior to the bronchus intermedius.
- The **left main bronchus** will appear as an air-containing circular structure on the left.

High Cardiac Level (Fig. 4-9)

- At this level, you should be able to identify the **left atrium, right atrium, aortic root,** and **right ventricular outflow tract.**
- The **left atrium** occupies the posterior and central portion of the heart. One or more **pulmonary veins** may be seen to enter the left atrium.
- The **right atrium** forms the right heart border and lies immediately to the right of the left atrium.

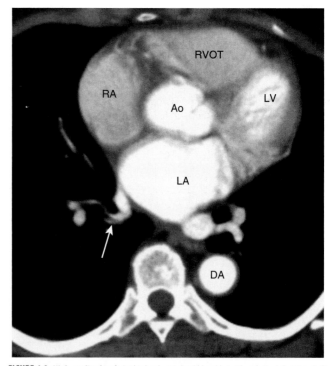

FIGURE 4-9 High cardiac level. At this level, you should be able to identify the left atrium (LA), right atrium (RA), aortic root (Ao), and right ventricular outflow tract (RVOT). The left atrium occupies the posterior and central portion of the heart posteriorly with one or more visible pulmonary veins draining into it *(white arrow)*. The right atrium produces the right heart border and lies anteriorly and to the right of the left atrium. The right ventricular outflow tract lies anterior, lateral, and superior to the root of the aorta. *DA,* Descending thoracic aorta; *LV,* left ventricle.

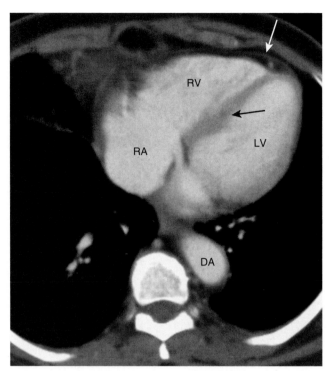

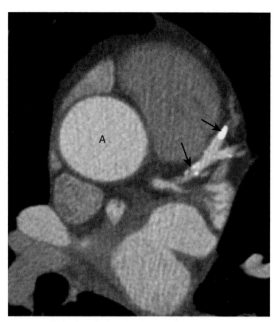

FIGURE 4-11 Coronary artery calcification. There is dense calcification *(black arrows)* in the left anterior descending coronary artery arising from the aorta (A). Using computed tomography, it is possible to image the coronary arteries and measure of the amount of coronary artery calcium, evaluate for vessel patency, and identify the presence of thrombus in the lumen or plaque in the vessel wall following the administration of intravenous contrast.

FIGURE 4-10 Low cardiac level. At this level, you should be able to identify the right atrium (RA), right ventricle (RV), left ventricle (LV), pericardium *(white arrow)*, and interventricular septum *(black arrow)*. The right atrium forms the right heart border. The right ventricle is anteriorly located, just behind the sternum. The left ventricle is more posterior and produces the left heart border. The interventricular septum is visible between the right and left ventricles *(black arrow)*. The normal pericardium is about 2 mm thick and is usually outlined by mediastinal fat *(outside the pericardium)* and epicardial fat *(on its inner surface)*. DA, Descending thoracic aorta.

> The **right ventricular outflow tract** lies anterior, lateral, and superior to the root of the aorta. The **P**ulmonic valve lies **A**nterior, **L**ateral, and **S**uperior to the aortic valve (you can remember that relationship if you can remember the acronym "PALS").

Low Cardiac Level (Fig. 4-10)

- At this level, you should be able to identify the **right atrium, right ventricle, left ventricle, pericardium,** and **interventricular septum.**
- The **right atrium** continues to form the right heart border. The **right ventricle** is anteriorly located, just behind the sternum, and demonstrates more trabeculation than the smoother-walled left ventricle. The **left ventricle** produces the left heart border and normally has a thicker wall than the right ventricle.
- With intravenous contrast you should be able to see the **interventricular septum** between the right and left ventricles.
- The **normal pericardium** is about **2 mm thick** and is usually outlined by **mediastinal fat** outside the pericardium and **epicardial fat** on its inner surface.

USES OF CARDIAC CT

- Cardiac CT is used to evaluate for the presence of cardiac masses, abnormalities of the aorta (including aortic dissection), and diseases of the pericardium.

- Cardiac CT also allows for the imaging and three-dimensional reconstruction of the **coronary arteries** and quantitative measurement of the amount of **coronary artery calcium.** The administration of intravenous contrast allows for evaluation of vessel patency with identification of a thrombus in the lumen or plaque in the vessel wall.
- **Calcium scoring** is based on the premise that the amount of calcium in the coronary arteries is related to the degree of atherosclerosis and that quantifying the amount of calcium may help predict future cardiac events related to coronary artery disease. The scoring is done by calculations that include the amount of calcium present in the coronary arteries visualized (Fig. 4-11). The **absence** of coronary artery calcification has a high **negative predictive value** for significant luminal narrowing.
- While calcium scoring is primarily used for risk analysis of asymptomatic patients, **coronary CT angiography (CCTA)** is primarily used in patients with **acute or chronic chest pain.** Like calcium scoring, a **negative CCTA** has a **high negative predictive value**, that is, a negative study effectively excludes obstructive coronary artery disease.
- One potential drawback to cardiac CT is the **radiation dose** delivered to the patient, which had previously been relatively high. Numerous methods are now being utilized to reduce that dose so that the procedure can now can be performed at a dose well below the average annual background radiation dose.

CCTA—Normal Anatomy

- CCTA compares favorably in accuracy with invasive (catheter) coronary angiography, which remains the reference standard in studying the coronary arteries (Video 4-1).

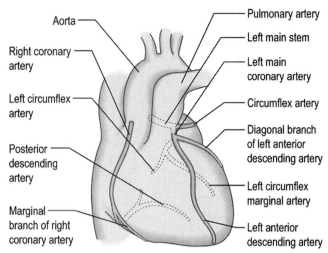

FIGURE 4-12 **The two main coronary arteries are the left (also known as the left main) and the right coronary artery.** The left coronary artery divides almost at once into the **circumflex artery** and **left anterior descending artery (LAD).** The LAD, in turn, gives rise to **diagonal branches** and **septal branches** (not shown). The circumflex artery has **marginal branches.** The **right coronary artery** courses between the right atrium and right ventricle to the inferior part of the septum. It gives rise to a large acute **marginal branch** and, in most people, the **posterior descending artery (PDA).** The PDA supplies the inferior wall of the left ventricle and inferior part of the septum. (From Camm AJ, Bunce NH: Cardiovascular disease. In Kumar P, Clark M, editors: *Clinical medicine*, ed 8, Oxford, 2013, Elsevier, p 673.)

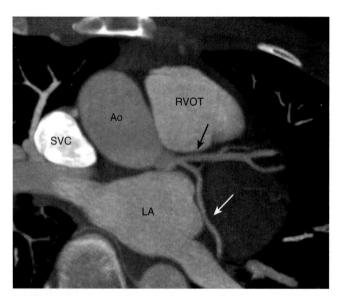

FIGURE 4-13 **CT coronary angiogram, left coronary artery.** The left coronary artery arises from the left coronary cusp at the aortic valve. It divides almost at once into the circumflex artery *(white arrow)* and left anterior descending artery (LAD) *(black arrow). Ao,* Aorta; *LA,* left atrium; *RVOT,* right ventricular outflow tract; *SVC,* superior vena cava.

- There are many variations of normal coronary artery anatomy. Only the most common branching is described here (Fig. 4-12).
- The two main coronary arteries are the **left** (also known as the left main) and the **right coronary artery.**
- The **left coronary artery (LCA)** arises from the left coronary cusp at the aortic valve. It divides almost at once into the **circumflex artery** and **left anterior descending artery (LAD).** The LAD, in turn, gives rise to **diagonal branches** and **septal branches.** The circumflex artery has **marginal branches** (Fig. 4-13).

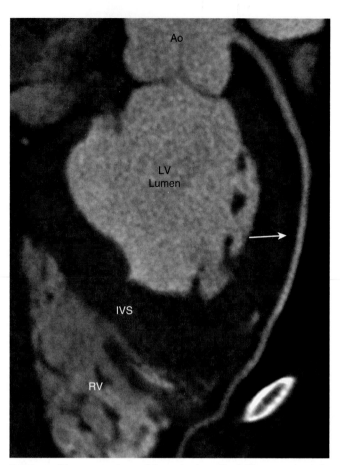

FIGURE 4-14 **CT coronary angiogram, left anterior descending (LAD) coronary artery.** The LAD travels in the **anterior interventricular groove** and continues to the apex of the heart. It supplies most of the left ventricle and also the AV-bundle. *Ao,* Aorta; *IVS,* interventricular septum; *LV lumen,* left ventricular lumen; *RV,* right ventricle.

- The LAD travels in the **anterior interventricular groove** and continues to the apex of the heart. It supplies most of the left ventricle and also the atrioventricular (AV)-bundle, serving the anterior part of the septum with **septal branches** and the anterior wall of the left ventricle with **diagonal branches** (Fig. 4-14).
- The **circumflex artery** lies between the left atrium and left ventricle and supplies **obtuse marginal** vessels to the lateral wall of the left ventricle (see Fig. 4-12).
- The right aortic sinus gives rise to the **right coronary artery (RCA),** which courses between the right atrium and right ventricle to the inferior part of the septum (Fig. 4-15).
- In most people, the first branch of the RCA is the **conus branch** that supplies the right ventricular outflow tract. In most people, a **sinus node artery** arises as a second branch of the RCA. The next branches are **diagonals** that supply the anterior wall of the right ventricle.
- The large **acute marginal branch (AM)** supplies the lateral wall of the right ventricle and runs along the *margin* of the right ventricle above the diaphragm. The RCA continues in the AV groove posteriorly and gives off a branch to the AV node (see Fig. 4-12).
- In most people, the **posterior descending artery (PDA)** is a **branch of the RCA.** The PDA supplies the inferior wall of the left ventricle and inferior part of the septum (see Fig. 4-15).

- **Coronary artery dominance**
  - ◆ The artery that supplies the PDA determines **coronary artery dominance.**
  - ◆ If the **PDA** is supplied by the **RCA,** then the coronary circulation is said to be *right-dominant.*
  - ◆ If the **PDA** is supplied by the **circumflex artery,** a branch of the left coronary artery, then the coronary circulation is called *left-dominant.*
  - ◆ If the **PDA** is supplied by **both** the **right coronary artery** and the **circumflex artery,** then the coronary circulation is called *co-dominant.*

 The **overwhelming majority** of the population is **right-dominant,** about 10% are left-dominant, and the remainder are co-dominant. A left-dominant coronary artery system is associated with an increased risk of nonfatal myocardial infarction and increased overall mortality.

- It is possible to perform an ultrafast CT scan in the emergency department that will allow for the simultaneous evaluation of **coronary artery disease, aortic dissection,** and **pulmonary thromboembolic disease,** the so-called *triple scan (triple rule-out)* for patients who present with **acute chest pain.** Such scans have been shown to improve clinical decision making and allow for earlier discharge from the hospital.

CARDIAC MRI

- MRI can be used to obtain **anatomic** and **functional** images of the heart using a combination of ECG-gating and rapid acquisition of images. Respiratory motion, which would also contribute to blurring the image, can be reduced by having the patient hold his or her breath for short periods of time while the images are acquired (Video 4-2).
- **Cardiac MRI** can depict scarring from a myocardial infarction, perfusion of the heart, anatomic defects, or masses and can assess the function of the valves and cardiac chambers.
- Cardiac MRI can be performed without intravenous contrast or with intravenous contrast (Gadolinium—see Chapter 22). Cardiac MRI is particularly useful in children as a method of evaluating congenital heart disease when other studies (such as echocardiography) produce inconclusive or conflicting information.

Normal Cardiac MRI Anatomy

- One of the benefits of MRI is that its images can be displayed in any plane. Besides the axial, sagittal, and coronal planes, there are several specific views that are typically used in cardiac MRI that allow for the best visualization of the heart. They are called the *horizontal long axis* (otherwise known as the *four-chamber view*), *vertical long axis, short axis,* and *three-chamber views.*
- The anatomy of the heart in the **axial, sagittal,** and **coronal** planes is the same as that seen on CT (Fig. 4-16).

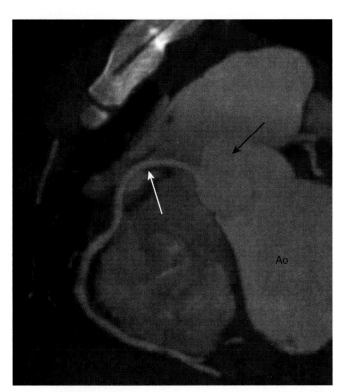

FIGURE 4-15 CT coronary angiogram, right coronary artery. The right aortic sinus *(black arrow)* gives rise to the right coronary artery *(RCA) (white arrow)* which courses between the right atrium and right ventricle to the inferior part of the septum. In most people, as here, the RCA continues to the posterior descending artery. *Ao,* Aorta.

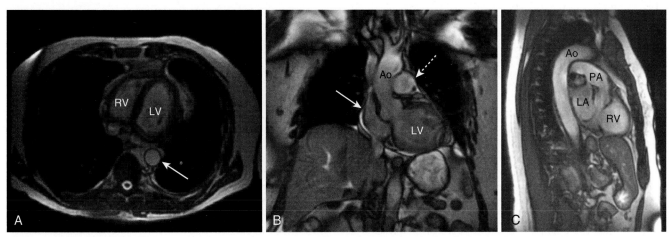

FIGURE 4-16 Cardiac MRI, axial, coronal, and sagittal planes. These three planes produce images the same as those on CT (see Fig. 3-11). **A,** The axial view at this level shows the right (RV) and left ventricles (LV) and the descending aorta *(arrow).* **B,** This coronal image demonstrates the right atrium *(solid arrow),* left ventricle (LV), aorta (Ao), and main pulmonary artery *(dotted arrow).* **C,** The sagittal image at this level shows the right ventricle (RV), pulmonary artery *(PA),* the left atrium *(LA),* and aorta *(Ao).* In all of these images, the blood is depicted as "bright" *(white).*

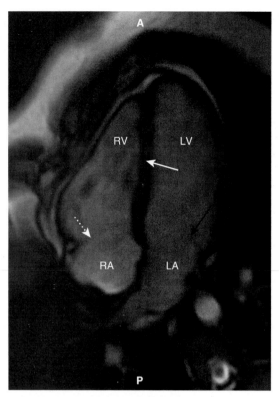

FIGURE 4-17 Cardiac MRI, horizontal long axis view. This is another standard view of the heart using MRI called the ***horizontal long axis*** or ***four-chamber view.*** The right (RV) and left ventricles (LV) are separated by the interventricular septum *(solid white arrow)*. Posterior to each of them are the right atrium (RA) and left atrium (LA) separated by the regions of the tricuspid *(dotted white arrow)* and mitral valves *(solid black arrow)*, respectively. This view is best for evaluating the left ventricle's septal and lateral walls and apex, the right ventricular free wall, and the size of the cardiac chambers. *A,* Anterior; *P,* posterior.

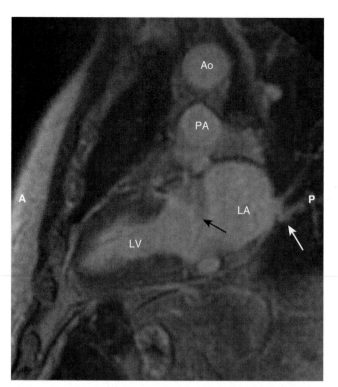

FIGURE 4-18 Cardiac MRI, vertical long axis view. The **vertical long axis** or **two-chamber view** demonstrates the left ventricle (LV) separated from the more posterior left atrium (LA) by the mitral valve area *(black arrow)*. Pulmonary veins drain into the left atrium *(white arrow)*. The aorta (Ao) sits atop the pulmonary artery (PA). This view is best in the evaluation of the anterior and inferior walls and apex of the left ventricle. *A,* Anterior; *P,* posterior.

- The **horizontal long axis (four-chamber) view** resembles an axial view and is best used for evaluating the left ventricle's septal and lateral walls and apex, the right ventricular free wall, and the size of the cardiac chambers. The mitral and tricuspid valves are especially well visualized in this view (Fig. 4-17).
- The **vertical long axis view** resembles a sagittal view and is best used in the evaluation of the anterior and inferior walls and apex of the left ventricle (Fig. 4-18).
- The **short axis view** depicts the left and right ventricles in a way that is useful for making volumetric measurements (Fig. 4-19).
  - Because MRI images of the heart are already obtained with three-dimensional volumes in both end-systole and end-diastole, computer-based measurements of ventricular mass, end-diastolic volume, and end-systolic volume can be made, and from them, stroke volume and ejection fraction can be calculated without intervention.

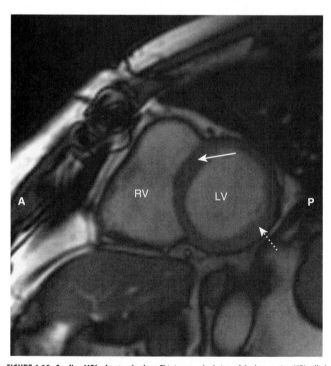

FIGURE 4-19 Cardiac MRI, short axis view. This is a standard view of the heart using MRI called the *short axis view*. The right ventricle (RV) lies anterior to the left ventricle (LV), separated by the interventricular septum *(solid white arrow)*. Note the normally thicker wall of the left ventricle *(dotted white arrow)* than the right ventricle. This view is best at depicting the left and right ventricles in a way that is useful for making volumetric measurements and, from them, calculating stroke volume and ejection fraction. *A,* Anterior; *P,* posterior.

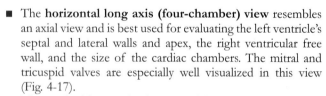

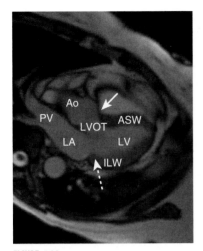

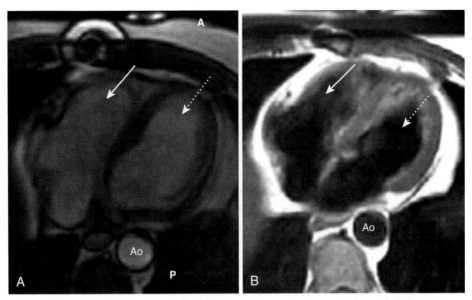

FIGURE 4-20 Cardiac MRI, three-chamber view. The **three-chamber view** is similar to a coronal view and shows the aorta (Ao), left ventricular outflow tract (LVOT), aortic valve *(solid arrow)*, left ventricle (LV), mitral valve *(dotted arrow)*, left atrium (LA), pulmonary veins (PV), and the anteroseptal (ASW) and inferolateral (ILW) walls of the left ventricle *(which are abnormally thickened in this person)*.

FIGURE 4-21 Cardiac MRI; *bright blood* and *black blood* images. Using different imaging algorithms, MRI is capable of displaying the same tissues with differing appearances. **(A)** and **(B)** are both axial sections through the heart showing the right ventricle *(solid white arrows)*, the left ventricle *(dotted white arrows)*, and the aorta (Ao). The **bright blood** technique is utilized to assess cardiac function **(A)**, whereas the **black blood** technique usually is better at depicting cardiac morphology **(B)**. *A,* Anterior; *P,* posterior.

- The **three-chamber view** is similar to a coronal view and shows the aortic root and aortic valve, left ventricular outflow tract, mitral valve, and the anteroseptal and inferolateral walls of the left ventricle (Fig. 4-20).
- Depending on the MRI pulse sequence used to obtain the images, blood can be depicted as either **black** (usually using something called a *spin echo* pulse sequence) and **most often used for anatomic evaluation,** or **bright, that is, white** (usually using something called a *gradient echo* pulse sequence), and most often used for **functional evaluation** (Fig. 4-21).

TAKE HOME POINTS

RECOGNIZING NORMAL CARDIAC ANATOMY

In adults, a quick assessment of heart size can be made using the cardiothoracic ratio, which is the ratio of the widest transverse diameter of the heart compared with the widest internal diameter of the rib cage. In normal adults, the cardiothoracic ratio is usually <50%.

The normal contours of the heart are reviewed.

The ventricles respond to obstruction to their outflow by first undergoing hypertrophy rather than dilatation. On plain films, cardiomegaly is primarily produced by ventricular enlargement. The most marked chamber enlargement will occur from volume overload rather than pressure overload.

Normal anatomy of the major structures is described at six levels in the chest (from top to bottom): five-vessel view, aortic arch, aortopulmonary window, main pulmonary artery, upper cardiac, and lower cardiac levels.

Cardiac CT scanning uses a fast, multislice CT scanner, usually with intravenous iodinated contrast and electrocardiographic (ECG)-gated acquisition to reduce motion artifact.

Cardiac CT scanning is used for evaluation of the coronary arteries, the presence of cardiac masses, abnormalities of the aorta (including aortic dissection) and pericardial diseases.

Normal coronary artery anatomy is described. The artery that supplies the posterior descending artery (PDA) determines coronary

artery dominance. The overwhelming majority of the population is right-dominant.

For patients who present with acute chest pain, it is possible to perform an ultrafast CT scan in the emergency department for the simultaneous evaluation of coronary artery disease, aortic dissection, and pulmonary thromboembolic disease *(triple rule-out scan)*

MRI can be used to obtain anatomic and functional images of the heart. Cardiac MRI can show scarring from a myocardial infarction, depict perfusion of the heart, show anatomic defects or masses, and assess the function of the valves and cardiac chambers.

Several specific views that are typically used in cardiac MRI that allow for the best visualization of the heart are described. They are called the *horizontal long axis* (otherwise known as the *four-chamber view*), *vertical long axis, short axis,* and *three-chamber* views.

Cardiac function is usually evaluated using MRI sequences producing "bright blood" images, so-called because the blood is depicted with increased signal intensity.

Cardiac morphology is usually evaluated using MRI sequences producing "black blood" images. These images allow for anatomic assessment of the cardiac structures without interference from the bright blood signal.

WEBLINK

Visit StudentConsult.Inkling.com for videos and more information.

For your convenience, the following QR Code may be used to link to **StudentConsult.com**. You must register this title using the PIN code on the inside front cover of the text to access online content.

CHAPTER 5
Recognizing Airspace versus Interstitial Lung Disease

We have reviewed the normal imaging anatomy of the heart and lungs in the last two chapters. Feel free to come back to them for a brush up whenever needed. Recognizing the difference between **normal anatomy** and what is abnormal is critical to your ability to make the correct diagnosis. In this chapter, we are going to examine the major patterns of parenchymal lung disease.

CLASSIFYING PARENCHYMAL LUNG DISEASE

- **Diseases** that affect the lung parenchyma **can be arbitrarily divided into two main categories** based in part on their pathology and in part on the pattern they typically produce on a chest imaging study.
    - **Airspace (alveolar) disease**
    - **Interstitial (infiltrative) disease**
- Why learn the difference?
    - While many diseases produce abnormalities that display both patterns, recognition of these patterns frequently helps narrow the disease possibilities so that you can form a reasonable differential diagnosis (Box 5-1).

CHARACTERISTICS OF AIRSPACE DISEASE

➡ Airspace disease characteristically produces opacities in the lung which can be described as *fluffy, cloudlike,* or *hazy.*

- These fluffy opacities tend to be **confluent,** meaning they blend into one another with imperceptible margins.
- The **margins of airspace disease are indistinct,** meaning it is frequently difficult to identify a clear demarcation point between the disease and the adjacent normal lung.
- Airspace disease may be **distributed throughout the lungs,** as in pulmonary edema (Fig. 5-1), **or** it may **appear more localized,** as in a segmental or lobar pneumonia (Fig. 5-2).
- Airspace disease may contain **air bronchograms.**
    - The **visibility of air in the bronchus because of surrounding airspace disease** is called an *air bronchogram.*
    - An air bronchogram is a **sign of airspace disease.**
        - Bronchi are normally not visible because their walls are very thin, they contain air, and they are surrounded by air. When something like fluid or soft tissue replaces the air normally surrounding the bronchus,

BOX 5-1   CLASSIFICATION OF PARENCHYMAL LUNG DISEASES

AIRSPACE DISEASES

Acute

Pneumonia	Aspiration
Pulmonary alveolar edema	Near-drowning
Hemorrhage	

Chronic

Bronchoalveolar cell carcinoma	Sarcoidosis
Alveolar cell proteinosis	Lymphoma

INTERSTITIAL DISEASES

Reticular

Pulmonary interstitial edema	Scleroderma
Interstitial pneumonia	Sarcoid

Nodular

Bronchogenic carcinoma	Miliary tuberculosis
Metastases	Sarcoid
Silicosis	

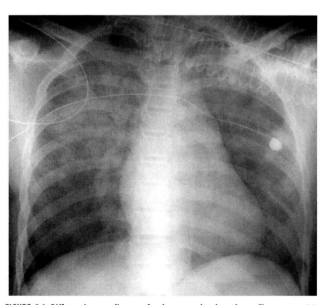

FIGURE 5-1 Diffuse airspace disease of pulmonary alveolar edema. There are opacities throughout both lungs, primarily involving the upper lobes that can be described as *fluffy, hazy,* or *cloudlike* and are confluent and poorly marginated, all pointing to airspace disease. This is a typical example of pulmonary alveolar edema (due to a heroin overdose in this patient).

then the air inside of the bronchus becomes visible as **a series of black, branching tubular structures—** this is the **air bronchogram** (Fig. 5-3).

♦ What can fill the airspaces besides air?
  • **Fluid,** such as occurs in pulmonary edema
  • **Blood,** for example, pulmonary hemorrhage
  • **Gastric juices,** as occurs with aspiration
  • **Inflammatory exudate,** for example, pneumonia
  • **Water,** which can be seen with near-drowning

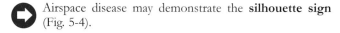

Airspace disease may demonstrate the **silhouette sign** (Fig. 5-4).

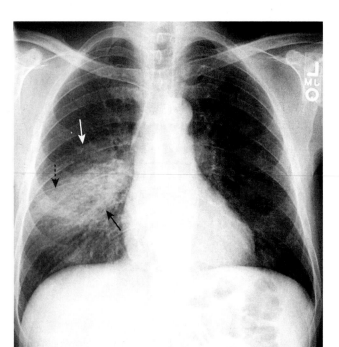

FIGURE 5-2 Right lower lobe pneumonia. There is an area of increased opacification in the right midlung field *(solid black arrow)* that has indistinct margins *(solid white arrow)*, characteristic of airspace disease. The minor fissure *(dotted black arrow)* appears to bisect the disease, locating this pneumonia in the superior segment of the right lower lobe. The right heart border and the right hemidiaphragm are still visible because the disease is not in anatomic contact with either of those structures.

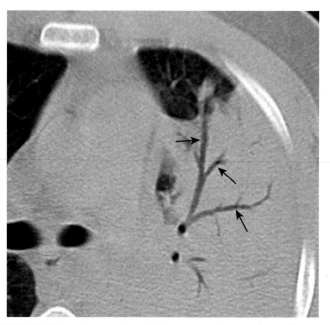

FIGURE 5-3 Air bronchograms demonstrated on computed tomography scan. There are numerous black, branching structures *(black arrows)* representing air that is now visible inside the bronchi because the surrounding airspaces are filled with inflammatory exudate in this patient with an obstructive pneumonia from a bronchogenic carcinoma. Normally on conventional radiographs, air inside bronchi is not visible because the bronchial walls are very thin, they contain air, and they are surrounded by air.

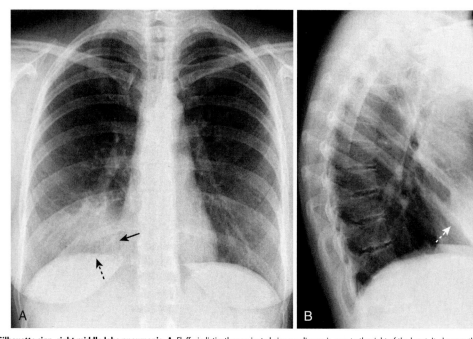

FIGURE 5-4 Silhouette sign, right middle lobe pneumonia. A, Fluffy, indistinctly marginated airspace disease is seen to the right of the heart. It obscures the right heart border *(solid black arrow)* but not the right hemidiaphragm *(dotted black arrow)*. This is called the *silhouette sign* and establishes that the disease (1) is in contact with the right heart border *(which lies anteriorly in the chest)* and (2) that the disease is the same radiographic density as the heart (fluid or soft tissue). Pneumonia fills the airspaces with an inflammatory exudate of fluid density. **B,** The area of the consolidation is indeed anterior, located in the right middle lobe, which is bounded by the major fissure below *(dotted white arrow)* and the minor fissure above *(solid white arrow)*.

BOX 5-2 CHARACTERISTICS OF AIRSPACE DISEASE

Produces opacities in the lung that can be described as *fluffy, cloudlike,* and *hazy.*

The opacities tend to be confluent, merging into one another.

The margins of airspace disease are fuzzy and indistinct.

Air bronchograms or the silhouette sign may be present.

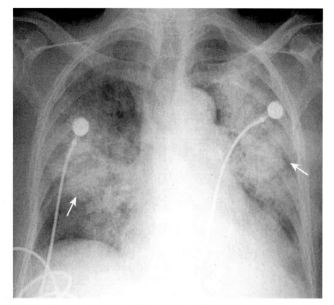

FIGURE 5-6 Acute pulmonary alveolar edema. Fluffy, bilateral, perihilar airspace disease with indistinct margins is present *(white arrows)*, sometimes described as having a *bat-wing* or *angel-wing* configuration. No air bronchograms are seen. The heart is enlarged. This represents pulmonary alveolar edema secondary to congestive heart failure.

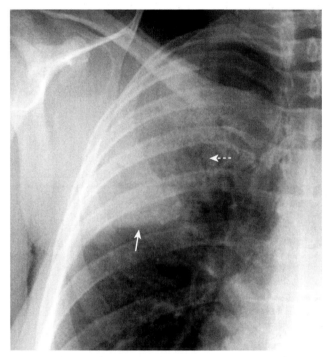

FIGURE 5-5 Right upper lobe pneumococcal pneumonia. Close-up view of the right upper lobe demonstrates confluent airspace disease with air bronchograms *(dotted white arrow)*. The inferior margin of the pneumonia is more sharply demarcated because it is in contact with the minor fissure *(solid white arrow)*. This patient had *Streptococcus pneumoniae* cultured from the sputum.

■ The silhouette sign occurs when two objects **of the same radiographic density** (such as water and soft tissue) **touch each other so that the edge or margin between them disappears.** It will be impossible to tell where one object begins and the other ends. **The silhouette sign is valuable not only in the chest but also as an aid in the analysis of imaging studies throughout the body.**

■ The characteristics of airspace disease are summarized in Box 5-2.

SOME CAUSES OF AIRSPACE DISEASE

■ There are many causes of airspace disease, three of which will be highlighted here. Each is described in greater detail later in the text.

■ **Pneumonia (see also Chapter 9)**
   ♦ About 90% of the time, community-acquired lobar or segmental pneumonia is caused by *Streptococcus pneumoniae* (formerly known as *Diplococcus pneumoniae*) (Fig. 5-5). Pneumonia usually manifests as patchy,

segmental, or lobar airspace disease. Pneumonias may contain air bronchograms. **Clearing** usually occurs **in less than 10 days** (pneumococcal pneumonia may clear within 48 hours).

■ **Pulmonary alveolar edema (see also Chapter 13)**
   ♦ Acute, pulmonary alveolar edema classically produces bilateral, perihilar airspace disease, sometimes described as having a *bat-wing* or *angel-wing configuration* (Fig. 5-6).
   ♦ It may be **asymmetric but** is usually **not unilateral.** Pulmonary edema that is **cardiac** in origin is frequently associated with **pleural effusions** and fluid that thickens the **major** and **minor fissures.**
   ♦ Because fluid fills not only the airspaces but also the bronchi themselves, there are **usually no air bronchograms** seen in pulmonary alveolar edema. Classically, **pulmonary edema clears rapidly** after treatment (<48 hours).

■ **Aspiration (see also Chapter 9)**
   ♦ Aspiration tends to affect whatever part of the lung is most dependent at the time the patient aspirates, and its manifestations depend on the substance(s) aspirated. For most bedridden patients, aspiration usually occurs in either the **lower lobes** or the **posterior portions of the upper lobes.**
   ♦ Because of the course and caliber characteristics of the right main bronchus, **aspiration occurs more often in the right lower lobe** than the left lower lobe (Fig. 5-7).
   ♦ The radiographic appearance of aspiration and how quickly the airspace disease resolves is determined by the type of aspirate and whether or not it becomes infected. **Aspiration** of bland (neutralized) gastric juice or **water usually clears rapidly** within 24 to 48 hours,

whereas aspiration that becomes infected can take weeks to resolve.

CHARACTERISTICS OF INTERSTITIAL LUNG DISEASE

■ The lung's interstitium consists of **connective tissue, lymphatics, blood vessels,** and **bronchi.** These are the structures that surround and support the airspaces.

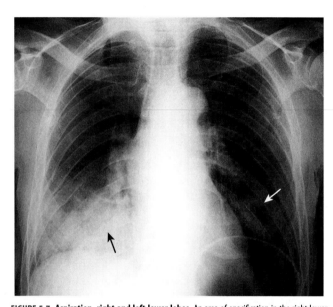

FIGURE 5-7 Aspiration, right and left lower lobes. An area of opacification in the right lower lobe is fluffy and confluent with indistinct margins characteristic of airspace disease *(black arrow)*. To a much lesser extent, there is a similar density in the left lower lobe *(white arrow)*. The bibasilar distribution of this disease should raise the suspicion of aspiration as an etiology. This patient had a recent stroke, and aspiration was demonstrated on a video swallowing study.

■ Interstitial lung disease (sometimes referred to as *infiltrative lung disease*) has the following characteristics:

 Interstitial lung disease produces what can be thought of as **discrete "particles" of disease** that develop in the abundant interstitial network of the lung (Fig. 5-8).

♦ These "particles" of disease can be further characterized as having **three patterns of presentation:**
  • **Reticular interstitial disease** appears as a network of lines (see Fig. 5-8, *A*).
  • **Nodular interstitial disease** appears as an assortment of dots (see Fig. 5-8, *B*).
  • **Reticulonodular interstitial disease** contains both lines and dots (see Fig. 5-8, *C*).

■ These "particles" or "packets" of interstitial disease tend to be **inhomogeneous,** separated from each other by visible areas of normally aerated lung.

■ The **margins of interstitial lung disease are sharper** than are the margins of airspace disease, whose boundaries tend to be indistinct.

■ Interstitial lung disease **can be focal** (as in a solitary pulmonary nodule) **or diffusely distributed** in the lungs (Fig. 5-9).

■ There are **usually no air bronchograms present,** as there may be with airspace disease.

 Pitfall: Sometimes there is so much interstitial disease present that the overlapping elements of disease may superimpose and mimic airspace disease on conventional chest radiographs. Remember that conventional radiographs are two-dimensional representations of three-dimensional objects (humans) so all of the densities in the lung, for example, are superimposed on themselves on any one projection. This may make the tiny packets of

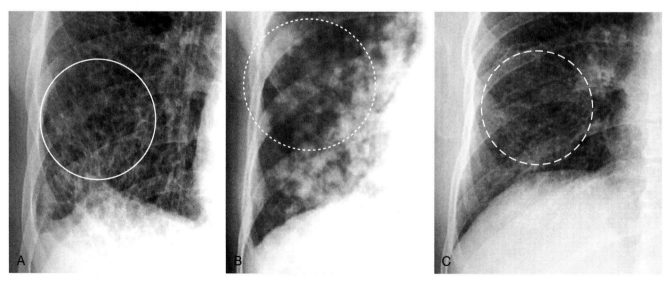

FIGURE 5-8 The patterns of interstitial lung disease. A, The disease is primarily **reticular** in nature, consisting of crisscrossing lines *(solid white circle)*. This patient had advanced sarcoidosis. **B,** The disease is predominantly **nodular** *(dotted white circle)*. The patient was known to have thyroid carcinoma, and these nodules represent innumerable small metastatic foci in the lungs. **(C)** Interstitial disease of the lung, **reticulonodular.** Most interstitial diseases of the lung have a mixture of both a reticular *(lines)* and nodular *(dots)* pattern, as does this case, which is a close-up view of the right lower lobe in another patient with sarcoidosis. The disease *(dashed white circle)* consists of both an intersecting, lacy network of lines and small nodules.

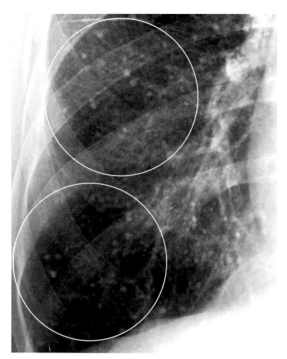

FIGURE 5-9 **Varicella pneumonia.** There are innumerable calcified granulomas that occur in the lung interstitium, seen here as small, discrete, nodules in the right lung *(white circles)*. This patient had a history of varicella (chicken pox) pneumonia years earlier. Varicella pneumonia clears with multiple small calcified granulomas remaining.

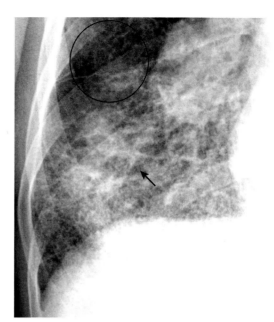

FIGURE 5-10 **The edge of the lesion.** Notice how a portion of this disease appears confluent, like airspace disease *(black arrow)*. Always look at the peripheral margins of parenchymal lung disease to best determine the nature of the "packets" of abnormality and to help in differentiating airspace disease from interstitial disease. At the periphery of this disease *(black circle)*, it is more clearly seen to be reticular interstitial disease.

interstitial disease seem coalescent and more like airspace disease.

♦ **Solutions:** Look at the **periphery** of such confluent shadows in the lung to help in determining whether they are actually caused by airspace disease or a superimposition of numerous reticular and nodular densities (Fig. 5-10).

♦ Obtain a computed tomography (CT) scan of the chest for even better characterization of the disease.

■ The characteristics of interstitial lung disease are summarized in Box 5-3.

SOME CAUSES OF INTERSTITIAL LUNG DISEASE

■ Just as with the airspace pattern, there are many diseases that produce an interstitial pattern in the lung. Several will be discussed briefly here. They are roughly divided into those diseases that are predominantly **reticular** and those that are predominantly **nodular.**

➡ Keep in mind that **many diseases have patterns that overlap and many interstitial lung diseases have mixtures of both reticular and nodular changes (reticulonodular disease).**

Predominantly Reticular Interstitial Lung Diseases

■ **Pulmonary interstitial edema**
♦ Pulmonary interstitial edema can occur because of increased capillary pressure (congestive heart failure),

> **BOX 5-3 CHARACTERISTICS OF INTERSTITIAL LUNG DISEASE**
>
> Interstitial disease has discrete reticular, nodular, or reticulonodular patterns.
>
> "Packets" of disease are separated by normal-appearing, aerated lung.
>
> Margins of "packets" of interstitial disease are usually sharp and discrete.
>
> Disease may be focal or diffusely distributed in the lungs.
>
> Usually, there are no air bronchograms present.

increased capillary permeability (allergic reactions), or decreased fluid absorption (lymphangitic blockade from metastatic disease).

♦ Considered the precursor of alveolar edema, pulmonary interstitial edema classically manifests **four key radiologic findings: fluid in the fissures (major and minor), peribronchial cuffing (from fluid in the walls of bronchioles), pleural effusions, and Kerley B lines** (Fig. 5-11).

♦ Classically, the patient may have few physical findings in the lungs (rales), even though the chest radiograph demonstrates considerable pulmonary interstitial edema, because almost all of the fluid is in the interstitium of the lung rather than in the airspaces.

♦ With appropriate therapy, pulmonary interstitial edema usually clears rapidly (<48 hours).

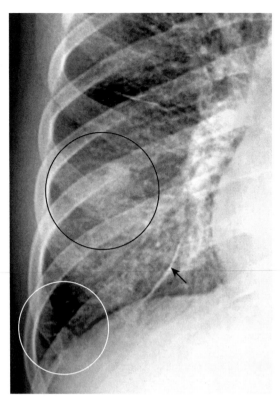

FIGURE 5-11 Pulmonary interstitial edema secondary to congestive heart failure. A close-up view of the right lung shows an accentuation of the pulmonary interstitial markings *(black circle)*. There are multiple Kerley B lines *(white circle)* representing fluid in thickened interlobular septa. There is fluid in the inferior accessory fissure *(black arrow)*.

- **Interstitial pneumonia**
  - ♦ There are many forms of interstitial pneumonia: **usual interstitial pneumonia (UIP), the most common form, desquamative interstitial pneumonia (DIP), and lipoid interstitial pneumonia (LIP),** among others. Another form, called *nonspecific interstitial pneumonia (NSIP),* by definition, does not fall easily into any of the other categories.
  - ♦ UIP is sometimes used to mean *idiopathic interstitial fibrosis.* UIP is more **common in older men** and has associations with **cigarette smoking** and **gastroesophageal reflux.**
  - ♦ Chest radiographs may be normal in UIP. The earliest manifestation on chest radiographs consists of a **fine reticular pattern, particularly at the lung bases.** The pattern then becomes coarser as the disease progresses and ends with a pattern that, for obvious reasons, is called *honeycomb lung* (other diseases produce the honeycomb pattern). There may also be **progressive volume loss.**
  - ♦ Diagnosis is made on CT where **honeycombing, traction bronchiectasis, and ground-glass opacities at the bases** are present (Fig. 5-12).
  - ♦ Idiopathic pulmonary fibrosis is considered the end-stage disease along the spectrum of these interstitial pneumonias.

Predominantly Nodular Interstitial Diseases

- **Bronchogenic carcinoma (see Chapter 12)**
  - ♦ There are four major cell types of bronchogenic carcinoma: **adenocarcinoma, squamous cell carcinoma, small cell carcinoma, and large cell carcinoma.**
  - ♦ Adenocarcinomas, in particular, can present as a **solitary peripheral pulmonary nodule.**

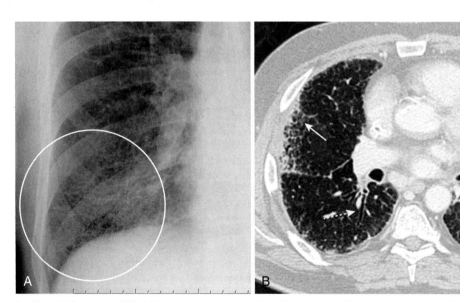

FIGURE 5-12 Usual interstitial pneumonia (UIP). A, There are coarse reticular interstitial markings representing fibrosis, seen here in a close-up of the right lung base *(white circle)*. The findings in UIP occur predominantly at the lung bases. **B,** An axial CT scan of the chest shows abnormalities at the lung bases in a subpleural location, the typical distribution for UIP. There are small cystic spaces called *honeycombing (white arrows)* with evidence of bronchiectasis, as manifest by thickened bronchial walls *(dotted white arrow)*.

- As a rule, on conventional chest radiographs, nodules or masses in the lung are more **sharply marginated** than airspace disease, producing a relatively clear demarcation between the nodule and the surrounding normal lung tissue.
- CT scans may demonstrate **spiculation** or **irregularity** of the lung nodule that may not be apparent on conventional radiographs (Fig. 5-13).

■ **Metastases to the lung**
- Metastases to the lung can be divided into three categories depending on the pattern of disease demonstrated in the lung.
- **Hematogenous metastases** arrive via the bloodstream and usually produce two or more nodules in the lungs, which, when they achieve a large size, are sometimes called *cannonball metastases.* Primary tumors that classically produce nodular metastases to the lung include **breast, colorectal, renal cell, bladder, testicular, head and neck carcinomas, soft tissue sarcomas,** and **malignant melanoma** (Fig. 5-14, *A*).
- The second form of tumor dissemination is **lymphangitic spread.** The pathogenesis of lymphangitic spread to the lungs is somewhat controversial but most likely involves blood-borne spread to the pulmonary capillaries and then invasion of adjacent lymphatics. An alternative means of

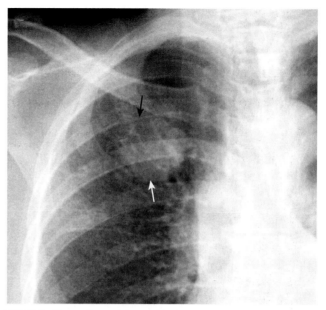

FIGURE 5-13 Adenocarcinoma, right upper lobe. There is a mass in the right upper lobe *(white arrow)*. Its margin is slightly indistinct along the superolateral border *(black arrow)*. CT scan of the chest confirmed the presence of the mass and also demonstrated paratracheal and right hilar adenopathy. The mass was biopsied and was an adenocarcinoma, primary to the lung. Adenocarcinoma of the lung most commonly presents as a peripheral nodule.

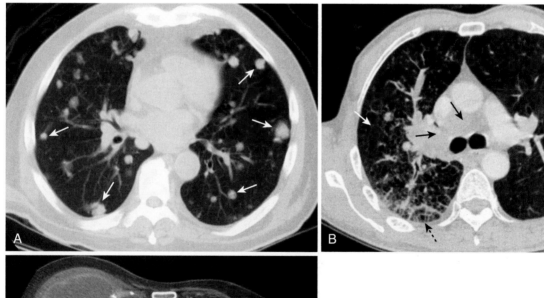

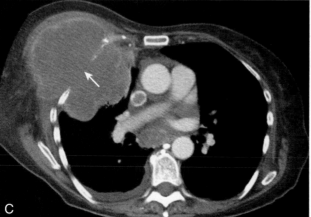

FIGURE 5-14 Metastases to the lung, CT scans. A, Hematogenous spread. There are multiple discrete nodules of varying size throughout both lungs *(white arrows)*. The diagnosis of exclusion, whenever multiple nodules are found in the lungs, is metastatic disease. In this case, the metastases were from colon carcinoma. **B,** Lymphangitic spread. The interstitial markings in the right lung are prominent *(solid white arrow)*, there are septal lines *(dotted black arrow)*, and lymphadenopathy *(solid black arrows)* representing lymphangitic spread of a bronchogenic carcinoma. **C,** Direct extension. In this case, the lung cancer has grown through the chest wall *(white arrow)* and invaded it by direct extension. The pleura usually serves as a strong barrier to the direct spread of tumor.

lymphangitic spread is obstruction of central lymphatics, usually in the hila, with retrograde dissemination through the lymphatics in the lung.

♦ Regardless of the mode of transmission, lymphangitic spread to the lung tends to **resemble pulmonary interstitial edema** from congestive heart failure except, unlike congestive heart failure, it tends to be **localized to a lung segment or involves only one lung.**

♦ Findings include **Kerley lines, fluid in the fissures,** and **pleural effusions** (see Fig. 5-14, *B*).

♦ Primary tumors that classically produce the lymphangitic pattern of metastases to the lung include **breast, lung, stomach, pancreatic,** and **infrequently, prostate carcinoma.**

♦ **Direct extension** is the least common form of tumor spread involving the lungs because the **pleura** is surprisingly **resistant to the spread of malignancy** through direct violation of its layers. Direct extension would most likely produce a **localized subpleural mass** in the lung, frequently with **adjacent rib destruction** (see Fig. 5-14, *C*).

Mixed Reticular and Nodular Interstitial Disease (Reticulonodular Disease)

■ **Sarcoidosis**

♦ In addition to the **bilateral hilar** and **right paratracheal adenopathy** characteristic of this disease, about half of patients with thoracic sarcoid also demonstrate **interstitial lung disease.** The interstitial lung disease is frequently a mixture of both reticular and nodular components.

♦ There is a progression of disease in sarcoid that tends to start with adenopathy **(stage I),** proceeds to a combination of both interstitial lung disease and adenopathy **(stage II),** and then progresses to a stage in which the adenopathy regresses while the interstitial lung disease remains **(stage III).**

♦ Most patients with parenchymal lung disease will undergo complete resolution of the disease (Fig. 5-15).

Mixed Airspace and Interstitial Disease

■ Unfortunately, not all diseases follow the rule of producing either airspace or interstitial disease. Some produce a mixture of both at the same time or may present as airspace disease followed by interstitial disease. **Tuberculosis** is one such disease.

Tuberculosis

■ Up to a third of the world's population is believed to have been infected with *Mycobacterium tuberculosis* (TB). While the incidence of TB is decreasing in the United States, it has been increasing in developing countries. Most infections cause no symptoms, but about one in ten go on to active tuberculosis.

Primary Pulmonary Tuberculosis

■ Relatively few patients with primary TB have clinical manifestations. The **upper lobes** are affected slightly more than the lower. Classical manifestations include **lobar pneumonia** (Fig. 5-16), especially with associated **adenopathy,** unilateral hilar or mediastinal adenopathy without parenchymal disease (more common in children), and large and

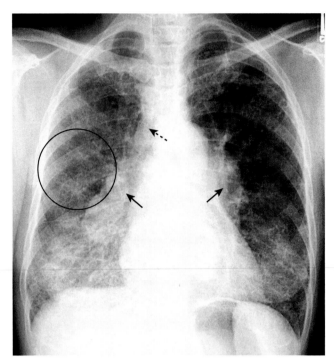

FIGURE 5-15 Sarcoidosis. A frontal radiograph of the chest reveals bilateral hilar *(solid black arrows)* and right paratracheal adenopathy *(dotted black arrow)*, a classical distribution for the adenopathy in sarcoidosis. In addition, the patient has diffuse, bilateral interstitial lung disease *(black circle)* that is reticulonodular in nature. In some patients with this stage of disease, the adenopathy regresses while the interstitial disease remains. In the overwhelming majority of patients with sarcoidosis, the disease completely resolves.

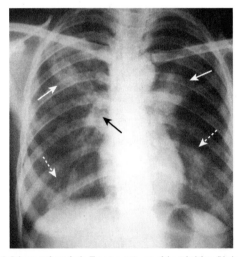

FIGURE 5-16 Primary tuberculosis. There is prominence of the right hilum *(black arrow)* caused by hilar adenopathy. Unilateral hilar adenopathy may be the only manifestation of primary infection with *Mycobacterium tuberculosis*, especially in children. When it produces pneumonia, primary TB affects the upper lobes *(solid white arrows)* slightly more than the lower lobes *(dotted white arrows)*.

typically asymptomatic **pleural effusions** (more common in adults) (Fig. 5-17). **Cavitation is rare.**

Postprimary Tuberculosis ("Reactivation TB")

■ Most cases of TB in adults occur as reactivation of a primary focus of infection acquired in childhood. The infection is limited mainly to the **apical and posterior segments of the upper lobes and the superior segments of the lower**

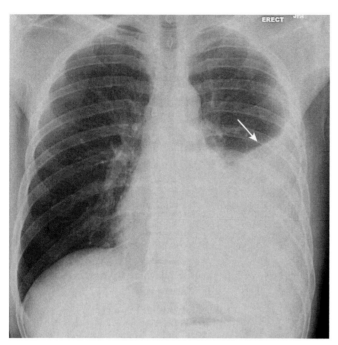

FIGURE 5-17 Tuberculous effusion. There is a large, left pleural effusion present. Tuberculous effusions are exudates which can occur in either primary or postprimary forms of the disease but are more common in the primary form. They are usually unilateral and have a tendency to loculate **(see Chapter 8).** They may grow large, as in this case, while the patient is still relatively symptom-free.

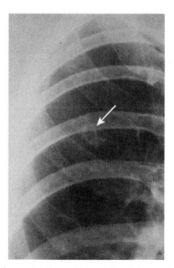

FIGURE 5-18 Tuberculous cavity. There is a thin-walled upper lobe cavity with no air–fluid level *(white arrow)*. The characteristics are consistent with tuberculosis. In the presence of a cavity, activity is best excluded on a clinical basis. This patient had clinical findings of active tuberculosis.

lobes. **Caseous necrosis** and the **tubercle** (accumulations of mononuclear macrophages, Langhans giant cells surrounded by lymphocytes and fibroblasts) are the pathologic hallmarks of postprimary TB.

■ **Healing** typically occurs with **fibrosis** and **contraction.**

Patterns of Distribution of Postprimary Tuberculosis

■ Bilateral upper lobe **cavitary disease** is very common. The cavity is usually thin walled, smooth on the inner margin, and contains no air–fluid level (Fig. 5-18).

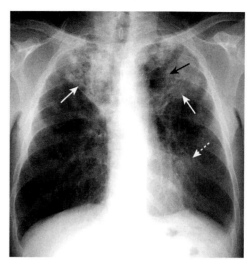

FIGURE 5-19 Post-primary TB with transbronchial spread. There is a cavitary pneumonia in both upper lobes *(solid white arrows)*. Numerous lucencies (cavities) are seen in the airspace disease in both upper lobes *(solid black arrow on the left)*. A cavitary upper lobe pneumonia is presumptively TB, until proven otherwise. In addition, there is airspace disease in the lingula *(dotted white arrow)*, another finding suggestive of TB, a disease which can spread via a transbronchial route to the opposite lower lobe or another lobe in either lung.

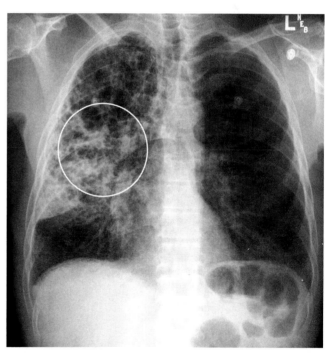

FIGURE 5-20 Tuberculous bronchiectasis. Bronchiectasis secondary to tuberculosis, both primary and postprimary, is relatively common. It may occur as a result of endobronchial infection or adjacent fibrosis *(traction bronchiectasis)*. The apical and posterior segments of the upper lobes are the most common sites. Multiple, small cystic structures creating an appearance that resembles a honeycomb may be seen *(white circle)*.

■ It may present as **pneumonia.**
■ **Transbronchial spread** may occur from one upper lobe to the opposite lower lobe or to another lobe in either lung (Fig. 5-19).
■ **Bronchiectasis**—usually asymptomatic (Fig. 5-20).

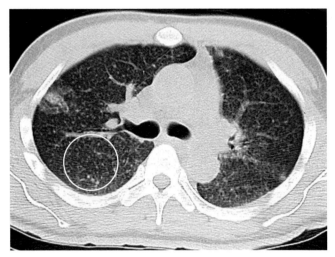

FIGURE 5-21 Miliary tuberculosis. There are innumerable small round nodules present in this axial computed tomography scan of the chest in a patient with miliary tuberculosis *(white circle)*. At the start of the disease, the nodules are so small they are frequently difficult to detect on conventional radiographs. When they reach about 1 mm or more is size, they begin to become visible. Miliary tuberculosis represents the widespread hematogenous dissemination of the tubercle bacillus.

- **Bronchostenosis** is due to fibrosis and stricture. Fibrosis may cause distortion of a bronchus and atelectasis many years after the initial infection—**"middle lobe syndrome."**
- Solitary pulmonary nodule—the **tuberculoma**—may occur in either primary or postprimary disease. These are round or oval lesions frequently associated with small, discrete shadows in the immediate vicinity of the lesion—the **"satellite"** lesion.
- Formation of a pleural effusion in postprimary TB almost always means direct spread of the disease into the pleural cavity and should be regarded as an **empyema**—this carries a graver prognosis than the pleural effusion of the primary form.

Miliary Tuberculosis

- The onset is insidious. Fever, chills, and night **sweats are common.** It may take weeks between the time of dissemination and the radiographic appearance of disease. Miliary TB may occur as a manifestation of either primary TB or postprimary TB, although the clinical appearance of miliary TB may not occur for many years after initial infection.
- When first visible, the miliary nodules **measure about 1 mm in size**; they can grow to 2 to 3 mm if left untreated. When treated, clearing is rapid. Miliary TB seldom, if ever, heals with calcifications (Fig. 5-21).

🏠 **TAKE-HOME POINTS**

AIRSPACE VERSUS INTERSTITIAL LUNG DISEASE

Parenchymal lung disease can be divided into airspace (alveolar) and interstitial (infiltrative) patterns.

Recognizing the pattern of disease can help in reaching the correct diagnosis.

Characteristics of airspace disease include: fluffy, confluent densities that are indistinctly marginated and may demonstrate air bronchograms.

An *air bronchogram* is typically a sign of airspace disease and occurs when something other than air (such as inflammatory exudate or blood) surrounds the bronchus allowing the air inside the bronchus to become visible.

When two objects of the same radiographic density are in contact with each other, the normal edge or margin between them will disappear. The disappearance of the margin between these two structures is called the *silhouette sign* and is useful throughout radiology in identifying either the location or the density of the abnormality in question.

Examples of airspace disease include pulmonary alveolar edema, pneumonia, and aspiration.

Characteristics of interstitial lung disease include discrete "particles" or "packets" of disease with distinct margins that tend to occur in a pattern of lines (reticular), dots (nodular), or very frequently, a combination of lines and dots (reticulonodular).

Examples of interstitial lung disease include pulmonary interstitial edema, pulmonary fibrosis, metastases to the lung, bronchogenic carcinoma, sarcoidosis, and interstitial pneumonia.

Tuberculosis is an example of a disease process that can demonstrate both airspace and interstitial lung patterns.

 WEBLINK

Visit StudentConsult.Inkling.com for more information and quizzes on Basic Radiographic Principles and Airspace vs. Interstitial Disease.

For your convenience, the following QR Code may be used to link to **StudentConsult.com**. You must register this title using the PIN code on the inside front cover of the text to access online content.

CHAPTER 6
Recognizing the Causes of an Opacified Hemithorax

- Mr. Smith, age 31, comes to the emergency department acutely short of breath. His frontal chest radiograph is shown in Figure 6-1.
- As you can see, Mr. Smith's left hemithorax is completely opaque.
  - Mr. Smith's treatment will vary greatly depending on whether he has **atelectasis** (which may require emergent bronchoscopy), a large **effusion** (which may require emergent thoracentesis), or **pneumonia** (which would require starting antibiotics).
  - You have to determine the cause of the opacification to be able to treat him correctly. The answer to that question is demonstrated on the radiograph, if you know how to approach the problem.

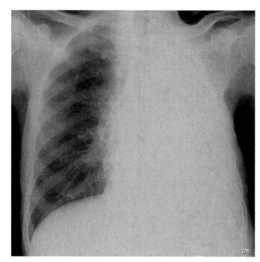

FIGURE 6-1 Mr. Smith, a 31-year-old male, comes into the emergency department acutely short of breath. This is his frontal chest radiograph. Would you recommend bronchoscopy for atelectasis, emergent thoracentesis for a large pleural effusion, or a course of antibiotics for his large pneumonia? The answer is on the radiograph (and in this chapter).

- There are three major causes of an opacified hemithorax and one other that is less common. They are:
  - **Atelectasis of the entire lung**
  - **A very large pleural effusion**
  - Pneumonia of an entire lung
  - *And a fourth cause:* **Pneumonectomy**—removal of an entire lung

ATELECTASIS OF THE ENTIRE LUNG

- **Atelectasis of an entire lung** usually results from **complete obstruction of the right or left main bronchus.** With bronchial obstruction, no air can enter the lung. The remaining air in the lung is absorbed into the bloodstream through the pulmonary capillary system. This leads to **loss of volume** of the affected lung.
- In an older individual, this might be caused by an **obstructing neoplasm** such as a bronchogenic carcinoma. In younger individuals, **asthma** may produce mucus plugs that obstruct the bronchi, or a **foreign body** may have been aspirated (in children, peanuts are a frequent culprit). Critically ill patients also develop atelectasis from **mucus plugs.**
- **In obstructive atelectasis,** even though there is volume loss within the affected lung, **the visceral and parietal pleura almost never separate from each other.** That is an important fact about atelectasis and is sometimes confusing to beginners who try to picture **atelectasis** and a **pneumothorax** as both producing collapse of a lung, without understanding why they look completely different (Table 6-1 and Fig. 6-2).
  - Because the visceral and parietal pleura **do not separate** from each other in atelectasis, mobile structures in the thorax are "pulled" toward the side of the atelectasis, producing a *shift* (movement) of these mobile thoracic structures **toward the side of opacification.**
  - The most visible mobile structures in the thorax are the **heart,** the **trachea,** and the **hemidiaphragms.**

TABLE 6-1 PNEUMOTHORAX VERSUS OBSTRUCTIVE ATELECTASIS

Feature	Pneumothorax	Obstructive Atelectasis
Pleural space	Air in the pleural space separates the visceral from the parietal pleura.	The visceral and parietal pleura do not separate from each other.
Density	The pneumothorax itself will appear "black" (air density). The hemithorax may appear more lucent than normal.	Atelectasis is the absence of air in the lung. The hemithorax will appear more opaque ("whiter") than normal.
Shift	There is never a shift of the heart or trachea toward the side of a pneumothorax.	There is almost always a shift of the heart and trachea toward the side of the atelectasis.

TABLE 6-2   RECOGNIZING A "SHIFT" IN ATELECTASIS/PNEUMONECTOMY

Structure	Normal Position	Right-Sided Atelectasis or Pneumonectomy	Left-Sided Atelectasis or Pneumonectomy
Heart	Midline	Heart moves rightward; left heart border may come to lie near left side of spine	Heart moves leftward; right heart border overlaps the spine
Trachea	Midline	Shifts toward right	Shifts toward left
Hemidiaphragm	Right slightly higher than left	Right hemidiaphragm moves upward and may disappear (silhouette sign)	Left hemidiaphragm moves upward and may disappear (silhouette sign)

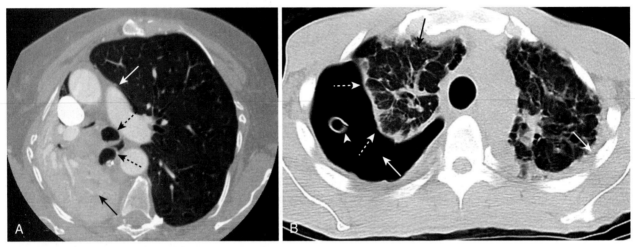

FIGURE 6-2 Obstructive atelectasis versus a pneumothorax. Two different causes of lung collapse and the difference in their radiologic appearance are shown. **A,** There is atelectasis of the entire right lung *(solid black arrow)* from an obstructing endobronchial lesion. The visceral and parietal pleura remain in contact with each other. Other mobile structures in the mediastinum, such as the trachea and right main bronchus *(dotted black arrows)*, shift toward the atelectasis. The left lung overexpands and crosses the midline *(solid white arrow)*. **B,** This is another patient who has a large right-sided pneumothorax. Air *(solid white arrow)* interposes between the visceral *(dotted white arrows)* and parietal pleurae, causing the lung to undergo passive atelectasis *(solid black arrow)*. There is a chest tube in the right hemithorax *(arrowhead)* that had been removed from suction.

> In obstructive atelectasis, one or all of these structures will shift toward the side of opacification (toward side of volume loss) (Fig. 6-3).

- ♦ Table 6-2 summarizes the movement of the mobile structures in the thorax in patients with atelectasis.

MASSIVE PLEURAL EFFUSION

- If fluid such as blood, exudate, or a transudate fills the pleural space so as to opacify almost the entire hemithorax, then the **fluid may act like a mass** compressing the underlying lung tissue.

> When enough pleural fluid accumulates, the **large effusion "pushes" mobile structures away** and there is a shift of the heart and trachea away from the side of opacification (Fig. 6-4).

- Massive pleural effusions are frequently the result of malignancy, either in the form of a **bronchogenic carcinoma** or secondary to **metastases to the pleura** from a distant organ. Trauma can produce a **hemothorax,** and **tuberculosis** is notorious for causing large, clinically silent effusions (see Fig. 5-17). The effusions from congestive heart failure are very common and are most often *bilateral (but asymmetrical),* but they rarely grow large enough to occupy an entire hemithorax.

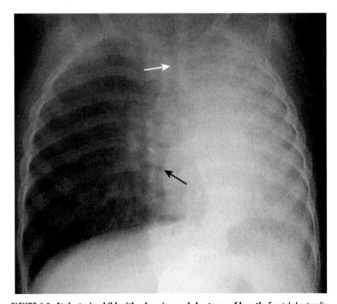

FIGURE 6-3 Atelectasis: child with wheezing and shortness of breath. Frontal chest radiograph shows opacification of the entire left hemithorax. There is a shift of the heart toward the left such that the right heart border no longer projects to the right of the spine. The heart now overlies the spine *(black arrow)*. The trachea *(white arrow)* has moved leftward from the midline toward the side of the opacification. These findings are characteristic of atelectasis of the entire lung. The child had asthma. Bronchoscopy was performed, and a large mucus plug that was obstructing the left main bronchus was removed.

At times, there may be a perfect **balance** between the **push** of a malignant effusion and the **pull** of underlying obstructive atelectasis from the malignancy itself.

■ In an adult patient with an opacified hemithorax, no air bronchograms, and little or no shift of the mobile thoracic structures, it is important to suspect an obstructing bronchogenic carcinoma, perhaps with metastases to the pleura. A computed tomography (CT) scan of the chest will reveal the abnormalities (Fig. 6-5).

■ Table 6-3 summarizes the movement of the mobile structures in the thorax in patients with a large pleural effusion.

PNEUMONIA OF AN ENTIRE LUNG

■ Inflammatory exudate fills the air spaces with pneumonia, causing consolidation and opacification of the lung.

➡ The hemithorax becomes opaque because the lung no longer contains air, but there is **neither a pull toward** the side of the pneumonia by volume loss **nor a push away** from the side of the pneumonia by a large effusion. There is **no shift of the heart or trachea.**

♦ There **may be air bronchograms** present (Fig. 6-6).

■ Table 6-4 summarizes the movement of the mobile structures in the thorax in patients with pneumonia of the entire lung.

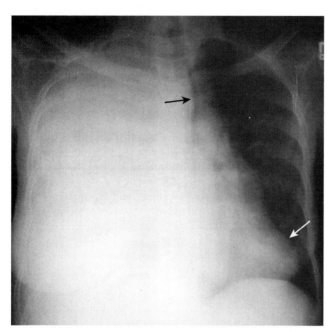

FIGURE 6-4 Large pleural effusion. There is complete opacification of the right hemithorax. The trachea is deviated to the left *(black arrow)*, and the apex of the heart is also displaced to the left, close to the lateral chest wall *(white arrow)*. These findings are characteristic of a large pleural effusion which is producing a mass effect. Almost 2 L of serosanguinous fluid were removed at thoracentesis. The fluid contained malignant cells from a primary bronchogenic carcinoma.

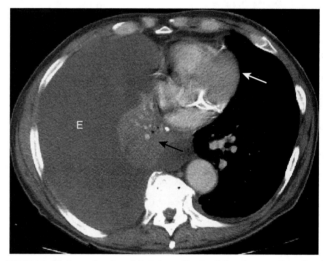

FIGURE 6-5 Effusion and atelectasis in balance. There is a balance between a large pleural effusion (E) and atelectasis of the right lung *(black arrow)* so that there is no significant resultant shift of the mobile midline structures. The heart remains in essentially its normal position *(white arrow)*. This combination of findings is highly suggestive of a central bronchogenic malignancy with a malignant effusion.

TABLE 6-3   RECOGNIZING A "SHIFT" IN PLEURAL EFFUSION

Structure	Normal Position	Right-Sided Effusion	Left-Sided Effusion
Heart	Midline	Heart moves leftward; apex may lie near chest wall	Heart moves rightward; more of heart protrudes to right of spine
Trachea	Midline	Shifts toward left	Shifts toward right
Hemidiaphragm	Right higher than left	Right hemidiaphragm disappears on chest radiograph (silhouette sign)	Left hemidiaphragm disappears on chest radiograph (silhouette sign)

TABLE 6-4   RECOGNIZING A "SHIFT" IN PNEUMONIA

Structure	Normal Position	Right-Sided Pneumonia	Left-Sided Pneumonia
Heart	Midline	There is usually no shift of the heart from its normal position	There is usually no shift of the heart from its normal position
Trachea	Midline	Midline	Midline
Hemidiaphragm	Right higher than left	Right hemidiaphragm may disappear on chest radiograph (silhouette sign)	Left hemidiaphragm may disappear on chest radiograph (silhouette sign)

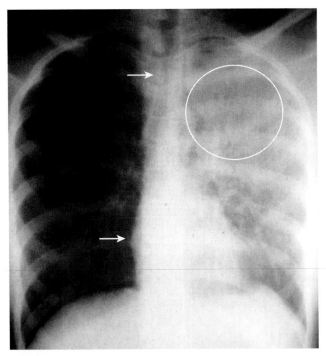

FIGURE 6-6 Pneumonia of the left upper lobe. There is near-complete opacification of the left hemithorax with no shift of the heart and little shift of the trachea *(white arrows)*. There are air bronchograms suggested within the upper area of opacification *(circle)*. These findings suggest pneumonia rather than atelectasis or pleural effusion. The patient had *Streptococcus pneumoniae* present in the sputum and improved quickly on antibiotics.

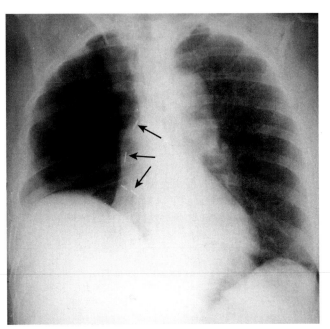

FIGURE 6-7 Postpneumonectomy day 1, right hemithorax. A pneumonectomy is the removal of the entire lung. This postoperative radiograph was obtained less than 24 hours after this patient underwent a pneumonectomy on the right side for a bronchogenic carcinoma. There are surgical clips in the region of the right hilum *(black arrows)*, and the right 5th rib has been surgically removed in order to perform the pneumonectomy. Over the next several weeks, the right hemithorax will fill with fluid, and there will be a gradual shift of the heart and mediastinal structures toward the side of the pneumonectomy (see Fig. 6-8).

POSTPNEUMONECTOMY

■ Pneumonectomy means the removal of an entire lung.

 In order to perform this procedure, **either the 5th or 6th rib on the affected side is almost always removed.** In most cases, **metallic surgical clips will be visible in the region of the hilum** on the side of the pneumonectomy.

■ For about 24 hours following the surgery, only air occupies the hemithorax from which the lung has been removed (Fig. 6-7).
■ Over the course of the next 2 weeks, the hemithorax gradually **fills with fluid.**
■ By about 4 months after surgery, the pneumonectomized hemithorax should be **completely opaque.**
■ Eventually, **fibrous tissue forms in the pneumonectomized hemithorax,** and in most patients the **entire hemithorax is completely opaque.** The heart and trachea shift toward the side of opacification.
  ◆ The chest examination looks identical to that of a patient with atelectasis of the entire lung. To tell the difference, look for the missing 5th or 6th rib and look for the surgical clips in the hilum to indicate a pneumonectomy has been performed (Fig. 6-8).

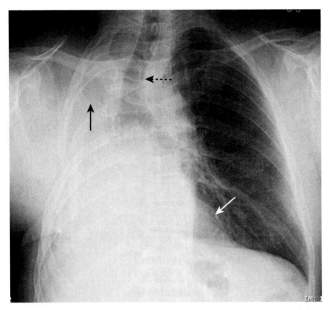

FIGURE 6-8 One year after pneumonectomy. There is complete opacification of the right hemithorax. The right 5th rib *(solid black arrow)* is surgically absent. The heart *(solid white arrow)* and trachea *(dotted black arrow)* are deviated toward the side of opacification. These signs are characteristic of volume loss. The surgery had been performed 1 year earlier for a bronchogenic carcinoma. The fluid that gradually filled the right hemithorax immediately following the pneumonectomy has probably fibrosed, leading to a permanent shift toward the pneumonectomized side.

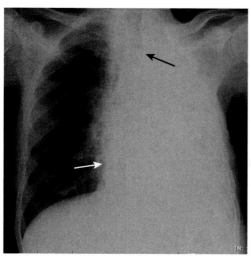

FIGURE 6-9 Mr. Smith's frontal chest radiograph. There is opacification of the entire left hemithorax. There is also a shift of the trachea toward the left *(black arrow)*, and the heart is displaced toward the left as well *(white arrow)*. Both of these mobile structures have moved toward the side of opacification. These signs are characteristic of atelectasis of the entire left lung, and in a patient like Mr. Smith, who is 31 and has a history of asthma, a mucus plug is the most likely cause. The plug was removed bronchoscopically.

 TAKE-HOME POINTS

RECOGNIZING THE CAUSES OF AN OPACIFIED HEMITHORAX

The differential possibilities for an opacified hemithorax should include atelectasis of the entire lung, a very large pleural effusion, pneumonia of the entire lung, or postpneumonectomy.

The trachea, heart, and hemidiaphragms are mobile structures that have the capability of moving (**shifting**) if there is either something pushing them or something pulling them.

With atelectasis, there is a shift **toward** the side of the opacified hemithorax because of volume loss in the affected lung.

With a large pleural effusion, there is a shift **away** from the side of opacification because the large pleural effusion can act as if it were a mass.

With pneumonia of an entire lung, there is usually **no shift,** but air bronchograms may be present.

Occasionally, the shift of a malignant effusion may be balanced by the opposite shift of atelectasis caused by an underlying, obstructing bronchogenic carcinoma so that the hemithorax will be completely opaque, but there will be no shift of the midline structures.

In the postpneumonectomy patient, there is eventually volume loss on the side from which the lung has been removed, and the clues to such surgery may include surgical absence of the 5th or 6th rib on the affected side or metallic surgical clips in the hilum.

- So let us return to the frontal radiograph of Mr. Smith with the opacified hemithorax, who has been waiting patiently in the emergency department while you finished reading this chapter.
  - ♦ Now, how would you proceed with his abnormality?
    - You will notice there is a shift of the heart and trachea **toward** the side of opacification (Fig. 6-9). This is characteristic of **atelectasis of the entire lung.** Because of his age (31) and his prior history of asthma (an historic question you asked him), this most likely represents atelectasis from an obstructing mucus plug.
    - Mr. Smith had a CT scan performed, which showed the obstruction was in the left main bronchus and the plug was then removed using a bronchoscope.

 WEBLINK

Visit StudentConsult.Inkling.com for more information and a quiz on The Opacified Hemithorax.

For your convenience, the following QR Code may be used to link to **StudentConsult.com**. You must register this title using the PIN code on the inside front cover of the text to access online content.

CHAPTER 7
Recognizing Atelectasis

WHAT IS ATELECTASIS?

- Common to all forms of atelectasis is a **loss of volume in some or all of the lung, usually leading to increased density of the lung involved.**
  - ◆ The lung normally appears "black" on a chest radiograph because it contains air. When fluid or soft tissue density is substituted for that air or when the air in the lung is resorbed (as it can be in atelectasis), that part of the lung becomes **whiter (denser** or **more opaque).**
- Unless mentioned otherwise, statements in this chapter that refer to *atelectasis* are referring to *obstructive atelectasis.* This might be a good time to review Table 6-1, highlighting the markedly different appearances of the thorax in a large pneumothorax and atelectasis of the entire lung (see Fig. 6-2).

➡ **Signs of atelectasis**

- **Displacement (shift) of the interlobar fissures** (major and minor) **toward** the area of atelectasis
- **Increase in the density of the affected lung** (Fig. 7-1)

- **Displacement (shift) of the mobile structures of the thorax.** The **mobile structures** are those capable of movement due to changes in lung volume:
  - ◆ **Trachea**
    - • **Normally midline in location** and centered on the spinous processes of the vertebral bodies (also midline structures) on a nonrotated, frontal chest x-ray. There is **always a slight rightward deviation of the trachea at the site of the left-sided aortic knob.**
    - • With atelectasis, especially of the upper lobes, the trachea may shift **toward** the side of the volume loss (Fig. 7-2).
  - ◆ **Heart**
    - • **At least 1 cm of the right heart border normally projects to the right of the spine** on a nonrotated, frontal radiograph.
    - • With atelectasis, especially of the lower lobes, the heart may shift to one side or the other. When the heart shifts toward the **left,** the right heart border will overlap the spine (Fig. 7-3). When the heart shifts toward the **right,** the left heart border will approach the midline (Fig. 7-4).

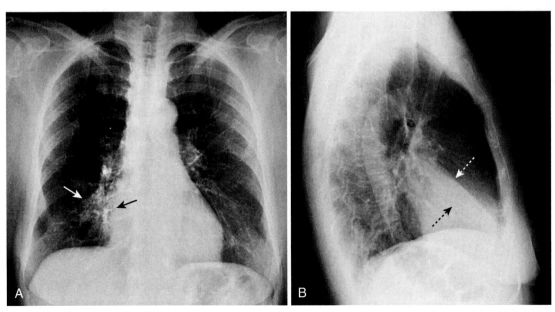

FIGURE 7-1 Right middle lobe atelectasis. A, Frontal view of the chest shows an area of increased density *(solid white arrow),* which is silhouetting the normal right heart border *(solid black arrow),* indicating its anterior location in the right middle lobe. **B,** On the lateral view, the minor fissure is displaced downward *(dotted white arrow),* and the major fissure is displaced slightly upward *(dotted black arrow).* Note the anterior location of the middle lobe.

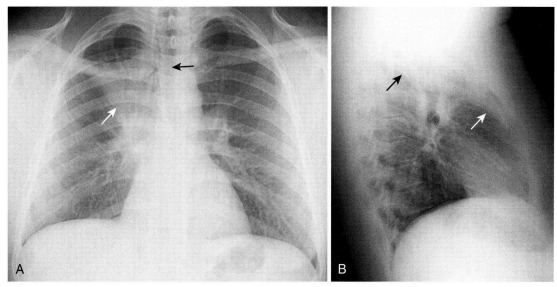

FIGURE 7-2 Right upper lobe atelectasis. A fan-shaped area of increased density is seen on the frontal projection **(A)** representing the airless right upper lobe. The minor fissure is displaced upward *(white arrow)*. The trachea is shifted to the right *(black arrow)*. **B,** The lateral view demonstrates a similar wedge-shaped density near the apex of the lung. The minor fissure *(white arrow)* is pulled upward and the major fissure is pulled forward *(black arrow)*. This is a child who had asthma, leading to formation of a mucus plug obstructing the right upper lobe bronchus.

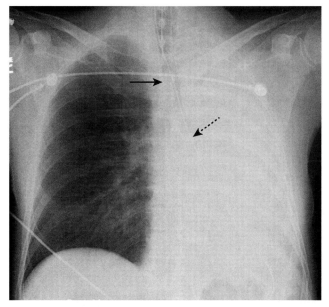

FIGURE 7-3 Atelectasis of the left lung. There is complete opacification of the left hemithorax with shift of the trachea *(solid black arrow)* and the esophagus (marked here by a nasogastric tube, *dotted black arrow*) toward the side of the atelectasis. The right heart border, which should project about a centimeter to the right of the spine, has been pulled to the left side and is no longer visible. The heart itself is not visible because it is no longer bordered by an air-filled lung. The patient had an obstructing bronchogenic carcinoma in the left main bronchus.

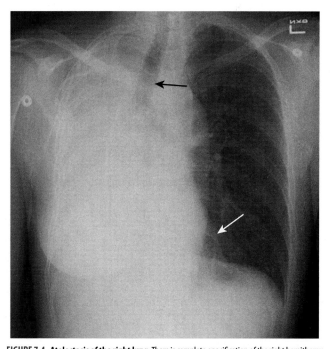

FIGURE 7-4 Atelectasis of the right lung. There is complete opacification of the right hemithorax with shift of the trachea *(black arrow)* toward the side of the atelectasis. The left heart border is displaced far to the right and now almost overlaps the spine *(white arrow)*. This patient had an endobronchial metastasis in the right main bronchus from her left-sided breast cancer. Did you notice that the left breast is surgically absent?

♦ **Hemidiaphragm**
   • The **right hemidiaphragm is almost always higher than the left** by about half the interspace distance between two adjacent ribs. In about 10% of normal people, the left hemidiaphragm is higher than the right.
   • In the presence of atelectasis**, especially of the **lower lobes,** the hemidiaphragm on the affected side will usually be **displaced upward** (Fig. 7-5).

■ **Overinflation of the unaffected ipsilateral lobes or the contralateral lung**
   ♦ The greater the volume loss and the more chronic its presence, the more the lung on the side **opposite** the atelectasis or the **unaffected lobe(s) in the ipsilateral lung** will attempt to **overinflate** to compensate for the volume loss. This may be noticeable on the lateral

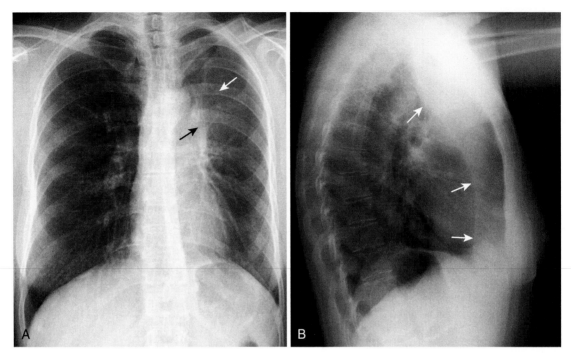

FIGURE 7-5 Left upper lobe atelectasis. A, On the frontal projection there is a hazy density surrounding the left hilum *(white arrow)*, and there is a soft tissue mass in the left hilum *(black arrow)*. Notice how the left hemidiaphragm has been pulled up to the same level as the right. **B,** The lateral projection shows a bandlike zone of increased density *(white arrows)* representing the atelectatic left upper lobe sharply demarcated by the major fissure, which has been pulled anteriorly. The patient had a squamous cell carcinoma of the left upper lobe bronchus that was producing complete obstruction of that bronchus.

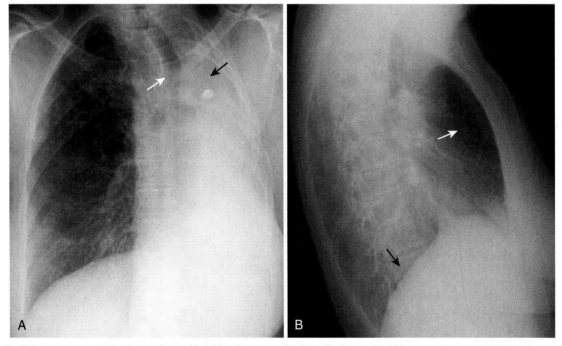

FIGURE 7-6 Left-sided pneumonectomy. A, Complete opacification of the left hemithorax is most likely from a fibrothorax produced following complete removal of the lung. There is associated marked volume loss with shift of the trachea to the left *(white arrow)*. The left 5th rib was surgically removed during the pneumonectomy *(black arrow)*. **B,** The right lung has herniated across the midline in an attempt to "fill up" the left hemithorax, which is seen by the increased lucency behind the sternum *(white arrow)*. Notice that because only the right hemithorax has an aerated lung remaining, only the right hemidiaphragm is visible on the lateral projection *(black arrow)*. The left hemidiaphragm has been silhouetted by the airless hemithorax above it.

projection by an **increase in the size of the retrosternal clear space** and on the frontal projection by **extension of the overinflated contralateral lung across the midline** (Fig. 7-6).

■ The signs of atelectasis are summarized in Box 7-1.

TYPES OF ATELECTASIS

■ **Subsegmental atelectasis** (also called *discoid atelectasis* or *platelike atelectasis*) (Fig. 7-7)

♦ **Subsegmental atelectasis produces linear densities of varying thickness usually parallel to the**

BOX 7-1 SIGNS OF ATELECTASIS

Displacement of the major or minor fissure*

Increased density of the atelectatic portion of lung

Shift of the mobile structures in the thorax (i.e., the heart, trachea and/or hemidiaphragms*)

Compensatory overinflation of the unaffected segments, lobes or lung

*Toward the atelectasis.

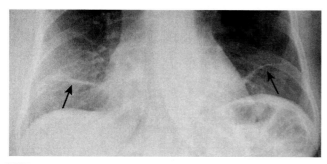

FIGURE 7-7 Subsegmental atelectasis. Subsegmental atelectasis. Close-up view of the lung bases demonstrates several linear densities extending across all segments of the lower lobes, paralleling the diaphragm *(black arrows)*. This is a characteristic appearance of subsegmental atelectasis, sometimes also called *discoid atelectasis* or *platelike atelectasis*. The patient was postoperative from abdominal surgery and was unable to take a deep breath. The atelectasis disappeared within a few days after surgery.

diaphragm, most commonly **at the lung bases.** It does not produce a sufficient amount of volume loss to cause a shift of the mobile thoracic structures.

♦ It occurs **mostly in patients who are "splinting,"** that is, not taking a deep breath, such as **postoperative patients** or patients with **pleuritic chest pain.**

• **Subsegmental atelectasis is not due to bronchial obstruction.** It is most likely related to **deactivation of surfactant,** which leads to collapse of airspaces in a nonsegmental or nonlobar distribution.

♦ On a single study, without prior examinations for comparison, subsegmental atelectasis and **chronic, linear scarring can look identical.** Subsegmental atelectasis **typically disappears within a matter of days** with resumption of normal, deep breathing, whereas **scarring remains.**

■ **Compressive atelectasis**

♦ Loss of volume due to **passive compression of the lung** can be caused by:

• A poor inspiratory effort in which there is passive atelectasis of the lung at the bases (Fig. 7-8, *A*)

• A **large pleural effusion, large pneumothorax,** or a **space-occupying lesion** (such as a large mass in the lung) (see Fig. 7-8, *B*)

♦ When caused by a poor inspiratory effort, passive atelectasis may mimic airspace disease at the bases.

⚠ **Pitfall:** Be suspicious of compressive atelectasis if the patient has taken less than an 8 posterior-rib breath.

• **Solution:** Check the lateral projection for confirmation of the presence of real airspace disease at the base.

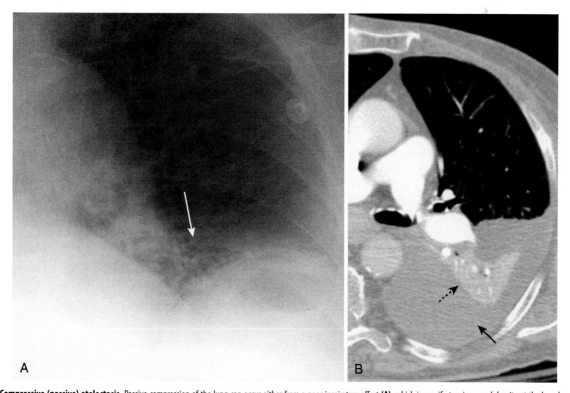

FIGURE 7-8 Compressive (passive) atelectasis. Passive compression of the lung can occur either from a poor inspiratory effort **(A)**, which is manifest as increased density at the lung bases *(solid white arrow)*, or secondary to a large pleural effusion or pneumothorax. **B,** Axial computed tomography scan of the chest in another patient showing only the left hemithorax demonstrates a large left pleural effusion *(solid black arrow)*. The left lower lobe *(dotted black arrow)* is atelectatic, having been compressed by the pleural fluid surrounding it.

♦ When caused by a large effusion or pneumothorax, the loss of volume associated with compressive atelectasis may balance the increased volume produced by either fluid (as in pleural effusion) or air (as in pneumothorax). In an adult patient with an **opacified hemithorax, no air bronchograms,** and **little or no shift of the mobile thoracic structures,** it is important to **suspect an obstructing bronchogenic carcinoma,** perhaps with metastases to the pleura (Fig. 7-9).

■ **Round atelectasis**
  ♦ This form of compressive atelectasis is **usually seen at the periphery of the lung base** and develops from a **combination of prior pleural disease** (such as from asbestos exposure or tuberculosis) **and the formation of a pleural effusion that produces adjacent compressive atelectasis.**
  ♦ When the pleural effusion recedes, the underlying pleural disease leads to a portion of the **atelectatic lung becoming "trapped."** This **produces a masslike lesion** that can be confused for a tumor.
    • On a computed tomography (CT) scan of the chest, the bronchovascular markings characteristically lead

from the **round atelectasis** back to the hilum, producing a **comet-tail** appearance (Fig. 7-10).

■ **Obstructive atelectasis** (see Fig. 7-3)
  ♦ Obstructive atelectasis is associated with the **resorption of air from the alveoli,** through the pulmonary capillary bed, **distal to an obstructing lesion** of the bronchial tree.
    • The rate at which air is absorbed and the lung collapses depends on its gas content when the bronchus is occluded. It takes about **18 to 24 hours for an entire lung to collapse** with the patient breathing room air but **less than an hour** with the patient breathing near 100% oxygen.
  ♦ The **affected segment, lobe,** or **lung collapses** and becomes more opaque (whiter) because it contains no air. The **collapse leads to volume loss** in the affected segment/lobe/lung. Because the visceral and parietal pleura invariably remain in contact with each other as the lung loses volume, there is a pull **on the mobile structures of the thorax toward the area of atelectasis.**

■ The types of atelectasis are summarized in Table 7-1.

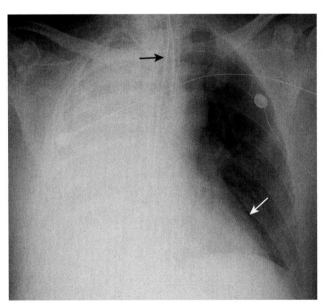

FIGURE 7-9 Atelectasis and effusion in balance, an ominous combination. There is complete opacification of the right hemithorax. There are neither air bronchograms to suggest pneumonia, nor any shift of the trachea *(black arrow)* or heart *(white arrow).* The absence of any shift suggests the possibility of atelectasis and pleural effusion in balance, a combination that should raise suspicion for a central bronchogenic carcinoma (producing obstructive atelectasis) with metastases (producing a large pleural effusion).

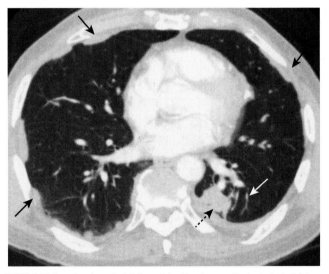

FIGURE 7-10 Round atelectasis, left lower lobe. There is a masslike density in the left lower lobe *(dotted black arrow).* The patient has underlying pleural disease in the form of pleural plaques from asbestos exposure *(solid black arrows).* There are comet-tail shaped bronchovascular markings that emanate from the "mass" and extend back to the hilum *(solid white arrow).* This combination of findings is characteristic of round atelectasis and should not be mistaken for a tumor.

TABLE 7-1 TYPES OF ATELECTASIS

Type	Associated With	Remarks
Subsegmental atelectasis	Splinting, especially in postoperative patients and those with pleuritic chest pain	May be related to deactivation of surfactant; does not usually lead to volume loss; disappears in days
Compressive atelectasis	Passive external compression of the lung from poor inspiration, pneumothorax, or pleural effusion	Volume loss of compressive atelectasis can balance volume increase from effusion or pneumothorax resulting in no shift; *round atelectasis* is a form of compressive atelectasis
Obstructive atelectasis	Obstruction of a bronchus from malignancy or mucus plugging	Visceral and parietal pleura maintain contact; mobile structures in the thorax are pulled toward the atelectasis

PATTERNS OF COLLAPSE IN LOBAR ATELECTASIS

- Obstructive atelectasis produces **consistently recognizable patterns of collapse** depending on the location of the atelectatic segment or lobe and the degree to which such factors as collateral airflow between lobes and obstructive pneumonia allow the affected lobe to collapse.
- In general, lobes collapse in a fanlike configuration with the **base** of the fan-shaped triangle **anchored at the pleural surface** and the **apex of the triangle anchored at the hilum.**
- Other unaffected lobes will undergo compensatory hyperinflation in an attempt to "fill" the affected hemithorax, and this hyperinflation may limit the amount of shift of the mobile chest structures.

 Pitfall: The **more atelectatic a lobe or segment becomes** (i.e., the smaller its volume), the **less visible it becomes on the chest radiograph.** This can lead to the false assumption of improvement when, in fact, the atelectasis is worsening.

- ◆ **Solution:** This can usually be resolved with a careful analysis of the study to check for the degree of displacement of the interlobar fissures or hemidiaphragms or with a CT scan of the chest.
- **Right upper lobe atelectasis** (see Fig. 7-2)
  - ◆ On the frontal radiograph:
    - There is an upward shift of the minor fissure.
    - There is a rightward shift of the trachea.
  - ◆ On the lateral radiograph:
    - There is an upward shift of the minor fissure and forward shift of the major fissure.
  - ◆ If there is a large enough mass in the right hilum, producing right upper lobe atelectasis, the combination of the **hilar mass** and the **upward shift of the minor fissure** produces a characteristic appearance on the frontal radiograph called the *S sign of Golden* (Fig. 7-11).
- **Left upper lobe atelectasis** (see Fig. 7-5)
  - ◆ On the frontal radiograph:
    - There is a hazy area of increased density surrounding the left hilum.
    - There is a leftward shift of the trachea.
    - There may be elevation with **"tenting" (peaking)** of the left hemidiaphragm.
    - Compensatory overinflation of the lower lobe may cause the superior segment of the left lower lobe to extend to the apex of the thorax on the affected side.
  - ◆ On the lateral radiograph:
    - There is forward displacement of the major fissure, and the opacified upper lobe forms a band of increased density running roughly parallel to the sternum.
- **Lower lobe atelectasis** (Fig. 7-12, *A* and Fig. 7-12, *B*)
  - ◆ On the frontal radiograph:
    - Both the right and left lower lobes collapse to form a triangular density that extends from its apex at the hilum to its base at the medial portion of the affected hemidiaphragm.

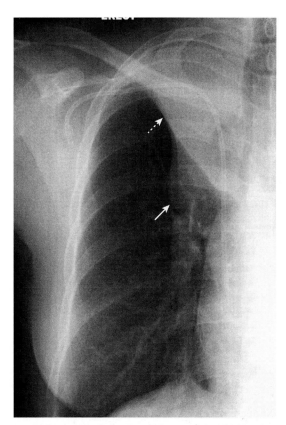

FIGURE 7-11 Right upper lobe atelectasis and hilar mass: S sign of Golden. There is a soft tissue mass in the right hilum *(solid white arrow)* which is causing opacification of the right upper lobe from atelectasis. The minor fissure is displaced upward toward the area of increased density *(dotted white arrow)* indicating right upper lobe volume loss. The shape of the curved edge formed by the mass and the elevated minor fissure is called the *S sign of Golden.* The patient had a large squamous cell carcinoma obstructing the right upper lobe bronchus.

- There is elevation of the hemidiaphragm on the affected side.
- The heart may shift toward the side of the volume loss.
- There is a downward shift of the major fissure (see Fig. 7-12, *C*).
  - ◆ On the lateral radiograph:
    - There is both downward and posterior displacement of the major fissure until the completely collapsed lower lobe forms a small triangular density at the posterior costophrenic angle (see Fig. 7-12, *B*).

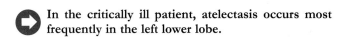 In the critically ill patient, atelectasis occurs most frequently in the left lower lobe.

- **Always check the left hemidiaphragm to be sure it is seen in its entire extent** through the heart shadow since **left lower lobe atelectasis will manifest by disappearance (silhouetting) of all or part of the left hemidiaphragm** (see Fig. 7-12, *A*).
- **Right middle lobe atelectasis** (see Fig. 7-1)
  - ◆ On the frontal radiograph:
    - There is a triangular density with its base silhouetting the right heart border and its apex pointing toward the lateral chest wall.
    - The minor fissure is displaced downward.

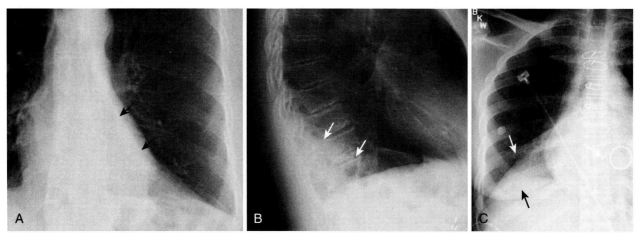

FIGURE 7-12 **Left lower lobe and right lower lobe atelectasis. A,** A fan-shaped area of increased density behind the heart is sharply demarcated by the medially displaced major fissure *(black arrows)* representing the characteristic appearance of left lower lobe atelectasis. **B,** On the lateral view, the major fissure *(white arrows)* is displaced posteriorly. The small triangular density in the posterior costophrenic sulcus is in the characteristic location for left lower lobe atelectasis on the lateral projection. **C,** In a different patient there is a fan-shaped triangular density in the right lower lobe bounded superiorly by the major fissure *(white arrow)*. Notice how the unaerated lower lobe silhouettes the right hemidiaphragm *(black arrow)*.

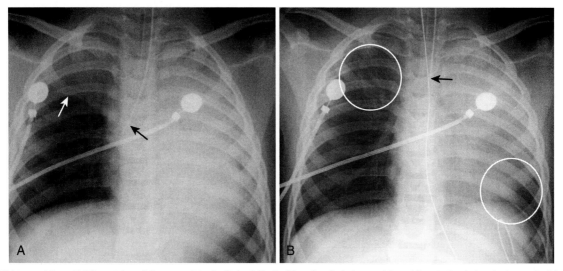

FIGURE 7-13 **Right upper lobe and left lung atelectasis from an endotracheal tube. A,** The tip of the endotracheal tube extends beyond the carina into the bronchus intermedius *(black arrow)*, which aerates only the right middle and lower lobes. The right upper lobe and entire left lung are opaque from atelectasis. The minor fissure is elevated *(white arrow)*. **B,** One hour later, the tip of the endotracheal tube has been retracted above the carina *(black arrow)*, and the right upper lobe and a portion of the left lower lobe are again aerated *(white circles)*.

◆ On the lateral radiograph:
   • There is a triangular density with its base directed anteriorly and its apex at the hilum.
   • The minor fissure may be displaced inferiorly and the major fissure superiorly.
■ **Endotracheal tube too low** (Fig. 7-13)
   ◆ If the tip of an endotracheal tube enters the right lower lobe bronchus, only the right lower lobe tends to be aerated and remains expanded. Within a short time, atelectasis of the entire left lung and the right upper and middle lobes will develop.
   ◆ Once the tip of the endotracheal tube is withdrawn above the carina, the atelectasis usually clears quite rapidly.
■ **Atelectasis of the entire lung** (see Figs. 7-3 and 7-4)
   ◆ On the frontal radiograph:
   • There is **opacification of the atelectatic lung** due to loss of air.

• The **hemidiaphragm** on the side of the atelectasis will be **silhouetted** by the nonaerated lung above it.
• There is a **shift of all of the mobile structures** of the thorax **toward the side of the atelectatic lung.**
◆ On the lateral radiograph:
   • The **hemidiaphragm on the side of the atelectasis will be silhouetted by the nonaerated lung above it.** Look closely and you will see **only one hemidiaphragm** on the lateral image, instead of two.

HOW ATELECTASIS RESOLVES

■ Depending in part on the rapidity with which the segment, lobe, or lung became atelectatic, atelectasis has the capacity to **resolve within hours or last for many days once the obstruction has been removed.**

- Slowly resolving lobar or whole-lung atelectasis may manifest patchy areas of airspace disease surrounded by progressively increasing zones of aerated lung until the atelectasis has completely cleared.
- The most common causes of obstructive atelectasis are summarized in Table 7-2.

TAKE-HOME POINTS

RECOGNIZING ATELECTASIS

Common to all forms of atelectasis is volume loss, but the radiographic appearance of atelectasis will differ depending on the type of atelectasis.

The three most commonly observed types of atelectasis are subsegmental atelectasis (also known as *discoid* or *platelike atelectasis*), compressive or passive atelectasis, and obstructive atelectasis.

Subsegmental atelectasis usually occurs in patients who are not taking a deep breath (splinting) and produces linear densities, usually at the lung bases.

Compressive atelectasis occurs passively when the lung is collapsed by a poor inspiration (at the bases) or from a large, adjacent pleural effusion or pneumothorax. When the underlying abnormality is removed, the lung usually expands.

Round atelectasis is a type of passive atelectasis in which the lung does not reexpand when a pleural effusion recedes, usually due to preexisting pleural disease. Round atelectasis may produce a masslike lesion on chest radiographs, which can mimic a tumor.

Obstructive atelectasis occurs distal to an occluding lesion of the bronchial tree because of reabsorption of the air in the distal airspaces via the pulmonary capillary bed.

Obstructive atelectasis produces consistently recognizable patterns of collapse based on the assumptions that the visceral and parietal pleura invariably remain in contact with each other and every lobe of the lung is anchored at or near the hilum.

Signs of obstructive atelectasis include displacement of the fissures, increased density of the affected lung, shift of the mobile structures of the thorax toward the atelectasis, and compensatory overinflation of the unaffected ipsilateral or contralateral lung.

Atelectasis tends to resolve quickly if it occurs acutely; the more chronic the process, the longer it usually takes to resolve.

TABLE 7-2 MOST COMMON CAUSES OF OBSTRUCTIVE ATELECTASIS

Cause	Remarks
Tumors	Includes bronchogenic carcinoma (especially squamous cell), endobronchial metastases, carcinoid tumors
Mucus plug	Especially in bedridden individuals, postoperative patients, those with asthma or cystic fibrosis
Foreign body aspiration	Especially peanuts, toys, following a traumatic intubation
Inflammation	As in scarring caused by tuberculosis

WEBLINK

Visit StudentConsult.Inkling.com for more information and a quiz on Recognizing Atelectasis.

For your convenience, the following QR Code may be used to link to **StudentConsult.com**. You must register this title using the PIN code on the inside front cover of the text to access online content.

CHAPTER 8
Recognizing a Pleural Effusion

NORMAL ANATOMY AND PHYSIOLOGY OF THE PLEURAL SPACE

Normal Anatomy

- The **parietal pleura lines the inside of the thoracic cage,** and the **visceral pleura adheres to the surface of the lung** parenchyma, including its interface with the mediastinum and diaphragm (see Chapter 3). The **enfolded visceral pleura forms the interlobar fissures**—the **major (oblique)** and **minor (horizontal)** on the right and only the major on the left. The space between the visceral and parietal pleura (i.e., the **pleural space) is a potential space normally containing only about 2 to 5 mL of pleural fluid.**

Normal Physiology

- Normally, several hundred milliliters of pleural fluid are produced and reabsorbed each day. Fluid is **produced primarily at the parietal pleura** from the pulmonary capillary bed and is **resorbed both at the visceral pleura and by lymphatic drainage through the parietal pleura.**

MODALITIES FOR DETECTING PLEURAL EFFUSIONS

- **Conventional radiography** is usually the first step in the detection of a pleural effusion. Other modalities used include **computed tomography (CT)** and **ultrasonography (US).** CT and US are both sensitive in detecting small amounts of fluid. CT is best for evaluating the disease underlying the effusion or in complete opacification of the hemithorax by effusion. US is especially helpful for guiding an intervention to remove the pleural fluid. The fundamental appearances of pleural effusion are similar, regardless of the modality used.

CAUSES OF PLEURAL EFFUSIONS

- Fluid accumulates in the pleural space when the rate at which the fluid **forms** exceeds the rate of by which it is **cleared.**
  - ◆ The **rate of formation may be increased** by:
    - **Increasing hydrostatic pressure,** as in left heart failure.
    - **Decreasing colloid osmotic pressure,** as in hypoproteinemia.
    - **Increasing capillary permeability,** as can occur in toxic disruption of the capillary membrane in pneumonia or hypersensitivity reactions.
  - ◆ The **rate of resorption can decrease** by:
    - **Decreased absorption** of fluid by **lymphatics, due either to lymphangitic blockade by tumor or to**

TABLE 8-1   SOME CAUSES OF PLEURAL EFFUSIONS

Cause	Examples
Excess formation of fluid	Congestive heart failure Hypoproteinemia Parapneumonic effusions Hypersensitivity reactions
Decreased resorption of fluid	Lymphangitic blockade due to tumor Elevated central venous pressure Decreased intrapleural pressure
Transport from peritoneal cavity	Ascites

   increased venous pressure that decreases the rate of fluid transport via the thoracic duct.
  - **Decreased pressure in the pleural space,** as in atelectasis of the lung due to bronchial obstruction.
- Pleural **effusions can also form when there is transport of peritoneal fluid from the abdominal cavity** through the diaphragm or via lymphatics from a subdiaphragmatic process (Table 8-1).

TYPES OF PLEURAL EFFUSIONS

- Pleural effusions are **divided into transudates** or **exudates** depending on their **protein content** and their **lactate dehydrogenase concentrations.**
- **Transudates** tend to form when there is **increased capillary hydrostatic pressure** or **decreased osmotic pressure.** Their causes include:
  - ◆ **Congestive heart failure,** primarily left-heart failure, is the most common cause of a transudative pleural effusion.
  - ◆ **Hypoalbuminemia**
  - ◆ **Cirrhosis**
  - ◆ **Nephrotic syndrome**
- **Exudates** tend to be the result of inflammation.
  - ◆ **The most common cause of an exudative pleural effusion is malignancy.**
  - ◆ An **empyema** is an exudate containing pus.
  - ◆ In a **hemothorax,** the fluid has a hematocrit that is >50% of the blood hematocrit.
  - ◆ A **chylothorax** contains increased triglycerides or cholesterol.

SIDE-SPECIFICITY OF PLEURAL EFFUSIONS

■ Certain diseases tend to produce pleural effusions in one hemithorax or the other, or in both.

■ Diseases that usually produce **bilateral effusions:**
  - ◆ **Congestive heart failure** usually produces about the same amount of fluid in each hemithorax. If there are **markedly** different amounts in each hemithorax, suspect a **parapneumonic effusion** or **malignancy** on the side with the greater volume of fluid.
  - ◆ **Lupus erythematosus** usually produces bilateral effusions, but when unilateral, it is usually left-sided.

■ Diseases that can produce **effusions on either side (but usually unilateral) are:**
  - ◆ **Tuberculosis** and other exudative effusions associated with infectious agents, including viruses
  - ◆ **Pulmonary thromboembolic disease**
  - ◆ **Trauma**

■ Diseases that usually produce **left-sided effusions are:**
  - ◆ **Pancreatitis**
  - ◆ **Distal thoracic duct obstruction**
  - ◆ **Dressler syndrome** (Box 8-1, Fig. 8-1)

BOX 8-1   DRESSLER SYNDROME

Also known as *postpericardiotomy/postmyocardial infarction syndrome.*

Typically occurs 2 to 3 weeks after a transmural myocardial infarct, producing a left pleural effusion, pericardial effusion, and patchy airspace disease at the left lung base.

Associated with chest pain and fever, it usually responds to high-dose aspirin or steroids.

■ Diseases that usually produce **right-sided effusions are:**
  - ◆ **Abdominal disease related to the liver or ovaries** (some ovarian tumors can be associated with a right pleural effusion and ascites [*Meigs syndrome*])
  - ◆ **Rheumatoid arthritis,** which can produce an effusion that remains unchanged for years
  - ◆ **Proximal thoracic duct obstruction**

RECOGNIZING THE DIFFERENT APPEARANCES OF PLEURAL EFFUSIONS

■ Forces that influence the appearance of pleural fluid on a chest radiograph depend on the position of the patient, the force of gravity, the amount of fluid, and the degree of elastic recoil of the lung.

■ The descriptions that follow, unless otherwise indicated, assume the patient is in the upright position.

Subpulmonic Effusions

■ It is believed that **almost all pleural effusions first collect in a subpulmonic location beneath the lung** between the parietal pleura lining the superior surface of the diaphragm and the visceral pleura under the lower lobe.

■ If the effusion remains entirely subpulmonic in location, it can be difficult to detect on conventional radiographs, except for contour alterations in what appears to be the hemidiaphragm but is actually the fluid–lung interface beneath the lung.

■ The different appearances of subpulmonic effusions are summarized in Table 8-2 (Fig. 8-2 and Fig. 8-3).

 Subpulmonic does **not** mean **loculated.**

  - ◆ **Most subpulmonic effusions flow freely** as the patient changes position.

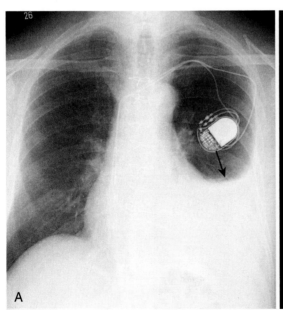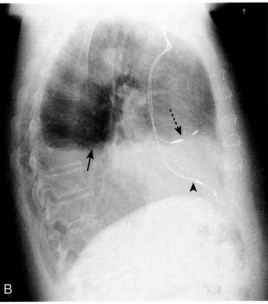

FIGURE 8-1   Dressler syndrome (postpericardiotomy/postmyocardial infarction syndrome). There is a left pleural effusion present *(solid black arrows* in **A** and **B**). This syndrome typically occurs 2 to 3 weeks after a transmural myocardial infarct. It also can occur following pericardiotomy, such as that which occurs in patients undergoing coronary artery bypass surgery, as in the case shown here. The combination of chest pain and fever, left pleural effusion, patchy left lower lobe airspace disease, and pericardial effusion several weeks following a myocardial infarction or open-heart surgery should suggest the syndrome. It usually responds to high-dose aspirin or steroids. This patient has a dual-lead pacemaker in place, and on the lateral projection **(B)**, the leads are seen in the region of the right atrium *(dotted black arrow)* and right ventricle *(arrowhead).*

TABLE 8-2   RECOGNIZING A SUBPOLMONIC EFFUSION

Right-Side Findings (see Fig. 8-2)		Left-Side Findings (see Fig. 8-3)	
Frontal View	**Lateral View**	**Frontal View**	**Lateral View**
The highest point of the apparent hemidiaphragm* is displaced more laterally than would the highest point of a normal hemidiaphragm. More difficult to recognize than left-sided subpulmonic effusions because the liver is the same density as the pleural fluid above it.	Posteriorly, the apparent hemidiaphragm has a curved arc, but as it meets the junction with the major fissure, the apparent hemidiaphragm* assumes a flat edge that drops sharply to the anterior chest wall.	Increased distance between the stomach bubble and the apparent left hemidiaphragm (should normally be only about 1 cm from top of stomach bubble to bottom of aerated left lower lobe). The highest point of the apparent hemidiaphragm* is displaced more laterally than would the highest point of a normal hemidiaphragm.	Posteriorly, the apparent hemidiaphragm has a curved arc but as it meets the junction with the major fissure, the apparent hemidiaphragm* assumes a flat edge that drops sharply to the anterior chest wall.

*The term **apparent hemidiaphragm** is used because the shadow being cast is actually from the subpulmonic fluid interfacing with the lung. The actual hemidiaphragm is invisible, silhouetted by the soft tissue in the abdomen below it and the pleural fluid above it.

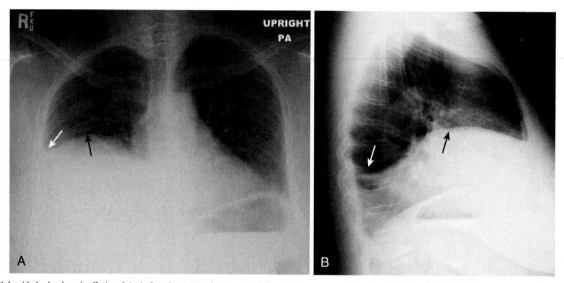

FIGURE 8-2   Right-sided subpulmonic effusion. A, In the frontal projection, the apparent right hemidiaphragm appears to be elevated *(black arrow)*. This edge does not represent the actual right hemidiaphragm, which has been rendered invisible by the pleural fluid that has accumulated above it; it is the interface between the effusion and the base of the lung (thus the term *apparent hemidiaphragm*). There is blunting of the right costophrenic sulcus *(white arrow)*. **B,** On the lateral projection, there is blunting of the posterior costophrenic sulcus *(white arrow)*. The apparent hemidiaphragm is rounded posteriorly, but then changes its contour as the effusion interfaces with the major fissure on the right side *(black arrow)*.

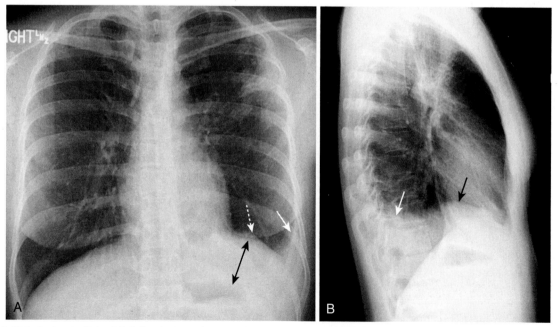

FIGURE 8-3   Left-sided subpulmonic effusion. A, In the frontal projection, there is more than 1 cm distance between the air in the stomach and the apparent left hemidiaphragm *(double black arrow)*. The edge between the aerated lung and the *dotted white arrow* does not represent the actual left hemidiaphragm, which has been rendered invisible by the pleural fluid that has accumulated above it; it is the interface between the effusion and the base of the lung. There is blunting of the left costophrenic sulcus *(solid white arrow)* on both projections. **B,** On the lateral projection, the apparent hemidiaphragm is rounded posteriorly, but then changes its contour as the effusion interfaces with the major fissure *(solid black arrow)*.

Blunting of the Costophrenic Angles

As the subpulmonic effusion grows in size, it first fills and thus blunts the posterior costophrenic sulcus, which is visible on the lateral view of the chest. This occurs with approximately 75 mL of fluid (Fig. 8-4).

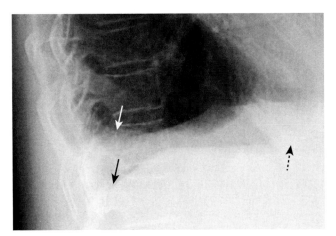

FIGURE 8-4 Blunting of the right posterior costophrenic sulcus on the lateral projection. When approximately 75 mL of fluid has accumulated in the pleural space, the fluid will typically ascend in the thorax and blunt the posterior costophrenic sulcus angle first *(solid white arrow)*. This can be visualized only on the lateral projection. There is a normal, sharp posterior costophrenic angle on the opposite side *(solid black arrow)*. Notice how the normal left hemidiaphragm is silhouetted by the heart anteriorly *(dotted black arrow)*, indicating which is the left hemidiaphragm. The pleural effusion is therefore on the right side.

- **When the effusion reaches about 300 mL in size, it blunts the lateral costophrenic angle,** which is visible on the frontal chest radiograph (Fig. 8-5).

  Pitfall: Pleural thickening caused by fibrosis can also produce blunting of the costophrenic angle.

  - ♦ **Solutions:** Scarring sometimes produces a characteristic **ski-slope appearance** of blunting, unlike the meniscoid appearance of a pleural effusion (Fig. 8-6).
  - ♦ **Pleural thickening will not change in location** with a change in patient position, as most effusions will.

The Meniscus Sign

- Because of the natural elastic recoil of the lungs, **pleural fluid appears to rise higher along the lateral margin of the thorax than it does medially** in the frontal projection. This produces a characteristic **meniscus** shape to the effusion that is higher on the sides and lower in the middle.
- In the lateral projection, the fluid assumes a U shape, ascending equally high both anteriorly and posteriorly (Fig. 8-7). **Identifying an abnormal lung density that demonstrates a meniscoid shape is strongly suggestive of a pleural effusion.**
- **Effect of patient positioning on the appearance of pleural fluid:**
  - ♦ When the patient is in the **upright position**, pleural fluid falls to the base of the thoracic cavity because of the force of gravity. When the patient is in the **supine**

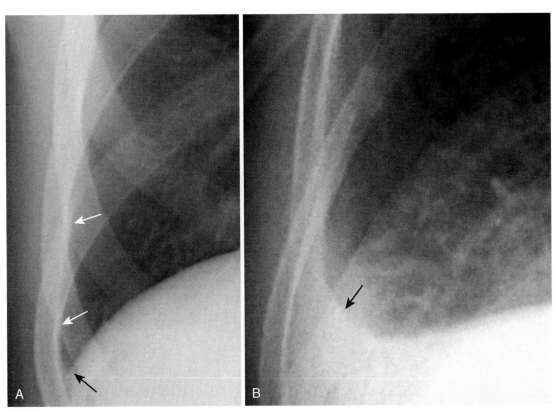

FIGURE 8-5 Normal and blunted right lateral costophrenic angle. A, The hemidiaphragm usually makes a sharp and acute angle as it meets the lateral chest wall on the frontal projection to produce the lateral costophrenic sulcus *(black arrow)*. Notice how normally aerated lung extends to the inner margin of each of the ribs *(white arrows)*. **B,** When an effusion reaches about 300 mL in volume, the lateral costophrenic sulcus loses its acute angulation and becomes **blunted** *(black arrow)*.

position, the same free-flowing effusion will layer along the posterior pleural space and produce a homogeneous "haze" over the entire hemithorax when viewed *en face* (Fig. 8-8).

♦ When the patient is **semirecumbent**, pleural fluid will layer in such a way that it forms a triangular density of varying thickness at the lung base with the apex, or thinnest part, of the triangle ascending to varying heights in the thorax, depending on how recumbent the patient is and how much fluid is present.

⚠ **Pitfall:** Depending on the patient's degree of recumbence, the upper lung fields may appear clearer (blacker) if the patient is upright and the fluid settles to the base of the thorax or denser (whiter) as the patient becomes more recumbent and the effusion begins to layer posteriorly. This change in appearance can occur with the same volume of pleural fluid redistributed because of patient positioning.

♦ **Solution:** In the best of all worlds, each portable chest radiograph would be exposed with the patient in the same position.

■ **The lateral decubitus view of the chest**
 ♦ The effect of patient positioning on the location of pleural fluid can be used for diagnostic advantage by having the patient lie on the side containing the effusion while taking a chest exposure using an x-ray beam directed horizontally. If the patient lies on the right side, it is called a *right lateral decubitus* view of the chest, and if the patient lies on the left side, it is called a *left lateral decubitus* view of the chest.
 ♦ **Decubitus views can be used to:**
  • **Confirm** the **presence of a pleural effusion.**
  • **Determine whether a pleural effusion flows freely** in the pleural space, **an important factor to know before attempting to drain** pleural fluid.
  • **"Uncover" a portion of the underlying lung** hidden by the effusion.
 ♦ If a pleural effusion can flow freely in the pleural space, the fluid will produce a characteristic **bandlike area of**

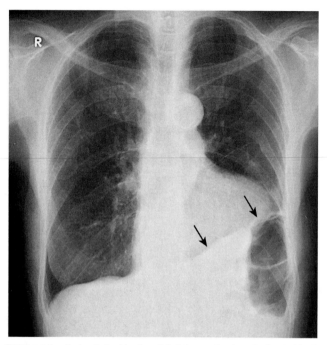

FIGURE 8-6 Scarring producing blunting of the left costophrenic angle. Scarring (due, for example, to previous infection, surgery, or blood in the pleural space) sometimes produces a characteristic "ski-slope appearance" of blunting *(black arrows)*, unlike the meniscoid appearance of a pleural effusion. This fibrosis would not change in appearance or location with changes in the patient's position, as free-flowing pleural fluid would.

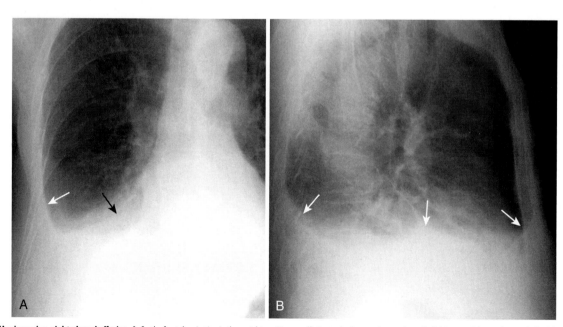

FIGURE 8-7 Meniscus sign, right pleural effusion. A, On the frontal projection in the upright position, an effusion typically ascends more laterally *(white arrow)* than it does medially *(black arrow)* because of factors affecting the natural elastic recoil of the lung. **B,** On the lateral projection, the fluid ascends about the same amount anteriorly and posteriorly, forming a U-shaped density called the *meniscus sign (white arrows).*

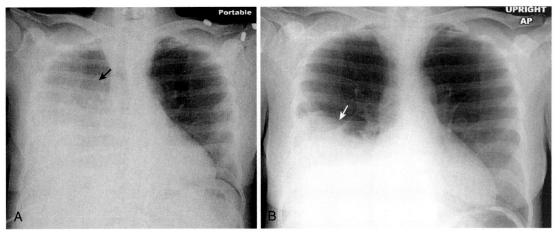

FIGURE 8-8 **Effect of patient positioning on the appearance of a pleural effusion. A,** With the patient in the recumbent position, the right-sided effusion layers along the posterior pleural surface and produces a "haze" over the entire hemithorax that is densest at the base and less dense toward the apex of the lung *(black arrow)*. **B,** On this x-ray taken a few minutes later, while the same patient was in a more upright position, pleural fluid falls to the base of the thoracic cavity as a result of the force of gravity *(white arrow)*. This simple change in position can produce the mistaken impression that an effusion has improved (or worsened if the supine study follows the upright examination) when there has actually been no change in the amount of pleural fluid. Ideally, the patient should be x-rayed in the same position each time for a meaningful comparison.

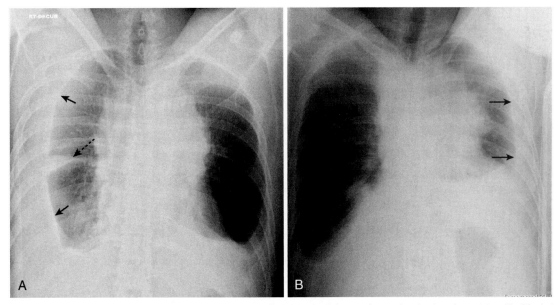

FIGURE 8-9 **Decubitus views of the chest. A,** In a right lateral decubitus view of the chest, the film is exposed with the patient lying on the right side on the examining table while a horizontal x-ray beam is directed posteroanteriorly. Because the patient's right side is dependent, any free-flowing pleural fluid will layer along the right side *(solid black arrows)*, forming a bandlike density. Notice how the fluid flows into the minor fissure *(dotted black arrow)*. **B,** In a left lateral decubitus view of the chest, the patient lies on the table with the left side down and free fluid on the left side layers along the left lateral chest wall *(solid black arrows)*. Parts **(A)** and **(B)** show the same patient who has bilateral pleural effusions due to lymphoma.

increased density along the inner margin of the chest cage on the dependent side of the body.

With a **right lateral decubitus view of the chest,** the patient's right side will be dependent and a **right pleural effusion will layer** along the dependent surface. With a **left lateral decubitus view of the chest,** the patient's left side will be dependent and a **left pleural effusion will layer** along the dependent surface (Fig. 8-9).

♦ There may be pleural fluid that **does not flow freely** in the pleural space **if there are adhesions present** that might impede the free flow of the fluid (see "Loculated Effusions").

♦ **Decubitus views** of the chest **can demonstrate effusions as small as 15 to 20 mL,** but CT scans of the chest have largely supplanted decubitus views in detecting very small amounts of pleural fluid (Fig. 8-10).

 Pitfall: Do not order a lateral decubitus view of the chest if the entire hemithorax is opacified, because there can be no change in the position of the fluid and the underlying lung will be no more visible in the decubitus position than it was with the patient upright. A CT scan of the

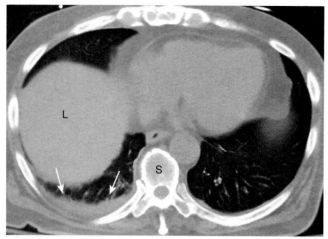

FIGURE 8-10 **Small pleural effusion on computed tomography (CT).** There is a small right pleural effusion *(white arrows)* that collects in the dependent portion of the hemithorax. The patient is supine in the CT scanner. This effusion would probably not be visible on conventional chest radiographs, because it would be obscured by the liver (L) on the frontal projection and the spine (S) on the lateral projection.

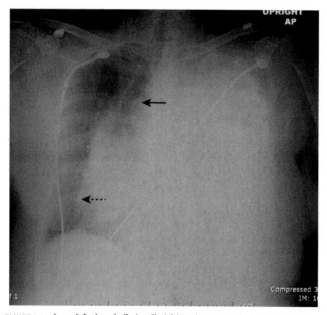

FIGURE 8-11 **Large left pleural effusion.** The left hemithorax is completely opacified, and there is a shift of the mobile mediastinal structures, such as the trachea *(solid black arrow)* and the heart *(dotted black arrow),* **away** from the side of opacification. This is characteristic of a large pleural effusion, which can act like a mass. In most adults, it requires about 2 L of fluid to fill or almost fill the entire hemithorax such as shown here.

chest is a better means of evaluating the underlying lung if the hemithorax is opacified.

Opacified Hemithorax

- When the **hemithorax of an adult contains about 2 L of fluid,** the **entire hemithorax will be opacified** (Fig. 8-11).
- As fluid fills the pleural space, the lung tends to undergo passive collapse (atelectasis) (see Chapter 7).
- **Large effusions are sufficiently opaque** on conventional chest radiographs **so as to envelop and mask whatever disease may be present in the lung. CT is the modality**

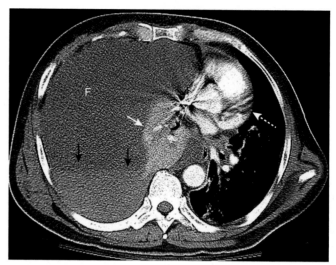

FIGURE 8-12 **Large hemothorax on CT scan.** There is a large fluid collection on the right (F), which is causing complete atelectasis of the right lung *(solid white arrow)* and a shift of the heart away from the side of the collection *(dotted white arrow).* Upon careful inspection, the fluid is denser (whiter) at the bottom of the collection *(solid black arrows)* than at the top, a **fluid-fluid** level that raises suspicion for the separation of blood products such as that which would occur in a hemothorax. This patient was bleeding heavily from a lacerated intercostal artery.

usually employed to visualize the underlying lung, which is rendered impenetrable by a large effusion (Fig. 8-12).
- **Large effusions** can act like a mass and **displace the heart and trachea away** from the **side of opacification.**
- For more on the opacified hemithorax, see Chapter 6.

Loculated Effusions

- **Adhesions in the pleural space,** caused most often by an old infection or hemothorax, **may limit the normal mobility of a pleural effusion** so that it remains in the same location no matter what position the patient assumes.
- **Imaging findings**
  - Loculated effusions **can be suspected when an effusion has an unusual shape or location in the thorax** (e.g., the effusion defies gravity by remaining at the nondependent part of the thorax when the patient is upright on a conventional radiograph [Fig. 8-13] or supine on a CT scan of the chest).
  - **Loculation of pleural fluid has therapeutic importance** because such collections tend to be traversed by multiple adhesions that make it difficult to drain the noncommunicating pockets of fluid with a single pleural drainage tube in the same way that free-flowing effusions can be drained.

Fissural Pseudotumors

- **Pseudotumors** (also called *vanishing tumors*) are **sharply marginated collections of pleural fluid** contained either between the layers of an **interlobar pulmonary fissure** or in a subpleural location just beneath the fissure. They are **transudates** that **almost always occur in patients with congestive heart failure.**
- The imaging findings of a pseudotumor are characteristic, so they should not be mistaken for an actual tumor (from which they derive their name).
  - They are **lenticular in shape,** most often occur in the **minor fissure (75%),** and frequently have **pointed**

ends on each side where they insinuate into the fissure, much like the shape of a lemon. They **do not tend to flow freely** with a change in patient positioning.

- They **disappear when the underlying condition (usually congestive heart failure) is treated**, but they **tend to recur in the same location** each time the patient's failure recurs (Fig. 8-14).

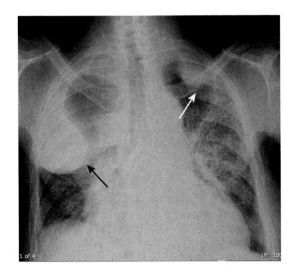

FIGURE 8-13 Loculated pleural effusions. There are bilateral fluid collections *(white and black arrows)* that have unusual shapes and seem to defy gravity, because they are trapped in the pleural space, usually by adhesions. Loculated effusions can be suspected when an effusion has something other than a meniscoid shape or collects in a location other than the base of the lung (e.g., if the effusion were to remain at the apex of the hemithorax even if the patient were upright).

Laminar Effusions

- A laminar effusion is a form of pleural effusion in which the fluid assumes a **thin, bandlike density along the lateral chest wall, especially near the costophrenic angle.** The lateral costophrenic angle tends to maintain its acute angle with a laminar effusion, unlike the blunting that occurs with a usual pleural effusion.
- Laminar effusions are **almost always the result of** elevated left atrial pressure, as in **congestive heart failure** or secondary to **lymphangitic spread of malignancy.** They are **usually not free-flowing.**
- They can be recognized by the band of increased density that separates the air-filled lung from the inside margin of the ribs at the lung base on the frontal chest radiograph. In healthy persons, an aerated lung should extend to the inside of each contiguous rib (Fig. 8-15).

Hydropneumothorax

- The presence of **both air** in the pleural space (pneumothorax) **and** abnormal amounts of **fluid** in the pleural space (pleural effusion or hydrothorax) is called a *hydropneumothorax.*
- Some of the more common causes of a hydropneumothorax are **trauma, surgery,** or a **recent thoracentesis to remove pleural fluid** in which air enters the pleural apace.
  - ♦ A **bronchopleural fistula**, an abnormal and relatively uncommon connection between the bronchial tree and the pleural space most often due to tumor, surgery, or infection, can also produce both air and fluid in the pleural space.
- Unlike pleural effusion alone, whose meniscoid shape is governed by the elastic recoil of the lung, a **hydropneumothorax produces an air-fluid level in the hemithorax** marked by a straight edge and a sharp, air-over-fluid interface when the exposure is made with a horizontal x-ray beam (Fig. 8-16).

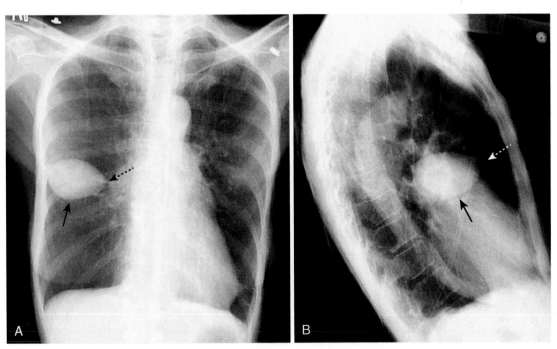

FIGURE 8-14 Pseudotumor in the minor fissure. A, A sharply marginated collection of pleural fluid contained between the layers of the minor fissure produces a characteristic lenticular shape *(solid black arrows in images **A** and **B**)* that frequently has pointed ends on each side, where it insinuates into the fissure so that pseudotumors look like a lemon on frontal **(A)** or lateral **(B)** chest radiographs *(dotted black arrow in [**A**]* and *dotted white arrow in [**B**]).* Pseudotumors always occur along the course of the minor or major fissure, which helps to distinguish them from an actual tumor of the lung.

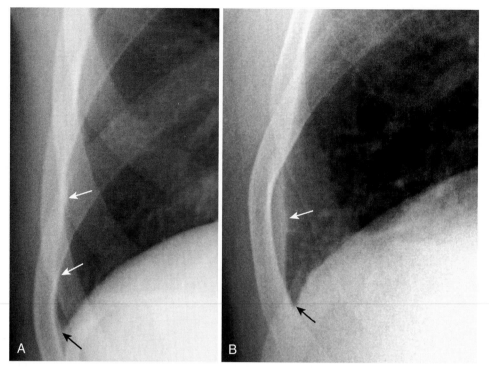

FIGURE 8-15 **Normal versus laminar pleural effusion. A,** A normal patient in whom the normally aerated lung extends to the inner margin of each of the ribs *(white arrows)*. The costophrenic sulcus is sharp *(black arrow)*. **B,** There is a thin band of increased density that extends superiorly from the lung base *(white arrow)* but does not appear to cause blunting of the costophrenic angle *(black arrow)*. This is the appearance of a **laminar pleural effusion**, which is most often associated with either congestive heart failure or lymphangitic spread of malignancy in the lung. This patient was in congestive heart failure.

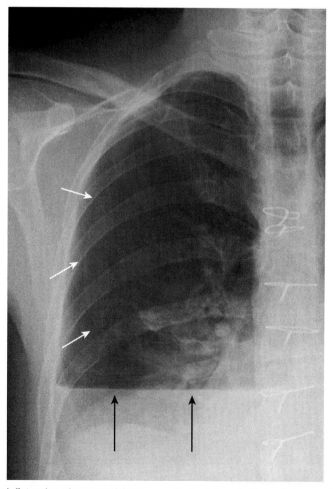

FIGURE 8-16 **Hydropneumothorax.** Unlike pleural effusions alone, whose meniscoid shape is governed by the elastic recoil of the lung, hydropneumothorax produces an air-fluid level in the hemithorax marked by a straight edge and a sharp, air-over-fluid interface when the exposure is made with a horizontal x-ray beam *(black arrows)*. This person was stabbed in the right side and there is a moderately large pneumothorax as shown by the visceral pleural white line *(white arrows)*. This actually represents a hemopneumothorax, but unlike the computed tomography scan shown in Figure 8-12, a conventional radiograph is unable to distinguish between blood and any other fluid.

TAKE-HOME POINTS

RECOGNIZING A PLEURAL EFFUSION

Pleural effusions collect in the potential space between the visceral and parietal pleura and are either transudates or exudates, depending on their protein content and lactate dehydrogenase concentration.

There are normally a few milliliters of fluid in the pleural space; about 75 mL is required to blunt the posterior costophrenic sulcus (seen on the lateral view) and about 200 to 300 mL is needed to blunt the lateral costophrenic sulcus (seen on the frontal view); approximately 2 L of fluid will cause opacification of the entire hemithorax in an adult.

Determining whether an effusion is unilateral or bilateral, mostly right-sided or mostly left-sided, can be an important clue as to its etiology.

Most pleural effusions begin by collecting in the pleural space between the hemidiaphragm and the base of the lung; these are called *subpulmonic effusions.*

As the amount of fluid increases, it forms a **meniscus** shape on the upright frontal chest radiograph because of the natural elastic recoil properties of the lung.

Very large pleural effusions may act like a mass and produce a shift of the mobile mediastinal structures (e.g., the heart) away from the side of the effusion.

In the absence of pleural adhesions, effusions will flow freely and change location with a change in the patient's position; with pleural adhesions (usually from old infection or hemothorax), the fluid may assume unusual appearances or occur in atypical locations; such effusions are said to be *loculated.*

A **pseudotumor** is a type of effusion that occurs in the fissures of the lung (mostly the minor fissure) and is most frequently secondary to congestive heart failure; it clears when the underlying failure is treated.

Laminar effusions are best recognized at the lung base just above the costophrenic angles on the frontal projection and most often occur as a result of either congestive heart failure or lymphangitic spread of malignancy.

A **hydropneumothorax** consists of both air and increased fluid in the pleural space and is recognizable on an upright view of the chest by a straight, air-fluid interface rather than the typical meniscus shape of pleural fluid alone.

■ CT is frequently necessary to distinguish between some presentations of hydropneumothorax and a **lung abscess,** both of which may have a similar appearance on conventional chest radiographs.

WEBLINK

Visit StudentConsult.Inkling.com for more information and a quiz on Recognizing a Pleural Effusion.

For your convenience, the following QR Code may be used to link to **StudentConsult.com**. You must register this title using the PIN code on the inside front cover of the text to access online content.

CHAPTER 9
Recognizing Pneumonia

GENERAL CONSIDERATIONS

- Pneumonia can be defined as **consolidation of the lung produced by inflammatory exudate, usually as a result of an infectious agent.**
- Most pneumonias produce airspace disease, either lobar or segmental. Other pneumonias demonstrate **interstitial disease** and others produce findings in both the airspaces and the interstitium.
- Most microorganisms that produce pneumonia are **spread to the lungs via the tracheobronchial tree, either through inhalation or aspiration** of the organisms.
- In some instances, microorganisms are spread via the bloodstream and, in even fewer cases, by direct extension.
- Because many **different microorganisms can produce similar imaging findings in the lungs, it is difficult to identify with certainty the causative organism from the radiographic presentation alone.** However, **certain patterns** of disease **are very suggestive** of a particular causative organism (Table 9-1).
- Some use the term *infiltrate* synonymously with pneumonia, although many diseases, from amyloid to pulmonary fibrosis, can infiltrate the lung.

TABLE 9-1 PATTERNS THAT MIGHT SUGGEST A CAUSATIVE ORGANISM

Pattern of Disease	Likely Causative Organism
Upper lobe cavitary pneumonia with spread to the opposite lower lobe	*Mycobacterium tuberculosis (TB)*
Upper lobe lobar pneumonia with bulging interlobar fissure	*Klebsiella pneumoniae*
Lower lobe cavitary pneumonia	*Pseudomonas aeruginosa* or anaerobic organisms *(Bacteroides)*
Perihilar interstitial disease or perihilar airspace disease	*Pneumocystis carinii (jiroveci)*
Thin-walled upper lobe cavity	*Coccidioides* (coccidiomycosis), TB
Airspace disease with effusion	*Streptococci, staphylococci*, TB
Diffuse nodules	*Histoplasma, Coccidioides, Mycobacterium tuberculosis* (histoplasmosis, coccidiomycosis, TB)
Soft-tissue, finger-like shadows in upper lobes	*Aspergillus* (allergic bronchopulmonary aspergillosis)
Solitary pulmonary nodule	*Cryptococcus* (cryptococcosis)
Spherical soft-tissue mass in a thin-walled upper lobe cavity	*Aspergillus* (aspergilloma)

GENERAL CHARACTERISTICS OF PNEUMONIA

- Because **pneumonia** fills the involved airspaces or interstitial tissues with some form of fluid, or inflammatory exudate, pneumonias **appear denser (whiter) than the surrounding, normally aerated lung.**
- **Pneumonia may contain air bronchograms** if the bronchi themselves are not filled with inflammatory exudate or fluid (see Fig. 5-3).
  - Air bronchograms are **much more likely to be visible when the pneumonia involves the central portion** of the lung near the hilum. Near the periphery of the lung, the bronchi are usually too small to be visible (Fig. 9-1).
  - Remember that anything of fluid or soft tissue density that replaces the normal gas in the airspaces may also

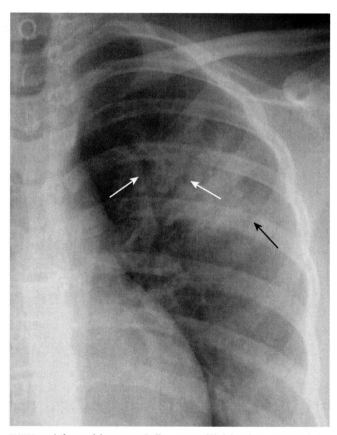

FIGURE 9-1 Left upper lobe pneumonia. There are several black, branching structures in this left upper lobe pneumonia *(white arrows)* that represent typical **air bronchograms** seen in airspace disease. This patient had pneumococcal pneumonia. The disease is homogeneous in density, except for the presence of the air bronchograms. Because this is airspace disease, its outer edges are poorly marginated, indistinct, and fluffy *(black arrow)*.

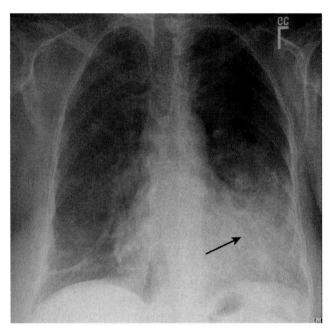

FIGURE 9-2 **Lingular pneumonia.** There is airspace disease in the lingular segments of the left upper lobe. The disease is of homogeneous density. The disease is in contact with the left lateral border of the heart, which is "silhouetted" by the fluid density of the consolidated upper lobe in contact with the soft tissue density of the heart *(black arrow)*. Because the pneumonia and the heart are the same radiographic density, the border between them disappears.

produce this sign; therefore **an air bronchogram is not specific for pneumonia** (see Chapter 5).

- Pneumonia that involves the airspaces **appears fluffy,** and its **margins are indistinct.**
  - ◆ **Where pneumonia abuts a pleural surface,** such as an interlobar fissure or the chest wall, **it will be sharply marginated.**
- **Interstitial** pneumonia, on the other hand, may produce prominence of the interstitial markings in the affected part of the lung or may spread to adjacent airways and resemble airspace disease.
- Except for the presence of air bronchograms, airspace **pneumonia is usually homogeneous in density** (Fig. 9-2).
- In some types of pneumonia (i.e., bronchopneumonia), **the bronchi as well as the airspaces contain inflammatory exudate.** This can lead to **atelectasis associated with the pneumonia.**
- Box 9-1 summarizes the keys to recognizing pneumonia.

PATTERNS OF PNEUMONIA

- Pneumonias may be distributed in the lung in several patterns described as *lobar, segmental, interstitial, round,* and *cavitary* (Table 9-2).
- Remember, these are terms that simply describe the distribution of the disease in the lungs; they aren't diagnostic of pneumonia because many other diseases can produce the same patterns of disease distribution in the lung.

LOBAR PNEUMONIA

- The **prototypical lobar pneumonia is pneumococcal pneumonia** caused by *Streptococcus pneumoniae* (Fig. 9-3).

BOX 9-1 RECOGNIZING A PNEUMONIA— KEY SIGNS

More opaque than surrounding normal lung.

In airspace disease, the margins may be fluffy and indistinct, except where they abut a pleural surface such as the interlobar fissures, in which case the margin will be sharp.

Interstitial pneumonias will cause a prominence of the interstitial tissues of the lung in the affected area; in some cases, the disease can spread to the alveoli and resemble airspace disease.

Pneumonia tends to be homogeneous in density.

Lobar pneumonias may contain air bronchograms.

Pneumonias may be associated with atelectasis in the affected portion of the lung.

TABLE 9-2 PATTERNS OF APPEARANCE OF PNEUMONIAS

Pattern	Characteristics
Lobar	Homogeneous consolidation of affected lobe with air bronchogram
Segmental (bronchopneumonia)	Patchy airspace disease frequently involving several segments simultaneously; no air bronchogram; atelectasis may be associated
Interstitial	Reticular interstitial disease usually diffusely spread throughout the lungs early in the disease process; frequently progresses to airspace disease
Round	Spherically shaped pneumonia usually seen in the lower lobes of children; may resemble a mass
Cavitary	Produced by numerous microorganisms, chief amongst them being *Mycobacterium tuberculosis*

- Although we are calling it *lobar pneumonia,* the patient may have symptoms before the disease involves the entire lobe. In its most classical form, the disease fills most or all of a lobe of the lung.
- Because lobes are bound by interlobar fissures, **one or more of the margins of a lobar pneumonia may be sharply marginated.**
- Where the disease is not bound by a fissure, it will have an **indistinct and irregular margin.**
- Lobar pneumonias almost always produce a **silhouette sign** where they come in contact with the heart, aorta, or diaphragm, and they almost always contain **air bronchograms** if they involve the central portions of the lung.

SEGMENTAL PNEUMONIA (BRONCHOPNEUMONIA)

- The **prototypical bronchopneumonia is caused by *Staphylococcus aureus.*** Many gram-negative bacteria, such as *Pseudomonas aeruginosa,* can produce the same picture.
- Bronchopneumonia is spread centrifugally via the tracheobronchial tree to **many foci in the lung at the same time.** Therefore **it frequently involves several segments** of the lung simultaneously.

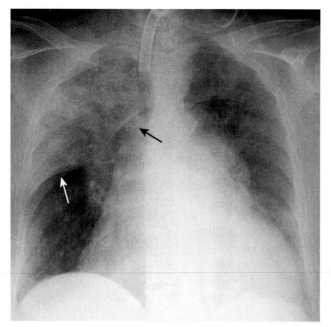

FIGURE 9-3 Right upper lobe pneumococcal pneumonia. There is airspace disease in the right upper lobe that occupies the entire lobe. Because lobes are bounded by interlobar fissures—in this case, the minor or horizontal fissure *(white arrow)*—the inferior margin of the pneumonia is sharply marginated. Where the disease contacts the ascending aorta *(black arrow)*, the border of the aorta is silhouetted by the fluid density of the pneumonia.

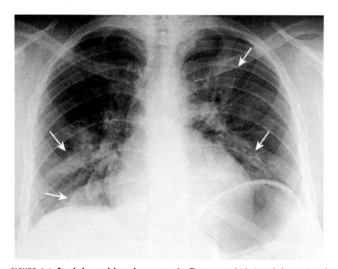

FIGURE 9-4 Staphylococcal bronchopneumonia. There are multiple irregularly marginated patches of airspace disease in both lungs *(white arrows)*. This is a characteristic distribution and appearance of bronchopneumonia. The disease is spread centrifugally via the tracheobronchial tree to many foci in the lung at the same time, and so it frequently involves several segments. Because lung segments are not bound by fissures, the margins of segmental pneumonias tend to be fluffy and indistinct. There are no air bronchograms present because inflammatory exudate fills the bronchi and the airspaces around them.

- Because lung segments are not bound by fissures, all of the **margins of segmental pneumonia tend to be fluffy and indistinct** (Fig. 9-4).
- Unlike lobar pneumonia, **segmental bronchopneumonia produces exudate that fills the bronchi.** Therefore **air bronchograms are usually not present** and there is frequently some **volume loss (atelectasis)** associated with bronchopneumonia.

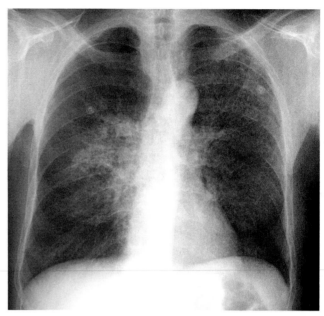

FIGURE 9-5 *Pneumocystis carinii (jiroveci)* pneumonia (PCP). There is bilateral, centrally-located interstitial lung disease that is primarily reticular in nature. Without the additional history that this patient had AIDS, this could be mistaken for pulmonary interstitial edema or for a chronic, fibrotic process such as sarcoidosis. There are, however, no pleural effusions present, as might be expected with pulmonary interstitial edema, and there is no evidence of hilar adenopathy, as might occur in sarcoidosis.

INTERSTITIAL PNEUMONIA

- The **prototypes for interstitial pneumonia are viral pneumonia,** *Mycoplasma pneumoniae,* and *Pneumocystis pneumonia* in patients with acquired immunodeficiency syndrome (AIDS).
- Interstitial pneumonia **tends to involve the airway walls and alveolar septa** and may produce, especially early in its course, a **fine, reticular pattern in the lungs.**
- Most **cases of interstitial pneumonia eventually spread to the adjacent alveoli** and produce patchy or confluent airspace disease, **making the original interstitial nature of the pneumonia impossible to recognize** radiographically.

 Pneumocystis carinii (jiroveci) **pneumonia (PCP)**

- PCP is the **most common** clinically recognized infection in patients with AIDS.
- It classically presents as a **perihilar, reticular interstitial pneumonia** or as **airspace disease that may mimic the central distribution pattern of pulmonary edema** (Fig. 9-5).
- Other presentations, such as unilateral airspace disease or widespread, patchy airspace disease are less common.
- There are usually **no pleural effusions** and **no hilar adenopathy.**
- Opportunistic infections **usually occur with CD4 counts under 200/mm³** of blood.

ROUND PNEUMONIA

- Some pneumonias, mostly in children, can assume a **spherical shape** on chest radiographs.

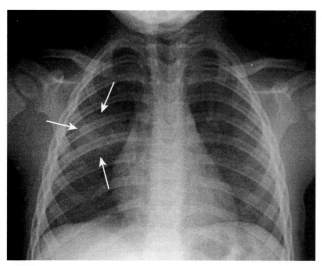

FIGURE 9-6 Round pneumonia. There is a soft tissue density in the right midlung field that has a rounded appearance *(white arrows)*. This is a 10-month-old child who had a cough and fever. This is a characteristic appearance of a round pneumonia, most common in children, and frequently due to either *Haemophilus,* streptococcal, or pneumococcal infection.

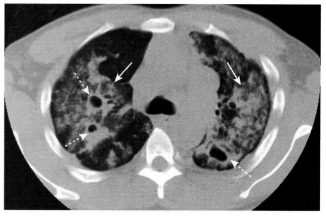

FIGURE 9-7 Cavitary pneumonia. A nonenhanced axial computed tomography image through the upper lobes demonstrates bilateral airspace disease *(white arrows)* containing multiple lucencies representing cavities *(dotted white arrows)*. The cavities contain no air–fluid levels. Upper lobe cavitary pneumonia is presumptively tuberculosis (TB), until proven otherwise. This patient had postprimary TB (reactivation tuberculosis).

- These **round pneumonias** are almost always **posterior** in the lungs, usually **in the lower lobes.**
- Causative agents include *Haemophilus influenzae, Streptococcus,* and *Pneumococcus.*
- A round pneumonia could be confused with a tumor mass except that symptoms associated with infection usually accompany the pulmonary findings and tumors are uncommon in children (Fig. 9-6).

CAVITARY PNEUMONIA

- The prototypical organism producing cavitary pneumonia is *Mycobacterium tuberculosis.* Tuberculosis is discussed in Chapter 5.

  Cavitation is common in postprimary tuberculosis (TB) (reactivation tuberculosis) but rare in primary TB. The cavities are usually located in the upper lobes, are bilateral and **thin-walled,** have a **smooth inner margin,** and contain **no air-fluid level** (Fig. 9-7). **Transbronchial spread** (from one upper lobe to the opposite lower lobe or to another lobe in the lung) **should make you think of infection with** *Mycobacterium tuberculosis.*

- **Other infectious agents** that produce cavitary disease:
  - **Staphylococcal pneumonia** can cavitate and produce thin-walled pneumatoceles.
  - *Streptococcal* **pneumonia,** *Klebsiella* **pneumonia, and coccidioidomycosis** can also produce cavitating pneumonias.

ASPIRATION

- There are many causes of aspiration of foreign material into the tracheobronchial tree, among them neurologic disorders (stroke, traumatic brain injury), altered mental status (anes-

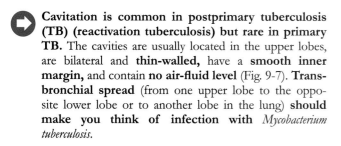

FIGURE 9-8 Aspiration, both lower lobes. Single, axial computed tomography image of the lungs demonstrates bilateral lower lobe airspace disease in a patient who had aspirated *(black arrows)*. Aspiration usually affects the most dependent portions of the lung. In the upright position, the lower lobes are affected. In the recumbent position, the superior segments of the lower lobes and the posterior segments of the upper lobes are most involved. Aspiration of water or neutralized gastric acid will usually clear in 24 to 48 hours depending on the volume aspirated.

thesia, drug overdose), gastroesophageal reflux, and postoperative effects from head and neck surgery.

- Aspiration almost always occurs in the **most dependent portions of the lung.**
  - When the person is **upright,** the most dependent portions of the lung will usually be the **lower lobes. The right side is more often affected than the left** because of the straighter and wider nature of the right main bronchus.
  - When a person is **recumbent,** aspiration usually occurs into the **superior segments of the lower lobes or the posterior segment of the upper lobes.**
  - **Acute aspiration** will produce radiographic findings of **airspace disease.**
    - Its location, the rapidity with which it appears, and the group of patients predisposed to aspirate are clues to its cause (Fig. 9-8).

TABLE 9-3 THREE PATTERNS OF ACUTE ASPIRATION

Pattern	Characteristics
Bland gastric acid or water	Rapidly appearing and rapidly clearing airspace disease in dependent lobe(s); not a pneumonia
Infected aspirate (aspiration pneumonia)	Usually lower lobes; frequently cavitates and may take months to clear
Unneutralized stomach acid (chemical pneumonitis)	Almost immediate appearance of dependent airspace disease that frequently becomes secondarily infected

TABLE 9-4 USING THE SILHOUETTE SIGN ON THE FRONTAL CHEST RADIOGRAPH

Structure That Is No Longer Visible	Disease Location
Ascending aorta	Right upper lobe
Right heart border	Right middle lobe
Right hemidiaphragm	Right lower lobe
Descending aorta	Left upper or lower lobe
Left heart border	Lingula of left upper lobe
Left hemidiaphragm	Left lower lobe

 Recognizing the different types of aspiration (Table 9-3)

- The clinical and radiologic course of aspiration depends on what was aspirated.
  - ♦ **Aspiration of bland (neutralized) gastric juices or water**
    - This is technically **not a pneumonia** because it does not involve an infectious agent, is handled by the lungs as if it were pulmonary edema fluid, and **classically remains for only a day or two before clearing** through resorption.
  - ♦ **Aspiration that produces pneumonia due to microorganisms in the lung**
    - Although **we routinely aspirate numerous microorganisms present in the normal oropharyngeal flora,** there are some patients in whom these microorganisms can develop into pneumonia, including older adults and those who are immunocompromised, debilitated, or have underlying lung disease.
    - Pneumonia caused by aspiration is usually due to **anaerobic organisms**, such as *Bacteroides*. These organisms produce **lower lobe airspace disease** that **frequently cavitates**. It may take **months to resolve.**
  - ♦ **Aspiration of unneutralized stomach acid** (*Mendelson syndrome*)
    - When large quantities of unneutralized gastric acid are aspirated, **chemical pneumonitis develops,** producing dependent lobe airspace disease or **pulmonary edema.** The disease may appear quickly, within a few **hours** of the aspiration. Clearing may take **days or longer,** and the chemical pneumonitis is prone to become **secondarily infected.**

LOCALIZING PNEUMONIA

- Although it is true that an antibiotic will travel to every lobe of the lung without regard for which lobe actually harbors the pneumonia, determining the location of a pneumonia may provide clues to the causative organism (e.g., upper lobes, think of TB) and the presence of associated pathology (e.g., lower lobes, think of recurrent aspiration).

- **On conventional radiographs,** it is always **best to localize disease using two views taken at 90° to each other (orthogonal views)** such as a frontal and lateral chest radiograph. Computed tomography (CT) may further localize and characterize the disease and demonstrate associated pathology, such as pleural effusions or cavities too small to see on conventional radiographs.
- Sometimes, only a frontal radiograph may be available, as with critically ill or debilitated patients who require a portable bedside examination. Nevertheless, it is still frequently **possible to localize the pneumonia using only the frontal radiograph** by analyzing which structure's edges are obscured by the disease (i.e., the silhouette sign) (Table 9-4).
- **Silhouette sign** (see also "Characteristics of Airspace Disease" in Chapter 5)
  - ♦ If two objects of the **same** radiographic density **touch** each other, then the **edge between them disappears** (see Fig. 9-2). The silhouette sign is valuable in localizing and identifying tissue types throughout the body, not just the chest.

 The spine sign (Fig. 9-9)

  - ♦ On the lateral chest radiograph, **the thoracic spine normally appears to get darker (blacker) as you survey it from the shoulder girdle to the diaphragm.**
    - This finding occurs because the x-ray beam normally needs to penetrate more tissue (more bones, more muscle) around the shoulders than it does just above the diaphragm, where it needs to pass through only the heart and aerated lungs.
  - ♦ **When disease** of soft tissue or fluid density **involves** the **posterior portion of the lower lobe,** more of the x-ray beam will be absorbed by the new, added density, and **the spine will appear to become "whiter" (more opaque) just above the posterior costophrenic sulcus.**
    - This is called the *spine sign* and it provides another way to localize disease in the lungs.
- The **lower lobe disease may not be apparent on the frontal projection** if the disease falls below the plane of the highest point of the affected side's hemidiaphragm. Therefore the **spine sign may indicate** the presence of lower lobe **disease,** such as lower lobe pneumonia, **which may be otherwise invisible on the frontal projection.**
- Figure 9-10 is a composite of the characteristic appearances of lobar pneumonia as seen on a frontal chest radiograph.

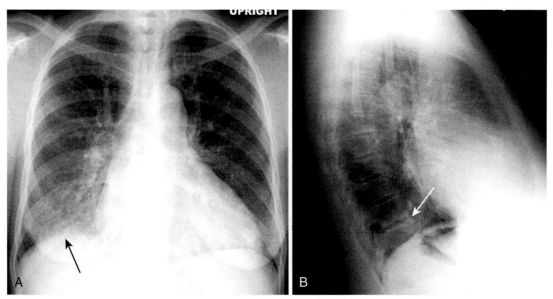

FIGURE 9-9 **The spine sign.** Frontal and lateral views of the chest demonstrate airspace disease on the lateral film **(B)** in the right lower lobe *(white arrow)* that may not be immediately apparent on the frontal film (you can see the pneumonia in the right lower lobe in **[A]** *[black arrow]*). Normally, the thoracic spine appears to get "blacker" as you view it from the neck to the diaphragm because there is less tissue for the x-ray beam to traverse just above the diaphragm than in the region of the shoulder girdle (see also Fig. 3-3). In this case, a right lower lobe pneumonia superimposed on the lower spine in the lateral view *(white arrow)* makes the spine appear "whiter" (more dense) just above the diaphragm. This is called the *spine sign*.

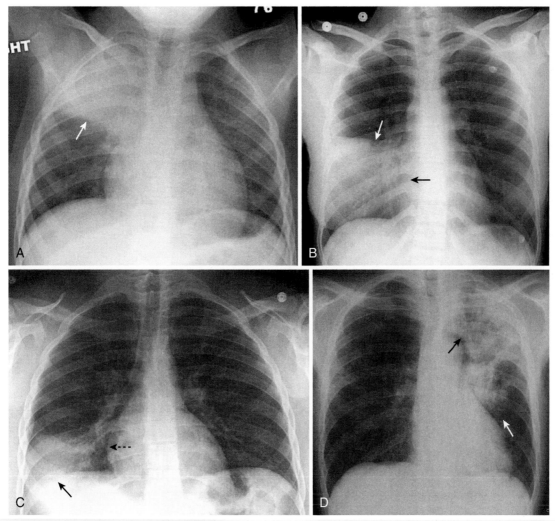

FIGURE 9-10 **Composite appearances of lobar pneumonias. A,** Right upper lobe. The disease obscures *(silhouettes)* the ascending aorta. Where it abuts the minor fissure, it produces a sharp margin *(white arrow)*. **B,** Right middle lobe. The disease silhouettes the right heart border *(solid black arrow)*. Where it abuts the minor fissure, it produces a sharp margin *(solid white arrow)*. **C,** Right lower lobe. The disease silhouettes the right hemidiaphragm *(solid black arrow)*. It spares the right heart border *(dotted black arrow)*. **D,** Left upper lobe. The disease is poorly marginated *(solid white arrow)* and obscures the aortic knob *(solid black arrow)*.

Continued

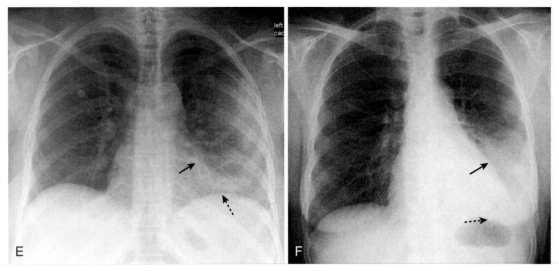

FIGURE 9-10, cont'd E, Lingula. The disease silhouettes the left heart border *(solid black arrow)* but spares the left hemidiaphragm *(dotted black arrow)*. **F,** Left lower lobe. The disease obscures the left hemidiaphragm *(dotted black arrow)* but spares the left heart border *(solid black arrow)*.

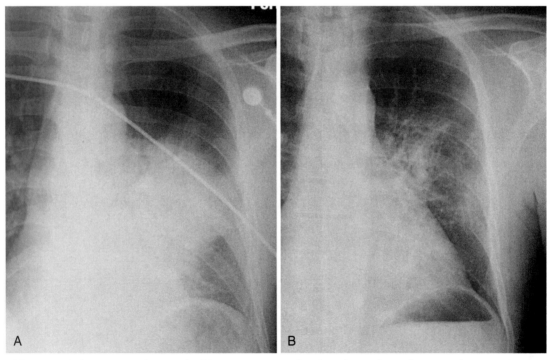

FIGURE 9-11 Resolving pneumonia. Pneumonia can resolve in 2 to 3 days if the organism is sensitive to the antibiotic administered, especially pneumococcal pneumonia. Most pneumonias, as shown in the radiographs of the left upper lobe taken 4 days apart **(A)** and **(B)**, typically resolve from within (vacuolization), gradually disappearing in a patchy fashion over days or weeks. If a pneumonia does not resolve in weeks, you should consider the presence of an underlying obstructing lesion, such as a neoplasm that might be preventing adequate drainage from that portion of the lung.

HOW PNEUMONIA RESOLVES

■ Pneumonia can resolve in 2 to 3 days if the organism is sensitive to the antibiotic administered, especially pneumococcal pneumonia.

■ **Most pneumonias typically resolve from within** (vacuolization), gradually disappearing in a patchy fashion over days, or weeks (Fig. 9-11).

■ If a pneumonia **does not** resolve in several weeks, consider the presence of an underlying **obstructing lesion**, such as a **neoplasm** that is preventing adequate drainage from that portion of the lung. A CT scan of the chest may help to demonstrate the obstructing lesion.

TAKE-HOME POINTS

RECOGNIZING PNEUMONIA

Pneumonia is more opaque than the surrounding normal lung; its margins may be fluffy and indistinct, except for where it abuts a pleural margin. Pneumonia tends to be homogeneous in density; it may contain air bronchograms and may be associated with atelectasis.

Although there is considerable overlap in the patterns of pneumonia that different organisms produce, there are some appearances that are highly suggestive of particular causes.

Lobar pneumonia (prototype: pneumococcal pneumonia) tends to be homogeneous, occupy most or all of a lobe, have air bronchograms centrally, and produce the silhouette sign.

Segmental pneumonia (prototype: staphylococcal pneumonia) tends to be multifocal, does not have air bronchograms, and can be associated with volume loss because the bronchi are also filled with inflammatory exudate.

Interstitial pneumonia (prototype: viral pneumonia or PCP) tends to involve the airway walls and alveolar septa and may produce, especially early in its course, a fine, reticular pattern in the lungs; later in its course, it tends to produce airspace disease.

Round pneumonia (prototype: *Haemophilus*) usually occurs in children in the lower lobes posteriorly and can resemble a mass, the clue being that masses in children are uncommon.

Cavitary pneumonia (prototype: tuberculosis) contains lucent cavities produced by lung necrosis as its hallmark. Postprimary tuberculosis usually involves the upper lobes; it can spread via a transbronchial route that can infect the opposite lower lobe or another lobe in the same lung.

Aspiration occurs in the most dependent portion of the lung at the time of the aspiration, usually the lower lobes, or the posterior segments of the upper lobes; aspiration can be bland and clear quickly, can be infected and take months to clear, or may be from a chemical pneumonitis, which can take weeks to clear.

Pneumonia can be localized by using the silhouette sign and the spine sign as aids.

Pneumonias frequently resolve by "breaking up" so that they contain patchy areas of newly aerated lung within the confines of the previous pneumonia (vacuolization).

 WEBLINK

Visit StudentConsult.Inkling.com for more information and a quiz on Recognizing Pneumonia.

For your convenience, the following QR Code may be used to link to **StudentConsult.com**. You must register this title using the PIN code on the inside front cover of the text to access online content.

CHAPTER 10
Recognizing Pneumothorax, Pneumomediastinum, Pneumopericardium, and Subcutaneous Emphysema

RECOGNIZING A PNEUMOTHORAX

- **A pneumothorax occurs when air enters the pleural space.**
  - ◆ The negative pressure normally present in the pleural space rises higher than the intraalveolar pressure, and the lung collapses.
  - ◆ The parietal pleura remains adjacent to the inner surface of the chest wall, but the visceral pleura retracts toward the hilum with the collapsing lung.
    - • The **visceral pleura becomes visible as a thin white line** outlined by air on both sides, marking the outer border of the lung and indicating the presence of the pneumothorax. The visible visceral pleura is called the *visceral pleural white line* or simply the *visceral pleural line.*

➡ **You must be able to identify the visceral pleural line** (Fig. 10-1) **to make the definitive diagnosis of a pneumothorax!**

- Even as the lung collapses, it tends to maintain its usual lunglike shape so that the curvature of **the visceral pleural line parallels the curvature of the chest wall**; that is, the visceral pleural line is **convex outward toward the chest wall** (Fig. 10-2).
  - ◆ Most other linear densities that mimic a pneumothorax do not demonstrate this spatial relationship with the chest wall.
- There is usually, but not always, an **absence of lung markings peripheral to the visceral pleural line.**

⚠ **Pitfall:** Pleural adhesions may keep part, but not all, of the visceral pleura adherent to the parietal pleura, even in the presence of a pneumothorax. On conventional radiographs it may be possible to visualize lung markings in front or in back of the pneumothorax, and to overlook the presence of a pneumothorax because lung markings appear to extend to the chest wall (Fig. 10-3).

  - ◆ **Absence of lung markings alone is not sufficient for the diagnosis of a pneumothorax, nor is the presence of lung markings distal to the visceral**

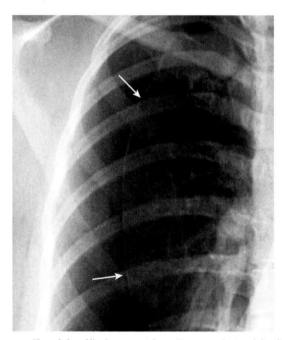

FIGURE 10-1 **Visceral pleural line in a pneumothorax.** You must see the visceral pleural line to make the definitive diagnosis of a pneumothorax *(white arrows)*. The visceral and parietal pleurae are normally not visible, both normally lying adjacent to the chest wall. When air enters the pleural space, the visceral pleura retracts toward the hilum along with the collapsing lung and becomes visible as a very thin white line, with air outlining it on either side. Notice how the contour of the pneumothorax parallels the curvature of the adjacent chest wall.

pleural line sufficient to eliminate the possibility of a pneumothorax.

- The **presence of an air–fluid interface in the pleural space** is, by definition, an indication that there is a pneumothorax present (see Fig. 8-16).
  - ◆ For more about recognizing a hydropneumothorax, see Chapter 8.
- In the supine position, air in a relatively large pneumothorax may collect anteriorly and inferiorly in the thorax and manifest itself by **displacing the costophrenic sulcus inferiorly,** while at the same time, producing **increased lucency of that costophrenic sulcus.** This is called the *deep sulcus*

sign, and it is presumptive evidence for the presence of a pneumothorax on a supine chest radiograph (Fig. 10-4).

■ The key signs for recognizing a pneumothorax are summarized in Box 10-1.

Recognizing the Pitfalls in Overdiagnosing a Pneumothorax

■ Several pitfalls can lead to the mistaken diagnosis of a pneumothorax.

 Pitfall 1: Absence of lung markings mistaken for a pneumothorax

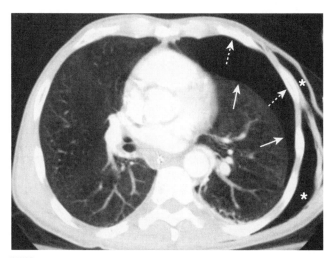

FIGURE 10-2 Pneumothorax seen on CT. As the lung collapses, it tends to maintain its usual shape so that the curve of the visceral pleural line *(solid white arrows)* parallels the curve of the chest wall *(dotted white arrows)*. This is important in differentiating a pneumothorax from artifacts or other diseases that can mimic a pneumothorax. As it collapses, the lung on the side of the pneumothorax also tends to remain lucent until the lung loses almost all of its normal volume, at which point it becomes more opaque. This patient also has *subcutaneous emphysema*–air in the soft tissues–of the left lateral chest wall *(white stars)*. The patient had been stabbed by a friend.

♦ The simple absence of lung markings is not sufficient to warrant the diagnosis of a pneumothorax as there are other diseases that produce such a finding.

♦ These diseases include:
  • **Bullous disease of the lung** (Fig. 10-5)
  • **Large cysts in the lung**
  • **Pulmonary embolism,** which can lead to a lack of perfusion and hence a decrease in the number of vessels visible in a particular part of the lung *(Westermark sign of oligemia)*

♦ In none of these diseases would the treatment ordinarily include the insertion of a chest tube, and in fact, insertion of a chest tube into a bulla might actually **produce** an intractable pneumothorax.

♦ **Solution:** Look at the contour of the structure you believe is the visceral pleural line. Unlike the margin of a bulla, the **visceral pleural line** will be **convex outward** toward the chest wall and will **parallel the curve of the chest wall** (Fig. 10-6).

 Pitfall 2: Mistaking a skinfold for a pneumothorax

♦ When the patient lies directly on the radiographic cassette (as for a portable supine radiograph), a **fold of**

> **BOX 10-1 RECOGNIZING A PNEUMOTHORAX— SIGNS TO LOOK FOR**
>
> Visualization of the visceral pleural line—a must for the diagnosis
>
> Convex curve of the visceral pleural line paralleling the contour of the chest wall
>
> Absence of lung markings distal to the visceral pleural line (most times)
>
> The **deep sulcus sign** of an inferiorly displaced costophrenic angle seen on a supine chest
>
> The presence of an air–fluid interface in the pleural space

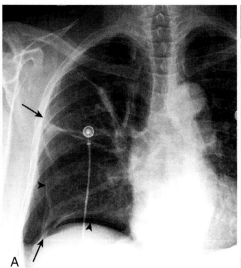
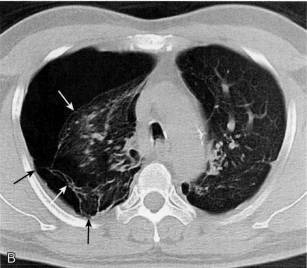

FIGURE 10-3 Pneumothorax with pleural adhesions. There may be lung markings visible on a conventional radiograph of the chest **distal** to the visceral pleural line if there are pleural adhesions. **A,** There is a pneumothorax *(arrowheads)* with pleural adhesions *(black arrows)* that prevent collapse of the lung. **B,** On a CT scan the pleural adhesions *(black arrows)* are seen tethering the partially collapsed lung *(white arrows)* to the parietal pleura. Adhesions most frequently result from prior infection or blood in the pleural space.

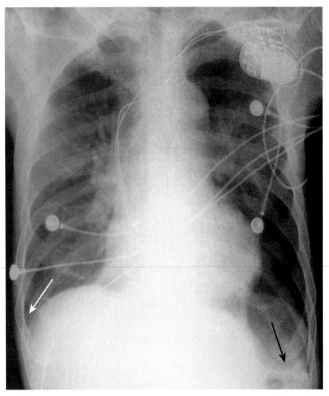

FIGURE 10-4 Deep sulcus sign. In the supine position, air in a relatively large pneumothorax may collect anteriorly and inferiorly in the thorax and manifest itself by displacing the costophrenic sulcus inferiorly, while at the same time producing increased lucency of that sulcus *(black arrow)*. This is called the *deep sulcus sign* and is a sign of a pneumothorax on a supine radiograph. Notice how much lower the left costophrenic sulcus appears than the right sulcus *(white arrow)*.

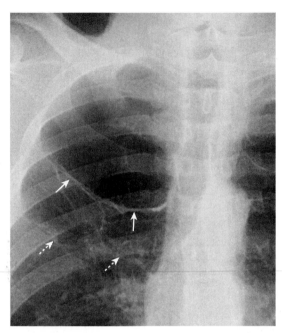

FIGURE 10-5 Bullous disease, right upper lobe. There is a thin white line visible on this close-up of the right upper lobe *(solid white arrows)*, and there are no lung markings peripheral to it. Unlike the visceral pleural line of a pneumothorax, this white line is convex *away* from the chest wall and does not parallel the curve of the chest wall. This is the classical appearance of a bulla in a patient with emphysema. It is important to differentiate between a pneumothorax and a bulla, because inadvertently placing a chest tube into a bulla will almost always **produce** a pneumothorax, which may be difficult to reexpand. The walls of several bullae are visible in this patient *(dotted white arrows)*. On rare occasions, the bullae can grow large enough to render the hemithorax to be seemingly devoid of visible lung tissue *(vanishing lung syndrome)*.

the patient's skin may become trapped between the patient's back and the surface of the cassette.

♦ This can produce an **edge** in the expected position of the visceral pleural line that **may, in fact, parallel the chest wall** just as you would expect the visceral pleural line in a pneumothorax (Fig. 10-7).

♦ **Solution:** Unlike the thin white line of the visceral pleura, **skinfolds produce a relatively thick white band of density.**

 Pitfall 3: Mistaking the medial border of the scapula for a pneumothorax

♦ Ordinarily the patient is positioned for an upright frontal chest radiograph in such a way that the scapulae are retracted lateral to the outer margin of the rib cage, thus preventing the medial border of the scapula from overlapping the lung fields.

♦ With supine radiographs, **the medial border of the scapulae may superimpose on the upper lobes and mimic the visceral pleural line** of a pneumothorax (Fig. 10-8).

♦ **Solution:** Before you diagnose a pneumothorax because you think you see the visceral pleural line, make sure you can **trace the outlines of the scapula** on the side in question and identify its medial border as being separate from the suspected pneumothorax.

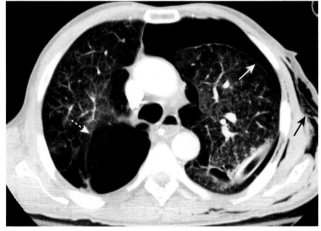

FIGURE 10-6 Bullous disease on right; pneumothorax on left. This axial section from a chest CT demonstrates the different appearances of bullous disease, seen on the right as a rounded cystic lucency *(dotted white arrow)*, and a pneumothorax, seen here on the left with its border convex paralleling the chest wall *(solid white arrow)*. This patient also has subcutaneous emphysema on the left *(solid black arrow)*.

Types of Pneumothorax

■ Pneumothorax can be categorized as *primary*, that is, occurring in what appears to have been normal lung (*spontaneous pneumothorax* being an example), or *secondary*, that is, it occurring in diseased lung (as in *emphysema*).

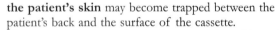

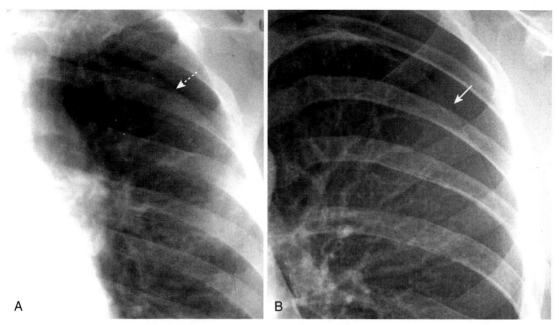

A   **B**

FIGURE 10-7 Skinfold mimicking a pneumothorax and a true pneumothorax. A, When patients lie directly on the radiographic cassette as they might for a portable, supine radiograph, a fold of the patient's skin may become trapped between the patient's back and the surface of the cassette. This can produce an *edge (dotted white arrow)* in the expected position of a pneumothorax, and that edge may, in fact, parallel the chest wall just as you would expect a pneumothorax to do. **B,** While skinfolds produce relatively thick white bands of density, this patient demonstrates the thin white line of the visceral pleura *(solid white arrow)*. A skinfold is an edge; the visceral pleura produces a line.

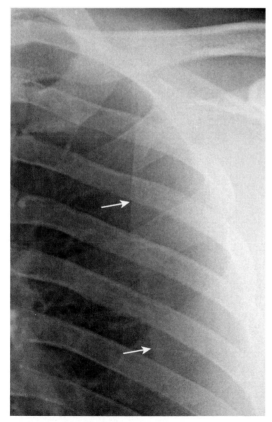

FIGURE 10-8 Scapular edge mimicking a pneumothorax. The patient is usually positioned for an upright frontal chest radiograph in such a way that the medial edges of the scapulae are retracted lateral to the outer edges of the rib cage, thus reducing the risk that the scapulae will produce superimposed densities on the chest. On supine radiographs, the medial border of the scapula *(white arrows)* will frequently superimpose on the upper lung field and may mimic the visceral pleural line of a pneumothorax. Before you diagnose a pneumothorax, make sure you can identify the medial border of the scapula as being separate from the suspected pneumothorax.

- Pneumothorax has also been classified based on the presence or absence of a **"shift" of the mobile mediastinal structures, such as the heart and trachea.**
  - **Simple**—there is usually **no shift of the mediastinal structures** (Fig. 10-9).
  - **Tension**—there is **frequently a shift of the mediastinal structures away from the side of the pneumothorax** associated with cardiopulmonary compromise (Fig. 10-10).
    - Progressive loss of air into the pleural space through a one-way, check-valve mechanism may cause a shift of the heart and mediastinal structures **away** from the side of the pneumothorax. The air may be entering through a rent in the parietal pleura or visceral pleura or from the tracheobronchial tree.
    - The continuously **increasing intrathoracic pressure may lead to cardiopulmonary compromise** by **impairing venous return to the heart.**
    - Besides a shift of the mobile mediastinal structures away from the side of the pneumothorax, there may be **inversion of the hemidiaphragm** (especially on the left side), and there may be **flattening of the heart contour on the side under tension.**
- Tension pneumothorax is associated with a shift of the mediastinal structures **away** from the side of the pneumothorax; simple pneumothorax is not associated with any shift. There is **never** a shift of the heart or mediastinal structures **toward** the side of a pneumothorax.
- The question, **"How large is the pneumothorax?"** is common and oft-repeated, but the actual issue is, "Does this patient require a chest tube to drain the pneumothorax?" Box 10-2 summarizes the answer.

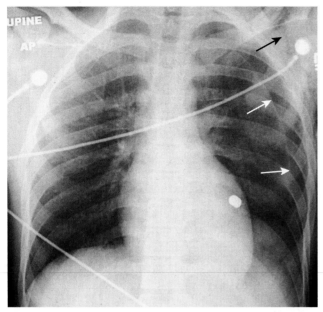

FIGURE 10-9 Simple pneumothorax with no shift. There is a large left-sided pneumothorax *(white arrows)* with no shift of the heart or trachea to the right. There is subcutaneous emphysema seen in the region of the left shoulder *(black arrow)*. Can you detect why the patient had all of these findings? Yes, that's a bullet superimposed on the heart (but on the CT it was posterior to the heart in the left lower lobe).

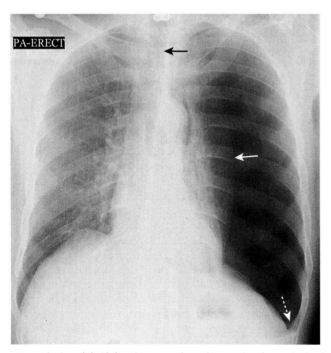

FIGURE 10-10 Large left-sided tension pneumothorax. Progressive loss of air into the pleural space through a one-way check-valve mechanism may cause a shift of the heart and mediastinal structures away from the side of the pneumothorax and lead to cardiopulmonary compromise by impairing venous return to the heart. In this patient with a spontaneous pneumothorax, the left lung is almost totally collapsed *(solid white arrow)*, and there is a shift of the trachea *(closed black arrow)* and heart to the right. The left hemidiaphragm is depressed because of the elevated left intrathoracic pressure *(dotted white arrow)*.

BOX 10-2 HOW LARGE IS THE PNEUMOTHORAX?

Size measurements of pneumothorax on conventional radiographs correlate poorly with computed tomography scans of the actual size.

There is poor correlation between the size of the pneumothorax and the degree of clinical impairment.

The 2 cm rule: if the distance between the lung margin and the chest wall at the apex is <2 cm, a chest tube is usually not needed; a distance >2 cm usually requires chest tube drainage.

Assessment of the patient's clinical status is the most important determinant in deciding whether chest tube drainage is required.

Causes of a Pneumothorax

- **Spontaneous**
  - ♦ Spontaneous pneumothorax is common and often develops from rupture of an apical, subpleural bleb or bulla. It **characteristically occurs in tall, thin males between 20 to 40 years of age.**
- **Traumatic**
  - ♦ **The most common cause of a pneumothorax, either accidental or iatrogenic**
    - • Through the chest wall, for example, stab wound
    - • Internal, for example, rupture of a bronchus from a motor vehicle collision
    - • Iatrogenic, for example, following transbronchial biopsy
- **Diseases that decrease lung compliance**
  - ♦ Chronic fibrotic diseases, for example, eosinophilic granuloma
- **Diseases that stiffen the lung,** for example, hyaline membrane disease in infants
- **Rupture of an alveolus or bronchiole,** for example, asthma

Other Ways to Diagnose a Pneumothorax

- **Computed tomography of the chest**
  - ♦ Today computed tomography (CT) of the chest has essentially replaced expiratory and decubitus views of the chest if there is a strong clinical suspicion of a pneumothorax that is not demonstrated on conventional radiographs. **CT is able to detect extremely small amounts of air in the pleural space** (Fig. 10-11).
- **Expiratory chest x-rays**
  - ♦ Based on the theory that the **volume of air in the lung will normally decrease** on expiration but the **size of a pneumothorax will not,** a small pneumothorax not visible on an inspiratory radiograph may become visible on a radiograph exposed in **full expiration.** Routine expiratory views are no longer recommended for the detection of a pneumothorax.
- **Decubitus chest x-rays**
  - ♦ Because air rises to the highest point, lateral decubitus films of the chest with the affected side up and the x-ray beam directed horizontally (parallel to the floor) may detect a small pneumothorax not seen in the supine position. This method may be helpful in demonstrating a pneumothorax in an infant.
- **Delayed films** are sometimes obtained about **6 hours after a penetrating injury** to the chest in patients with no visible

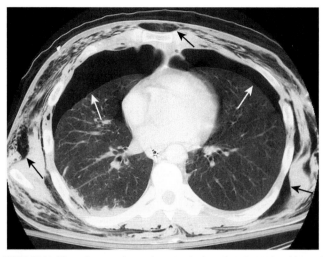

FIGURE 10-11 Bilateral pneumothorax. Conventional radiography is the initial modality used for detecting pneumothorax, but a smaller pneumothorax may be visible only on computed tomography (CT) scans of the chest. This patient has a bilateral pneumothorax *(white arrows)*. Air will rise to the highest point (the patient is supine in the CT scanner). There is also extensive subcutaneous emphysema present *(black arrows)*, which developed because of an air leak from a chest tube that had been inserted earlier.

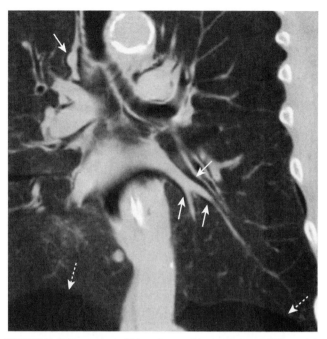

FIGURE 10-12 Pulmonary interstitial emphysema. This coronal reformatted CT scan of the chest demonstrates air *(solid white arrows)* surrounding the pulmonary arteries *(white branching structures)* in the lung. This air arose from a ruptured alveolus in a patient with asthma and is tracking back to the hilum, where it also produced pneumomediastinum and subcutaneous emphysema. The patient also has a bilateral, basilar pneumothorax *(dotted white arrows)*.

pneumothorax on the initial examination, but there may be the occasional appearance of a delayed traumatic pneumothorax.

Pulmonary Interstitial Emphysema

- When the pressure or volume in the alveolus becomes sufficiently elevated, the **alveolus may rupture,** leading to extraalveolar air.
- This extraalveolar air may take one of two main paths:
  - If the alveolus is in close proximity to a pleural surface, the **air may burst outward into the pleural space and create a pneumothorax.**
  - Alternatively, the air can **track backward along the bronchovascular bundles in the lung to the mediastinum,** then into the neck and out to the subcutaneous tissues of the chest and abdominal wall. The air can eventually track downward into the abdomen and retroperitoneum.
  - It is also possible for air to dissect **both** into the **pleural space** and backward along the **bronchovascular sheaths** at the same time.
- The air that tracks backward toward the hilum does so along the **perivascular connective tissue of the lung,** forming **small cystic collections** of extraalveolar air, which **dissect retrograde through the bronchovascular sheaths** to the hila. When the extraalveolar air is confined to the interstitial network of the lung, it is called *pulmonary interstitial emphysema,* or *perivascular interstitial emphysema (PIE).*
- Presumably because of the looser connective tissue in the lungs of children and young adults, **pulmonary interstitial emphysema is more likely to occur in those under the age of 40.**
- **Assisted, mechanical ventilation increases the risk for developing pulmonary interstitial emphysema,** and its formation **heralds a significant risk for the imminent appearance of a pneumothorax,** frequently within a matter of a few hours or days.

- Other causes of increased intraalveolar pressure and rupture include asthma and barotrauma.
- **Pulmonary interstitial emphysema may not be recognizable** on conventional radiographs because the air **collections are small,** and because there is usually a **considerable amount of coexisting disease** in the lung that obscures it. It may, however, be more visible on CT scans of the chest (Fig. 10-12).

RECOGNIZING PNEUMOMEDIASTINUM

- About **one in three patients** with **pulmonary interstitial emphysema will develop pneumomediastinum** (more than three in four develop pneumothorax). Pneumomediastinum occurs because the air tracks backward along the bronchovascular bundles until it enters the mediastinum
- Pneumomediastinum may also develop **when there is perforation of an air-containing viscus in the mediastinum,** such as the esophagus or tracheobronchial tree.
  - **Rupture of the distal esophagus,** usually the left posterolateral wall, can occur with increased intraesophageal pressure from retching or vomiting in *Boerhaave syndrome.*
  - **Rupture of the tracheobronchial tree** is most often **secondary to significant trauma,** either **iatrogenic** such as during a traumatic **intubation,** or **accidental** such as from a penetrating wound, or from severe blunt trauma.

 Radiographic findings of pneumomediastinum:

- **Linear, streaklike lucency associated with a thin white line paralleling the left heart border**

♦ **Streaky air outlining the great vessels** (aorta, superior vena cava, carotid arteries)
♦ Linear streaks of **air parallel to the spine in the upper thorax extending into the neck** and surrounding the esophagus and trachea (Fig. 10-13)

♦ ***Continuous diaphragm sign***—With pneumomediastinum, **air can outline the central portion of the diaphragm** beneath the heart, producing an unbroken superior surface of the diaphragm, which extends from one lateral chest wall to the other (Fig. 10-14).

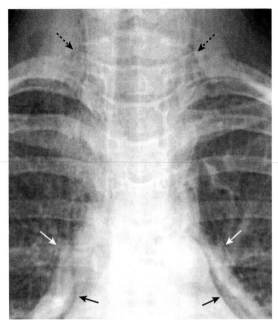

FIGURE 10-13 Pneumomediastinum, pneumopericardium, and subcutaneous emphysema. This patient with asthma developed spontaneous pneumomediastinum, most likely from rupture of an alveolus followed by formation of pulmonary interstitial emphysema. The air tracked back to the hila, then into the mediastinum, where it produced streaky white linear densities *(solid white arrows)* extending to the neck. In the neck, there is subcutaneous emphysema *(dotted black arrows)*. In adults air does not usually enter the pericardium, except by direct penetration, and so it is somewhat unusual that this patient also developed pneumopericardium *(solid black arrows)*. Notice how the air in the pericardial space does not extend above the reflections of the aorta and pulmonary artery, unlike pneumomediastinum, which does extend above the great vessels.

RECOGNIZING PNEUMOPERICARDIUM

■ **Pneumopericardium is usually caused by direct penetrating injuries** to the pericardium, **either caused iatrogenically** (during cardiac surgery) **or accidentally** (from penetrating trauma). Pneumopericardium is **more common in pediatric patients** than adults and may develop in neonates with hyaline membrane disease.

■ It is **rare for air in the pleural space to enter the pericardium,** except in those who have a pericardial defect, such as a surgical window incised in the pericardium, which allows free exchange between the pleural and pericardial spaces.

➡ Pneumopericardium produces a **continuous** band of **lucency that encircles the heart,** bound by the parietal pericardial layer, **which extends no higher than the root of the great vessels** (corresponding to the level of the main pulmonary artery) (Fig. 10-15).

♦ Pneumomediastinum, in contrast, does extend above the root of the great vessels into the uppermost thorax.

■ **CT is usually necessary to demonstrate the findings of pneumopericardium.**

RECOGNIZING SUBCUTANEOUS EMPHYSEMA

■ **Air can extend into the soft tissues of the neck, chest, and abdominal walls** from the mediastinum, or it can

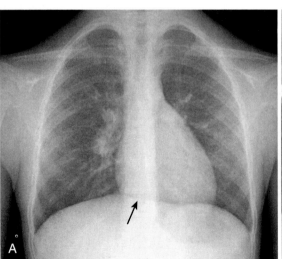
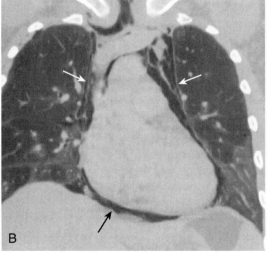

FIGURE 10-14 Continuous diaphragm sign of pneumomediastinum. A, With pneumomediastinum, air can outline the central portion of the diaphragm beneath the heart, producing an unbroken diaphragmatic contour that extends from one lateral chest wall to the other on conventional radiograph *(black arrow)*. This is called the *continuous diaphragm sign*. Normally, the diaphragm is not visible in the center of the chest because there is no air in the mediastinum, and the soft tissue density of the heart rests on and silhouettes the soft tissue density of the diaphragm in its central portion. **B,** A coronal reformatted CT scan of the chest in another patient shows pneumomediastinum outlining the central portion of the diaphragm *(black arrow)* and the remainder of the pneumomediastinum extending superior to the great vessels *(white arrows)*.

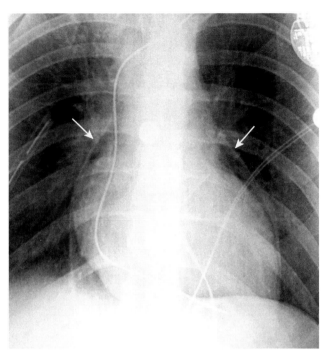

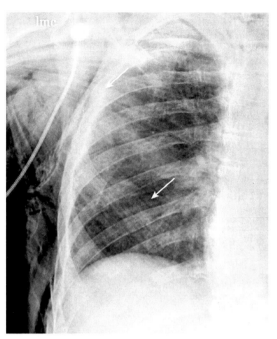

FIGURE 10-15 Pneumopericardium. This patient underwent a procedure to produce a pericardial window for recurrent pericardial effusions. Postoperatively, there is a pneumopericardium, shown by the visible parietal pericardium *(white arrows)*, outlining air around the heart in the pericardial space. Notice how the air does not extend above the reflection of the aorta and main pulmonary artery. Pneumopericardium usually occurs from direct violation of the pericardium by trauma.

FIGURE 10-16 Subcutaneous emphysema. Air can extend into the subcutaneous tissues of the neck, chest, and abdominal walls from the mediastinum, or it can dissect in the soft tissues from a thoracotomy drainage tube or a penetrating injury to the chest wall. Air dissecting along muscle bundles produces this characteristic striated appearance *(white arrows)*. Although dramatic radiographically, subcutaneous emphysema usually produces no serious clinical effects by itself.

dissect in the subcutaneous tissues from a thoracotomy drainage tube or a penetrating injury to the chest wall.

- **Air dissecting along muscle bundles** produces a **characteristic comblike, striated appearance** that superimposes on the underlying lung, often making it difficult to evaluate the lungs by conventional radiography (Fig. 10-16).

- Although radiographically dramatic, **subcutaneous emphysema usually produces no serious clinical effects on its own.**
- Depending on the volume of subcutaneous air present, it may require **several days or even up to a week** or longer for the air to reabsorb.

TAKE-HOME POINTS

RECOGNIZING PNEUMOTHORAX, PNEUMOMEDIASTINUM, PNEUMOPERICARDIUM, AND SUBCUTANEOUS EMPHYSEMA

There is normally no air in the pleural space; air in the pleural space is called a *pneumothorax*.

You must identify the visceral pleural white line to diagnose a pneumothorax.

Beware of the pitfalls that resemble pneumothorax: bullae, skinfolds, and the medial border of the scapula.

Simple pneumothorax has no shift of the heart or mobile mediastinal structures; most pneumothoraces are simple.

Tension pneumothorax (usually associated with cardiorespiratory compromise) produces a shift of the heart and mediastinal structures **away** from the side of the pneumothorax by virtue of a check-valve mechanism that allows air to enter the pleural space, but not leave.

Most pneumothoraces have traumatic causes, either accidental or idiopathic.

Conventional chest radiographs are poor at estimating the size of a pneumothorax; computed tomography (CT) is better; the most important assessment to be made is the clinical status of the patient.

Besides the conventional upright chest radiograph, other ways to diagnose a pneumothorax include expiratory exposures, decubitus views, and delayed images. CT remains the most sensitive test for detecting a small pneumothorax.

Spontaneous pneumothorax most often occurs as a result of rupture of a small apical, subpleural bleb; they most often occur in younger men.

Pulmonary interstitial emphysema results from an increase in the intraalveolar pressure, which leads to rupture of an alveolus and dissection of air back toward the hila along the bronchovascular bundles; it is frequently difficult to visualize on radiographs.

Pneumomediastinum can occur when air tracks back to the mediastinum from a ruptured alveolus or from perforation of an air-containing viscus such as the esophagus or trachea; it can produce the *continuous diaphragm sign* on a frontal chest radiograph.

Pneumopericardium usually requires direct penetration of the pericardium to occur rather than dissection of air from a pneumomediastinum; it can be difficult to differentiate from a pneumomediastinum, but a key is that pneumopericardium does not extend above the roots of the great vessels, whereas pneumomediastinum does.

Air dissecting into the neck and chest wall can produce **subcutaneous emphysema,** which because of its superimposition on the lungs can make evaluation of the underlying lung more difficult. By itself it is usually of no clinical significance and usually clears in a few days, depending on its volume.

WEBLINK

Visit StudentConsult.Inkling.com for more information and a quiz: Pneumothorax or Not?

For your convenience, the following QR Code may be used to link to **StudentConsult.com**. You must register this title using the PIN code on the inside front cover of the text to access online content.

CHAPTER 11
Recognizing the Correct Placement of Lines and Tubes and Their Potential Complications: Critical Care Radiology

- Patients in critical or intensive care units (ICU) are monitored on a frequent basis with portable chest radiography, both to check on the position of their multiple assistive devices and to assess their cardiopulmonary status.
- Diseases commonly seen in critically ill patients are discussed in other chapters (Table 11-1).
- In this chapter, you will get practical advice for evaluating the successful (or unsuccessful) insertion and ultimate position of multiple tubes, lines, catheters, and other supportive apparatus used in the ICU.
- It is customary to obtain a conventional radiograph after the insertion or attempted insertion of one of these devices, to check on its position, and to rule out any unintended complications.
- Therefore, **for each tube** or **device, you will learn:**
  - **Why it is used**
  - **Where it belongs** when properly placed
  - Where the device can be **malpositioned** and what **complications** may occur from the device

TABLE 11-1 COMMON DISEASES IN CRITICALLY ILL PATIENTS

Finding or Disease	Discussed in
Acute respiratory distress syndrome	Chapter 13
Aspiration	Chapter 9
Atelectasis	Chapter 7
Congestive heart failure (pulmonary edema)	Chapter 13
Pleural effusion	Chapter 8
Pneumomediastinum	Chapter 10
Pneumonia	Chapter 9
Pneumothorax	Chapter 10
Pulmonary embolic disease	Chapter 12

ENDOTRACHEAL AND TRACHEOSTOMY TUBES

Endotracheal Tubes

- **Why endotracheal tubes (ETTs) are used**
  - To assist ventilation
  - To isolate the trachea to permit control of the airway
  - To prevent gastric distension
  - To provide a direct route for suctioning
  - To administer medications
- **Correct placement of an ETT** (Box 11-1)
  - Endotracheal tubes are usually **wide-bore tubes** (about 1 cm) with **a radiopaque marker stripe** and **no side holes. The tip is frequently diagonally shaped.**
  - With the patient's head in the neutral position (i.e., bottom of mandible is at the level of C5 to C6), the **tip of ETT should be about 3 to 5 cm from the carina.** This is roughly **half the distance between the medial ends of clavicles** and **the carina** (Fig. 11-1).
- Ideally, the **diameter of the endotracheal tube should be one half to two thirds the width of the trachea.** An inflated cuff (balloon), if present, may fill—but should not distend—the lumen of the trachea (Fig. 11-2).

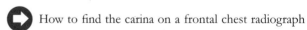 How to find the carina on a frontal chest radiograph

- Follow the right or left main bronchus backward until it meets the opposite main bronchus. The carina projects

BOX 11-1 ENDOTRACHEAL TUBES

The tip should be about 3 to 5 cm above the carina.

The inflated cuff should not distend the lumen of the trachea.

They are most commonly malpositioned in the right main or right lower lobe bronchi.

If positioned with their tip in the neck, damage to vocal cords can occur.

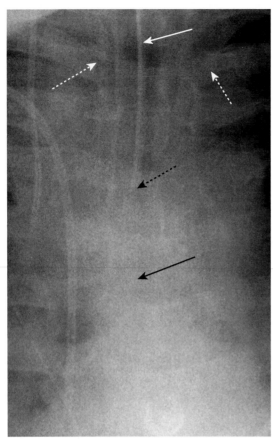

FIGURE 11-1 Endotracheal tube in a satisfactory position. Endotracheal tubes (ETTs) are usually wide-bore tubes (about 1 cm) with a radiopaque marker stripe *(solid white arrow)* and no side holes. The tip is frequently diagonally angled *(dotted black arrow)*. With the patient's head in the neutral position, the tip of ETT should be about 3 to 5 cm from the carina *(solid black arrow)*, which is roughly half the distance between the medial ends of clavicles *(dotted white arrows)* and the carina.

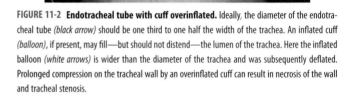

FIGURE 11-2 Endotracheal tube with cuff overinflated. Ideally, the diameter of the endotracheal tube *(black arrow)* should be one third to one half the width of the trachea. An inflated cuff *(balloon)*, if present, may fill—but should not distend—the lumen of the trachea. Here the inflated balloon *(white arrows)* is wider than the diameter of the trachea and was subsequently deflated. Prolonged compression on the tracheal wall by an overinflated cuff can result in necrosis of the wall and tracheal stenosis.

over the T5, T6, or T7 vertebral bodies in 95% of people.

- **Movement of tip with flexion and extension**
  - Neck **flexion** may cause **2 cm of descent** of the tube tip. This is why the tip should be 3 to 5 cm from the carina.
  - Neck **extension** from neutral **may cause 2 cm of ascent** of the tip.
- **Incorrect placement** and **complications of an ETT**
  - **Most common malposition:** because of the shallower angle and wider diameter of the **right main bronchus** or bronchus intermedius, the **tip of the ETT will tend to slide into the right-sided bronchial tree** preferentially to the left.
    - This can lead to **atelectasis** (especially of the nonaerated right upper lobe and left lung) (see Fig. 7-13).
    - Intubation of the right main bronchus could also lead to a **right-sided tension pneumothorax.**
  - Inadvertent **esophageal intubation** will produce a grossly dilated stomach.
  - The tip of the tube should not be positioned in the larynx or pharynx—the tip should be at least 3 cm distal to the level of vocal cords so that damage to the vocal cords and aspiration does not occur (Fig. 11-3).

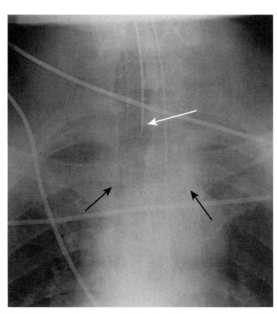

FIGURE 11-3 Endotracheal tube too high. The tip of the tube *(white arrow)* should not be positioned in the larynx or pharynx. The tip should be at least 3 cm distal to the level of the vocal cords so that damage to the vocal cords and aspiration does not occur. The medial ends of the clavicles are marked by the black arrows.

Tracheostomy Tubes

■ **When tracheostomies are used**
  ◆ In patients with airway obstruction at or above the level of the larynx
  ◆ In respiratory failure requiring long-term intubation (>21 days)
  ◆ For airway obstruction during sleep apnea
  ◆ When there is paralysis of the muscles that affect swallowing or respiration

■ **Correct placement of a tracheostomy tube**
  ◆ The **tip should be about halfway between the stoma in which the tracheostomy tube was inserted and the carina.** This is usually around the **level of T3** (Fig. 11-4).
  ◆ Unlike an ETT, the placement of the tip of a tracheostomy tube is not affected by flexion and extension of the head.
  ◆ **The width of the tracheostomy tube should be about two thirds the width of trachea.**

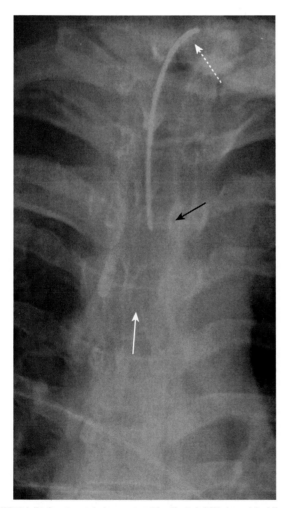

FIGURE 11-4 Tracheostomy tube in correct position. The tip *(solid black arrow)* should be about half-way between the stoma in which the tracheostomy tube was inserted *(dotted white arrow)* and the carina *(solid white arrow)*. This is usually around the level of T3. Unlike the tip of an endotracheal tube, the placement of the tip of a tracheostomy tube is not affected by flexion and extension of the neck.

■ **Incorrect placement** and **complications of a tracheostomy tube** (Box 11-2)
  ◆ Immediately after insertion, **look for signs of** inadvertent **perforation of the trachea** such as pneumomediastinum, pneumothorax, and subcutaneous emphysema.
  ◆ If the tracheostomy tube is equipped with a cuff, the **cuff should generally be inflated to a diameter that fills, but does not distend, the normal tracheal contour.**
  ◆ **Long-term complication of tracheostomies:**
    • **Tracheal stenosis is the most common late-occurring complication** of a tracheostomy tube and can occur at the entrance stoma, level of the cuff, or at the tip of the tube, but is **most common at the stoma.**

INTRAVASCULAR CATHETERS

Central Venous (Pressure) Catheters

■ **Why they are used**
  ◆ For venous access to instill chemotherapeutic and hyperosmolar agents not suitable for peripheral venous administration
  ◆ Measurement of central venous pressure
  ◆ To maintain and monitor intravascular blood volume

■ **Correct placement of central venous catheters (CVC)** (Box 11-3)
  ◆ **Central venous catheters are small** (3 mm) and **uniformly opaque without a marker stripe.**
  ◆ They are usually inserted either by the **subclavian** or **internal jugular** route. The internal jugular veins join the subclavian veins to form the brachiocephalic (innominate) veins, which in turn drain into the superior vena cava. The junctions of the subclavian and brachiocephalic veins generally occur posterior to the **medial ends** of the clavicles.

BOX 11-2 TRACHEOSTOMY TUBES

The tip should be halfway between the entrance stoma and the carina.

If so equipped, the cuff is generally not inflated to a size greater than the tracheal lumen.

Short-term complications may include perforation of the trachea.

Tracheal stenosis is the most common long-term complication, usually at the site of the entrance stoma.

BOX 11-3 CENTRAL VENOUS CATHETERS

The tip should lie in the superior vena cava.

All bends in the catheter should be smooth curves, not sharp kinks.

Most common malpositions are the right atrium and internal jugular vein (for subclavian inserted catheters).

Always check for pneumothorax after successful or unsuccessful insertion attempt.

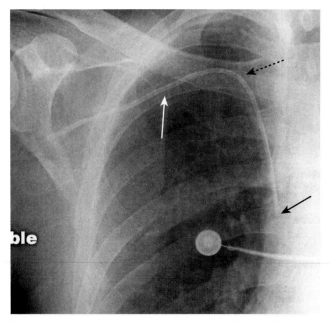

FIGURE 11-5 Subclavian central venous catheter in correct position. Central venous catheters are small in diameter (3 mm) and uniformly opaque without a marker stripe *(solid white arrow)*. The subclavian vein joins the brachiocephalic vein behind the medial end of the clavicle. A central venous catheter should reach the medial end of the clavicle *(dotted black arrow)* before descending. The catheter should descend to the right of the thoracic spine, and the tip should be in the superior vena cava *(solid black arrow)*.

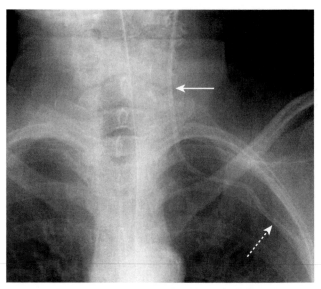

FIGURE 11-6 Central venous catheter malpositioned in internal jugular vein. Central venous catheters, especially those placed by the subclavian route *(dotted white arrow)*, are often malpositioned. They are most often malpositioned with their tips in the right atrium or internal jugular vein *(solid white arrow)*. When central venous catheters are malpositioned, they may provide inaccurate central venous pressure measurements.

♦ A central venous catheter should reach the **medial end of the clavicle before descending,** and its **tip** should **lie medial** to the **anterior end of the 1st rib.**

♦ The catheter should **descend lateral to the right side of the spine,** and the **tip should lie in the superior vena cava** (Fig. 11-5). You should be able to recognize the indentation of the cardiac contour that marks the junction between the superior vena cava and the right atrium (see Fig. 3-1).

♦ **All bends in the catheter should be smooth curves,** not sharp kinks.

■ **Incorrect placement and complications of central venous catheters**

♦ Central venous catheters (more frequently, those placed by the subclavian route) can be malpositioned.

♦ They are **most often malpositioned with their tips in the right atrium or internal jugular vein** (if inserted via the subclavian vein) (Fig. 11-6). In the **right atrium** they can **produce cardiac arrhythmias.** When central venous catheters are malpositioned, they may provide **inaccurate central venous pressure readings.**

♦ **Pneumothorax** can occur in up to 5% of CVC insertions, more often with the subclavian approach than the internal jugular route.

♦ **Occasionally CVCs may perforate the vein** and lie outside of the blood vessel. Look for sharp bends in the catheter as a clue to a potential perforation.

♦ Sometimes they may be **inadvertently inserted in the subclavian artery** rather than the subclavian vein. Suspect **arterial placement** if the **flow is pulsatile** and the course of the catheter **parallels the aortic arch** or **fails to descend to the right of the spine** (Fig. 11-7).

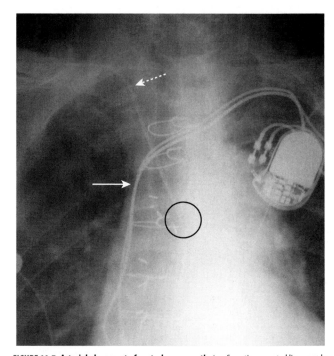

FIGURE 11-7 Arterial placement of central venous catheter. Sometimes, central lines may be inadvertently inserted in the subclavian artery rather than the subclavian vein. This catheter does not reach the medial end of the clavicle *(dotted white arrow)* before descending, and its tip *(black circle)* is oriented over the spine directed away from the shadow of the superior vena cava *(solid white arrow)*. Suspect arterial placement if the return flow is pulsatile.

 Two or more attempts at inserting a CVC

• A frontal chest radiograph is obtained following placement of a CVC.

• **Should initial placement fail,** you should **obtain a chest radiograph** before **trying insertion on the**

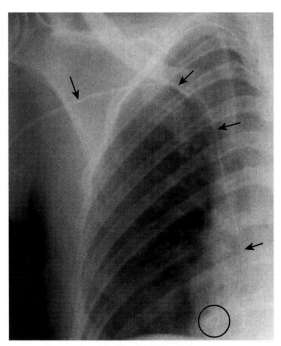

FIGURE 11-8 Peripherally inserted central catheter (PICC) in right atrium. A PICC *(black arrows)* can be used for long-term venous access. Because the lines are so small, they may be difficult to visualize. The tip should lie within the superior vena cava but may be placed in an axillary vein. In this case, the tip extends to the region of the right atrium *(black circle)*.

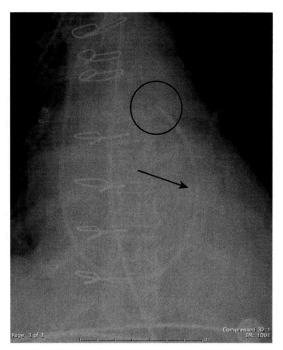

FIGURE 11-9 Swan-Ganz catheter in correct position. Swan-Ganz catheters have the same appearance as central venous lines but are longer *(black arrow)*. They are inserted via the subclavian vein or internal jugular vein, and their tips *(black circle)* are floated out into the proximal right or proximal left pulmonary artery. The tip should not be more than 2 cm distal to either hilar shadow.

BOX 11-4 PERIPHERALLY INSERTED CENTRAL CATHETERS

The tip should lie in the superior vena cava or axillary vein.

They may be difficult to visualize because of their small size.

Thrombosis of the line may occur over time.

BOX 11-5 PULMONARY ARTERY CATHETERS (SWAN-GANZ CATHETERS)

Tip should be no more than 2 cm from the hilum in either the right or left pulmonary artery.

Balloon should be inflated only when pressure measurements are performed.

The tip of the catheter should not lie within a peripheral pulmonary artery.

---

other side to avoid the possibility of producing bilateral pneumothoraces.

Peripherally Inserted Central Catheters

- **Why they are used**
  - For long-term venous access (months)
    - To administer medications such as chemotherapy or antibiotics
    - For frequent blood sampling
      - Because of their small size, they can be inserted into an antecubital vein.
- **Correct placement of peripherally inserted catheters (PICCs)** (Box 11-4)
  - The **tip** should lie **within** the **superior vena cava** but may be placed in an axillary vein. Because the lines are so small, they may be difficult to visualize (Fig. 11-8).
- **Incorrect placement** and **complications of PICCs**
  - Tips may become malpositioned over time.
  - Thrombosis of the line may occur because of its small lumen size.

Pulmonary Artery Catheters—Swan-Ganz Catheters

- **Why they are used**
  - Monitor hemodynamic status of critically ill patients
  - Help in differentiating cardiac from noncardiac pulmonary edema

- **Correct placement of Swan-Ganz catheters** (also known as *pulmonary capillary wedge pressure catheters*)
  - Swan-Ganz catheters have the **same appearance as central venous catheters** but are **longer.**
  - Inserted via the subclavian vein or internal jugular vein, their tips are floated out into the **proximal right** or **proximal left pulmonary artery. The tip of the Swan-Ganz catheter should be no more than about 2 cm from the hila** (Fig. 11-9).
  - The catheter's balloon is temporarily inflated only when pressure measurements are made and should then be deflated.
- **Incorrect placement** and **complications of Swan-Ganz catheters** (Box 11-5)
  - **Serious complications are uncommon.**
  - The most common significant complication is **pulmonary infarction** from occlusion of a pulmonary artery by the catheter or from emboli arising from catheter.
  - Peripherally placed catheters may produce a localized, confined perforation, or **pseudoaneurysm,** which can be recognized by consolidation or a mass that forms at

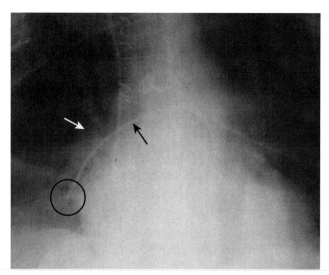

FIGURE 11-10 **Swan-Ganz catheter with tip too peripheral.** This Swan-Ganz catheter is directed toward the right pulmonary artery *(black arrow)*. The tip of a Swan-Ganz catheter should lie within 2 cm of the hilar shadow *(white arrow)*. The tip of this catheter *(black circle)* lies in a peripheral branch of the right descending pulmonary artery. This increases the risk of complication such as pulmonary infarction or pseudoaneurysm formation.

BOX 11-6   MULTIPLE LUMEN CATHETERS

The tip should either be in the superior vena cava or right atrium: some catheters are designed with separate lumens so that one tip is in the superior vena cava and the other is in the right atrium.

The right internal jugular vein has the lowest incidence of clotting so it is the preferred access route.

Complications include pneumothorax, thrombosis, and infection.

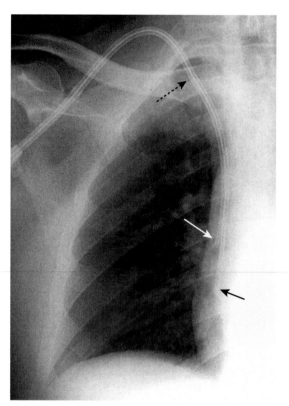

FIGURE 11-11 **Double lumen catheter in correct position.** These large-bore catheters are typically marked with a central stripe *(dotted black arrow)*. All have at least two lumens, with the tip from which blood is withdrawn from the patient *(solid white arrow)* more proximal than the tip through which blood is returned to the patient *(solid black arrow)* in order to minimize recirculation. Some catheters, such as this one, are designed as two, separate single-lumen catheters with one tip in the superior vena cava and the other in the right atrium.

the site of the catheter tip in a critical care patient, who also develops hemoptysis.
- **Make sure the catheter tip does not lie in a distal branch of a pulmonary artery** because this increases the risk of complication (Fig. 11-10).

Multiple Lumen Catheters—Quinton Catheters, Hemodialysis Catheters

- **Why multiple lumen catheters are used** (Box 11-6)
  - Hemodialysis
  - Simultaneous ports for administration of medication and blood sampling
- **Correct placement of multiple-lumen catheters for hemodialysis**
  - These are **large-bore catheters** that are **typically marked with a central stripe.**
  - There are many variations in design among different commercial brands, but all have at least two lumens arranged coaxially inside a single catheter with the goal to minimize the amount of recirculation that occurs between the two ports.
  - The **"arterial" port** from which blood is **withdrawn from the patient is proximal to the "venous" port** through which blood is **returned to the patient** to minimize recirculation of blood. Some catheters are designed as two separate single-lumen catheters, with one catheter tip in the superior vena cava and the other catheter tip in the right atrium (Fig. 11-11).

- The **right internal jugular route is most often used for access.**
- Those that are used temporarily (2 to 3 weeks) usually have their tips in the superior vena cava, whereas the tips of the more permanent catheters may be in the right atrium.
- **Incorrect placement** and **complications of multiple lumen catheters**
  - **Immediate complications** can include **pneumothorax, malposition,** or **perforation of the tip.**
  - **Long-term complications** include **infection** and **thrombosis** of the vein containing the catheter or **occlusion** of the catheter itself.

Pleural Drainage Tubes (Chest Tubes, Thoracotomy Tubes)

- **Why thoracotomy tubes are used**
  - To remove either air or abnormal collections of fluid from the pleural space
- **Correct placement of pleural drainage tubes** (Box 11-7)
  - Chest tubes are **wide-bore tubes with a radiopaque stripe** used as a marker. The **stripe "breaks" at the site of a side hole.**
  - Ideal position is **anterosuperior for evacuating a pneumothorax** and **posteroinferior for draining an effusion,** but chest tubes usually work well no matter where positioned (Fig. 11-12).

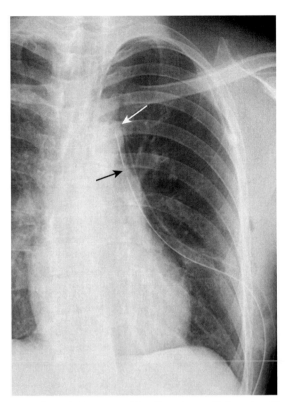

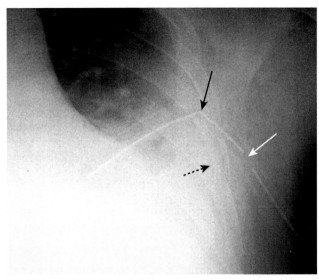

FIGURE 11-13 Side hole of chest tube extends outside thorax. Chest tubes typically have one or more side holes marked by a discontinuity in the marker stripe. No side hole *(solid white arrow)* should lie outside the thoracic cavity *(dotted black arrow)* as it does in this patient. An air leak can develop, leading to persistence of the underlying problem, in this case a pleural effusion. This tube is also kinked as it enters the chest *(solid black arrow)*, which may further reduce its efficiency.

FIGURE 11-12 Chest tube in correct position. Chest tubes are wide-bore tubes with a radi-opaque stripe used as a marker. The stripe *breaks* at the site of the side hole *(black arrow)*. The ideal position is anterosuperior *(white arrow)*, such as in this patient, for evacuating a pneumothorax and posteroinferior for draining an effusion. Chest tubes usually work well no matter where positioned.

BOX 11-7   PLEURAL DRAINAGE TUBES (CHEST TUBES)

In general, chest tubes work well no matter where they are positioned, but malpositioning can result in inadequate drainage.

For pleural effusions, they work best with their tip placed posteriorly and inferiorly.

For pneumothorax, they work best with their tip placed anteriorly and superiorly.

Rapid drainage of a large pleural effusion or large pneumothorax can produce reexpansion pulmonary edema in the underlying lung.

BOX 11-8   PACEMAKERS

Usually placed in the left anterior chest wall, at least one lead should be in the right ventricular apex.

Remember that the right ventricle projects to the left of the spine on the frontal view and anteriorly on the lateral view of the chest.

Complications are infrequent but include fractures in the lead wires and pneumothorax.

Ectopically placed leads may result in failure of the pacemaker to function properly.

- ▪ **Rapid reexpansion of a collapsed lung** caused either by a large pneumothorax or a large pleural effusion may lead to unilateral **reexpansion pulmonary edema** (see Fig. 13-12).

CARDIAC DEVICES—PACEMAKER, AICD, IABP

Pacemakers (Box 11-8)

- ■ **Why they are used**
  - ◆ For cardiac conduction abnormalities
  - ◆ Certain conditions refractory to medical treatment (e.g., congestive heart failure)
- ■ **Correct placement of cardiac pacemakers**
  - ◆ All **pacemakers consist of a pulse generator** usually implanted subcutaneously in the left anterior chest wall and **at least one lead (electrode)** inserted percutaneously, most often into the subclavian vein.

➡ The **tip of one lead is almost always located in the apex of the right ventricle.** Remember that **in the frontal projection, the apex of the right ventricle lies to the left of the spine,** and on the lateral film the **apex of the right ventricle is anterior** (Fig. 11-15).

- ◆ **None of the side holes should lie outside of the thoracic wall** (Fig. 11-13).
- ◆ **Incorrect placement** and **complications of pleural drainage tubes**
  - • Most malpositions lead to **inadequate drainage** rather than serious complication. This includes malpositions in which the tube is inadvertently placed in the major fissure (Fig. 11-14).
  - • If the **side hole extends outside of the chest wall**, this can lead to an *air-leak* causing both inadequate drainage and subcutaneous emphysema.
  - • **Serious complications are uncommon**
    - ▪ Bleeding secondary to laceration of the intercostal artery
    - ▪ Laceration of the liver or spleen on insertion

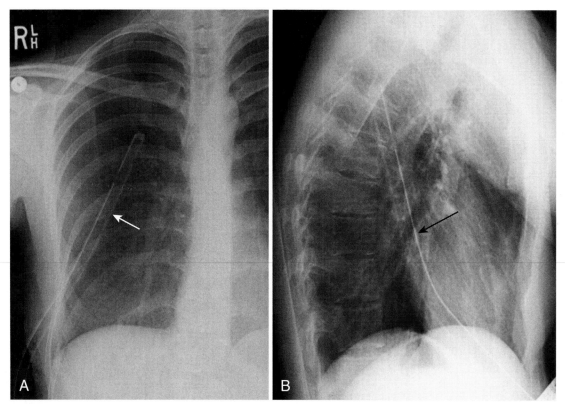

FIGURE 11-14 Chest tube in fissure. Suspect insertion of a chest tube into one of the interlobar fissures when the tube is oriented along the course of the fissure. This tube lies within the right major fissure (*white arrow* on the frontal view **[A]** and *black arrow* on the lateral view **[B]**). Malpositions like this can lead to inadequate drainage rather than a serious complication.

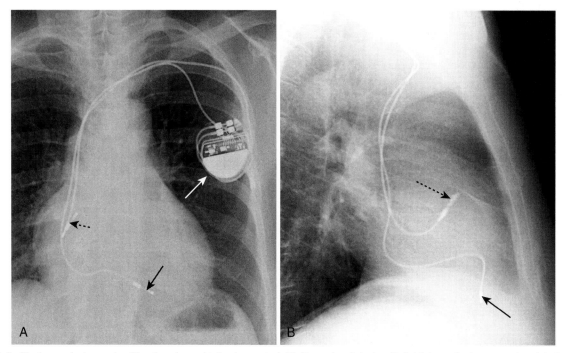

FIGURE 11-15 Dual-lead pacemaker in correct position. Pacemakers consist of a pulse generator *(solid white arrow)* usually implanted in the left chest wall and one or more electrode leads usually inserted into the subclavian vein. This patient has a dual-lead pacemaker. One of the leads is in the apex of the right ventricle (solid black arrow in **[A]** and **[B]**) and the other lead is in the right atrium (*dotted black arrow* in **[A]** and **[B]**). Notice that the right ventricular lead *(solid black arrows)* projects to the left of the midline **(A)** and anteriorly **(B)**, and the right atrial lead typically curls upward **(A)** and **(B)**.

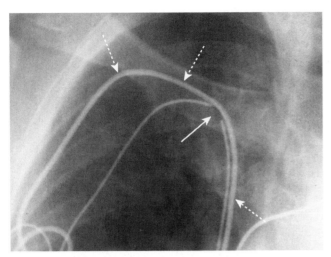

FIGURE 11-16 Fractured pacemaker lead. Fractures of pacemaker or automatic implantable cardiac defibrillator (AICD) leads may occur at any of three places: the generator itself, the tip of the lead, or the site of venous access. Breaks in the lead can be recognized by discontinuity in the wire lead itself *(solid white arrow)*, which in this patient occurred at the site of venous access to the subclavian vein. The broken lead had been discovered earlier, and a second, intact lead is already in place *(dotted white arrows)*.

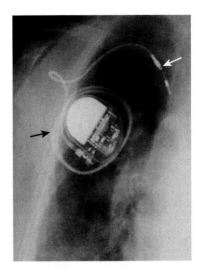

FIGURE 11-17 Twiddler's syndrome. Some patients inadvertently "twiddle" with their subcutaneous pulse generator and, if the subcutaneous tissue allows, may rotate the generator many times on its own axis, curling the lead(s) around the device *(black arrow)*. This can retract the tip of the electrode from the inner wall of the right ventricle, rendering the pacemaker useless. This lead has retracted to the superior vena cava *(white arrow)*.

- Some pacemakers have two leads (usually their tips are in the right atrium and right ventricle), whereas others may have three leads (with their tips usually in the right atrium, right ventricle, and coronary sinus).
- **All leads should have gentle curves.** There should be no sharp kinks in the electrodes.

■ **Incorrect placement and complications of cardiac pacemakers**
- **Pneumothoraces** are infrequent complications of either pacemaker or AICD insertion.
- **Fracture of the leads** may occur at any of three places: the pacer itself, the tip of the lead, or the site of venous access. Breaks in the lead can be **recognized by discontinuity in the wire lead itself** (Fig. 11-16).
- **Leads can perforate the heart** producing cardiac tamponade. **Look for sharp bends** in leads secondary to perforation of a blood vessel.
- **Leads may retract from normal contact with the ventricular wall** because the patient "twists" or "twiddles" the pacemaker generator under the skin, unknowingly winding the leads around the pacer and causing retraction of the tips (*twiddler's syndrome*) (Fig. 11-17) or from subcutaneous migration of the pacer.
- **Leads may be ectopically placed,** for example, in the hepatic vein.

Automatic Implantable Cardiac Defibrillators
■ **Why they are used**
- To prevent sudden death, usually from tachyarrhythmias such as ventricular fibrillation or ventricular tachycardia
■ **Correct placement of automatic implantable cardiac defibrillators (AICDs)** (Box 11-9)
- AICDs can usually be **differentiated from pacemakers by the wider** and **more opaque segment of at least one of the electrodes** (Fig. 11-18). One electrode is usually placed in the superior vena cava or

> ### BOX 11-9   AUTOMATIC IMPLANTABLE CARDIAC DEFIBRILLATOR
>
> Can be differentiated from pacemakers by the presence of a thicker electrode on at least one lead.
>
> May have one (right ventricle), two (right atrium and right ventricle), or three leads (right atrium, right ventricle and coronary sinus).
>
> Bends in the leads should be smooth curves, not sharp kinks.
>
> Visible complications can include lead breakage and dislodgement.

brachiocephalic vein. If present, the other electrode tip is placed in the apex of the right ventricle.
- **All bends in the leads should be smooth curves,** not sharp kinks.
■ **Incorrect placement** and **complications of automatic implantable cardiac defibrillators**
- Leads may migrate and become dislodged.
- Leads may fracture.

Intraaortic Counterpulsation Balloon Pump
■ **Why they are used**
- Used to improve cardiac output and perfusion of the coronary arteries following surgery and in patients with cardiogenic shock or refractory ventricular failure. Placed in the proximal descending thoracic aorta, **the balloon is inflated in diastole, increasing blood flow to the coronary arteries,** and is **deflated in systole, which reduces the cardiac afterload.**
■ **Correct placement of intraaortic balloon pumps (IACB or IABP)** (Box 11-10)
- **The tip can be identified by a small, linear metallic marker** (Fig. 11-19).
  - **The tip should lie distal to the origin of the left subclavian artery** so as not to occlude it.
  - The metallic **marker may point slightly toward the right** in the region of aortic arch.

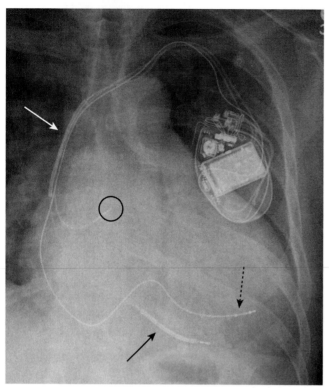

FIGURE 11-18 Automatic implantable cardiac defibrillator (AICD). AICDs can usually can be differentiated from pacemakers by the wider and more opaque segment of at least one of the electrodes *(solid white and black arrows)*. One electrode is usually placed in the superior vena cava *(solid white arrow)* or brachiocephalic vein with its tip in the right atrium *(black circle)*, and another electrode tip is placed in the apex of the right ventricle *(solid black arrow)*. This AICD has a third lead that entered the coronary sinus and lies in a coronary vein *(dotted black arrow)*.

BOX 11-10   INTRAAORTIC BALLOON PUMPS

The tip has a metallic marker, which should lie distal to origin of left subclavian artery.

When inflated, the balloon will be visible as an air-containing "sausage" in thoracic aorta.

Catheters placed too proximally may occlude the great vessels.

Catheters placed too distally may be ineffective.

♦ When inflated, the sausage-shaped balloon may be visualized as an air-containing structure in the descending thoracic aorta.

■ **Incorrect placement** and **complications of IABPs**
  ♦ If the catheter is **too proximal, the inflated balloon may occlude the great vessels,** leading to stroke.
  ♦ If the balloon is **too distal,** the **device has decreased effectiveness.**
  ♦ Aortic dissection and arterial perforation may occur infrequently.

GI TUBES AND LINES—NASOGASTRIC TUBES, FEEDING TUBES

Nasogastric Tubes

■ **Why they are used**
  ♦ Short-term feeding
  ♦ Gastric sampling and decompression through suction
  ♦ Administering medication

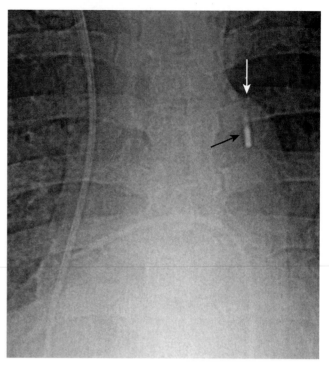

FIGURE 11-19 Intraaortic counterpulsation balloon pump (IABP). The tip of this assistive device can be identified by a small, linear metallic marker in the region of the descending thoracic aorta *(black arrow)*. The tip should lie distal to origin of the left subclavian artery so as not to occlude it. This is usually about 2 cm from the top of the aortic arch *(white arrow)*.

BOX 11-11   NASOGASTRIC TUBE (LEVIN TUBE)

The tip of a nasogastric tube should extend into the stomach about 10 cm past the EG junction.

NG tubes are the most commonly malpositioned of all tubes; always check their positioning with a radiograph.

When malpositioned, they most frequently coil in the esophagus.

If inserted in the trachea, they can extend into a bronchus to periphery of lung, more often on the right side.

■ **Correct placement of nasogastric tubes (NGTs)** (Box 11-11)
  ♦ **Nasogastric tubes** are **wide tubes** (about 1 cm) marked with a **radiopaque stripe** that "breaks" at a **side hole,** usually about **10 cm from the tip.**
  ♦ **The tip** and **all side holes** of the tube should **extend about 10 cm into the stomach beyond the esophagogastric (EG) junction** to prevent aspiration into the esophagus from administration of a feeding (Fig. 11-20). If the NGT is to be used only for suction, the position of the side holes is less important.

 How to recognize the location of the EG junction

  • The **EG junction** is usually **located at the junction of the left hemidiaphragm** and **the left side of the thoracic spine** (this is called the *left cardiophrenic angle*).

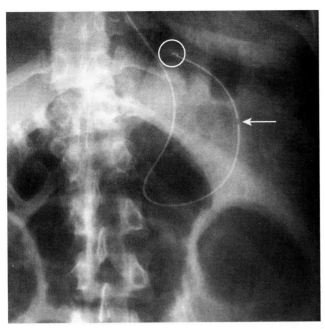

FIGURE 11-20 Nasogastric tube in stomach. Nasogastric tubes are wide-bore tubes marked with a radiopaque stripe that "breaks" in the position of the side hole, usually about 10 cm from the tip *(white arrow)*. The tip and all side holes of the tube should extend about 10 cm into the stomach beyond the esophagogastric (EG) junction to prevent aspiration from administration of the feeding into esophagus. This tube is coiled back on itself, and the tip *(white circle)* is too near the EG junction.

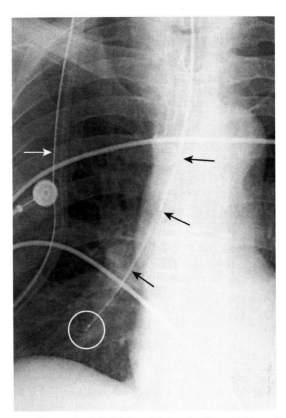

FIGURE 11-21 Nasogastric tube in right lower lobe bronchus. Nasogastric tubes are the most commonly malpositioned of all tubes and lines. Coiling of the NG tube in the esophagus is the most common malposition. In this patient, the nasogastric tube *(black arrows)* entered the trachea instead of the esophagus, and its tip extends to the right lower lobe *(white circle)*. It is important to obtain a radiograph to confirm positioning of a nasogastric tube before using it for feeding. A portion of the NG tube that lies outside the patient is superimposed on the chest *(white arrow)*, as are several heart monitor leads.

■ **Incorrect placement** and **complications of nasogastric tubes**
   ♦ The **nasogastric tube is the most commonly malpositioned of all tubes** and **lines.**
     • **Coiling of the NGT in the esophagus is the most common malposition.**
     • It may be **inadvertently inserted into the trachea** and enter a bronchus (Fig. 11-21).
     • **Perforation caused by an NG tube is rare,** but when it occurs, it is usually in the cervical esophagus.
     • Long-term indwelling NG tube **can lead to gastroesophageal reflux,** which may, in turn, produce **esophagitis** and **stricture.**
   ♦ Always obtain a confirmatory radiograph before feeding the patient or administering any medication through the tube.

Feeding Tubes (Dobbhoff Tubes)

■ **Why they are used**
   ♦ For nutrition
■ **Correct placement of feeding tubes** (Box 11-12)
   ♦ The ideal position of the **tip of the feeding tube is considered to be postpyloric, that is, in the duodenum** or **jejunum,** the rationale being to reduce risk of aspiration after feeding (Fig. 11-22). In fact, placement in the stomach is very common.
   ♦ The tip of a Dobbhoff tube (DHT) is recognizable by a weighted, metallic linear density.
     • Did you know that the Dobbhoff tube was not named after a person called *Dobbhoff?* The tube is

---

BOX 11-12 FEEDING TUBES (DOBBHOFF TUBES)

The tip should ideally be in the duodenum, although most lie in the stomach.

The tip is recognizable by a metallic marker.

If inadvertently inserted into the trachea, the tip may extend into the lung.

Always obtain a confirmatory radiograph before using the tube for feedings.

---

named after two physicians, Drs. Dobbie and Hoffmeister.
■ **Incorrect placement** and **complications of feeding tubes**
   ♦ **Placement in the trachea** rather than the esophagus may lead to the tip entering the lung. Always obtain a confirmatory radiograph before feeding the patient (Fig. 11-23).
   ♦ **Perforation of the esophagus by the guide wire** is an uncommon complication.
     • Once the guide wire is removed, it is not reinserted.

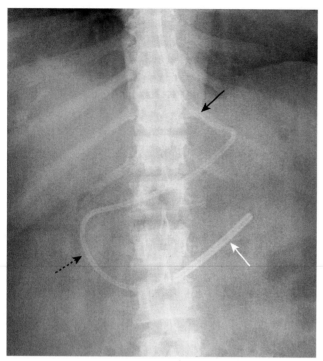

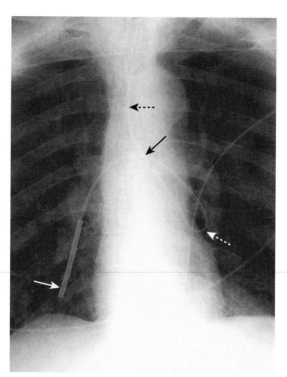

FIGURE 11-22 Dobbhoff tube in duodenum. Correctly placed, the tip of a Dobbhoff feeding tube should be in the duodenum so as to reduce risk of aspiration after feeding. The tip is recognizable by a weighted, metallic end *(solid white arrow)*. This Dobbhoff tube enters the stomach *(solid black arrow)*, courses around the duodenal sweep *(dotted black arrow)*, and ends at the junction between the fourth portion of the duodenum and the jejunum *(solid white arrow)*. Placement in the stomach rather than the duodenum is very common. Note that the EG junction is usually located at the junction of the left hemidiaphragm and the left side of the thoracic spine (the **left cardiophrenic angle)** *(solid black arrow)*.

FIGURE 11-23 Dobbhoff tube in left and right lower lobe bronchi. In this case, the Dobbhoff tube inadvertently entered the trachea *(dotted black arrow)*, entered the left lower lobe bronchus *(dotted white arrow)*, and then coiled back on itself *(solid black arrow)* to cross the midline and end in the right lower lobe bronchus *(solid white arrow)*. It is important to obtain a confirmatory radiograph before using the tube for feedings.

TAKE-HOME POINTS

RECOGNIZING THE CORRECT PLACEMENT OF LINES AND TUBES AND THEIR POTENTIAL COMPLICATIONS: CRITICAL CARE RADIOLOGY

Tube or Line	Desired Position
Endotracheal tube (ETT)	Tip 3-5 cm from carina; usually, halfway between medial clavicles and carina
Tracheostomy tube tip	Halfway between stoma and carina
Central venous catheter (CVC)	Tip in superior vena cava
Peripherally inserted central catheters (PICCs)	Tip in superior vena cava
Swan-Ganz catheter	Tip in proximal right or left pulmonary artery, within 2 cm of hilum
Double-lumen (Quinton) catheters	Tips in either superior vena cava or right atrium (or both), depending on type of catheter
Pleural drainage tube	Anterosuperior for pneumothorax; posteroinferior for pleural effusion
Pacemaker	Tip at apex of right ventricle; other(s) in right atrium and/or coronary sinus
Automatic implantable cardiac defibrillator (AICD)	One lead in superior vena cava; other(s) in right ventricle and/or coronary sinus
Intraaortic balloon pump (IABP)	Tip about 2 cm from top of aortic arch in descending thoracic aorta
Nasogastric (Levin) tube (NGT)	Tip in stomach 10 cm from esophagogastric junction
Feeding (Dobbhoff) tube (DHT)	Tip ideally in the duodenum but more frequently in stomach

 **WEBLINK**

Visit StudentConsult.Inkling.com for additional figures of lines, tubes, and devices and their correct placement (eFigures 11-1 to 11-10) and a quiz on Lines and Tubes in the ICU.

For your convenience, the following QR code may be used to link to **StudentConsult.com**. You must register this title using the PIN code on the inside front cover of the text to access online content.

CHAPTER 12

Recognizing Diseases of the Chest

In this chapter, you will learn how to recognize mediastinal masses, benign and malignant pulmonary neoplasms, pulmonary thromboembolic disease, and selected airway diseases.

- Several chest abnormalities are discussed in other chapters (Table 12-1).
- A complete discussion of all of the abnormalities imaged in the chest is beyond the scope of this text. We will begin here with mediastinal masses and work our way outward to the lungs.

MEDIASTINAL MASSES

- The mediastinum is an area whose **lateral margins are defined by the medial borders of each lung,** whose **anterior margin** is the **sternum** and **anterior chest wall,** and whose **posterior margin is the spine,** usually including the paravertebral gutters.
- The mediastinum can be arbitrarily subdivided into three compartments: the **anterior, middle,** and **posterior** compartments, and each contains its favorite set of diseases. The **superior mediastinum,** roughly the area above the plane of the aortic arch, is a division which is now usually combined with one of the other three compartments mentioned above (Fig. 12-1).

 Pitfall: Since these compartments have no true anatomic boundaries, **diseases from one compartment may extend into another compartment.** When a mediastinal abnormality becomes extensive or a mediastinal mass becomes quite large, it is often impossible to determine which compartment was its site of origin.

TABLE 12-1 CHEST ABNORMALITIES DISCUSSED ELSEWHERE IN THIS TEXT

Topic	Appears in
Atelectasis	Chapter 7
Pleural effusion	Chapter 8
Pneumonia	Chapter 9
Pneumothorax, pneumomediastinum, and pneumopericardium	Chapter 10
Cardiac and thoracic aortic abnormalities	Chapter 13
Chest trauma	Chapter 19

- **Differentiating a mediastinal from a parenchymal lung mass on frontal and lateral chest radiographs** can be difficult.
    - If a mass is completely surrounded by lung tissue in both the frontal and lateral projections, it lies within the lung, not the mediastinum.
    - In general, **the margin of a mediastinal mass** is **sharper** than a mass originating in the lung.
    - Mediastinal masses frequently **displace, compress,** or **obstruct** other mediastinal structures.
- Ultimately, computed tomography (CT) scans of the chest are more accurate in determining the location and nature of a mediastinal mass than conventional radiographs.

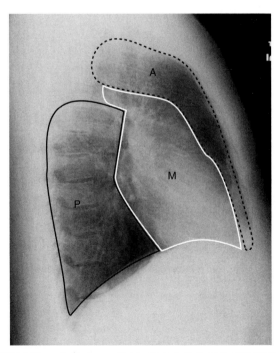

FIGURE 12-1 The mediastinal compartments. The mediastinum can be arbitrarily subdivided into three compartments: anterior, middle, and posterior with each containing its favorite set of diseases. The anterior mediastinum is the compartment that extends from the back of the sternum to the anterior border of the heart and great vessels (A). The middle mediastinum is the compartment that extends from the anterior border of the heart and aorta to the posterior border of the heart and origins of the great vessels (M). The posterior mediastinum is the compartment that extends from the posterior border of the heart to the anterior border of the vertebral column (P). For practical purposes, however, it is considered to extend into the paravertebral gutters.

TABLE 12-2 ANTERIOR MEDIASTINAL MASSES (3 *T*'S AND AN *L*)

Mass	What to Look for
Thyroid goiter	The only anterior mediastinal mass that routinely deviates the trachea
Lymphoma (lymphadenopathy)	Lobulated, polycyclic mass, frequently asymmetric, that may occur in any compartment of the mediastinum
Thymoma	Look for a well-marginated mass that may be associated with myasthenia gravis
Teratoma	Well-marginated mass that may contain fat and calcium, especially seen on computed tomography scans

ANTERIOR MEDIASTINUM

- The **anterior mediastinum** is the compartment that extends from the **back of the sternum** to the **anterior border of the heart** and **great vessels.**
- There is a **differential diagnosis for anterior mediastinal masses** that most often includes:
  - Substernal **thyroid masses**
  - **Lymphoma**
  - **Thymoma**
  - **Teratoma**
  - Lymphoma is sometimes called *terrible lymphoma* in this list, so that all of the diseases in this differential "start" with the letter "T" (Table 12-2).

Thyroid Masses

- In **everyday practice, enlarged substernal thyroids** are the **most frequently encountered anterior mediastinal masses.** The vast majority of these masses are **multinodular goiters,** and this type of mass is called a *substernal goiter, substernal thyroid,* or *substernal thyroid goiter.*
- On occasion the isthmus or lower pole of either lobe of the thyroid may enlarge but project **downward** into the upper thorax rather than **anteriorly** into the neck. About three out of four thyroid masses extend anterior to the trachea; the remaining (almost all of which are right-sided) descend posterior to the trachea.
- **Substernal goiters characteristically displace the trachea** either to the left or right **above the level of the aortic arch,** a tendency the other anterior mediastinal masses do not typically demonstrate. Classically, **substernal goiters do not extend below the top of the aortic arch** (Fig. 12-2).

- ⮕ Therefore you should **think of an enlarged substernal thyroid** whenever you see **an anterior mediastinal mass that displaces the trachea.**

- Radioisotope **thyroid scans are the study of first choice** in confirming the diagnosis of a substernal thyroid, because virtually all of them will display some uptake of the radioactive tracer, which can be imaged and recorded with a special camera (see online chapter, Nuclear Medicine: Understanding the Principles and Recognizing the Basics).

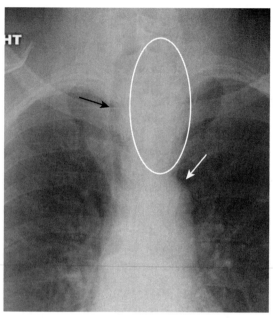

FIGURE 12-2 Substernal thyroid mass. The lower pole of the thyroid may enlarge but project downward into the upper thorax *(white oval)* rather than anteriorly into the neck. Classically substernal thyroid goiters produce mediastinal masses that do not extend below the top of the aortic arch *(white arrow).* Substernal goiters characteristically displace the trachea *(black arrow)* either to the left or right above the aortic knob, a tendency the other anterior mediastinal masses do not typically demonstrate. Therefore you should think of an enlarged substernal thyroid goiter whenever you see an anterior mediastinal mass that displaces the trachea.

- **On CT scans** substernal thyroid masses are **contiguous with the thyroid gland, frequently contain calcification,** and **avidly take up intravenous contrast,** but **with a mottled, inhomogeneous appearance** (Fig. 12-3).

Lymphoma

- **Lymphadenopathy,** whether from lymphoma, metastatic carcinoma, sarcoid, or tuberculosis, is the most common cause of a mediastinal mass overall.
- **Anterior mediastinal lymphadenopathy is most common in Hodgkin disease, especially the nodular sclerosing variety.** Hodgkin disease is a malignancy of the lymph nodes, more common in females, which most often presents with painless, enlarged lymph nodes in the neck.
- Unlike teratomas and thymomas, which are presumed to expand outward from a single abnormal cell, lymphomatous masses are frequently composed of several contiguously enlarged lymph nodes. **Lymphadenopathy frequently presents with a border that is lobulated** or **polycyclic in contour** owing to the conglomeration of enlarged nodes that make up the mass.

- ⮕ On chest radiographs this lobulation **may help differentiate lymphadenopathy from other mediastinal masses.**

- **Mediastinal lymphadenopathy** in **Hodgkin disease** is usually **bilateral** and **asymmetric** (Fig. 12-4). In addition, **asymmetric hilar adenopathy** is associated with mediastinal adenopathy in many patients with Hodgkin disease.

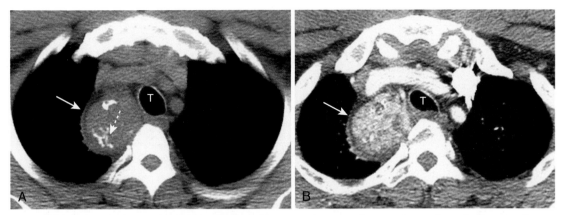

FIGURE 12-3 Computed tomography (CT) of a substernal thyroid goiter without and with contrast enhancement. These are two images at the same level in a patient who was scanned both before **(A)** and then after intravenous contrast administration **(B)**. **A,** On CT scans substernal thyroid masses *(solid white arrow)* are contiguous with the thyroid gland, frequently contain calcification *(dotted white arrow)*, and **(B)** avidly take up intravenous contrast but with a mottled, inhomogeneous appearance *(solid white arrow)*. This mass is displacing the trachea (T) slightly to the left.

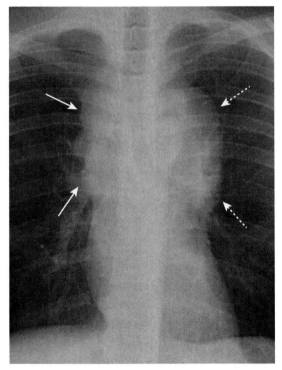

FIGURE 12-4 Mediastinal adenopathy from Hodgkin disease. Lymphadenopathy frequently presents with a lobulated or polycyclic border as a result of the conglomeration of enlarged nodes that produce the mass *(solid white arrows)*. This finding may help differentiate lymphadenopathy from other mediastinal masses. Mediastinal lymphadenopathy in Hodgkin disease is usually bilateral *(dotted white arrows)* and frequently asymmetric.

- In general, **mediastinal lymph nodes that exceed about 1 cm,** measured along their **short axis** on CT scans of the mediastinum, are **considered to be enlarged.**
- Lymphoma will **produce multiple, lobulated soft-tissue masses,** or one **large, soft-tissue mass** from lymph node aggregation.
- The **mass is usually homogeneous** in density on CT but **may be heterogeneous** when it achieves a sufficient size to undergo **necrosis (areas of lower attenuation, i.e., blacker),** or hemorrhage (areas of higher attenuation, i.e., whiter)** (Fig. 12-5).

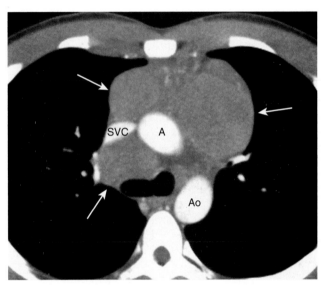

FIGURE 12-5 Computed tomography (CT) of anterior mediastinal adenopathy in Hodgkin disease. On CT, lymphomas will produce multiple, lobulated soft-tissue masses or a large soft-tissue mass from lymph node aggregation *(white arrows)*. The mass is usually homogeneous in density, as in this case, but may be heterogeneous when the nodes achieve a sufficient size to undergo necrosis *(areas of low attenuation, i.e., blacker),* or hemorrhage *(areas of high attenuation, i.e., whiter)*. The superior vena cava (SVC) is compressed by the nodes, whereas the ascending (A) and descending aorta (Ao) are typically less so.

- Some findings of lymphoma may mimic those of a sarcoid since both produce thoracic adenopathy. Table 12-3 contains several key points to differentiate them.

Thymic Masses

- **Normal thymic tissue can be visible on CT throughout life,** although the gland begins to **involute after age 20.**
- **Thymomas** are **neoplasms of thymic epithelium and lymphocytes.** They occur most often in **middle-aged adults,** generally at an **older** age than those with **teratomas.** Most thymomas are **benign.**

Thymomas are **associated with myasthenia gravis about 35% of the time** that they are present. Conversely, about **15% of patients with clinical myasthenia gravis**

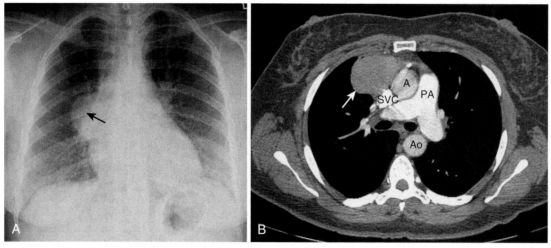

FIGURE 12-6 Thymoma, chest radiograph, and computed tomography (CT) scan. Thymomas are neoplasms of thymic epithelium and lymphocytes that occur most often in middle-aged adults, generally at an older age than those with teratomas. **A,** The chest radiograph shows a smoothly marginated anterior mediastinal mass *(black arrow).* **B,** Contrast-enhanced CT scan confirms the anterior mediastinal location and homogeneous density of the mass *(white arrow).* The patient had myasthenia gravis and improved following resection of the thymoma. Ascending aorta (A), descending aorta (Ao), main pulmonary artery (PA), superior vena cava (SVC).

TABLE 12-3 SARCOIDOSIS VERSUS LYMPHOMA

Sarcoidosis	Lymphoma
Bilateral hilar and right paratracheal adenopathy are classic combination	More often mediastinal adenopathy, associated with asymmetric hilar enlargement
Bronchopulmonary nodes more peripheral	Hilar nodes more central
Pleural effusion in about 5%	Pleural effusion more common in 30%
Anterior mediastinal adenopathy is uncommon	Anterior mediastinal adenopathy is common

will be found to **have a thymoma.** The importance of identifying a thymoma in patients with myasthenia gravis lies in the **favorable prognosis** for patients with myasthenia **after thymectomy.**

- On **CT scans thymomas** classically present as a **smooth or lobulated mass** that arises near the **junction of the heart** and **great vessels** and which, like a teratoma, **may contain calcification** (Fig. 12-6).
- Other lesions that can produce enlargement of the thymus include thymic cysts, thymic hyperplasia, thymic lymphoma, carcinoma, or lipoma.

Teratoma

- Teratomas are **germinal tumors** that typically **contain all three germ layers** (ectoderm, mesoderm, and endoderm). Most teratomas are benign and occur earlier in life than thymomas. Usually asymptomatic and discovered serendipitously, about **30% of mediastinal teratomas are malignant** and have a poor prognosis.
- The most common variety of teratoma is **cystic,** producing a **well-marginated mass** near the origin of the great vessels that **characteristically contains fat, cartilage,** and **possibly bone on CT examination** (Fig. 12-7).

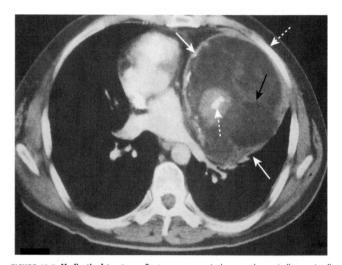

FIGURE 12-7 Mediastinal teratoma. Teratomas are germinal tumors that typically contain all three germ layers. They tend to be discovered at a younger age than those with thymomas. The most common variety of teratoma is cystic, as in this case *(solid white arrows).* As shown here they usually produce a well-marginated mass near the origin of the great vessels. On CT they characteristically contain fat *(solid black arrow),* cartilage, and sometimes bone *(dotted white arrow).*

MIDDLE MEDIASTINAL MASSES

- The **middle mediastinum** is the compartment that extends from the **anterior border of the heart** and **aorta** to the **posterior border of the heart** and **contains the heart, the origins of the great vessels, trachea,** and **main bronchi,** along with **lymph nodes** (see Fig. 12-1).
- **Lymphadenopathy produces the most common mass in this compartment.** Although Hodgkin disease is the most likely cause of mediastinal adenopathy, other malignancies and several benign diseases can produce such findings.
  - Other malignancies that produce mediastinal lymphadenopathy include **small cell lung carcinoma** and **metastatic disease** such as from primary **breast carcinoma** (Fig. 12-8).

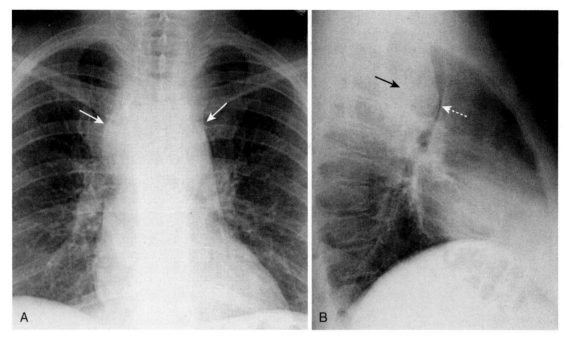

FIGURE 12-8 Middle mediastinal lymphadenopathy. Although lymphoma is the most likely cause of adenopathy in the middle mediastinum, other malignancies such as small cell lung carcinoma, metastatic disease, and several benign diseases can produce these findings. This patient has a mediastinal mass demonstrated on both the frontal **(A)** *(solid white arrows)* and lateral **(B)** views *(solid black arrow)*. The mass is pushing the trachea forward *(dotted white arrow)* on the lateral view. The biopsied lymph nodes in this patient demonstrated small cell carcinoma of the lung.

♦ Benign causes of mediastinal lymphadenopathy include **infectious mononucleosis** and **tuberculosis,** the latter usually producing **unilateral mediastinal adenopathy.**

POSTERIOR MEDIASTINAL MASSES

■ The **posterior mediastinum** is the compartment that extends from the **posterior border of the heart** to the **anterior border of the vertebral column.** For practical purposes, however, it is **considered to extend to either side of the spine into the paravertebral gutters** (see Fig. 12-1).

■ It contains the **descending aorta, esophagus, and lymph nodes;** is the site of masses representing **extramedullary hematopoiesis;** and most important, is the home of **tumors of neural origin.**

Neurogenic Tumors

■ Although neurogenic tumors produce the largest percentage of posterior mediastinal masses, none of these lesions is particularly common. Neurogenic tumors include such entities as **neurofibroma, Schwannoma, ganglioneuroma,** and **neuroblastoma.**

■ **Nerve sheath tumors** (Schwannoma) are the **most common** and are almost always **benign.** They usually affect persons 20 to 50 years of age and, since they are slow growing, may produce no symptoms until late in their course.

■ Neoplasms that arise from nerve elements **other than the sheath,** such as **ganglioneuromas** and **neuroblastomas,** are usually **malignant.**

■ **Neurogenic tumors** will produce a soft tissue **mass, usually sharply marginated,** in the paravertebral gutter (Fig. 12-9). Both benign and malignant tumors **may erode adjacent ribs** (Fig. 12-10, *A*). They may **enlarge neural foramina,** producing **dumbbell**-shaped lesions that arise from the spinal canal but project through the neural foramen into the mediastinum (see Fig. 12-10, *B*).

■ **Neurofibromas** can occur as an **isolated tumor** arising from the Schwann cell of the nerve sheath or **as part of a syndrome** called *neurofibromatosis.* As part of the latter, they are a component of a neurocutaneous bone dysplasia that can cause numerous abnormalities, including **subcutaneous nodules, erosion of adjacent bone (rib notching), scalloping** of the posterior aspect of the vertebral bodies (Fig. 12-11), **absence** of the **sphenoid wings, pseudarthrosis,** and **sharp-angled kyphoscoliosis** at the thoracolumbar junction.

SOLITARY NODULE/MASS IN THE LUNG

■ **The difference between a nodule** and **a mass is size:** in general, under 3 cm it is called a *nodule,* and over 3 cm it is called a *mass.*

■ Much has been written about the workup of a patient in whom a single nodular density is discovered in the lung on imaging of the thorax, (i.e., the **solitary pulmonary nodule**). It is estimated that as many as 50% of smokers have a nodule discovered by chest CT; yet less than 1% of the nodules <5 mm in size show any malignant tendencies (growth or metastases) when followed for 2 years.

■ In evaluating a solitary pulmonary nodule, the **critical question** to be answered is this: **Is the nodule most likely benign** or **malignant?** If most likely benign, it can be watched in follow-up, whereas if most likely malignant, it will almost certainly be treated aggressively with therapies that carry some risk of morbidity and mortality.

■ The answer to the question of benign versus malignant will depend on many factors, including the availability of **prior**

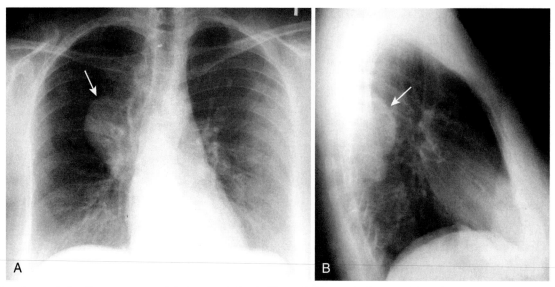

FIGURE 12-9 **Neurofibromatosis.** Neurofibromas can occur as an isolated tumor arising from the Schwann cell of the nerve sheath or as part of the syndrome **neurofibromatosis**, as in this case. There is a large posterior mediastinal neurofibroma *(white arrows)* seen in the right paravertebral gutter on the frontal **(A)** and lateral **(B)** views.

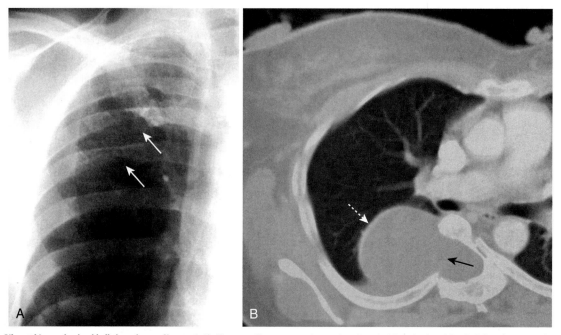

FIGURE 12-10 **Rib-notching and a dumbbell-shaped neurofibroma. A,** Plexiform neurofibromas can produce erosions along the inferior borders of the ribs (where the intercostal nerves are located) and produce either notching or a wavy appearance called **ribbon ribs** *(solid white arrows).* **B,** Another patient demonstrates a large neurofibroma that is enlarging the neural foramen, eroding half of the vertebral body *(solid black arrow),* and producing a dumbbell-shaped lesion that arises from the spinal canal but projects through the foramen into the mediastinum *(dotted white arrow).*

imaging studies, which **can help greatly in establishing stability of a lesion over time.**

■ In 2005 an international society of chest radiologists (Fleischner Society) approved a set of evidence-based criteria setting guidelines for the follow-up of **noncalcified nodules** found **incidentally during chest CT.**

♦ **Nodules 4 mm or smaller** do not require follow-up for low-risk patients (patients with minimal or no history of smoking or other risk factors). High-risk

patients (patients with a history of smoking or other known risk factors) require follow-up after 12 months. If no change, further imaging is not needed.

♦ **Nodules >4 to 6 mm** require follow-up after 12 months for low-risk patients (no further imaging needed if there is no change). High-risk patients should receive an initial follow-up CT at 6 to 12 months (and at 18 to 24 months if there is no change).

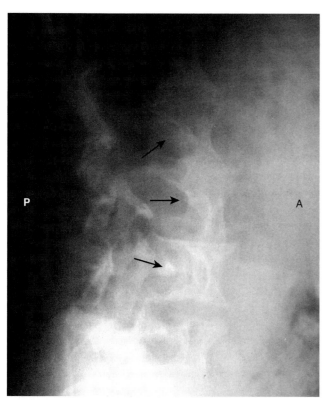

FIGURE 12-11 Scalloping of the vertebral bodies in neurofibromatosis. Neurofibromatosis is a neurocutaneous disorder associated with a skeletal dysplasia. There may be numerous skeletal abnormalities associated with the disease, including scalloping of posterior vertebral bodies *(black arrows)*, especially in the thoracic or lumbar spine (as shown here in this lateral view). This is produced by diverticula of the thecal sac caused by dysplasia of the meninges that leads to erosion of adjacent bone through the pulsations transmitted via the spinal fluid. *A,* Anterior; *P,* posterior.

♦ **Nodules >6 to 8 mm** require an initial follow-up CT at 6 to 12 months (and at 18 to 24 months if there is no change) for low-risk patients. High-risk patients require an initial follow-up CT at 3 to 6 months (and at 9 to 12 and 24 months if there is no change).

♦ **Nodules >8 mm** require follow-up CTs at 3, 9, and 24 months with consideration for the use of dynamic-contrast enhanced CT, positron emission tomography (PET), and/or biopsy for both high- and low-risk patients.

Signs of a Benign Versus Malignant Solitary Pulmonary Nodule

■ **Size of the lesion.** Nodules smaller than 4 mm rarely show malignant behavior. Masses **larger than 5 cm have a 95% chance of malignancy** (Fig. 12-12).

■ **Calcification.** The presence of calcification is usually determined by CT. **Lesions containing central, laminar,** or **diffuse patterns of calcification are invariably benign.**

■ **Margin.** Lobulation, spiculation, and shagginess all suggest malignancy (Fig. 12-13).

■ **Change in size over time.** Requires a previous study or sufficient confidence so as to obtain a follow-up study that will provide a basis of comparison of size over time. **Malignancies** tend to **increase in size at a rate that is neither brief enough to suggest an inflammatory cause** (changes in weeks) nor **prolonged enough to suggest benignity** (no change over a year or more).

♦ **Large cell carcinomas,** as a cell type, **grow the most rapidly.**

♦ **Squamous cell carcinomas** and **small cell carcinomas** tend to grow slightly less **rapidly.**

♦ **Adenocarcinomas** grow the **most slowly.**

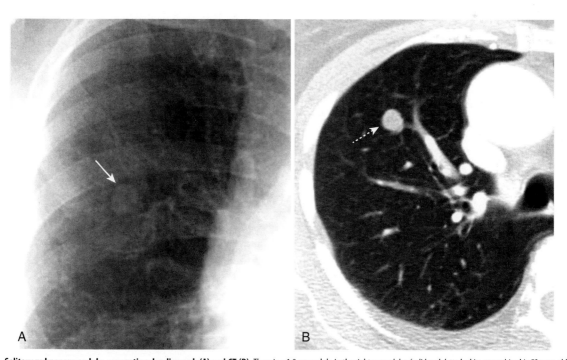

FIGURE 12-12 Solitary pulmonary nodule, conventional radiograph (A) and CT (B). There is a 1.8 cm nodule in the right upper lobe *(solid and dotted white arrows)* in this 53 year-old male with an episode of hemoptysis. The critical question to be answered in evaluating any solitary pulmonary nodule is whether the lesion is benign or malignant. The answer to the question will depend on many factors including the lesion's size and availability of prior imaging studies, which can help greatly in establishing growth of a lesion over time. A positron emission tomography scan in a lesion of this size might help in indicating if the nodule was benign or not. This patient had a biopsy that revealed an adenocarcinoma of the lung.

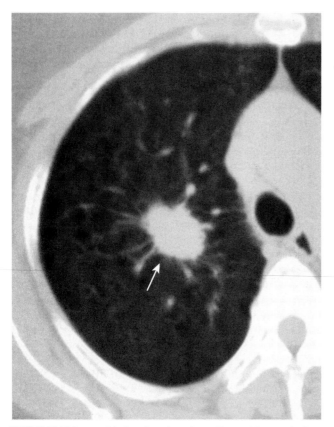

FIGURE 12-13 Right upper lobe bronchogenic carcinoma. There is a 3.2-cm spiculated mass in the right upper lobe *(white arrow)*. The relatively large size and irregular margins of this mass point toward a malignant process. A percutaneous biopsy revealed an adenocarcinoma of the lung.

 When there are clinical signs or symptoms present, the chance of malignancy rises. Solitary pulmonary nodules that are surgically removed from those who showed **clinical signs** or **symptoms** and imaging findings that suggested malignancy are **malignant 50% of the time in men over the age of 50.**

Benign Causes of Solitary Pulmonary Nodules

- **Granulomas.** Tuberculosis and histoplasmosis usually produce calcified nodules <1 cm in size, although tuberculomas and histoplasmomas can reach up to 4 cm.
    - **When calcified, they are clearly benign.** Tuberculous granulomas are usually **homogeneously calcified.** **Histoplasmomas** may contain a **central** or **target calcification** or may have a **laminated calcification,** which is diagnostic (Fig. 12-14).
    - **PET** scans can also help in differentiating benign from malignant nodules and provide insight into the site of any metastatic disease. Using a glucose analog (fluorodeoxyglucose [FDG]), the test is based on the enhanced glucose metabolism and uptake of lung cancer cells. Usually the nodules must be >1 cm in size for accurate characterization as being PET-avid (i.e., most likely malignant). (See online chapter, Nuclear Medicine: Understanding the Principles and Recognizing the Basics.)
- **Hamartomas.** Peripherally located lung tumors of **disorganized lung tissue** that characteristically **contain fat and calcification** on CT scan. The **classical calcification** of a hamartoma is called *popcorn calcification* (Fig. 12-15).

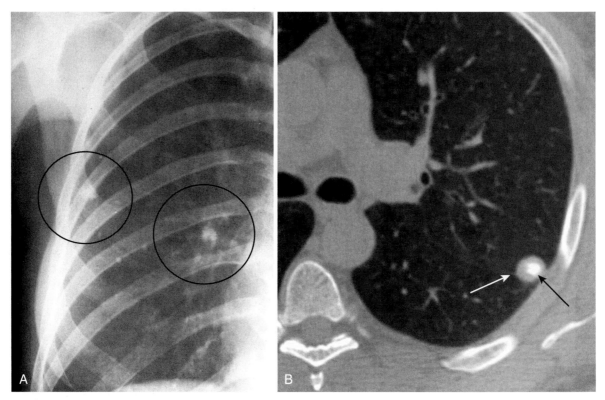

FIGURE 12-14 Calcified tuberculous granulomas and histoplasmomas. When a pulmonary nodule is heavily calcified, it is almost always benign. **A,** Tuberculous granulomas are common sequelae of prior, usually subclinical, tuberculous infection and are usually homogeneously calcified *(black circles)*. **B,** Histoplasmomas *(white arrow)* may contain a central or *target* calcification *(black arrow)* or may have a laminated calcification, which is diagnostic. CT can be used to differentiate between a calcified and a noncalcified pulmonary nodule with greater sensitivity than conventional radiographs.

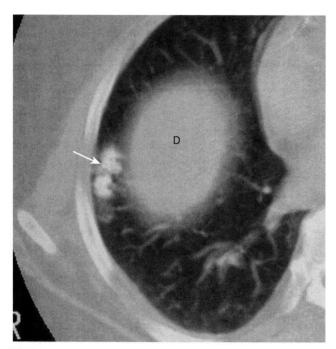

FIGURE 12-15 Hamartoma of the lung. Hamartomas of the lung are peripherally located tumors of disorganized lung tissue that classically contain fat and calcification on CT scans. The classical calcification of a hamartoma is called *popcorn calcification* (*white arrow*). The small island of soft tissue in the middle of the right lung (D) is the uppermost part of the right hemidiaphragm.

- Other uncommon benign lesions that can produce solitary pulmonary nodules include rheumatoid nodules, fungal diseases such as nocardiosis, arteriovenous malformations, and granulomatosis with polyangiitis (Wegener granulomatosis).
- Round atelectasis may mimic a solitary pulmonary nodule and is discussed in Chapter 7.

BRONCHOGENIC CARCINOMA

- In the United States **lung cancer is the most common fatal malignancy in men** and the **second most common in women** (next to breast cancer).
- The number of nodules in the lung can help direct the workup. For malignant pulmonary nodules, **primary lung cancer** usually presents as a **solitary nodule,** whereas **metastatic disease** to the lung from another organ characteristically produces **multiple nodules.**
- Table 12-4 summarizes the classical manifestations and growth tendencies of the four types of bronchogenic carcinoma by cell type.
- **Recognizing a bronchogenic carcinoma**
  - They may be recognized by visualization of the **tumor itself** (i.e., a **nodule/mass in the lung**).
  - They may be suspected by recognizing the **effects of bronchial obstruction:** (i.e., **pneumonitis** and/or **atelectasis**).
  - They may be suspected by recognizing the results of either their **direct extension** or **metastatic spread** to the lung, ribs, or other organs.

TABLE 12-4 CARCINOMA OF THE LUNG: CELL TYPES

Cell Type and Clinical Manifestations	Graphic Representation	Cell Type and Clinical Manifestations	Graphic Representation
Squamous Cell Carcinoma • Primarily central in location • Arise in segmental or lobar bronchi • Invariably produce bronchial obstruction leading to obstructive pneumonitis or atelectasis • Tend to grow rapidly		**Small Cell, Including Oat Cell Carcinoma** • Primarily central in location • Many contain neurosecretory granules that lead to an association of small cell carcinoma with paraneoplastic syndromes such as Cushing syndrome, inappropriate secretion of antidiuretic hormone	
Adenocarcinoma • Primarily peripheral in location • Usually solitary except in the case of diffuse adenocarcinoma, which can present as multiple nodules • Slowest growing		**Large Cell Carcinoma** • Diagnosis of exclusion for lesions that are nonsmall cell and not squamous or adenocarcinoma • Larger peripheral lesions • Grow extremely rapidly	

Bronchogenic Carcinomas Presenting as a Nodule/Mass in the Lung

■ These are most often **adenocarcinomas.**
■ The nodule may have **irregular** and **spiculated margins** (see Fig. 12-13).
■ The mass **may cavitate**, more often if it is of squamous cell origin (although cavitation also occurs with adenocarcinoma), usually producing a **relatively thick-walled** cavity with a **nodular** and **irregular inner margin** (Fig. 12-16).

Bronchogenic Carcinoma Presenting with Bronchial Obstruction

■ **Bronchial obstruction is most often caused by a squamous cell carcinoma.** Endobronchial lesions produce varying degrees of bronchial obstruction, which can lead to pneumonitis or atelectasis.
  ♦ **Obstructive pneumonitis and atelectasis** (see Fig. 7-11).
    • It is called *pneumonitis* because the obstructed lung is consolidated but frequently not infected (although it can be).
    • **Atelectasis** secondary to an endobronchial obstructing lesion features the usual shifts of the fissures or mobile mediastinal structures **toward** the side of the atelectasis (see Chapter 7) and occasional visualization of the obstructing mass itself.

Bronchogenic Carcinoma Presenting with Direct Extension or Metastatic Lesions

■ **Rib destruction by direct extension.** *Pancoast tumor* is the eponym for a tumor arising from the superior sulcus of the lung, frequently producing destruction of one or more of the first three ribs on the affected side (Box 12-1; Fig. 12-17).
■ **Hilar adenopathy.** Usually unilateral on the same side as the tumor (Fig. 12-18).

BOX 12-1   PANCOAST TUMOR—APICAL LUNG CANCER

Manifests as a soft tissue mass in the apex of the lung

Most often squamous cell carcinoma or adenocarcinoma

Frequently produces adjacent rib destruction

May invade brachial plexus or cause Horner syndrome on affected side

On the right side, it may produce superior vena caval obstruction

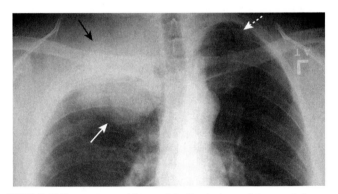

FIGURE 12-17 Pancoast tumor, right upper lobe. There is a large, soft tissue mass in the apex of the right lung *(white arrow)*. It is associated with rib destruction *(black arrow)*. On the normal left side, the ribs are intact *(dotted white arrow)*. The finding of an apical soft tissue mass with associated rib destruction is classical for a Pancoast or superior sulcus tumor.

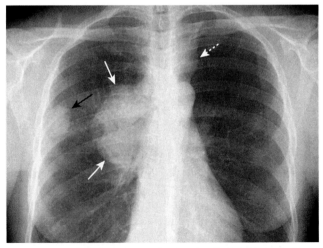

FIGURE 12-18 Bronchogenic carcinoma with hilar and mediastinal adenopathy. There is a peripheral lung mass *(solid black arrow)* with evidence of ipsilateral hilar and mediastinal adenopathy *(solid white arrows)* and contralateral mediastinal adenopathy *(dotted white arrow)*. Bronchogenic carcinoma may present with metastatic lesions that can manifest in distant organs or in the thorax itself. This was an adenocarcinoma of the lung.

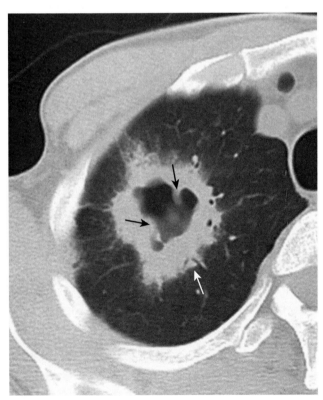

FIGURE 12-16 Cavitary bronchogenic carcinoma. There is a large cavitating neoplasm in the right upper lobe with a thick wall *(white arrow)*. The outer margin of the lesion is spiculated. The internal contour of the cavity is nodular *(black arrows)*. These features point toward a cavitating malignancy. This was a squamous cell carcinoma.

- **Mediastinal adenopathy.** May be the sole manifestation of a small cell carcinoma with the peripheral lung nodule being invisible (see Fig. 12-8).
- **Other nodules in the lung.** One of the manifestations of diffuse adenocarcinoma may be multiple nodules throughout both lungs, and as such, may mimic metastatic disease.
- **Pleural effusion.** Frequently there is associated lymphangitic spread of tumor when there is a pleural effusion.
- **Metastases to bones.** These tend to be mixed osteolytic and osteoblastic.

METASTATIC NEOPLASMS IN THE LUNG
Multiple Nodules

- Multiple nodules in the lung are most often metastatic lesions that have traveled through the bloodstream from a distant primary **(hematogenous spread).** Multiple metastatic nodules are **usually of slightly differing sizes,** indicating tumor embolization that occurred at different times.
- They are **frequently sharply marginated,** varying in size from **micronodular** to **"cannonball" masses** (see Fig. 5-8, *B*).
- For all practical purposes, it is **impossible to determine the primary site** by the **appearance of the metastatic nodules** (i.e., all metastatic nodules appear similar).

➡ Tissue sampling, whether by bronchoscopic or percutaneous biopsy, is the best means of determining the organ of origin of the metastatic nodule.

- Table 12-5 summarizes the primary malignancies most likely to metastasize to the lung hematogenously.

Lymphangitic Spread of Carcinoma

- In lymphangitic spread of carcinoma, a tumor grows in and obstructs the lymphatics in the lung, producing a pattern that is **radiologically similar to pulmonary interstitial edema** from heart failure, **including Kerley B lines, thickening of the fissures,** and **pleural effusions.**

➡ The findings **may be unilateral** or **may involve only one lobe,** a pattern that should alert you to suspect the possibility of lymphangitic spread rather than congestive heart failure, which is usually bilateral (Fig. 12-19).

- The **most common primary malignancies to produce lymphangitic spread** to the lung are those that arise around the thorax: **breast, lung,** and **pancreatic carcinoma.**

TABLE 12-5 SOME COMMON PRIMARY SITES OF METASTATIC LUNG NODULES

Males	Females
Colorectal carcinoma	Breast cancer
Renal cell carcinoma	Colorectal carcinoma
Head and neck tumors	Renal cell carcinoma
Testicular and bladder carcinoma	Cervical or endometrial carcinoma
Malignant melanoma	Malignant melanoma
Sarcomas	Sarcomas

PULMONARY THROMBOEMBOLIC DISEASE

- **Over 90% of pulmonary emboli (PE) develop from thrombi in the deep veins of the leg,** especially above the level of the popliteal veins. They are usually a **complication of surgery** or **prolonged bed rest** or **cancer.** Because of the **dual** circulation of the lungs (pulmonary and bronchial), **most pulmonary emboli do not result in infarction.**
- Although conventional chest radiographs **are** frequently abnormal in patients with PE, they can demonstrate nonspecific findings, such as **subsegmental atelectasis, small pleural effusions,** or **elevation of the hemidiaphragm.** Conventional chest radiography has a **high false-negative rate in detecting PE.**
- Chest radiographs **infrequently manifest one of the "classic" findings for pulmonary embolism, which can include:**
  - Wedge-shaped peripheral air-space disease **(Hampton hump)** (Fig. 12-20).
  - Focal oligemia **(Westermark sign).**
  - **A prominent central pulmonary artery (knuckle sign).**

➡ If the chest radiograph is normal, a nuclear medicine ventilation-perfusion scan (V/Q scan) may be diagnostic. If, however, the chest radiograph is abnormal, CT is usually performed.

- **CT pulmonary angiography (CT-PA)** is made possible by the **fast data acquisition** of **spiral CT** scanners (one breath hold) combined with **thin slices** and **rapid bolus intravenous injection of iodinated contrast** that produce maximal opacification of the pulmonary arteries with little or no motion artifact.

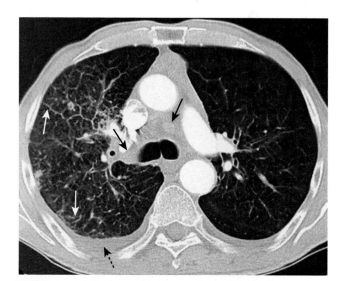

FIGURE 12-19 Bronchogenic carcinoma with lymphangitic spread of tumor. In lymphangitic spread of carcinoma, a tumor grows in and obstructs lymphatics in the lung, producing a pattern that is radiologically similar to pulmonary interstitial edema from heart failure. The findings may be unilateral, as in this case, which should alert you to the possibility of lymphangitic spread rather than congestive heart failure. There is extensive hilar and mediastinal adenopathy *(solid black arrows)* from a carcinoma of the lung. The interstitial markings are prominent in the right lung compared with the left, and there are thickened septal lines *(Kerley B lines)* present *(solid white arrows),* along with a right pleural effusion *(dotted black arrow).*

■ Another benefit of CT-PA studies is the ability to acquire images of the veins of the pelvis and legs by obtaining slightly delayed images following the pulmonary arterial phase of the study. In this way a deep venous thrombus may be detected even if the angiogram is nondiagnostic.

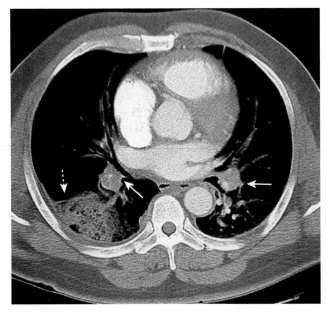

FIGURE 12-20 Hampton hump. There is a wedge-shaped, peripheral air-space density present *(dotted white arrow)* associated with filling defects in both the left and right pulmonary arteries *(solid white arrows)*. The wedge-shaped infarct is called a *Hampton hump*. Without the associated emboli being present, the pleural-based airspace disease could have a differential diagnosis that includes pneumonia, lung contusion, or aspiration.

■ CT-PA has a sensitivity **in excess of 90%** and has replaced the use of V/Q scans in patients with **chronic obstructive pulmonary disease** or a **positive chest radiograph** in whom a V/Q scan is known to be less sensitive.
■ On CT-PA, acute **pulmonary emboli** appear as **partial or complete filling defects centrally located** within the **contrast-enhanced** lumina of the **pulmonary arteries** (Fig. 12-21).
  ◆ CT-PA has the **additional benefit** of **demonstrating other diseases** that may be present, such as pneumonia, even if the study is negative for pulmonary embolism. CT-PA is also part of the diagnostic imaging examination called the *triple rule-out,* in which patients presenting with chest pain can be simultaneously evaluated for coronary artery disease, aortic dissection, and pulmonary embolism.

CHRONIC OBSTRUCTIVE PULMONARY DISEASE

■ Chronic obstructive lung disease (COPD) is defined as a disease **of airflow obstruction due to chronic bronchitis or emphysema.**
■ **Chronic bronchitis** is defined **clinically** by **productive cough,** whereas **emphysema** is defined **pathologically** by the presence of **permanent** and **abnormal enlargement** and **destruction of the air spaces distal to the terminal bronchioles.**
■ **Emphysema has three pathologic patterns:**
  ◆ **Centriacinar (centrilobular) emphysema** features focal destruction limited to the respiratory bronchioles and the central portions of the acinus. It is **associated with cigarette smoking** and is most severe in the **upper lobes** (Fig. 12-22, *A*).

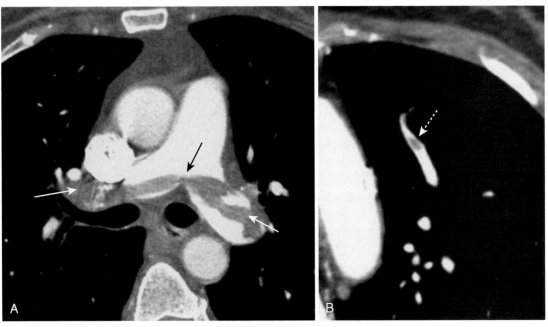

FIGURE 12-21 Saddle and peripheral pulmonary emboli. Acute pulmonary emboli appear as partial or complete filling defects centrally located within the contrast-enhanced lumina of the pulmonary arteries. **A,** A large pulmonary embolus almost completely fills both the left and right pulmonary arteries *(solid white and black arrows)*. This is a **saddle embolus. B,** A small central filling defect is seen in a more peripheral pulmonary artery *(dotted white arrow)*. This pulmonary artery seems to be floating disconnected in the lung because the plane of this particular image does not display its connection to the pulmonary artery.

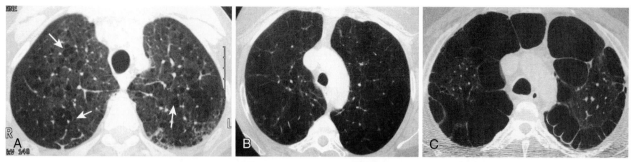

FIGURE 12-22 **Types of emphysema. A, Centriacinar (centrilobular) emphysema** features focal destruction limited to the respiratory bronchioles and the central portions of the acinus *(white arrows)*. It is associated with cigarette smoking and is most severe in the upper lobes. **B, Panacinar (panlobular) emphysema** involves the entire alveolus distal to the terminal bronchiole, is most severe in the lower lung zones, and generally develops in patients with homozygous alpha₁-antitrypsin deficiency. **C, Paraseptal emphysema** is the least common form; it involves distal airway structures, alveolar ducts, and sacs, tends to be subpleural, and may cause pneumothorax.

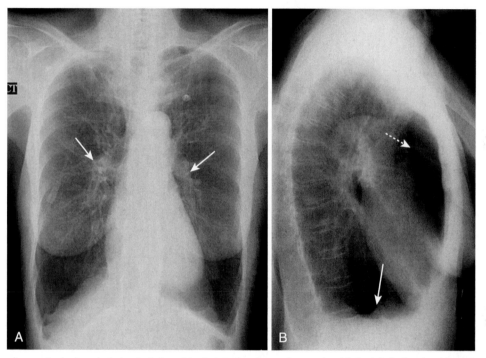

FIGURE 12-23 **Emphysema.** On conventional radiographs, the imaging findings of chronic obstructive pulmonary disease are hyperinflation, including flattening of the diaphragm, especially on the lateral exposure *(solid white arrow in B)*, increase in the retrosternal clear space *(dotted white arrow)*, hyperlucency of the lungs with fewer than normal vascular markings, and prominence of the pulmonary arteries secondary to pulmonary arterial hypertension *(solid white arrows in A)*.

♦ **Panacinar emphysema** involves the entire alveolus distal to the terminal bronchiole. It is most severe in the **lower lung zones** and generally develops in patients with homozygous **alpha₁-antitrypsin deficiency** (see Fig. 12-22, *B*).

♦ **Paraseptal emphysema** is the **least common** form. It involves distal airway structures, alveolar ducts, and sacs. Localized to fibrous septa or to the pleura, it can lead to formation of bullae, which **may cause pneumothorax.** It is not associated with airflow obstruction (see Fig. 12-22, *C*).

■ On conventional radiographs, **the most reliable finding of COPD is hyperinflation,** including **flattening of the diaphragm,** especially on the **lateral** exposure (Fig. 12-23). Other findings may include an **increase** in the **retrosternal clear space, hyperlucency** of the lungs with fewer than normal vascular markings visible, and **prominence of the pulmonary arteries** from pulmonary arterial hypertension.

■ **With CT, findings of COPD** may include **focal areas of low density** in which the cystic areas lack visible walls, except where bounded by interlobular septa. CT is helpful in evaluating the extent of emphysematous disease and in planning for surgical procedures designed to remove bullae to reduce lung volume.

BLEBS AND BULLAE, CYSTS AND CAVITIES

■ Blebs, bullae (singular: bulla), cysts, and cavities are all **air-containing lesions in the lung** of differing **size, location,** and **wall composition.**

■ Almost any of these lesions can contain **fluid** instead of, or in addition to, air.

■ Fluid usually develops as a result of **infection, hemorrhage,** or **liquefaction necrosis.**

■ When these lesions are **completely filled with fluid,** they **will appear solid** on conventional radiographs and CT

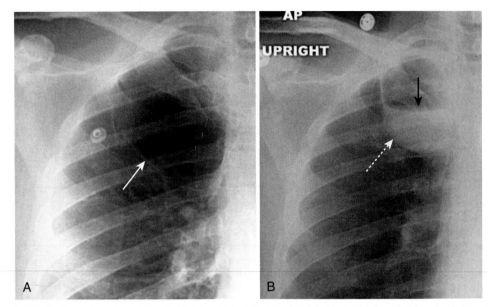

FIGURE 12-24 Infected bulla. A, In this close-up, there are several thin-walled, but air-containing, bullae in the right upper lobe *(solid white arrow)*. **B,** Several weeks later, one of the bullae *(dotted white arrow)* contains both fluid and air *(solid black arrow)*. Bullae normally contain air but can become partially or completely fluid-filled from infection or hemorrhage. Patients with infected bullae tend to be less sick than those with a lung abscess, which this lesion can mimic.

scans, but they will typically demonstrate a **low CT number,** which will differentiate them from a solid tumor. When they contain some fluid and some air, they will manifest an **air–fluid level** on a conventional radiograph (exposed with a horizontal x-ray beam) or on CT scans (Fig. 12-24).

Blebs

- Blebs are **very small, blister-like lesions that form in the visceral pleura,** usually at the **apex** of the lung. They are **very thin-walled,** and although they may be seen by CT, they are too small to be visible on chest radiographs. They are thought to be associated with spontaneous pneumothoraces.

Bullae

- Bullae measure **more than 1 cm** in size. They are usually **associated with emphysema.** They occur in the **lung parenchyma** and have a **very thin wall** (<1 mm) that is frequently **only partially visible** on conventional radiography but well seen on CT. On **conventional radiographs** their **presence is often inferred** by a localized paucity of lung markings (Fig. 12-25).
- They can **grow to fill the entire hemithorax** and compress the lung on the affected side to such an extent that the lung seems to disappear *(vanishing lung syndrome).*

Cysts

- Cysts are either congenital or acquired. They can occur in either the **lung parenchyma** or the **mediastinum.** They have a **thin wall** that is usually thicker than that of a bulla **(<3 mm).**
  - ♦ **Pneumatocoeles** represent **thin-walled cysts** that usually develop after a lung infection caused by such organisms as *Staphylococcus* or *Pneumocystis* (Fig. 12-26).

Cavities

- Cavities can **vary in size** from a few millimeters to many centimeters. They occur in the **lung parenchyma** and usually **result from a process** that produces necrosis of the central portion of the lesion.

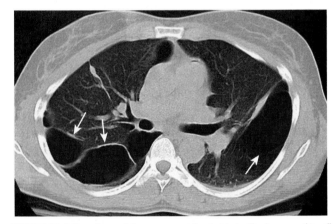

FIGURE 12-25 Bullous disease. Bullae measure more than 1 cm in size. They have a very thin wall that is often only partially visible on conventional radiography. CT demonstrates them more easily *(white arrows)*. Characteristically, they contain no blood vessels, but there may be septae that appear to traverse the bulla. On conventional radiographs their presence is often inferred by a localized paucity of lung markings (see Fig. 12-24, *A*).

- Cavities usually have **the thickest wall** of any of these four air-containing lesions **with a wall thickness from 3 mm to several cm** (Fig. 12-27).
- Differentiation between three of the most frequent causes of cavities (carcinoma, pyogenic abscess, and tuberculosis) is summarized in Table 12-6.

BRONCHIECTASIS

- Bronchiectasis is defined as **localized irreversible dilatation of part of the bronchial tree.** Although it can be associated with many other diseases, it is **usually caused by necrotizing bacterial infections** like those of *Staphylococcus* and *Klebsiella* and it usually affects the lower lobes.
- Bronchiectasis may also occur with cystic fibrosis, primary ciliary dyskinesia (**Kartagener syndrome**), allergic bronchopulmonary aspergillosis, and **Swyer-James syndrome** (unilateral hyperlucent lung).

- Clinically, **chronic productive cough,** frequently associated with **hemoptysis,** is the major symptom.
- **Conventional radiographs** may demonstrate findings suggestive of the disease but are **usually not specific.**

TABLE 12-6 DIFFERENTIATING THREE CAVITATING LUNG LESIONS

Lesion	Thickness* of the Cavity Wall	Inner Margin of Cavity
Bronchogenic carcinoma (Fig. 12-27, A)	Thick	Nodular
Tuberculosis (Fig. 12-27, B)	Thin	Smooth
Lung abscess (Fig. 12-27, C)	Thick	Smooth

*Thick means >5 mm; thin means <5 mm.

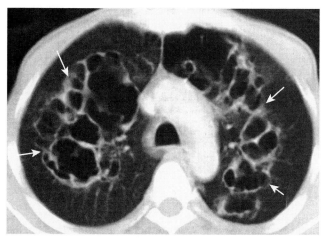

FIGURE 12-26 Cysts (pneumatocoeles) in *Pneumocystis* pneumonia (PCP). Cysts are visible on chest radiographs in 10% of patients with PCP and far more frequently with computed tomography scans (up to one in three patients). They may occur in the acute or in the postinfective phase of the disease. They have a predilection for the upper lobes and are commonly multiple *(white arrows)*. Their cause is unclear.

- Findings include parallel-line opacities (**"tram-tracks"**) resulting from thickened, dilated bronchial walls, **cystic lesions** as large as 2 cm in diameter caused by cystic bronchiectasis, and **tubular densities** from fluid-filled bronchi (Fig. 12-28). **Today CT is the study of choice in diagnosing bronchiectasis.**
- The hallmark lesion on CT is the **signet ring sign,** in which the **bronchus,** frequently with a thickened wall, **becomes larger than its associated pulmonary artery,** which is the opposite of the normal relationship between the two. The bronchus may also show a failure to taper normally (Fig. 12-29).

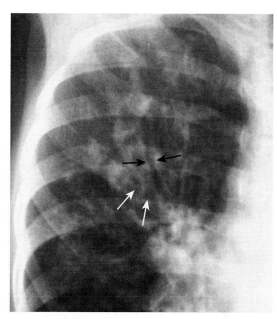

FIGURE 12-28 Bronchiectasis in cystic fibrosis. Conventional radiographs may demonstrate parallel line opacities called ***tram-tracks,*** thickened walls of dilated bronchi *(black arrows)*, and cystic lesions as large as 2 cm in diameter due to cystic bronchiectasis *(white arrows)*. Bilateral upper lobe bronchiectasis in children is highly suggestive of cystic fibrosis.

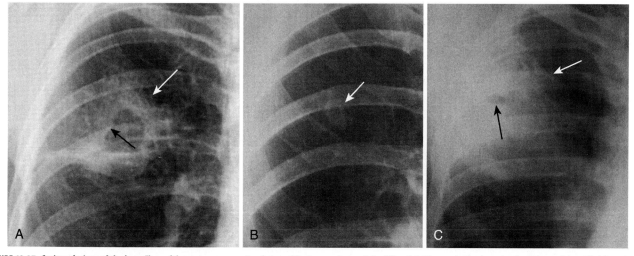

FIGURE 12-27 Cavitary lesions of the lung. Three of the most common cavitary lesions of the lung can frequently be differentiated from each other by noting the thickness of the wall of the cavity and the smoothness or nodularity of its inner margin. **A,** Squamous cell bronchogenic carcinoma produces a cavity with a thick wall *(white arrow)* and a nodular interior margin *(black arrow)*. **B,** Tuberculosis usually produces a relatively thin-walled, upper lobe cavity with a smooth inner margin *(white arrow)*. **C,** A staphylococcal lung abscess demonstrates a characteristically thickened wall *(white arrow)*, which in this case has a very small cavity with a smooth inner margin *(black arrow)*.

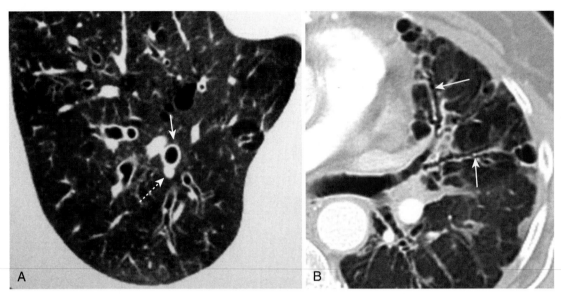

FIGURE 12-29 Bronchiectasis. CT is the study of choice in diagnosing bronchiectasis. **A,** The hallmark lesion is called the ***signet ring sign,*** in which the bronchus with a thickened wall *(solid white arrow)* becomes larger than its associated pulmonary artery *(dotted white arrow)*, which is the opposite of the normal relationship between the two. **B,** The bronchus may also show **tram-tracking,** thickened walls, and a failure to taper normally *(solid white arrows)*.

TAKE-HOME POINTS

RECOGNIZING DISEASES OF THE CHEST

The mediastinum lies in the central portion of the thorax between the two lungs and is arbitrarily divided into anterior, middle, and posterior compartments.

Masses in the anterior mediastinum include substernal thyroid goiters, lymphoma, thymoma, and teratoma.

The middle mediastinum is home primarily to lymphadenopathy from lymphoma and metastatic disease, such as from small cell carcinoma of the lung.

The posterior mediastinum is the location of neurogenic tumors, which originate either from the nerve sheath (mostly benign) or from tissues other than the sheath (mostly malignant).

Incidental solitary pulmonary nodules (SPNs) < 4 mm in size are rarely malignant; in those in whom clinical or imaging findings suggest malignancy, 50% of SPNs in those over the age of 50 are malignant. The key question is to determine whether a nodule is most likely benign or most likely malignant in any given individual.

Criteria on which an evaluation of benignity can be made include absolute size of the nodule upon discovery, presence of calcification within it, the margin of the nodule, and change in the size of the nodule over time.

Bronchogenic carcinomas present in one of three ways: visualization of the tumor itself; recognition of the effects of bronchial obstruction such as pneumonitis and/or atelectasis; or by identification of either their direct extension or metastatic spread to the chest or to distant organs.

Bronchogenic carcinomas presenting as a solitary nodule/mass in the lung are most often adenocarcinomas; adenocarcinomas may present with multiple nodules, mimicking metastatic disease.

Bronchogenic carcinoma presenting with bronchial obstruction is most often caused by squamous cell carcinoma; squamous cell carcinomas are most likely to cavitate.

Small cell carcinomas are highly aggressive, centrally located, peribronchial tumors, the majority of which have already metastasized at the time of initial presentation of symptoms; they can be associated with paraneoplastic syndromes such as inappropriate secretion of antidiuretic hormone and Cushing syndrome.

Multiple nodules in the lung are most often metastatic lesions that have traveled through the bloodstream from a distant primary site **(hematogenous spread);** common primary sites for such metastases include colorectal, breast, renal cell, head and neck, bladder, uterine and cervical carcinomas, soft tissue sarcomas, and melanoma.

In lymphangitic spread of carcinoma, a tumor grows in and obstructs lymphatics in the lung, producing a pattern that is radiologically similar to pulmonary interstitial edema; primaries that metastasize to the lung in this fashion include breast, lung, and pancreatic cancer.

Conventional radiography has a high false-negative rate in pulmonary thromboembolic disease because demonstration of classic findings like Hampton hump, Westermark sign, and the knuckle sign is infrequent.

CT-pulmonary angiography is now widely used for the diagnosis of pulmonary embolism, producing images of the pulmonary arteries with little or no motion artifact.

Chronic obstructive pulmonary disease consists of emphysema and chronic bronchitis; of the two, chronic bronchitis is a clinical diagnosis, whereas emphysema is defined pathologically and has findings that can be seen on both conventional radiographs and CT scans.

Blebs, bullae, cysts, and cavities are all air-containing lesions in the lung that differ in size, location, and wall composition; bullae, cysts, and cavities are seen on CT and may also be seen on conventional radiographs.

Although bronchiectasis may be seen on conventional radiographs, CT is the study of choice. CT demonstrates the signet ring sign, tram-tracks, cystic lesions, or tubular densities.

 **WEBLINK**

Visit StudentConsult.Inkling.com for more information and quizzes on Chest Abnormalities and Pulmonary Abnormalities.

For your convenience, the following QR Code may be used to link to **StudentConsult.com**. You must register this title using the PIN code on the inside front cover of the text to access online content.

CHAPTER 13
Recognizing Adult Heart Disease

This chapter will begin with how to assess heart size, then describe the normal and abnormal contours of the heart on the frontal radiograph, and finally, illustrate some imaging findings in common cardiac diseases.

RECOGNIZING AN ENLARGED CARDIAC SILHOUETTE

- The cardiac silhouette can appear enlarged for three main reasons (arranged in the order in which they will be discussed, not by frequency of occurrence):
  - **Pericardial effusion,** which mimics the appearance of cardiomegaly on conventional radiographs.
  - **Extracardiac** factors that produce apparent cardiac enlargement.
  - And most important, **cardiomegaly,** true cardiac enlargement.

PERICARDIAL EFFUSION

- Normally, there are 15 to 50 mL of fluid in the pericardial space between the parietal and visceral pericardial layers. Abnormal accumulations of fluid begin in the **dependent portions of the pericardial space,** which in the supine position is **posterior to the left ventricle** (Fig. 13-1, *A*). As the pericardial effusion increases in size, it tends to accumulate more **along the right heart border** until it fills the pericardial space and **encircles the heart** (see Fig. 13-1, *B*).

- Computed tomography (CT) scans can demonstrate small pericardial effusions, although pericardial ultrasonography is usually the imaging study of first choice. Conventional radiographs are poor at defining a pericardial effusion.
- Some of the causes of pericardial effusions are outlined in Box 13-1.

EXTRACARDIAC CAUSES OF APPARENT CARDIAC ENLARGEMENT

 Sometimes, there is an **extracardiac** cause of apparent cardiac enlargement on conventional radiography that may cause the cardiothoracic ratio to appear >50%, **while the heart itself may actually be normal in size.**

- The extracardiac causes of apparent cardiomegaly are outlined in Table 13-1. **Magnification** of the heart produced

---

BOX 13-1 CAUSES OF PERICARDIAL EFFUSION

Congestive heart failure
Infection (TB, viral)
Metastatic malignancy (lung and breast, especially)
Uremic pericarditis
Collagen-vascular disease (lupus)
Trauma
Postpericardiotomy syndrome

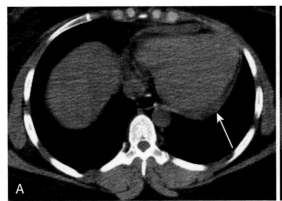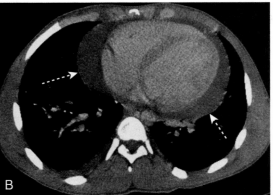

FIGURE 13-1 Pericardial effusions, small and large. A, Fluid first begins to accumulate in the dependent portions of the pericardial space, which is posterior to the left ventricle in the supine position *(white arrow).* **B,** As the effusion increases in size, it fills the pericardial space and encircles the heart *(dotted white arrows).* Conventional chest radiographs may show an enlarged cardiac silhouette but cannot differentiate the density of the heart from the effusion.

by projection, usually on an anteroposterior (AP), supine, portable chest examination, is the most common cause of apparent cardiomegaly (see Fig. 2-8).

IDENTIFYING CARDIAC ENLARGEMENT ON AN AP CHEST RADIOGRAPH

■ Is it possible to estimate the size of the heart on a portable chest radiograph? The answer is yes. A good rule of thumb:

TABLE 13-1 EXTRACARDIAC CAUSES OF APPARENT CARDIOMEGALY

Cause	Reason for Enlarged Appearance
AP portable supine chest—most common cause	Magnification due to AP projection.
Suboptimal inspiration	In expiration, the diaphragm moves upward and compresses the heart, making the heart appear larger than it would in full inspiration. If there are eight to nine posterior ribs visible on the frontal chest radiograph, the inspiration is adequate (see Fig. 2-4).
Obesity, pregnancy, ascites	These conditions prevent an adequate inspiration.
Pectus excavatum deformity, a congenital deformity of the lowermost section of the sternum, causes it to bow inward and compress the heart	The heart is compressed between the sternum and the spine.
Rotation	Especially when it occurs to the patient's left, rotation may make the heart appear larger.
Pericardial effusion	Other imaging modalities (most commonly ultrasound) or electrocardiographic findings will help to identify pericardial fluid.

if the heart appears enlarged on a well-inspired, portable chest radiograph, it probably is enlarged (Table 13-2 and Fig. 13-2).

RECOGNIZING CARDIOMEGALY ON THE LATERAL CHEST RADIOGRAPH

■ Generally speaking, evaluation of cardiac size is best made on the frontal chest radiograph. To evaluate for the presence of enlargement of the cardiac silhouette in the lateral projection, **look at the space posterior to the heart** and **anterior to the spine at the level of the diaphragm** (Fig. 13-3). In a normal person the **cardiac silhouette will usually not extend posteriorly and project over the spine** (see Fig. 13-3, *A*).
■ As the **heart enlarges,** whether that enlargement is the result of cardiomegaly or pericardial effusion, the **posterior border of the heart may extend to, or overlap, the anterior border of the thoracic spine.** This can be useful as a confirmatory sign of an enlarged cardiac silhouette (see Fig. 13-3, *B*).

RECOGNIZING COMMON CARDIAC DISEASES

■ In this section, several diseases will be discussed in more detail. Most of them produce abnormalities in size and contour of the heart or great vessels. For a systematic approach to recognizing the cardiac contour abnormalities

TABLE 13-2 RECOGNIZING CARDIOMEGALY ON AN AP CHEST RADIOGRAPH

Appearance of Heart on AP Study	Probable Heart Size
Borderline enlarged	Normal size
Significantly enlarged	Enlarged
Touching, or almost touching, the left lateral chest wall	Definitely enlarged

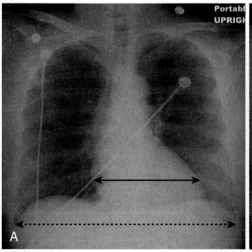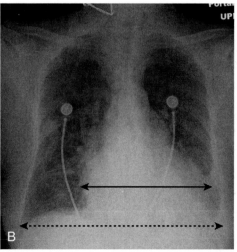

FIGURE 13-2 Cardiomegaly on a portable radiograph. It is possible to determine heart size on a portable supine chest radiograph. **A,** If the heart is about 50% of the cardiothoracic ratio *(ratio of length of the solid black arrows to dotted black arrows)*, the heart is not enlarged. **B,** If the heart is near or touching the lateral chest wall *(ratio of length of the solid black arrows to dotted black arrows)*, the heart is enlarged.

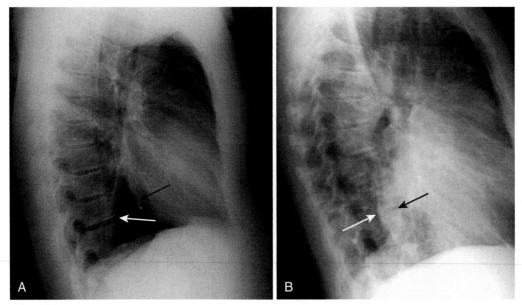

FIGURE 13-3 Normal and enlargement of the cardiac silhouette in the lateral projection. A, In most normal patients, the posterior border of the heart *(black arrow)* does not overlap the thoracic spine *(white arrow).* **B,** In this patient with cardiomegaly, the posterior border of the heart *(white arrow)* overlaps the anterior border of the thoracic spine *(black arrow).* Estimation of cardiac size is best made on the frontal projection, but the lateral projection can be used for a confirmatory sign of enlargement of the cardiac silhouette.

in adult cardiac disease, see The ABCs of Heart Disease available online to registered users at StudentConsult.com. The diseases to be discussed are:

- ◆ Congestive heart failure and pulmonary edema
  - • Cardiogenic versus noncardiogenic pulmonary edema
- ◆ Hypertensive cardiovascular disease
- ◆ Mitral stenosis
- ◆ Pulmonary arterial hypertension
- ◆ Aortic stenosis
- ◆ Cardiomyopathy
- ◆ Thoracic aortic aneurysm and aortic dissection
- ◆ Coronary artery disease

Congestive Heart Failure

- ▪ The incidence of congestive heart failure (CHF) has grown rapidly over the last two decades so that CHF is currently **the most common diagnosis in hospitalized patients over the age of 65.**
- ▪ **Causes of congestive heart failure**
  - ◆ In the United States, **the two most common causes of CHF** are **coronary artery disease** and **hypertension.**
  - ◆ Other causes of CHF:
    - • **Cardiomyopathy,** such as from long-standing alcohol abuse
    - • **Cardiac valvular lesions,** such as aortic stenosis and mitral stenosis
    - • **Arrhythmias**
    - • **Hyperthyroidism**
    - • **Severe anemia**
    - • **Left-to-right shunts**
- ▪ Typically, **congestive heart failure** presents with **one of two radiographic patterns: pulmonary interstitial edema** or **pulmonary alveolar edema.** Not every feature of each pattern is always present, and there is often overlap between the two patterns.

Pulmonary interstitial edema

 There are **four key radiographic signs of pulmonary interstitial edema.**

- ▪ **Thickening of the interlobular septa**
- ▪ **Peribronchial cuffing**
- ▪ **Fluid in the fissures**
- ▪ **Pleural effusions**

Thickening of the Interlobular Septa—The Kerley B Line

- ▪ The interlobular septae are not detectable on a normal chest radiograph but can become visible if they accumulate excessive fluid, usually at a pulmonary (venous) capillary wedge pressure of about **15 mm Hg.** The thickened septae are called **septal lines** or **Kerley B lines** (named after Peter James Kerley, an Irish neurologist and radiologist).
- ▪ **Recognizing Kerley B lines**
  - ◆ Kerley B lines actually do exist. They are visible on a frontal radiograph usually **at the lung bases, at** or **near the costophrenic angles.**
  - ◆ They are **very short** (1 to 2 cm long), **very thin** (about 1 mm), and **horizontal in orientation,** which means they are **perpendicular to the pleural surface.**
  - ◆ They **usually extend to** and **abut the pleural surface** (Fig. 13-4).
  - ◆ After repeated episodes of pulmonary interstitial edema, the **septal lines may fibrose** and therefore **remain, even after all other signs of pulmonary interstitial edema clear.** These are called *chronic Kerley B lines,* and they may be present even if the patient is not clinically in congestive failure.
- ▪ **Kerley A Lines**
  - ◆ Kerley named other lines seen in congestive heart failure besides the "B" line.
  - ◆ **Kerley A** lines appear when connective tissue around the bronchovascular sheaths in the lung distends with

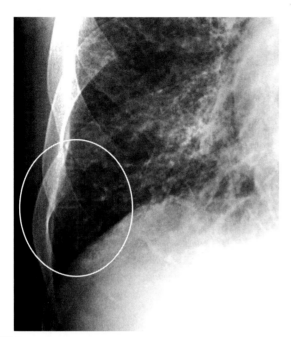

FIGURE 13-4 Kerley B lines. Interlobular septae are not visible on a normal chest radiograph but can become visible if they accumulate excessive fluid. First described by neurologist/radiologist Peter James Kerley, they are very short (1 to 2 cm long), very thin (about 1 mm) horizontal lines perpendicular to and abutting the pleural surface *(oval)*.

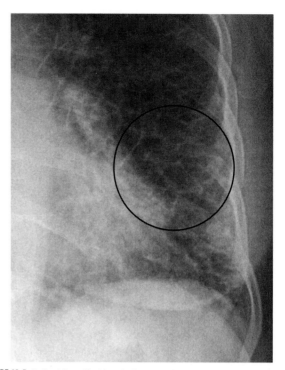

FIGURE 13-5 Kerley A lines. The A lines *(circle)* appear when connective tissue near the bronchovascular bundle distends with fluid. They extend from the hila for several centimeters in the midlung and do not reach the periphery of the lung like Kerley B lines do. A network of Kerley lines is produced in the lungs in patients with congestive heart failure producing the "prominence of the pulmonary interstitial markings" seen in that disease.

fluid. **Kerley A lines extend from the hila for several centimeters** but **do not reach the periphery of the lung** as Kerley B lines do (Fig. 13-5).

♦ Kerley also described "C" lines, but there is doubt that they exist as separate entities.

Peribronchial Cuffing

■ In adults, bronchi may **normally** be visible *en face* in the region of the hila, but their walls are usually too thin and **not visible** more peripherally in the lung.

■ When fluid accumulates in the interstitial tissue around and in the wall of a bronchus, such as in congestive heart failure, the **bronchial wall becomes thicker** and **may appear as a ringlike density that can be seen on-end** in radiographs.

■ When seen on-end, **peribronchial cuffing** appears as numerous, small, ringlike shadows that look like little **doughnuts** (Fig. 13-6).

Fluid in the Fissures

■ The major (or oblique) and minor (or horizontal) **fissures may be visible normally** but are almost never thicker than a line you could draw with the **point of a sharpened pencil** (Fig. 13-7, *A*).

■ **Fluid can collect between the two layers of visceral pleura that form the fissures or in the subpleural space,** between the visceral pleura and the lung parenchyma. This **fluid** distends the fissure and makes it **thicker,** more **irregular in contour,** and therefore **more visible** than normal (see Fig. 13-7, *B*).

Pleural Effusion

■ As a result of either increased production or decreased absorption of pleural fluid, fluid in excess of the normal 2

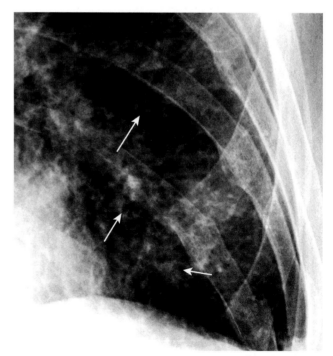

FIGURE 13-6 Peribronchial cuffing. Normally the bronchus is invisible when seen on-end in the periphery of the lung. When fluid accumulates in the interstitial tissue around and in the wall of a bronchus as it does in CHF, the bronchial wall becomes thicker and can appear as ringlike densities when seen on-end *(white arrows)*. Peribronchial cuffing may not always produce perfectly round circles.

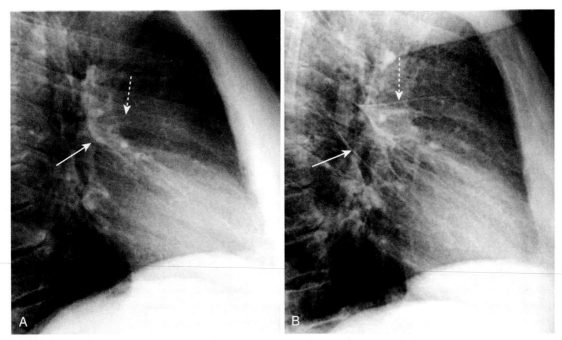

FIGURE 13-7 Normal fissures and fluid in the fissures. A, The major *(solid white arrow)* and minor fissures *(dotted white arrow)* may be barely visible normally but are almost never thicker than a line you could draw with the point of a sharpened pencil. Fluid can collect in the fissures in congestive heart failure and distend them, making them appear thicker and more irregular in contour and more visible than normal. **B,** The major fissure is enlarged *(solid white arrow)*, as is the minor fissure *(dotted white arrow)*. When the patient's heart failure clears, the fissures will return to normal appearance, but after repeated and prolonged bouts of failure, fibrosis may result in permanent thickening of the fissures.

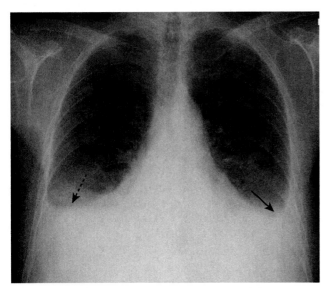

FIGURE 13-8 Pleural effusions in congestive heart failure. There are bilateral pleural effusions *(dotted and solid black arrows)*. Effusions in congestive heart failure are most often bilateral but may be asymmetric, the right side invariably being slightly larger. Although a unilateral, left pleural effusion may occur with congestive heart failure, a large, unilateral left effusion should draw suspicion to another possible cause, such as metastatic disease. This patient had congestive heart failure.

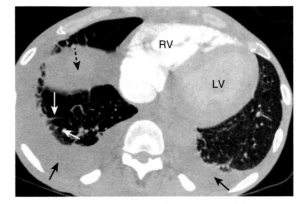

FIGURE 13-9 Congestive heart failure, computed tomography (CT). Axial CT scan of the chest following intravenous injection of contrast. Note that the right heart is opacified (RV), but the contrast has not yet passed through the lungs a sufficient number of times to fully opacify the left heart (LV). There are bilateral pleural effusions present, and they layer posteriorly because the patient is being scanned supine *(solid black arrows)*. Fluid has insinuated itself into the fissure *(dotted black arrow)*. The thickened interlobular septae *(solid white arrows)* are Kerley B lines. *LV,* Left ventricle; *RV,* right ventricle.

to 5 mL can collect in the pleural space, typically at a pulmonary capillary wedge pressure of about **20 mm Hg.**

■ **Pleural effusions accompanying congestive heart failure are usually bilateral** but can be asymmetric (Fig. 13-8).

 ◆ When **unilateral,** they are **almost always right-sided.**

 • About 15% of the time they can be unilateral and on the left, but if you see a unilateral left pleural effusion, you should think of causes other than CHF, such as metastases, tuberculosis, or pulmonary thromboembolic disease.

■ At times, pleural fluid accumulates in the form of a **laminar effusion,** in which the fluid assumes a **thin, bandlike density along the lateral chest wall, beginning near the costophrenic sulcus** but often preserving the sulcus itself (see Fig. 8-15).

■ For more information about pleural effusions, see Chapter 8.

■ All of these findings can be seen on both CT (Fig. 13-9) and conventional radiographs.

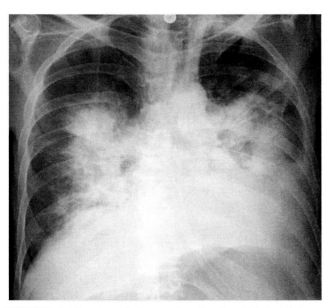

FIGURE 13-10 Batwing pattern of pulmonary edema. The radiographic findings of pulmonary alveolar edema include fluffy, indistinct, and patchy airspace densities frequently centrally located and sparing the outer third of the lung. This is called the *batwing (angel-wing)* or *butterfly* pattern, and it is suggestive of pulmonary edema versus other airspace diseases such as pneumonia. There is considerable overlap in the patterns of cardiogenic and noncardiogenic pulmonary edema, but the absence of pleural effusions, absence of fluid in the fissures, and the normal-sized heart favor a noncardiogenic cause in this case. The patient was in septic shock from an overwhelming urinary tract infection.

■ The key findings of **pulmonary interstitial edema** are summarized in Box 13-2.

Pulmonary alveolar edema

■ When the **pulmonary venous pressure is sufficiently elevated** (about 25 mm Hg), **fluid spills out of the interstitial tissues** of the lung **into the airspaces.** This results in **pulmonary alveolar edema** (most often shortened to *pulmonary edema* without including "alveolar").

 The **radiographic findings** of **pulmonary alveolar edema are:**

♦ **Fluffy, indistinct, patchy airspace densities that are usually centrally located.** The outer third of the lung is frequently spared, and the lower lung zones are more affected than the upper. This is called the *batwing, angel-wing,* or *butterfly* configuration of pulmonary edema (Fig. 13-10).

♦ **Pleural effusions** and **fluid in the fissures** are commonly found in pulmonary alveolar edema on a cardiogenic basis.

■ The key findings in pulmonary alveolar edema are summarized in Box 13-3.

■ **What happened to cardiomegaly** and **cephalization?**

♦ Although most patients with congestive heart failure have an enlarged heart, most patients with an enlarged heart are not in congestive heart failure. In any individual, **cardiomegaly is not a particularly sensitive indicator for the presence or absence of congestive heart failure.**

♦ **Cephalization,** defined as redistribution of flow in the lungs such that the upper lobe pulmonary vasculature becomes larger than the lower lobe vessels, **is difficult to identify for most beginners** and is meaningful only if you are certain the patient was upright at the time of the chest exposure.

■ **How pulmonary edema resolves**

♦ Pulmonary edema generally is both **abrupt in onset** and **quick to clear**—typically in a matter of a few hours to a few days (Fig. 13-11).

♦ Resolution frequently begins peripherally and moves centrally. Radiologic resolution may lag behind clinical improvement, especially if the patient has large pleural effusions.

NONCARDIOGENIC PULMONARY EDEMA—GENERAL CONSIDERATIONS

■ Although congestive heart failure accounts for the majority of the cases of pulmonary edema (i.e., **cardiogenic pulmonary edema**), there are other, **noncardiogenic causes of pulmonary edema.**

■ The causes of noncardiogenic pulmonary edema are a diverse group of diseases:

♦ **Increased capillary permeability**—includes all of the various causes of **acute (adult) respiratory distress syndrome** or **ARDS** such as:
  • **Sepsis**
  • **Uremia**
  • **Disseminated intravascular coagulopathy**
  • **Smoke inhalation**
  • **Near-drowning**

♦ **Volume overload**

♦ **Lymphangitic spread of malignancy**

♦ Other causes of noncardiogenic pulmonary edema may include:
  • **High-altitude pulmonary edema**
  • **Neurogenic pulmonary edema**
  • **Reexpansion pulmonary edema** (Fig. 13-12)
  • **Heroin or other overdoses**

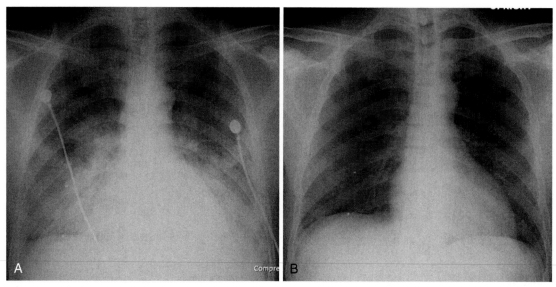

FIGURE 13-11 Rapidly clearing pulmonary edema. Pulmonary edema generally is both abrupt in its onset and quick to clear. **A,** This patient demonstrates bilateral, perihilar airspace disease with diffuse prominence of the interstitial markings characteristic of pulmonary edema. **B,** Four days later, the lungs are clear. Patients with acute respiratory distress syndrome are not likely to clear this quickly, nor are patients who have coexisting diseases such as renal or hepatic failure, or superimposed pneumonia.

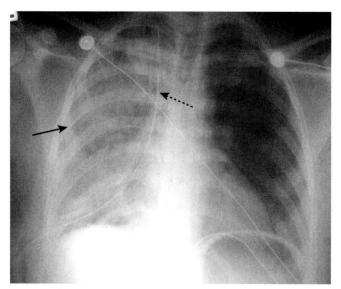

FIGURE 13-12 Reexpansion pulmonary edema. Unilateral airspace disease affects the entire right lung *(solid black arrow).* In addition, a chest tube *(dotted black arrow)* is seen on the same side. The chest tube was inserted for a large, right-sided, tension pneumothorax that was rapidly reexpanded. Reexpansion pulmonary edema results from the overly rapid expansion of a lung that has typically been chronically collapsed by pneumothorax or a large pleural effusion. Its exact cause is not known. Other causes of unilateral pulmonary edema can include either an abnormality on the same side as the pulmonary edema (e.g., prolonged positioning with the affected side dependent) or an abnormality on the opposite side (e.g., large pulmonary embolus occluding flow to the opposite lung).

NONCARDIOGENIC PULMONARY EDEMA—IMAGING FINDINGS

■ **Acute respiratory distress syndrome (ARDS)** represents one form of noncardiogenic pulmonary edema.
   ◆ Characteristically, **patients with ARDS are radiographically normal for 24 to 36 hours** after the initial insult. Then pulmonary abnormalities become evident in the form of **pulmonary interstitial edema, patchy airspace disease,** or typical bilateral **pulmonary alveolar edema.**
   ◆ Clinically, the **patient demonstrates severe hypoxia, cyanosis, tachypnea, and dyspnea.**
   ◆ Typically, the findings of ARDS stabilize after 5 to 7 days and begin improving in about 2 weeks. Complete clearing, when it occurs, may take months.

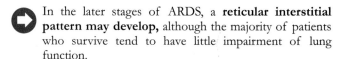
In the later stages of ARDS, a **reticular interstitial pattern may develop,** although the majority of patients who survive tend to have little impairment of lung function.

DIFFERENTIATING CARDIAC FROM NONCARDIAC PULMONARY EDEMA

■ The **patterns of cardiac (cardiogenic)** and **noncardiac (noncardiogenic) pulmonary edema overlap considerably,** and the patient's **history** and **clinical picture are keys** to establishing the most likely cause of pulmonary edema.
■ In general, **noncardiogenic pulmonary edema is:**
   ◆ **Less likely** to demonstrate **pleural effusions** and **Kerley B lines** than cardiogenic pulmonary edema.
   ◆ **More likely** to demonstrate a **normal pulmonary capillary wedge pressure (PCWP)** of <12 mm Hg than cardiogenic pulmonary edema.
   ◆ **More likely** to be associated with a **normal-sized heart** (Fig. 13-13).
   ◆ **More likely** to demonstrate airspace disease that is more patchy and peripheral than that in cardiogenic pulmonary edema, but this is highly variable.
   ◆ The key differences between cardiogenic and noncardiogenic pulmonary edema are summarized in Table 13-3.

HYPERTENSIVE CARDIOVASCULAR DISEASE

■ Chronic elevation of systemic blood pressure leads to left ventricular hypertrophy in about 20% of patients; double that incidence if the patient is obese. Most of the time (90%), the hypertension is **essential hypertension** with no identifiable cause. Heart failure, coronary artery disease, and cardiac arrhythmias are common complications of hypertension.

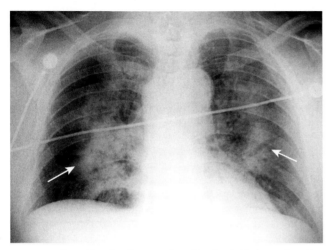

FIGURE 13-13 Noncardiogenic pulmonary edema. Even though this airspace disease has a perihilar distribution similar to cardiogenic pulmonary edema *(white arrows)*, there is no pleural fluid, fluid in the fissures, or cardiomegaly. In general, noncardiogenic pulmonary edema is less likely to demonstrate pleural effusions and Kerley B lines, more likely to demonstrate a normal pulmonary capillary wedge pressure (PCWP) of less than 12 mm Hg, and more likely to be associated with a normal-sized heart than cardiogenic pulmonary edema. The patient had just used crack cocaine.

■ Systemic hypertension can lead to left ventricular hypertrophy. Left ventricular hypertrophy occurs at the expense of the lumen, the wall becoming thicker while the lumen becomes smaller. Therefore, **the heart is usually normal** or **slightly increased in size** early in the disease. It is not until the muscle begins to decompensate that the heart increases dramatically in size (Fig. 13-14).

■ **The aorta,** under increased systemic pressure, pivots outward around the aortic valve and the aortic hiatus in the diaphragm and **gradually uncoils,** becoming more prominent in **both** its **ascending** and **descending** portions (see Fig. 4-3).

■ Prolonged systemic hypertension may eventually lead to congestive heart failure.

MITRAL STENOSIS

■ In developed nations, the incidence of mitral stenosis from rheumatic heart disease has declined markedly, but it is still

TABLE 13-3   CARDIOGENIC VERSUS NONCARDIOGENIC PULMONARY EDEMA

Imaging Finding	Cardiogenic	Noncardiogenic
Pleural effusions	Common	Infrequent
Kerley B lines	Common	Infrequent
Heart size	Frequently enlarged	May be normal
Pulmonary capillary wedge pressure	Elevated	Normal

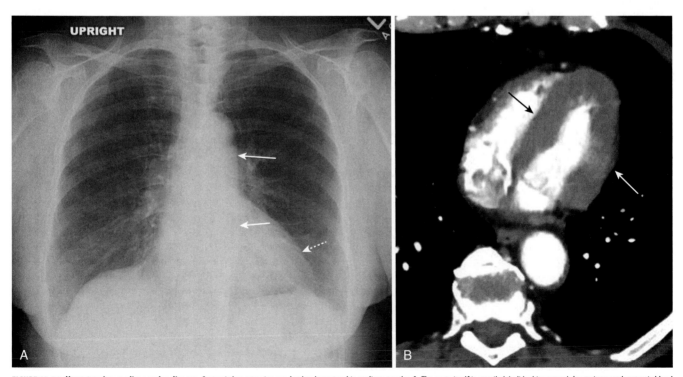

FIGURE 13-14 Hypertensive cardiovascular disease. Systemic hypertension can lead to hypertrophic cardiomyopathy. **A,** The aorta itself is uncoiled *(solid white arrows)* due to increased systemic blood pressure (see also Fig. 4-3). The left ventricle *(dotted white arrow)* is only slightly enlarged, but a computed tomography scan through the heart **(B)** demonstrates marked concentric hypertrophy of the left ventricular wall *(white and black arrows),* which has occurred at the expense of the lumen (which contains contrast).

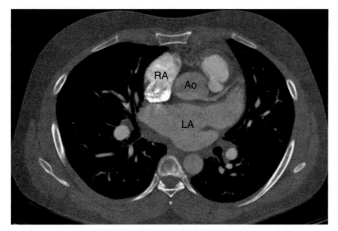

FIGURE 13-15 **Mitral stenosis.** This is a contrast-enhanced axial computed tomography image through the heart. The left atrium (LA) is markedly enlarged in this patient because of the obstruction of outflow from the left atrium by the stenotic mitral valve. Eventually, the increase in pressure is transmitted backward through the pulmonary venous and then pulmonary arterial circulations. *Ao,* Aorta; *RA,* right atrium.

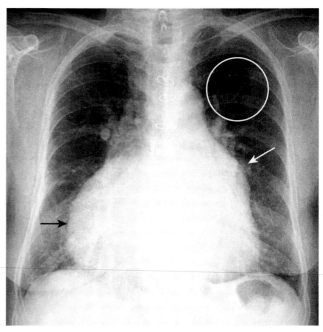

FIGURE 13-16 **Chronic mitral stenosis with tricuspid regurgitation.** The left atrium is enlarged *(white arrow).* Pulmonary venous hypertension has produced a redistribution of flow in the lungs so that the upper lobe vessels have become more prominent than the lower lobe **(cephalization)** *(white circle).* As a result of increased pulmonary vascular resistance and subsequent pulmonary arterial hypertension, the right heart also undergoes changes eventually, including tricuspid regurgitation with enlargement of the right atrium *(black arrow).*

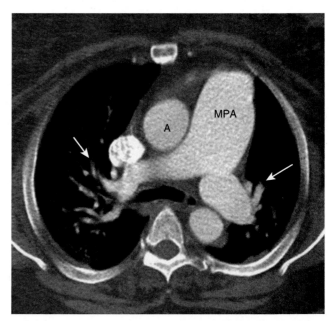

FIGURE 13-17 **Pulmonary arterial hypertension.** Normally, the main pulmonary artery (MPA) is about the same diameter as the ascending aorta (A). In this patient with pulmonary arterial hypertension, the MPA is much larger than the aorta. There is also a rapid attenuation in the size of the pulmonary arteries *(white arrows)* called *pruning,* which is also seen in pulmonary arterial hypertension.

seen in older adults and in younger individuals from developing countries. The **most common symptoms** are the result of **left heart failure:** dyspnea on exertion, orthopnea, and paroxysmal nocturnal dyspnea.

- Mitral stenosis leads to obstruction to the outflow of blood from the left atrium and usually becomes symptomatic when the **valve area falls below one third** of its normal size. As the left atrial pressure builds, the left atrium enlarges and the increased pulmonary venous pressure *(pulmonary venous hypertension)* is reflected retrograde into the pulmonary circulation (Fig. 13-15).

- Upper lobe vessels become as large as, or more prominent than, lower lobe vessels **(cephalization).** Pulmonary venous hypertension eventually leads to congestive heart failure. With prolonged elevation of pulmonary venous pressure, there may be physical changes in the pulmonary vasculature leading to escalating pulmonary vascular resistance requiring ever-increasing levels of pulmonary arterial pressure.

- Eventually there is pulmonary arterial hypertension and right-sided heart failure (Fig. 13-16).

PULMONARY ARTERIAL HYPERTENSION

- The normal mean pulmonary artery pressure is about 15 mm Hg. Pulmonary arterial hypertension may be **idiopathic (primary)** or **secondary** to another disease, usually emphysema. Mitral stenosis is another cause of pulmonary arterial hypertension.

- With primary pulmonary hypertension, the leading cause of death is progressive right heart failure. Secondary pulmonary hypertension shares comorbidities with the diseases that cause it: emphysema, recurrent thromboembolic disease, mitral stenosis, and CHF.

- The **imaging hallmark of pulmonary arterial hypertension is** a discrepancy in size between the central pulmonary vasculature (i.e., the main, right, and left pulmonary arteries are large) and the peripheral pulmonary vasculature. This discrepancy is called *pruning.*

- On CT scans, the **main pulmonary artery is normally about the same diameter as the ascending aorta,** but in **pulmonary arterial hypertension** the main pulmonary artery is usually larger than the ascending aorta and ≥**3 cm** in size (Fig. 13-17).

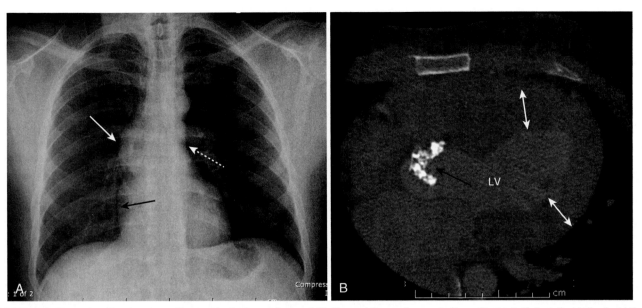

FIGURE 13-18 Aortic stenosis. A, There is poststenotic dilatation isolated to the ascending aorta *(solid white arrow).* The ascending aorta should normally not project farther to the right than the right heart border *(black arrow).* Notice that the heart is not enlarged and the descending aorta *(dotted white arrow)* is normal in appearance. **B,** This unenhanced axial CT scan of the heart shows dense calcification of the aortic valve *(black arrow)* and a markedly thickened left ventricular wall *(double white arrows)* from hypertrophy. *LV,* Left ventricle.

AORTIC STENOSIS

- The classic triad of clinical symptoms in aortic stenosis is **chest pain,** symptoms related to **heart failure,** and **syncope.**
- Aortic stenosis may be secondary to a **congenital bicuspid aortic valve,** from **degeneration of a tricuspid valve,** or less frequently now, **rheumatic heart disease.**
- Since there is obstruction to left ventricular outflow and since the ventricles respond to obstruction by undergoing hypertrophy of their walls, the heart is usually normal in size early in the course of the disease.
- The ascending aorta may be unusually prominent because of **poststenotic dilatation,** a hallmark of a significantly stenotic lesion in any major artery where there is increased intraluminal pressure for several centimeters **distal** to an obstructing lesion as a result of **eddy currents** and **turbulent flow.** Calcification of the aortic valve is easily seen on CT and has been shown to be predictive of increased likelihood of aortic stenosis and mortality from cardiovascular disease (Fig. 13-18).
- Eventually, when the heart begins to decompensate, it will become enlarged and congestive heart failure may ensue.

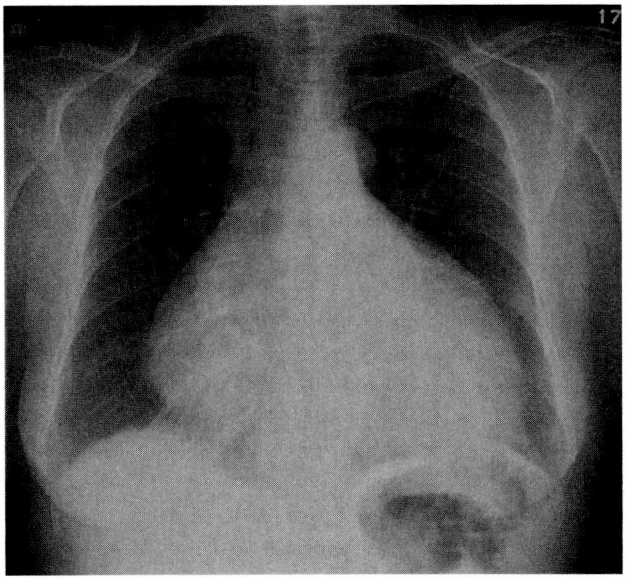

CARDIOMYOPATHY

Dilated Cardiomyopathy

- Dilated cardiomyopathy is a condition in which there is **increased systolic** and **diastolic volume of the ventricles** associated with a **decreased (<40%) ejection fraction.** It is the **most common** form of **cardiomyopathy** (90%).
- It can be idiopathic (primary) or associated with known diseases such as cardiac ischemia, diabetes, and alcoholism.

FIGURE 13-19 Dilated alcoholic cardiomyopathy. The cardiac silhouette is markedly enlarged, primarily as a result of biventricular enlargement. The patient had a long history of alcohol abuse. Dilated cardiomyopathy is frequently associated with congestive heart failure.

- **Decreased contractility and ventricular dilatation are hallmarks,** so it is usually characterized by an **enlarged heart** and frequently associated with the imaging signs of **congestive heart failure** (Fig. 13-19).
- The diagnosis can usually be made by echocardiography following the initial chest radiograph, in concert with the clinical findings.
- Magnetic resonance imaging (MRI) can provide the most accurate and reproducible findings for this disease. Using

electrocardiography (ECG)–gated, cine-magnetic resonance angiography (MRA), **cardiac ejection fraction** and **cardiac dimensions** can be accurately assessed.

■ **Radionuclide ventriculography,** using small quantities of intravenously injected radioisotope, can also determine **ejection fraction** and can be useful in differentiating between **ischemic** and **nonischemic causes of cardiomyopathy.**

Hypertrophic Cardiomyopathy

■ Hypertrophic cardiomyopathy is an abnormality (divided into primary [genetic] and secondary forms) that causes **asymmetric or concentric thickening of the myocardium,** sometimes with obstruction to left ventricular outflow caused by **systolic anterior motion (SAM)** of the anterior **mitral valve leaflet** (in the primary form). It **may lead to sudden cardiac death** and has been implicated as the cause of death in several high-profile athletes.

■ In the secondary form, its most common manifestation, it is caused by **hypertensive cardiovascular disease,** which produces concentric and diffuse hypertrophy of the left ventricle not associated with left ventricular outflow tract obstruction (see Fig. 13-14).

■ The primary form may be diagnosed with echocardiography or ECG-gated MRI of the heart in which the **asymmetric septal hypertrophy (ASH)** may be demonstrated.

Restrictive Cardiomyopathy

■ A **rare form** of cardiomyopathy characterized by **high diastolic filling pressures of the ventricles** in association with relatively well-preserved systolic function. It is usually secondary to an **infiltrative process** in the myocardium. Such diseases include **amyloid, autoimmune disease,** and **radiation.** The predominant presenting symptoms are related to **congestive heart failure.**

■ Although clinically similar to **constrictive pericarditis,** the key difference is that the **pericardium is normal in restrictive cardiomyopathy,** whereas it is **thickened in constrictive pericarditis.** The importance in differentiating the two is that constrictive pericarditis, unlike restrictive cardiomyopathy, is surgically curable.

■ In restrictive cardiomyopathy, the **heart is usually not enlarged.** There are pulmonary changes of **congestive heart failure.**

■ MRI can demonstrate the thickness of the pericardium, and if the pericardium is normal in size (<4 mm), constrictive pericarditis can effectively be excluded. If there is pericardial calcification, better seen on CT, restrictive cardiomyopathy can be excluded (Fig. 13-20).

AORTIC ANEURYSMS— GENERAL CONSIDERATIONS

■ Aneurysms are **defined as enlargements of a vessel greater than 50% of its original size. Atherosclerosis** is the **most common cause** of a descending thoracic aortic aneurysm. The majority of patients with aortic aneurysms are also hypertensive.

■ Most patients with aneurysms are **asymptomatic,** and the **aneurysm is discovered serendipitously.** When an aneurysm of the descending thoracic aorta expands, it may cause pain, which classically, but not always, radiates to the back.

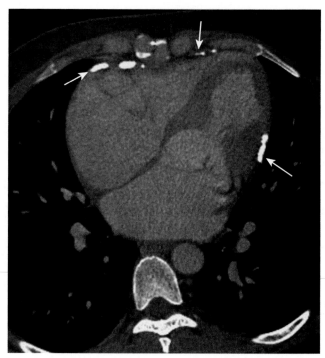

FIGURE 13-20 Constrictive pericarditis. There is extensive pericardial calcification *(white arrows),* most likely postinflammatory in etiology in this patient. Although restrictive cardiomyopathy and constrictive pericarditis can have identical clinical findings, the presence of pericardial calcifications excludes *restrictive cardiomyopathy.* If indicated, pericardiectomy holds the potential for cure for constrictive pericarditis.

■ As measured on CT or MRI scans, the **ascending aorta is normally <3.5 cm in diameter** and the **descending aorta is <3 cm.**

■ An aneurysm of the thoracic aorta is usually defined as a **persistent enlargement of >4 cm.**

■ In general, **aneurysms of 5 to 6 cm are at risk to rupture** and will require surgical intervention. The **rate of growth** of an aneurysm is also important in determining the need for surgical intervention and repair. Annual aneurysm growth rates should be <1 cm/year; otherwise, elective resection is considered.

RECOGNIZING A THORACIC AORTIC ANEURYSM

■ The **appearance** of a thoracic aortic aneurysm on chest radiographs **will depend on,** in part, **from which portion of the thoracic aorta it arises.** Aneurysms of the **ascending aorta** may extend **anteriorly and to the right.** Aneurysms of the **aortic arch** produce a **middle mediastinal mass.** Aneurysms of the **descending aorta** project **posteriorly, laterally,** and to the **left** (Fig. 13-21).

■ **Contrast-enhanced CT** is the modality **most often used** to diagnose a thoracic aortic aneurysm; MRI is also excellent at demonstrating aneurysms but is usually less available and more expensive.

■ On CT, aneurysms can appear as **fusiform (long)** or **saccular (globular)** in shape.

■ Their anatomy will be **more readily delineated on CT studies** using iodinated contrast material injected intravenously as a bolus, but they may also be visible on noncontrast (unenhanced) studies. Often, both unenhanced and contrast-enhanced CT studies are obtained to fully evaluate the aneurysm and its contained clot.

■ Frequently, **calcification is seen in the intima,** which may be **separated** from the **contrast-filled lumen by varying amounts of clot** (Fig. 13-22).

THORACIC AORTIC DISSECTION

■ **Aortic dissections most often originate in the ascending aorta (Stanford type A),** or they may involve only the **descending aorta (Stanford type B).**

■ They result from a tear that allows blood to dissect in the wall for varying lengths of the aorta, usually along the media.

■ In general, **patients with aortic dissection have been hypertensive** and may have an **underlying condition that can predispose to dissection,** such as cystic medial degeneration, atherosclerosis, Marfan syndrome, Ehlers-Danlos syndrome, trauma, syphilis, or crack cocaine abuse.

■ In many patients **abrupt onset of ripping or tearing chest pain,** which is **maximal at its time of origin,** is the characteristic history.

■ **Conventional radiographs** are **not significantly sensitive to be diagnostically reliable,** but they **may point to the diagnosis** when several imaging findings occur together, especially in the proper clinical setting.

  ♦ "Widening of the mediastinum" is a **poor means of establishing the diagnosis** because (1) it is commonly overinterpreted on portable supine radiographs and (2) it occurs in only about one in four cases of aortic dissection.

  ♦ **Left pleural effusion frequently represents a transudate** caused by pleural irritation, although transient hemorrhage from the aorta can also produce a hemothorax (Fig. 13-23).

  ♦ **Left apical pleural cap** of fluid or blood

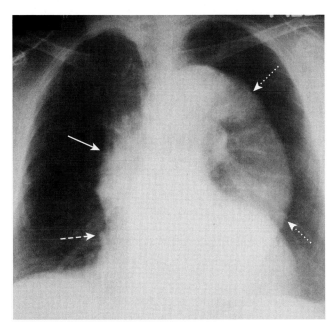

FIGURE 13-21 Aortic aneurysm. The entire thoracic aorta is enlarged in this 67-year-old man. The ascending aorta *(solid white arrow)* should normally not project farther to the right than the right heart border *(dashed white arrow)* on a nonrotated chest radiograph. The descending thoracic aorta should normally parallel and almost disappear with the thoracic spine; as it becomes larger, it swings farther away from the spine *(dotted white arrows)*.

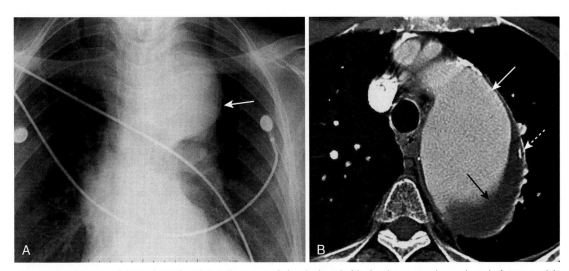

FIGURE 13-22 Aortic aneurysm, conventional chest radiograph, and CT. A, Close-up view of a frontal radiograph of the chest demonstrates a large mediastinal soft tissue mass *(white arrow)*. This soft tissue density represents a large aneurysm of the proximal descending aorta seen also in the CT scan **(B).** The aneurysm measured 6.7 cm, which placed it at significant risk for rupture. Calcification in the wall of an aneurysm is common *(dotted white arrow)*. Contrast material mixes with blood flowing in the lumen of the aorta *(solid white arrow)*, but the flowing blood is separated from the intimal calcification by a considerable amount of noncontrast-containing thrombus adherent to the wall *(black arrow)*.

- ◆ Loss of the normal shadow of the aortic knob
- ◆ Increased deviation of the trachea or esophagus to the right
- ■ In various studies, MRI and contrast-enhanced CT have been shown to be equally sensitive and specific for aortic

dissection; therefore the imaging modality used will depend in part on the patient's hemodynamic stability and the availability of resources. Transesophageal ultrasound is also used to establish the diagnosis.

➡ On both **MRI and CT,** the **diagnosis rests on identification** of the **intimal flap** that separates the **true (original)** from the **false lumen** (canal created by the dissection) (Fig. 13-24).

- ■ In general **type A (ascending aortic) dissections are treated surgically,** whereas **type B (descending aortic dissections)** are treated **medically.**

CORONARY ARTERY DISEASE

- ■ Coronary artery disease is the **leading cause of death worldwide.**
- ■ The coronary artery lumen becomes narrowed by varying amounts of atheromatous plaque. Calcium is deposited in the muscular layers of the artery's walls. Vulnerable plaque may rupture, there may be vasospasm, or there may be emboli, which produce enough stenosis to lead to *ischemia* and possibly *infarction* of cardiac muscle (Fig. 13-25).
- ■ MRI, CT, and nuclear medicine studies can be used in the evaluation of coronary artery disease. MRI can demonstrate **postinfarct scar formation** and **myocardial contractility.** Ventricular **function** can be quantitatively assessed.
- ■ Cardiac CT is used for imaging of the coronary arteries. It has a **very high negative predictive value;** a negative test virtually rules out coronary artery disease. One of the major drawbacks in the use of cardiac CT angiography is the relatively high x-ray dose delivered, but changes in equipment and algorithms are reducing this dose considerably. Cardiac CT angiography requires injection of iodinated contrast material (Video 13-1).

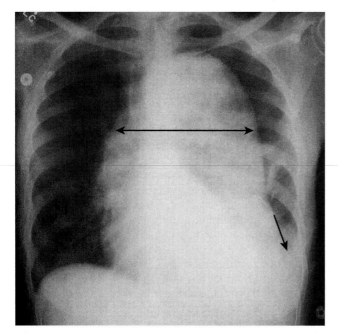

FIGURE 13-23 **Aortic dissection.** Conventional radiographs are not sensitive enough to be diagnostically reliable for aortic dissection, but they may point to the diagnosis when several imaging findings are seen together, especially in the proper clinical setting. "Widening of the mediastinum" is frequently not present and is a poor means of establishing the diagnosis, although in this patient the mediastinum is clearly widened by an enlarged aorta *(double black arrow)*. Also, a left pleural effusion is present *(single black arrow)*. The combination of a widened mediastinum and a left pleural effusion in a patient with chest pain should alert you to the possibility of an aortic dissection.

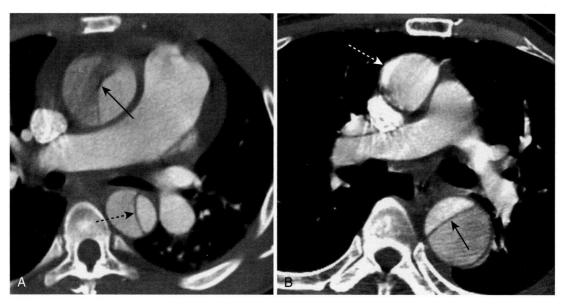

FIGURE 13-24 **Aortic dissections, types A and B. A,** An intimal flap is seen to traverse both the ascending *(closed black arrow)* and descending aorta *(dotted black arrow)*. This is a Stanford Type A dissection. **B,** There is a normal appearing ascending aorta *(dotted white arrow)* and there is an intimal flap noted by the black line traversing the descending aorta *(solid black arrow)*. The intimal flap is the characteristic lesion of an aortic dissection. The smaller lumen is usually the true (original) lumen, and the larger, false lumen is actually a channel that has been produced by blood dissecting through the media.

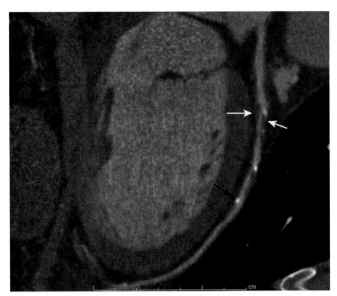

FIGURE 13-25 Cardiac CT, coronary artery stenosis, and calcification. Contrast fills the lumen of the left anterior descending coronary artery, except where atheromatous plaque *(white arrows)* narrows the lumen by about 70%, a significant stenosis. Rupture of this soft plaque, hemorrhage, or clot formation can further narrow or completely occlude the lumen. There is also calcified plaque present *(black arrows)*.

- CT may also be used in asymptomatic patients for **calcium scoring,** in which the calcium found in the coronary arteries is used as a marker for coronary artery disease. The amount of calcium detected on a cardiac CT scan and calculated by computer can be a helpful prognostic tool. The findings on cardiac CT are expressed as a calcium score—the higher the score, the more extensive the evidence for CAD and the higher the potential mortality from a cardiovascular event. Calcium scoring is performed without intravenous contrast (see Fig. 4-11).
- CT can also demonstrate complications of myocardial infarction, such as ventricular aneurysms and intracardiac clots.
- **Single-photon emission computed tomography (SPECT)** is an imaging technique that blends the intravascular injection of a **radioactive isotope** with acquisition of images using a rotating nuclear gamma camera capable of three-dimensional localization of disease. Stress and resting myocardial perfusion images using SPECT imaging can demonstrate areas of **ischemia,** especially compared with the same study done at rest. Nuclear medicine studies can also estimate left ventricular function.
- Coronary angiography remains the gold standard for detecting coronary artery stenosis (see Video 4-1).

TAKE-HOME POINTS

RECOGNIZING ADULT HEART DISEASE

In adults, a quick assessment of heart size can be made using the *cardiothoracic ratio*, which is the ratio of the widest transverse diameter of the heart compared with the widest internal diameter of the rib cage. In normal adults, the cardiothoracic ratio is usually <50%.

There are extracardiac causes that can make the heart appear to be enlarged, even if it is actually normal; these include AP portable studies, factors that inhibit a deep inspiration, abnormalities of the bony thorax, and the presence of a pericardial effusion.

The heart will appear slightly larger on an AP projection than on a PA projection of the chest because the heart is closer to the imaging surface on a PA exposure.

On the lateral projection, the heart usually does not extend posteriorly to overlap the spine unless it is enlarged or there is a pericardial effusion.

Two major patterns of congestive heart failure are pulmonary interstitial and pulmonary alveolar edema.

The four key findings of **pulmonary interstitial edema** are thickening of the interlobular septa, peribronchial cuffing, fluid in the fissures, and pleural effusions.

The key findings in **pulmonary alveolar edema** are fluffy, indistinct, patchy airspace densities; batwing or butterfly configuration, frequently sparing the outer third of lungs; and pleural effusions, especially with cardiogenic pulmonary edema.

Causes of pulmonary edema can be divided into two major categories: cardiogenic and noncardiogenic causes.

A patient with **cardiogenic pulmonary edema** is more likely to have pleural effusions, Kerley B lines, cardiomegaly, and an elevated pulmonary capillary wedge pressure than would a patient with noncardiogenic pulmonary edema.

The **noncardiogenic causes of pulmonary edema** are a diverse group of conditions including uremia, disseminated intravascular

coagulopathy, smoke inhalation, near-drowning, volume overload, and lymphangitic spread of malignancy.

Acute respiratory distress syndrome (ARDS) can be considered a subset of noncardiogenic pulmonary edema in which the clinical picture is one of severe hypoxia, cyanosis, tachypnea, and dyspnea.

Essential **hypertension** is a common disease that can lead to congestive heart failure, coronary artery disease, and secondary hypertrophic cardiomyopathy.

Mitral stenosis has become less common with antibiotic treatment of rheumatic fever, but it can lead to left, and then right, heart failure through chronic elevation of the pulmonary venous and arterial pressures with increased pulmonary vascular resistance.

Pulmonary arterial hypertension may be either idiopathic (primary) or secondary to emphysema or recurrent thromboembolic disease. It produces **pruning** of the pulmonary vasculature and might be suspected when the main pulmonary artery achieves a diameter of 3 cm or more on CT or MRI.

Aortic stenosis in older adults is most often secondary to degeneration of a tricuspid aortic valve and can lead to angina, syncope, or congestive heart failure. The ascending aorta may be prominent from **poststenotic dilatation.**

Cardiomyopathies are divided into dilated, hypertrophic, and restrictive forms. Restrictive cardiomyopathy must be differentiated from constrictive pericarditis, with which it shares its clinical findings.

Aortic aneurysms can be saccular or fusiform or can dissect. Most thoracic aortic dissections begin in the ascending aorta (Stanford Type A) and are treated surgically.

Coronary artery disease is the leading cause of death worldwide, and it or its sequelae can be imaged using a variety of techniques, including CT, MRI, and single-photon emission computed tomography (SPECT).

 **WEBLINK**

Visit StudentConsult.Inkling.com for videos and quizzes on Recognizing Cardiomegaly, Plain Film Cardiac Abnormalities, and The Faces of Congestive Heart Failure.

For your convenience, the following QR Code may be used to link to **StudentConsult.com**. You must register this title using the PIN code on the inside front cover of the text to access online content.

CHAPTER 14
Recognizing the Normal Abdomen: Conventional Radiology

CONVENTIONAL RADIOGRAPHY

■ Imaging of the abdomen is now largely performed with computed tomography (CT), ultrasonography (US), or magnetic resonance imaging (MRI); however, for many patients, "plain films" of the abdomen are obtained as a first step before other imaging studies are ordered or as a method of following up on findings demonstrated by other modalities. The principles that guide the interpretation of conventional radiographs apply to the modalities of CT, MRI, and US.

■ To recognize **abnormal** findings on conventional radiographs of the abdomen, you must familiarize yourself with the appearance of **normal** first.

WHAT TO LOOK FOR

 First, look at the **overall gas pattern** (Box 14-1).

♦ You are looking for the **overall** pattern, so do not spend too much time trying to identify every bubble of bowel gas you see.

■ **Second,** check to see if there is **extraluminal air** (see Chapter 17 for a discussion of extraluminal air).

■ **Third,** look for **abnormal abdominal calcifications.**

■ **Fourth,** look for any **soft tissue masses.**

NORMAL BOWEL GAS PATTERN

■ Virtually all gas in the bowel comes from swallowed air. Only a fraction comes from the bacterial fermentation of food. In the abdomen, the terms *gas* and *air* are used interchangeably to refer to the contents of the bowel.

■ Loops of bowel that contain a sufficient amount of air to fill the lumen completely are said to be *distended.* **Distension** of the bowel is **normal.**

■ Loops of bowel that are filled beyond their normal size are said to be *dilated.* **Dilatation** of the bowel is **abnormal.**

BOX 14-1 RECOGNIZING THE NORMAL ABDOMEN: WHAT TO LOOK FOR

The gas pattern
Extraluminal air
Calcifications
Soft tissues masses

■ **Stomach**
  ♦ There is almost **always air in the stomach,** unless:
    • The patient has recently vomited, or
    • There is a nasogastric tube in the stomach, and the tube is attached to a suction device.

■ **Small bowel**
  ♦ There is usually a small amount of **air in about two** or **three loops of nondilated small bowel** (Fig. 14-1).
  ♦ The **normal diameter of the small bowel is <2.5 cm,** which is about 1 inch, or the diameter of one U.S. quarter.

■ **Large bowel**
  ♦ There is **almost always air in the rectum** or **sigmoid colon.** There may be **varying amounts of gas** in the remainder of the **colon** (Fig. 14-2).

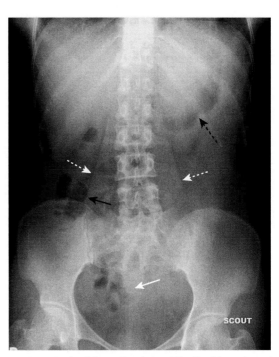

FIGURE 14-1 Normal supine abdomen. This is the "scout" film of the abdomen, the one that gives you a general idea of the bowel gas pattern and allows you to search for abnormal calcifications and detect organomegaly. There is usually a small amount of air in about two or three loops of nondilated small bowel *(solid black arrow).* There will almost always be air in the stomach *(dotted black arrow)* and in the rectosigmoid colon *(solid white arrow).* Depending on the amount of fat around the visceral organs, the outlines of these organs may be partially visible on conventional radiographs. The psoas muscles are outlined by fat *(dotted white arrows),* making them visible on this image.

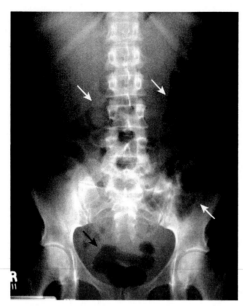

FIGURE 14-2 **Normal prone abdomen.** In the prone position, the ascending and descending colon, as well as the rectosigmoid colon, all of which are posterior structures, are the highest parts of the large bowel and thus most likely to fill with air. There is air in the S-shaped rectosigmoid colon *(black arrow)*. Air can also be seen throughout the remainder of the colon *(white arrows)*.

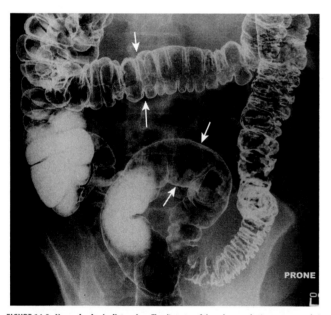

FIGURE 14-3 **Normal colonic distension.** The diameter of the colon on a barium enema study is the size to which the colon can normally distend *(white arrows)*, beyond which it would be considered dilated. This patient has had a **double-contrast barium enema** examination in which both air and barium were instilled as contrast agents. The combination of air and barium allows for excellent visualization of the mucosal surface of the colon.

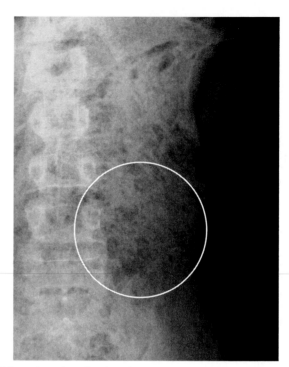

FIGURE 14-4 **Appearance of stool.** Stool is recognizable by the multiple, small bubbles of gas present within a semisolid-appearing soft tissue density *(white circle)*. Stool marks the location of the large bowel and can help in identification of individual loops of bowel on conventional radiographs. This patient has a markedly dilated sigmoid colon due to chronic constipation.

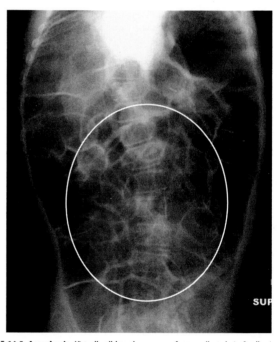

FIGURE 14-5 **Aerophagia.** Virtually all bowel gas comes from swallowed air. Swallowing large quantities of air may produce a picture called *aerophagia*, characterized by numerous polygon-shaped, air-containing loops of bowel, none of which is dilated *(white circle)*.

♦ Use this rule to decide if the large bowel is dilated or not:
  • **The large bowel can normally distend to about the same size as it does on a barium enema examination.** To give you an idea of how large that is, look at Figure 14-3.
  • Stool is recognizable by the **multiple small bubbles of gas** present within a semisolid-appearing mass.

Recognizing the appearance of stool will help in localizing the large bowel (Fig. 14-4).

 Individuals who swallow large quantities of air may develop **aerophagia,** characterized by numerous polygon-shaped, air-containing loops of bowel, none of which are dilated (Fig. 14-5).

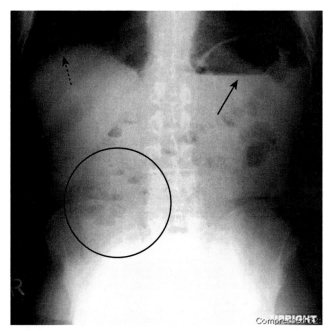

FIGURE 14-6 Normal upright abdomen. There are two things to look for on an upright view of the abdomen: air–fluid levels and free intraperitoneal air. Normally, there is an air–fluid level in the stomach *(solid black arrow)*. There may be short air–fluid levels in a few nondilated loops of small bowel *(black circle)*. There are usually very few or no air–fluid levels in the colon. Free air, if present, should be visible just below the hemidiaphragm *(dotted black arrow)* and would be easier to recognize on the right than on the left.

NORMAL FLUID LEVELS

- **Stomach**
  - ◆ There is almost always fluid in the stomach, so there is **almost always an air–fluid level in the stomach** on an upright abdominal image, a study done with the patient in the decubitus position, or an upright chest radiograph.
    - • **To see an air–fluid level, the x-ray beam must be directed horizontally,** parallel to the floor (see Terminology section in the online content).
- **Small bowel**
  - ◆ **Two** or **three air–fluid levels** in the small bowel may be seen normally on an upright or decubitus view of the abdomen.
- **Large bowel**
  - ◆ The large bowel functions, in part, to remove fluid, so **there are usually no** or **very few air–fluid levels in the colon** (Fig. 14-6).

⊙ There may be **many air–fluid levels** present in the colon if the patient has had a **recent enema** or if the patient is **taking medication with a strong anticholinergic, antiperistaltic effect.**

- The normal distribution of bowel gas and fluid is summarized in Table 14-1.

DIFFERENTIATING LARGE FROM SMALL BOWEL

- **Recognizing the large bowel**
  - ◆ The **large bowel is peripherally placed** around the perimeter of the abdominal cavity, except for the right

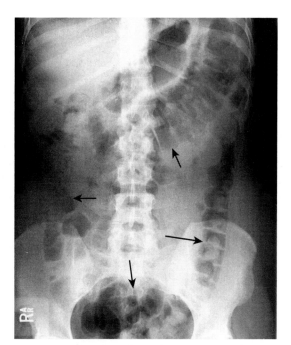

FIGURE 14-7 Location of large bowel. The large bowel usually occupies the periphery of the abdomen. The small bowel is located more centrally. Here, the large bowel *(black arrows)* contains a normal amount of air. The liver occupies the right upper quadrant and normally displaces all bowel from this area.

TABLE 14-1 NORMAL DISTRIBUTION OF GAS AND FLUID IN THE ABDOMEN

Organ	Normally Contains Gas	Normally Has Air–Fluid Levels
Stomach	Yes	Yes
Small bowel	Yes, 2-3 loops	Yes
Large bowel	Yes, especially rectosigmoid colon	No

upper quadrant, which is occupied by the liver (Fig. 14-7).
  - ◆ **Haustral markings usually do not extend completely across the large bowel** from one wall to the other. If they do connect one wall with another, **haustral markings are spaced more widely apart** than the valvulae conniventes of the small bowel (Fig. 14-8).
- **Recognizing the small bowel**
  - ◆ The **small bowel is centrally placed** in the abdomen. **Valvular markings typically extend across the lumen** of the small bowel from one wall to the other. The **valvulae are spaced much closer together** than the haustra of the large bowel (Fig. 14-9).
  - ◆ The **small bowel can achieve a maximum diameter, when abnormally dilated, of about 5 cm. The large bowel can dilate to many times that size.**

ACUTE ABDOMINAL SERIES: THE VIEWS AND WHAT THEY SHOW

- Almost every Department of Radiology has a series of radiographic images (a protocol) that are routinely obtained in

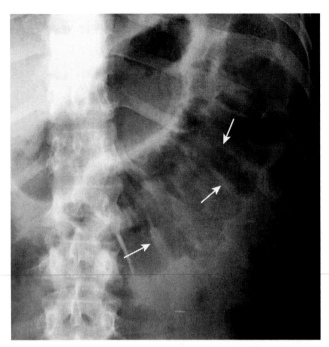

FIGURE 14-8 Normal large bowel haustral markings. Most haustral markings in the colon do not traverse the entire lumen to extend from one wall to the opposite wall *(white arrows)*. This appearance is unlike that of the valvulae conniventes in the small bowel, which do appear to traverse the entire lumen. The haustral markings are also spaced more widely apart than the valvulae of the small bowel (see Fig. 14-9).

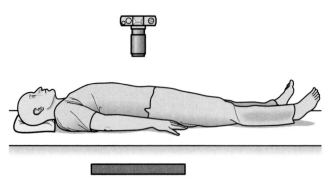

FIGURE 14-10 Positioning for supine view of the abdomen. The patient lies on his or her back on the x-ray table or stretcher, and the x-ray beam is directed vertically downward. The camera icon represents the x-ray tube, which would actually be positioned about 40 inches above the cassette, represented by the thick bar under the table.

TABLE 14-2 ACUTE ABDOMINAL SERIES: THE VIEWS AND WHAT TO LOOK FOR

View	Look For
Supine abdomen	Overall bowel gas pattern, calcifications, masses
Prone abdomen	Gas in the rectosigmoid colon
Upright abdomen	Free air, air–fluid levels in the bowel
Upright chest	Free air, pneumonia, pleural effusions

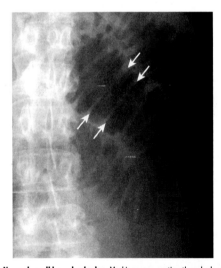

FIGURE 14-9 Normal small bowel valvulae. Markings representing the valvulae typically extend across the lumen of the small bowel to reach from one wall to the other. In addition, the valvulae are spaced much closer together than the haustra of the large bowel, even when the small bowel is dilated. The *white arrows* point to two valvulae that traverse the entire lumen in this close-up of dilated small bowel.

- ♦ **Prone** or **lateral rectal view**
  - The inclusion of these views is the most widely variable between hospitals.
- ♦ **Upright or left-side-down (left lateral) decubitus view**
  - One or the other is almost always included.
- ♦ **Chest: upright or supine view**
  - Inclusion of a chest radiograph depends on hospital practices.
- ▪ Table 14-2 summarizes **what to look for on each of the views** of an acute abdominal series.

Supine View ("Scout Film")
- ▪ **What it's good for**
  - ♦ **Overall appearance of the gas pattern**
    - The **overall appearance of the bowel gas pattern,** including how much air and fluid there are and their most likely locations, is **more important** than identifying every small bubble of air on the radiograph.
  - ♦ Identifying the **presence** or **absence of calcifications**
  - ♦ Identifying the **presence of soft tissue masses** (see Fig. 14-1)
- ▪ **How it's obtained**
  - ♦ The patient lies on his or her back on the x-ray table or stretcher, and the x-ray beam is directed vertically downward (Fig. 14-10).
- ▪ **Substitute view**
  - ♦ There is really no substitute for a supine view of the abdomen. Virtually all patients, regardless of their condition, can tolerate this part of the examination.

patients who have acute abdominal pain. These series are sometimes called *obstruction series* or *complete abdominal series* or *acute abdominal series* or something similar. For the purposes of this book, we will call such series *acute abdominal series*.
- ▪ **Acute abdominal series: what it may contain**
  - ♦ **Supine view** of the abdomen
    - This view is almost always obtained.

Prone View

- ■ **What it's good for**
  - ♦ **Identifying gas in the rectum** and/or **sigmoid colon**
    - • Because the rectum and sigmoid colon are the **highest points of the large bowel** when the person is lying prone on the x-ray table, air will rise into the rectosigmoid colon.
    - • Almost no air is introduced into the rectosigmoid colon during the course of a routine rectal examination.
  - ♦ **Identifying gas in the ascending and descending colon**
    - • Because these two parts of the large bowel, in addition to the rectosigmoid colon, are also posteriorly positioned, air will collect in them when the patient is lying prone (see Fig. 14-2).
- ■ **How it's obtained**
  - ♦ The patient lies on his or her abdomen on the x-ray table or stretcher, and the x-ray beam is directed vertically downward (Fig. 14-11).
- ■ **Substitute view**
  - ♦ Frequently, patients are **unable to lie prone** because of their physical condition (e.g., recent surgery, severe abdominal pain).
  - ♦ These patients can turn onto their **left side,** and thus a **lateral view of the rectum** can be exposed with a **vertical beam** to substitute for the prone radiograph (Fig. 14-12). The lateral view of the rectum will usually demonstrate the presence or absence of air in the rectum and/or sigmoid colon (Fig. 14-13).

Upright View of the Abdomen

- ■ **What it's good for**
  - ♦ Seeing **free air in the peritoneal cavity** (i.e., extraluminal air)
  - ♦ Seeing **air–fluid levels within the bowel lumen** (see Fig. 14-6)
- ■ **How it's obtained**
  - ♦ The patient stands or sits up, and the exposure is made with the **x-ray beam directed horizontally,** parallel to the plane of the floor (Fig. 14-14).

- ■ **Substitute view**
  - ♦ Frequently, patients with the signs and symptoms of an acute condition in the abdomen cannot tolerate standing or sitting up for an upright view of the abdomen.
  - ♦ In such cases, a **left lateral decubitus view** of the abdomen can be substituted for the upright radiograph. For a left lateral decubitus view, the **patient lies on his** or **her left side on the x-ray table.** This is done so that any **"free air"** will distribute itself at the **highest part of the abdominal cavity,** which will be the patient's **right side** (Fig. 14-15).

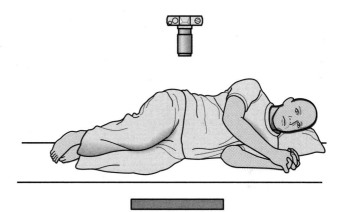

FIGURE 14-12 Positioning for the lateral rectum view. Patients who cannot lie prone can usually turn onto their left side and have a lateral view of the rectum exposed, with a vertical beam used to substitute for the prone radiograph. The camera icon represents the x-ray tube, which would actually be positioned about 40 inches above the cassette, represented by the thick bar under the table.

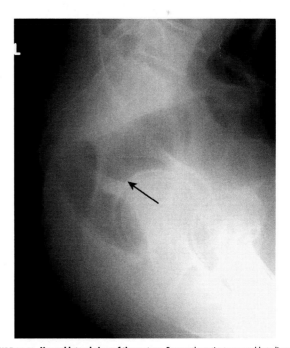

FIGURE 14-13 Normal lateral view of the rectum. Frequently, patients are unable to lie prone because of their physical condition (e.g., recent surgery, severe abdominal pain). These patients can turn onto their left side and have a lateral view of the rectum exposed, with a vertical beam used to substitute for the prone radiograph. The lateral view of the rectum will usually demonstrate the presence or absence of air in the rectum and/or sigmoid colon (black arrow).

FIGURE 14-11 Positioning for prone view of the abdomen. The patient lies on his or her abdomen on the x-ray table or stretcher, and the x-ray beam is directed vertically downward. The camera icon represents the x-ray tube, which would actually be positioned about 40 inches above the cassette, represented by the thick bar under the table.

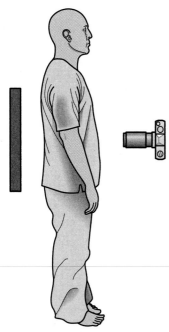

FIGURE 14-14 Positioning of patient for an upright view of the abdomen. The patient stands or sits up, and the x-ray beam is directed horizontally, parallel to the plane of the floor. The camera icon represents the x-ray tube, which would actually be positioned about 40 inches from the cassette, represented by the thick bar behind the patient.

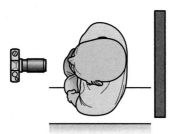

FIGURE 14-15 Positioning of the patient for a left lateral decubitus view of the abdomen. In patients who cannot tolerate an upright view of their abdomen, a left lateral decubitus view is usually used as a substitute. The patient lies on his or her left side on the examining table; the x-ray tube is usually positioned anteriorly *(camera icon)*; and the cassette *(thick bar)* is placed in back of the patient. The x-ray beam is directed horizontally, parallel to the floor, at a distance of about 40 inches from the patient.

- **Free air should be easily visible** over the **outside edge of the liver,** where there is normally no bowel gas present (Fig. 14-16).
- If a right lateral decubitus view is obtained, any free air, if present, will rise to the left side of the abdomen. The left side of the abdomen is the normal location of the stomach bubble as well as gas in the splenic flexure of the colon, either of which could be mistaken for free air.
- **To see** the **free air, the x-ray beam must be directed horizontally,** parallel to the floor, when a decubitus view is obtained.

 Box 14-2 summarizes the ingredients that must be present for visualization of air–fluid levels in the abdomen.

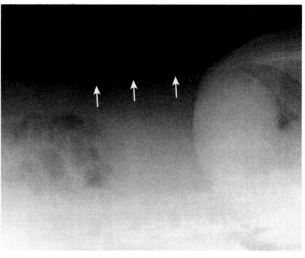

FIGURE 14-16 Normal left lateral decubitus view of the abdomen. For a left lateral decubitus view, patients lie on their left side on the examining table, and an exposure is made with a horizontal x-ray beam (parallel to the floor). This is done so that any "free air" will distribute itself at the highest part of the abdominal cavity, which will be the patient's right side. Free air, if present, should be easily visible as a black crescent over the outside edge of the liver *(white arrows)*, a location in which there is normally no bowel gas present. In this photo, the patient's head is positioned toward your right, with the feet pointing toward your left.

BOX 14-2 TO SEE AN AIR–FLUID LEVEL ON CONVENTIONAL RADIOGRAPHS, YOU MUST HAVE

Air

Fluid

A horizontal x-ray beam (parallel to the plane of the floor)

Air–fluid interfaces cannot be visualized on conventional radiographs taken with a vertical x-ray beam (e.g., supine studies)

Upright View of Chest

- **What it's good for**
  - Seeing **free air beneath the diaphragm**
  - Finding **pneumonia at the lung bases,** which might mimic the symptoms of an acute condition in the abdomen
  - Finding **pleural effusions,** which could be secondary to an intraabdominal process and could help to identify the presence of such a process.
    - **Pancreatitis,** for example, **may be associated with a left pleural effusion.**
    - Some **ovarian tumors** may occasionally be associated with **right-sided** or **bilateral pleural effusions.**
    - An abscess beneath the right hemidiaphragm **(subphrenic abscess) may be associated with a right pleural effusion.**
    - See Chapter 8 for more details on the laterality of pleural effusions.
- **How it's obtained**
  - The **patient stands** or **sits up,** and an exposure of the thorax is made using a **horizontal x-ray beam** (Fig. 14-17).

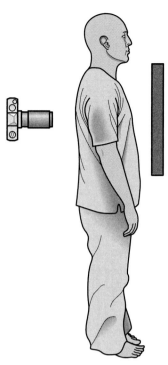

FIGURE 14-17 **Positioning of patient for an upright chest radiograph.** The patient sits upright or stands with their anterior chest wall closest to the cassette. The camera icon represents the x-ray tube, which is actually about 72 inches from the cassette, represented by the thick bar.

- ■ **Substitute view for an upright chest x-ray**
  - ◆ Frequently, patients with the signs and symptoms of an acute condition in the abdomen cannot tolerate standing for an upright view of the chest. In those cases, **a supine view of the chest may be obtained** with the patient lying on the stretcher or x-ray table.
  - ◆ In a supine view, the x-ray beam is directed **vertically downward,** and **free air, especially in small amounts, may not be visible.**

CALCIFICATIONS

- ■ Abdominal calcifications are discussed in more detail in Chapter 18.

 There are two abdominal calcifications that should not be confused with pathologic calcifications.

- ◆ **Phleboliths** are small, rounded calcifications that represent calcified venous thrombi and occur with increasing age, most often in the pelvic veins of women. They classically have a lucent center, which helps to differentiate them from ureteral calculi, with which phleboliths can be confused (Fig. 14-18).
- ◆ **Calcification of the rib cartilages** occurs with advancing age and, though not a true abdominal calcification, can sometimes be confused for renal or biliary calculi when these calcifications overlie the kidney or the region of the gallbladder. Calcified cartilage tends to have an amorphous, speckled appearance, and the calcified cartilage will occur in an arc corresponding to that of the anterior rib cartilage as it sweeps back toward articulation with the sternum (Fig. 14-19).

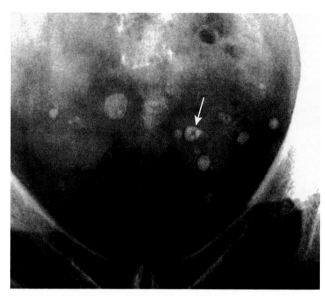

FIGURE 14-18 **Phleboliths.** Phleboliths are small, rounded calcifications that represent calcified venous thrombi that occur with increasing age, most often in the pelvic veins of women. They classically have a lucent center *(white arrow).* In the pelvic veins, they are considered incidental and nonpathologic calcifications, but they can be confused with ureteral calculi.

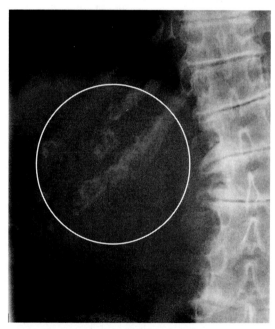

FIGURE 14-19 **Calcified rib cartilages.** Calcifications of the rib cartilage *(white circle)* occur with advancing age, and, though they are not true abdominal calcifications, they can sometimes be confused for calculi when they superimpose on the kidney or the region of the gallbladder. Calcified cartilage tends to have an amorphous, mottled appearance. Calcified rib cartilages occur along an arc corresponding to the sweep of the anterior ribs as they turn back toward the sternum.

ORGANOMEGALY

- ■ **Conventional radiographic evaluation of soft tissue structures in the abdomen** (such as the liver, spleen, kidneys, gallbladder, urinary bladder, or soft tissue masses such as tumors or abscesses) **is limited,** because these structures are **soft tissue densities** and are **surrounded by other soft tissues** or **fluid** of similar density. Only a **difference**

in density between two adjacent structures **will render their outlines visible** on conventional radiographs.

■ Still, conventional radiographs are easy to obtain and frequently are the first study ordered in a patient with abdominal symptoms.

➡ There are **two fundamental ways of recognizing the presence,** and **estimating the size, of soft tissue masses** or **organs** on conventional radiographs of the abdomen:

◆ The first is by **direct visualization of the edges of the structure,** which can only occur if the structure is surrounded by something with a density different from that of soft tissue, such as fat or free air.

◆ The second is to recognize **indirect evidence of the mass** or enlarged visceral organ by **recognizing pathologic displacement of air-filled loops of bowel.**

Liver

■ **Normal**

◆ The liver normally displaces all bowel gas from the right upper quadrant.

◆ Occasionally, a **tongue-like projection of the right lobe** of the liver may extend to the iliac crest, **especially in women.** This is called a *Riedel lobe* and is normal (Fig. 14-20).

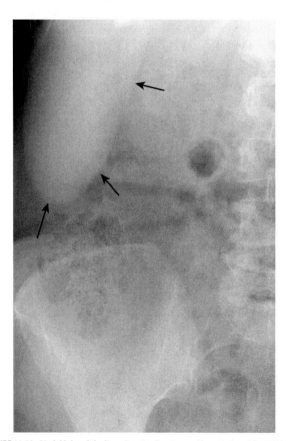

FIGURE 14-20 Riedel lobe of the liver. Occasionally, a tongue-like projection of the right lobe of the liver may extend to the iliac crest, especially in women. This is called a *Riedel lobe* and is normal *(black arrows)*. Conventional radiography is a notoriously poor tool for estimating the size of the liver. CT, MRI, and US give a more accurate picture of liver size.

■ **Enlarged liver**

◆ On conventional radiographs, **an enlarged liver might be suggested** if there is displacement of all bowel from the right upper quadrant **down to the iliac crest** and **across the midline** (Fig. 14-21). **Conventional radiography is a notoriously poor tool for estimating the size of the liver.** The best imaging modalities to use for estimating liver size are CT, MRI, and US.

Spleen

■ **Normal**

◆ The adult **spleen is about 12 cm in length and usually does not project below the 12th posterior rib.** As a general rule, the **spleen is about as large as the left kidney.**

■ The **stomach bubble** (i.e., air in the gastric fundus) **usually nestles beneath the highest part of the left hemidiaphragm,** about midway between the abdominal wall and the spine.

■ **Enlarged spleen**

◆ If the spleen projects well below the 12th posterior rib and/or displaces the stomach bubble toward or across the midline, the spleen is probably enlarged (Fig. 14-22).

Kidneys

■ **Normal**

◆ Portions of the kidney outlines may be visible on conventional radiographs if there is an adequate amount of perirenal fat present.

◆ The **kidney length is approximately the height of four lumbar vertebral bodies,** or about **10 to 14 cm in size in an adult.**

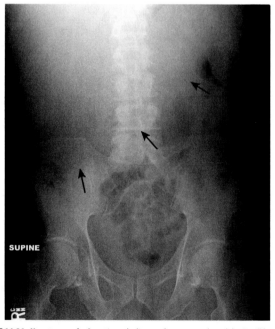

FIGURE 14-21 Hepatomegaly. Sometimes the liver can become so enlarged that it will be obvious even on conventional radiographs. Conventional radiographs may suggest an enlarged liver if there is displacement of all bowel loops from the right upper quadrant down to the iliac crest and across the midline *(black arrows)*, such as in this patient with cirrhosis.

- ♦ The liver depresses the right kidney such that the **right kidney is usually lower in the abdomen than the left kidney** (Fig. 14-23).
- ♦ The **left kidney is roughly the same length as the spleen.**
- ■ **Enlarged kidney**
  - ♦ Usually, only extremely enlarged kidneys or large renal masses are recognizable on conventional radiographs by displacement of bowel gas (Fig. 14-24).

Urinary Bladder

- ■ **Normal**
  - ♦ The bladder frequently is surrounded by **enough extravesical fat that at least the dome is visible** in most individuals as the top of an oval structure, with its

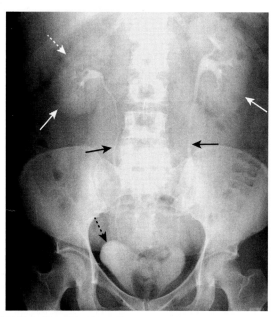

FIGURE 14-23 Position of the kidneys. This is one image from an **intravenous urogram,** also known as an **intravenous pyelogram** (**IVP**). For an IVP, the patient is given an intravenous injection of iodinated contrast dye, which is excreted by the kidneys. Both kidney outlines *(solid white arrows),* the ureters *(solid black arrows),* and the urinary bladder *(dotted black arrow)* can be seen. Using IVPs, other images of the kidneys, including oblique views, were often obtained to visualize the entire contour of the kidney. IVPs have largely been replaced by CT scans in the form of CT urograms. The liver *(dotted white arrow)* normally depresses the right kidney more inferiorly than the left kidney.

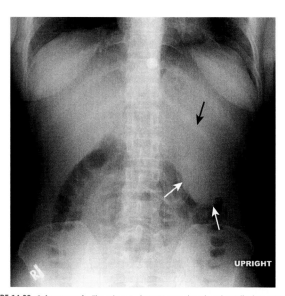

FIGURE 14-22 Splenomegaly. The spleen is about 12 cm in length and usually does not project below the 12th posterior rib. If the spleen *(white arrows)* projects well below the 12th posterior rib *(black arrow)* and/or displaces the stomach bubble toward or across the midline, it is probably enlarged, as it is in this patient with leukemia.

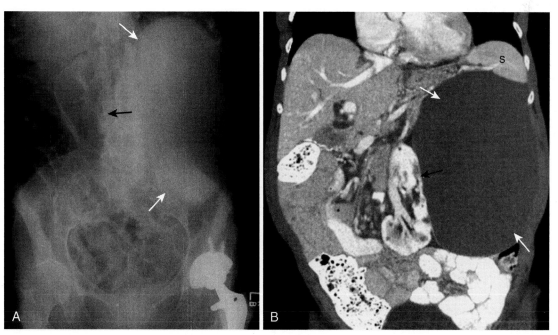

FIGURE 14-24 Enlarged kidney. Soft tissue masses or organomegaly can be diagnosed on the basis of a conventional radiograph, either by visualizing the edge of the mass if there is fat or air surrounding it or by displacement of bowel. **A,** On this conventional radiograph, there is a soft tissue mass in the left upper quadrant *(white arrows)* that is displacing bowel to the right *(black arrow).* **B,** A coronal reformatted CT scan of the same patient demonstrates a large renal cyst *(white arrows)* arising from the left kidney *(black arrow),* displacing it and surrounding bowel. The spleen (S) is being compressed by the cyst.

long axis parallel to the hips and the **base of the bladder just above the top of the symphysis pubis.**

♦ The urinary bladder is about the size of a small cantaloupe when distended and about the size of a lemon when contracted (Fig. 14-25).

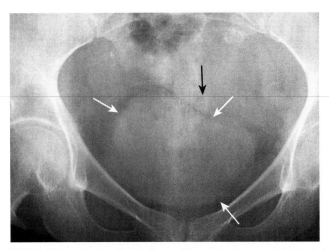

FIGURE 14-25 Normal urinary bladder. Close-up of the pelvis demonstrates enough perivesical fat present to make the outline of the urinary bladder visible *(white arrows)*. In men, the sigmoid colon usually occupies the space just above the bladder *(black arrow)*. In women, the soft tissue above the bladder may be either the uterus or the sigmoid colon.

■ **Enlarged urinary bladder**

♦ Bladder enlargement is usually recognizable by displacement of the bowel upward and out of the pelvis by a soft tissue mass. Bladder outlet obstruction is much more common in men due to enlargement of the prostate, so that a pelvic soft tissue mass is more likely to be a dilated bladder in a man than in a woman (Fig. 14-26).

Uterus

■ **Normal**

♦ The uterus usually sits atop the dome of the bladder. There is frequently a lucency produced by fat between the top of the bladder and the bottom the uterus. The **normal uterus** is about **8 cm × 4 cm × 6 cm.**

■ **Enlarged uterus**

♦ US is the **best** tool for **evaluating the size of the uterus** and **ovaries.**

♦ **Occasionally, uterine enlargement,** when marked, **may be visible** on **conventional radiographs.** The key to differentiating an enlarged uterus from a distended bladder is **identification of the lucency between the bladder** and **the uterus.** When the uterus is enlarged, the fat plane will be present; when the bladder is dilated, the fat plane will not be visible (see Fig. 14-26).

Psoas Muscles

■ One or both of the psoas muscles may be visible if there is adequate extraperitoneal fat surrounding them. Inability to visualize one or both psoas muscles is not a reliable indicator of retroperitoneal disease (see Fig. 14-1).

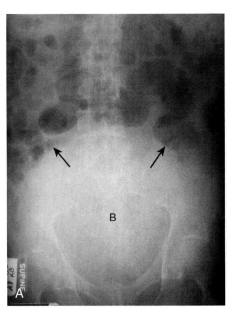
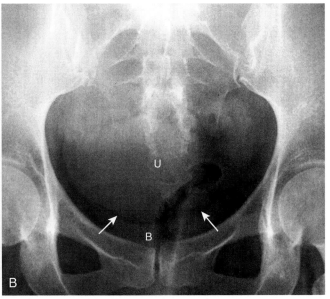

FIGURE 14-26 Distended urinary bladder and enlarged uterus. A, The distended bladder (labeled *B*) is a soft tissue mass that ascends from the pelvis into the lower abdomen, displacing the bowel into the mid abdomen *(black arrows).* This image was obtained from a 72-year-old man with bladder outlet obstruction due to benign prostatic hypertrophy. **B,** The uterus (labeled *U*) is slightly enlarged. It can be distinguished from the bladder because there is a fat plane *(white arrows)* between it and the urinary bladder (labeled *B*) below it.

TAKE-HOME POINTS

RECOGNIZING THE NORMAL ABDOMEN: CONVENTIONAL RADIOGRAPHS

Evaluation of the abdomen should be focused on four main areas: the gas pattern, free air, soft tissue masses or organomegaly, and abnormal calcifications.

There is normally air present in the stomach and colon, especially in the rectosigmoid colon, and a small amount of air (two or three loops) may normally be seen in the small bowel.

There is normally an air–fluid level in the stomach. There may be two or three air–fluid levels in nondilated small bowel, but there is usually no fluid visible in the colon.

An acute abdominal series may consist of supine abdomen, prone abdomen (or its substitute, which is a lateral rectal view), upright abdomen (or its substitute, which is a left lateral decubitus view), and upright chest (or its substitute, which is supine chest).

The supine view of the abdomen is the general scout view for the bowel gas pattern and is useful for identifying calcifications and detecting organomegaly or soft tissue masses.

The prone view allows air, if present, to be seen in the rectosigmoid colon, which is important in the evaluation of mechanical obstruction of the bowel.

The upright abdomen may demonstrate air–fluid levels in the bowel or peritoneal free air.

The upright chest radiograph may demonstrate free air beneath the diaphragm, pleural effusion (which may provide a clue as to the presence and the nature of intraabdominal pathology), or pneumonia (which can mimic an acute condition in the abdomen).

CT, US, and MRI have essentially replaced conventional radiography in the assessment of organomegaly and soft tissue masses.

WEBLINK

Visit StudentConsult.Inkling.com for more information and a quiz on Plain Films of the Abdomen.

For your convenience, the following QR Code may be used to link to **StudentConsult.com**. You must register this title using the PIN code on the inside front cover of the text to access online content.

CHAPTER 15
Recognizing the Normal Abdomen and Pelvis on Computed Tomography

INTRODUCTION TO ABDOMINAL AND PELVIC COMPUTED TOMOGRAPHY

- It is estimated that around 80 million computed tomography (CT) scans of all kinds are performed in the United States each year. Almost 10% of all visits to the emergency department are for abdominal pain of a nontraumatic nature. Many of those patients undergo a CT scan of the abdomen and pelvis, performed to detect or clarify their clinical findings.
- As with all imaging studies, one of the guiding principles in CT scanning of the abdomen and pelvis is to maximize the differences in density between tissues to best demonstrate their unique anatomy. Toward that end, CT studies make extensive use of **intravenous** and **oral contrast** agents.

Intravenous Contrast in CT Scanning

- CT scans **can be performed with** and/or **without the intravenous administration of iodinated contrast** material, but **in general, they yield more diagnostic information** that is more easily recognizable **when intravenous contrast can be used.**
- CT scans done **with intravenous contrast** are called *contrast-enhanced* or simply *enhanced*. Typically, the radiologist will choose the scanning parameters to optimize the CT study for the patient's particular clinical issues. For example, different rates of contrast administration and timing of the scan will allow diagnostic enhancement of hepatic vessels versus the liver parenchyma.
- Although it might sound like a wonderful idea to give everyone contrast, keep in mind that **iodinated contrast** can have **adverse effects** and produce **serious reactions** in **susceptible individuals** (Box 15-1).

Oral Contrast in CT Scanning

- For abdominal and pelvic CT imaging, **oral contrast** may also be administered to define the bowel, although its use has diminished as the quality of CT images has improved. Oral contrast is usually **not employed in chest CT scanning** unless there is a particular question concerning the esophagus.
- Orally administered contrast, frequently given in temporally divided doses to allow earlier contrast to reach the colon while later contrast opacifies the stomach, is utilized for many abdominal CT scans, except those performed for **trauma,** the **stone search study,** and studies specifically directed toward evaluating vascular structures like the **aorta.**

BOX 15-1 CONTRAST REACTIONS AND RENAL FAILURE

Intravenous contrast materials available today are nonionic, low-osmolar solutions containing a high concentration of iodine, which circulate through the bloodstream, opacify those tissues and organs with high blood flow, are absorbed by x-ray (and therefore appear *"whiter"* on images), and are finally excreted in the urine by the kidneys.

In some patients (e.g., those with diabetes, dehydration, multiple myeloma) who have compromised renal function evidenced by creatinine >1.5, iodinated contrast can produce a nephrotoxic effect, resulting in acute tubular necrosis. Though usually reversible, in a small number of patients with underlying renal insufficiency, renal dysfunction may permanently worsen. This effect is dose-related.

Iodinated contrast agents can sometimes produce mild side effects, including a feeling of warmth, nausea and vomiting, local irritation at the site of injection, itching, and hives; these side effects usually require no treatment. Occasional idiosyncratic, allergic-like reactions may also include itching, hives, and laryngeal irritation.

Asthmatics and those with a history of severe allergies or prior reactions to IV contrast have a higher likelihood of contrast reactions (but still very low overall) and may benefit from steroids, diphenhydramine (Benadryl), and cimetidine administered before and/or after injection. Prior shellfish allergy bears absolutely NO relationship to iodinated contrast reactions.

In about 0.01% to 0.04% of all patients, severe and idiosyncratic reactions to contrast can occur that can produce intense bronchospasm, laryngeal edema, circulatory collapse and, very rarely, death (1 in 200,000 to 300,000).

- One of two different types of oral contrast may be used. The most widely used is a dilute solution of **barium sulfate,** the same contrast agent employed in upper gastrointestinal studies and barium enemas. If there is concern for bowel perforation and the possibility that contrast may exit from the lumen of the bowel, an iodine-based, water-soluble contrast is sometimes used **(Gastrografin).** Contrast may also be introduced rectally to opacify the colon more quickly than it would take for orally administered contrast to reach the large bowel or through a Foley catheter to quickly opacify the urinary bladder.
- You will probably not be required to make the decision of when or if to use contrast; **the radiologist will usually tailor the examination to best answer the clinical question being asked,** so that it is always important to provide

TABLE 15-1 CT SCANS: WHEN CONTRAST IS USED

IV Contrast Used	IV Contrast Usually Not Used
CHEST	
CT–pulmonary angiogram (CT-PA) for pulmonary embolism	Evaluation of diffuse infiltrative lung diseases using HRCT
Evaluation of the mediastinum or hila for mass or adenopathy	Confirmation of the presence of a nodule suspected from conventional radiographs
Detection of aortic aneurysm or dissection	Detection of pneumothorax/ pneumomediastinum
Evaluation for blunt or penetrating trauma	Calcium scoring for the coronary arteries
Characterization for pleural disease (metastases, empyema)	Known allergies to contrast or renal failure
CT densitometry of pulmonary masses Evaluation for the coronary arteries	
ABDOMEN AND PELVIS	
IV Contrast Used	*IV Contrast Usually Not Used*
Evaluation for the presence of and/or characterization of a mass; staging of or follow-up on malignancies	CT colonography
Trauma	Search for a ureteral calculus
Abdominal pain (e.g., appendicitis)	
Detection of aortic aneurysm or dissection	
WHEN ORAL CONTRAST IS USED	
Most cases of nontraumatic abdominal pain	
Inflammatory bowel disease	
Abdominal or pelvic abscess	
Locating of the site of bowel perforation, including fistulae	

CT, Computed tomography; *HRCT,* high resolution computed tomography.

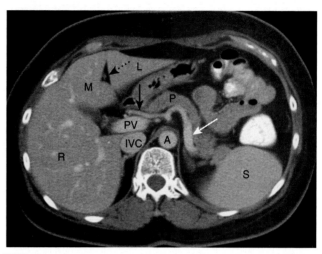

FIGURE 15-1 Normal liver anatomy. The ligamentum teres *(dotted black arrow)* divides the left lobe of the liver into a medial (M) and lateral (L) segment with the larger right (R) lobe lying more posterior. The portal vein (PV) lies just posterior to the hepatic artery *(solid black arrow).* The splenic artery *(solid white arrow)* follows the path of the pancreas (P) towards the spleen (S). The inferior vena cava (IVC) lies to the right of the aorta (A).

as much clinical information as possible when requesting a study.

- Table 15-1 summarizes, in general, **when intravenous and oral contrast are utilized** for particular problems.
- Table 15-2 outlines some of the common **patient preparations** that are generally suggested for a variety of imaging studies. Preparation instructions may vary depending on the facility and individual patient needs. In general, a patient is permitted to take medications with a small sip of water, even when instructions specify nothing to eat or drink before the study.

ABDOMINAL CT: GENERAL CONSIDERATIONS

- Conventional radiography, ultrasonography (US), CT, and magnetic resonance imaging (MRI) are all utilized in the imaging evaluation of abdominal abnormalities.
- Each has advantages and disadvantages inherent to its own particular technology, and the choice of modality is frequently based on the patient's clinical condition (Table 15-3).
- Although advances in CT imaging have resulted in better diagnostic studies, they may have come with an unintended consequence—potentially higher **radiation dose.**

The radiation dose delivered by CT studies is dependent on many factors, including the type of equipment, the energy of the x-rays used to produce the images, and the size of the patient. Dose-reducing measures are being employed, including the use of optimized CT settings, reduction in the x-ray energy used, limiting the number of repeat scans, and assuring, through appropriate consultation, that the **benefits** derived from obtaining the study **outweigh** any potential **risks** of the radiation exposure.

⮕ **History** and **physical examination continue to be an essential part of evaluating abdominal abnormalities** not only to suggest a cause but in helping to determine which, if any, imaging study (studies) will provide the best yield in establishing the correct diagnosis.

Liver

- By convention, **CT scans,** like most other radiologic studies, **are viewed** with the **patient's right on your left** and **the patient's left on your right.** If the patient is scanned in the supine position, as most usually are, the **top** of each image is **anterior** and the **bottom** of each image is **posterior.**
- The liver receives its blood **supply** from both **hepatic arteries** and **portal veins** and **drains** to the inferior vena cava via the **hepatic veins.** The normal parenchyma is supplied 80% by the portal vein and 20% by the hepatic artery, so it will enhance in the portal venous phase.
- For practical purposes, the **vascular distribution of the liver defines its anatomy** since the vascular anatomy is what directs the surgical approach to liver lesions.
- The liver is divided into **right, left, and caudate lobes** by its vessels.
  - ◆ The **right lobe** is subdivided into **two segments:** the **anterior** and **posterior.** The **left lobe** is also subdivided into **two segments:** the **medial and lateral.**
  - ◆ A prominent, fat-filled fissure that contains the **falciform ligament** and **ligamentum teres** (formerly, the umbilical vein) separates the **medial and lateral segments** of the **left lobe** of the liver (Fig. 15-1).

TABLE 15-2 INSTRUCTIONS FOR IMAGING STUDIES

Study	Before the Order Is Placed	Before the Start of the Study	After the Study Is Done
CT Head CT with or without contrast	Solicit history of contrast reaction in the past	Serum creatinine may be needed before contrast injections	Nothing
Body CT without contrast	Nothing	No preparation needed	Nothing
Body CT with oral contrast and/or IV contrast	Solicit history of contrast reaction in the past	Serum creatinine may be needed before contrast injections; oral contrast given just before study	Nothing
US Upper abdomen, general survey study: aorta, gall-bladder, inferior vena cava, liver, pancreas, renal stenosis, retroperitoneal, spleen	Nothing	Nothing to eat or drink for several hours before exam	Nothing
Renal or kidney	Nothing	Patient may be asked to drink a prescribed amount of water to distend bladder 1-2 h before procedure; patient should not empty bladder	Nothing
Male or female pelvis or lower abdomen; obstetrical/gynecologic US	Nothing	Patient may be asked to drink a prescribed amount of water to distend bladder 1-2 h before procedure; patient should not empty bladder	Nothing
Renal transplant; thyroid and vascular studies	Nothing	No preparation needed	Nothing
MRI Without contrast	Solicit history of working with metal, grinding, welding, or possible metal in eyes; patient may need an orbital x-ray; solicit history of pacemaker, aneurysm clips, neural stimulators, IUD, permanent makeup, cochlear implants, artificial heart valves, pregnancy, metallic fragments; claustrophobia	No preparation needed	Nothing
With contrast	Solicit history of working with metal, grinding, welding, or possible metal in eyes; patient may need an orbital x-ray; solicit history of pacemaker, aneurysm clips, neural stimulators, IUD, permanent makeup; cochlear implants, artificial heart valves, pregnancy, metallic fragments; claustrophobia	Serum creatinine may be needed before contrast injections or in patients with renal insufficiency	Nothing
BARIUM STUDY Esophagram	Nothing	No preparation needed	Nothing
Upper gastrointestinal series/small bowel series	Nothing	Nothing to eat or drink for several hours before study	Nothing
Barium enema; virtual colonography	Nothing	Bowel prep to cleanse colon before study may consist of oral laxatives, suppositories, fluids	Mild laxative if desired
MAMMOGRAPHY Mammogram	Nothing	Patient should not use any deodorant, perfume, powder, ointment, or any other skin products on chest, breast, or under arms on the day of appointment	Nothing
NUCLEAR MEDICINE Bone density	Is the patient pregnant?	No contrast or barium studies for 48 h before the procedure; no food restriction	Nothing
Bone scan	Is the patient pregnant?	No food restriction.	Nothing
Cardiac treadmill and adenosine stress test	Is the patient pregnant?	Nothing to eat or drink for several hours before the exam; no caffeine several hours prior to exam	Nothing

CT, Computed tomography; *IUD,* intrauterine device; *IV,* intravenous; *MRI,* magnetic resonance imaging; *US,* ultrasonography.

TABLE 15-3   IMAGING OF THE ABDOMEN AND PELVIS MODALITIES COMPARED

Uses	Advantages	Disadvantages
CONVENTIONAL RADIOGRAPHY		
Primarily used for screening in abdominal pain	Availability Cost Patients tolerate procedure well	Lower sensitivity Ionizing radiation
ULTRASONOGRAPHY		
Primary imaging mode for gallbladder and biliary tree; screening for aortic aneurysm; identification of vascular abnormalities and flow; detection of ascites; primary imaging mode for the female pelvis	Availability Cost No ionizing radiation Patients tolerate procedure well Portable	Operator dependent More difficult to interpret
CT		
Diagnostic modality of choice for most abdominal abnormalities, including trauma	Availability Cost High spatial resolution and image reconstruction Evaluates multiple organ systems simultaneously	Cost Ionizing radiation Contrast reactions Inability to use IV contrast in renal insufficiency Patient weight and size may affect scan results
MRI		
Problem solving for difficult diagnoses; extension of known disease into surrounding soft tissues (staging); vascular anatomy	Soft tissue contrast No ionizing radiation No iodinated contrast Image reconstruction	Cost Availability Longer scan times Claustrophobia Monitoring issues in acutely ill patients Patient weight and size may affect scan results Incompatible with aneurysm clips, pacemakers, etc.

CT, Computed tomography; *MRI,* magnetic resonance imaging.

- The outer surface of the liver is normally smooth. The normal liver usually appears homogeneous in density on CT, and its attenuation should always be **denser than** or **equal to the density of the spleen on noncontrast scans.**
- The adult liver usually measures 15 cm or less on coronal scans at its maximum height. Care must be taken not to measure the length at the site of a Riedel lobe of the liver, a normal inferior projection from the right lobe seen mainly in females (see Fig. 14-20). The liver's greatest transverse measurement is 20 to 26 cm.
- The diaphragmatic surface of the liver is affixed by connective tissue to a triangular section of the undersurface of the diaphragm, termed the **bare area.** This will have importance later in differentiating ascites from pleural effusion (see Chapter 20) (Fig. 15-2).
- **Liver volume** can be calculated using CT (as well as US and MRI). The volume of the liver will vary by gender and by patient weight. The adult liver volume is about 1500 cm³. Liver volume determinations may be used for liver resection, transplantation, and in evaluating the progression of various diseases, such as alcohol-related liver disease.

Spleen

 On early contrast-enhanced scans, the spleen may be inhomogeneous in its attenuation, a finding that should disappear over the course of the next several minutes.

- The normal adult spleen may have lobulations, not to be confused with lacerations. The splenic artery and vein enter and exit at the hilum of the spleen.
- The spleen is usually **about 12 cm long, does not project** substantially below the margin of the **12th rib,** and is about the **same size as the left kidney.**

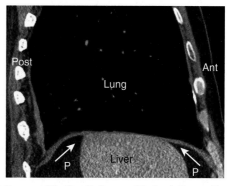

FIGURE 15-2   Bare area of the liver. The bare area of the liver *(white arrows)* has no peritoneal covering but is affixed directly to the undersurface of the diaphragm. As such, it will be impossible for ascitic fluid in the peritoneal cavity (P) to insert itself between the liver and the lung in this area, which will be important for differentiating pleural effusion from ascites (see Chapter 20). *Ant,* Anterior; *Post,* posterior.

Pancreas

- The pancreas is a retroperitoneal organ oriented obliquely so that the entire organ is not seen on any one axial image of the upper abdomen.
  - The **tail** is usually **most superior,** lying in the hilum of the spleen.
  - Proceeding inferiorly, the **body of the pancreas** crosses the midline and rests anterior to the **superior mesenteric artery.** The **head of the pancreas** is nestled in the duodenal loop (Fig. 15-3). The **uncinate process** is part of the head and curves around the **superior mesenteric vein.**
- The **splenic vein** courses along the posterior border of the pancreas to the **superior mesenteric vein,** and the **splenic**

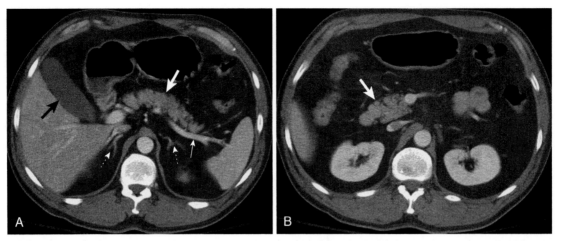

FIGURE 15-3 Normal pancreas. A, Body of pancreas *(thick white arrow)* and splenic artery *(thin white arrow)*. Additionally well visualized are both adrenal glands *(dotted white arrows)* and gallbladder *(black arrow)*. **B,** Normal head of pancreas *(solid white arrow)*. Because the pancreas is oriented obliquely, the entire organ is not seen on any one axial image of the upper abdomen. The tail is most superior and the body and then head are usually visualized on successively more inferior slices.

artery runs along the superior border of the pancreas from the **celiac axis** to the spleen. The main pancreatic duct empties into the duodenum as the **duct of Wirsung** and sometimes through an **accessory duct of Santorini.** The pancreatic duct may be visible and measures 3 to 4 mm in diameter.

- The **head** of the pancreas measures **3 cm** in maximum dimension, the **body 2.5 cm,** and the **tail 2 cm.** The gland is about **12 to 15 cm in length.** As a person ages, the gland may undergo fatty infiltration, giving it a "feathery" appearance.

Kidneys

- The kidneys are retroperitoneal organs, encircled by varying amounts of fat and enclosed within a fibrous capsule.
- They are surrounded by the **perirenal space,** which in turn, is delimited by the **anterior** and **posterior renal fasciae.** Certain fascial attachments, muscles, and other organs define a series of spaces that produce predictable patterns of abnormality when those spaces are filled with fluid, pus, blood, or air.
- In adults, the **left kidney** is **minimally larger** than the **right,** each kidney being about **11 cm** in size, or about the **same size** as the **spleen.**
- The **right renal artery** passes **posterior** to the inferior vena cava (IVC). The renal vein**s** lie **anterior to the arteries;** the longer left renal vein passes anterior to the aorta before draining into the inferior vena cava (Fig. 15-4).
- The **renal hilum** contains the renal pelvis and the renal artery and vein. The upper poles of the kidneys are more posterior than the lower poles and are tilted inward toward the spine. Although dependent on the timing of the scan and the patient's renal function, the outer **cortex** is frequently brighter and more homogeneous than the inner **medulla** (Fig. 15-5).
- As long as they are functioning properly, the **kidneys are the major route** for excretion of iodinated **contrast** material. They should therefore **enhance** whenever intravenous contrast is administered. Over time, the urine will become more opacified, increasing significantly from its normal water density, sometimes requiring delayed imaging of the urinary tract.

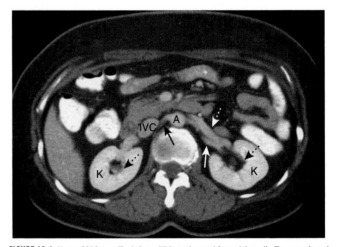

FIGURE 15-4 Normal kidneys. The kidneys (K) lie in the renal fossae bilaterally. The normal renal pelvis, containing fat, occupies the central portion of the kidneys *(dotted black arrows)*. The right renal artery *(solid black arrow)* runs posterior to the inferior vena cava (IVC). The left renal vein *(dotted white arrow)* here lies anterior to the left renal artery *(solid white arrow)*. *A,* Abdominal aorta.

- If the kidneys are not functioning properly, contrast is excreted through alternative pathways (bile, bowel), a process called *vicarious excretion* of contrast.

Small and Large Bowel

- **Opacification** and **distension** of the bowel lumen are helpful for proper evaluation of the bowel wall regardless of the modality used to evaluate it. Wall thickness may artificially seem increased in collapsed bowel.
- The **small bowel** is usually **2.5 cm** or less in **diameter** with a **wall thickness** usually less than **3 mm.** Adjacent loops of small bowel are usually in contact with each other, depending in part on the amount of intraperitoneal fat present.
- The **colonic wall** is usually **less than 3 mm** thick with the colon distended and less than 5 mm with the colon collapsed. The cecum is recognizable by the presence of the terminal ileum. The positions of the transverse and sigmoid segments of the colon are variable, depending on their degree of redundancy (Fig. 15-6).

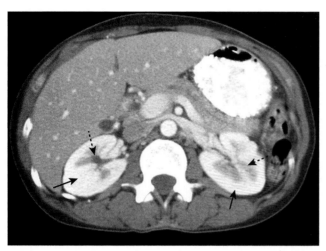

FIGURE 15-5 Renal cortex and medulla, computed tomography. The kidneys are the primary organ by which iodinated contrast is excreted. The kidneys will appear slightly different depending on the time the scan is performed following the injection. About 70 to 100 seconds after the injection, a **nephrographic (nephrogram) phase,** such as this, will demonstrate the corticomedullary junction between the outer, brighter cortex *(solid black arrows)* and the inner, less-dense medulla *(dotted black arrows).*

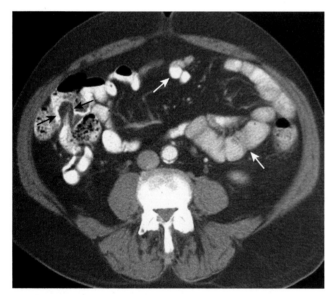

FIGURE 15-6 Normal bowel. Contrast fills a nondilated lumen (less than 2.5 cm). The small bowel wall is so thin as to normally appear almost invisible *(white arrows).* The terminal ileum can be recognized by the fat-containing "lips" of the ileocecal valve outlined with contrast *(black arrows).*

Urinary Bladder

- The bladder is an **extraperitoneal organ**, the extraperitoneal space being continuous with the retroperitoneum. The **dome of the bladder** is covered by the **inferior reflection of the peritoneum.**
- The bladder's location in males is superior to the prostate gland and anterior to the rectum. In females, the bladder is anterior to the vagina and anteroinferior to the uterus. The ureters enter the posterolateral aspect of the bladder at the trigone.
- The **bladder wall measures 5 mm** or **less with the bladder distended.**

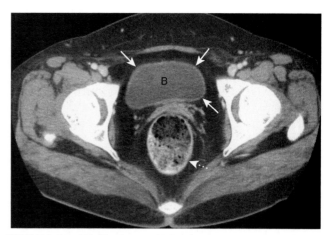

FIGURE 15-7 Normal bladder. The urinary bladder (B) contains unopacified urine in this early image of a contrast-enhanced computed tomography scan of the pelvis. The bladder wall *(solid white arrows)* is thin and of equal thickness around the circumference of the bladder. The rectum lies posterior to the bladder *(dotted white arrow).*

- The bladder is **best evaluated when distended** with either urine or urine-containing contrast, but the **bladder wall is usually visible** whether or not intravenous contrast has been administered (Fig. 15-7).

TAKE-HOME POINTS

RECOGNIZING THE NORMAL ABDOMEN AND PELVIS ON COMPUTED TOMOGRAPHY

CT studies make extensive use of intravenous and oral contrast agents to maximize the differences in density between structures so as to best demonstrate their anatomy.

In general, CT scans yield more diagnostic information that is more easily recognizable when IV contrast can be used.

Iodinated contrast agents may occasionally produce side effects such as warmth or nausea and vomiting; rarely, idiosyncratic allergic-like reactions have led to anaphylaxis and death.

Oral and/or rectal contrast may be used to define the bowel and help in differentiating bowel from adjacent lymph nodes or pathologic, fluid-containing lesions.

There are a number of clinical settings in which either iodinated and/or oral contrast improve the diagnostic accuracy of CT scanning. Imaging studies are usually tailored by the radiologist based on the clinical problem.

Advances in CT imaging have resulted in better diagnostic studies, but they may carry the potential for a higher radiation dose. Dose-reducing measures include the use of optimized CT settings, reduction in the x-ray energy used, limiting the number of repeat scans, and assuring that the benefits derived from obtaining a study outweigh any potential risks of the radiation exposure.

It is essential to provide an appropriate history to help in determining which, if any, imaging study (studies) will yield the best information in establishing the correct diagnosis.

Some of the common patient preparations that are generally suggested for a variety of imaging studies are outlined.

The normal CT appearance of the liver, spleen, pancreas, kidneys, bowel, and bladder is described.

WEBLINK

Visit StudentConsult.Inkling.com for more information.

For your convenience, the following QR Code may be used to link to **StudentConsult.com**. You must register this title using the PIN code on the inside front cover of the text to access online content.

CHAPTER 16
Recognizing Bowel Obstruction and Ileus

- In Chapters 14 and 15 we discussed how to recognize the normal intestinal gas pattern on conventional radiographs and computed tomography (CT). In this chapter you'll learn how to recognize and categorize the four most common abnormal bowel gas patterns and their causes. These patterns of abnormal bowel gas will appear the same, whether imaged initially by conventional radiography or by CT scanning. CT is superior in revealing the location, degree, and cause of an obstruction and in demonstrating any signs of reduced bowel viability.
- Abnormalities of bowel function are suspected by the history and clinical findings.
- The key questions in assessing the bowel gas pattern on imaging studies are:
  - Are there **dilated loops of small** and/or **large bowel?**
  - On CT, is there a **transition point?**
  - On plain films, is there **air in the rectum** or **sigmoid?**

ABNORMAL GAS PATTERNS

- Abnormal intestinal gas patterns can be divided into two main categories, each of which can be subdivided into two subcategories (Box 16-1).
- **Functional ileus** is one main category in which it is presumed that **one** or **more loops of bowel lose their ability to propagate the peristaltic waves of the bowel,** usually due to some **local irritation** or **inflammation,** and hence cause a **functional type of "obstruction"** proximal to the affected loop(s).
- There are **two kinds** of **functional ileus.**
  - **Localized ileus affects only one** or **two loops** of (usually **small**) **bowel** (also called *sentinel loops*).
  - **Generalized adynamic ileus** affects **all loops of large** and **small bowel** and frequently the stomach.

BOX 16-1   ABNORMAL BOWEL GAS PATTERNS
FUNCTIONAL ILEUS
Localized ileus (sentinel loops)

Generalized adynamic ileus

MECHANICAL OBSTRUCTION
Small bowel obstruction (SBO)

Large bowel obstruction (LBO)

- **Mechanical obstruction** is the other main category of abnormal bowel gas pattern.
- With mechanical obstruction, a **physical, organic, obstructing lesion prevents the passage of intestinal content** past the point of either the small or large bowel blockage.
- There are **two kinds** of **mechanical obstruction.**
  - **Small bowel obstruction** (SBO)
  - **Large bowel obstruction** (LBO)

LAWS OF THE GUT

- The bowel reacts to a mechanical obstruction in more or less predictable ways.
- After the obstruction occurs, **peristalsis continues** (except in the loops of bowel involved in a functional ileus) in an attempt to propel intestinal contents through the bowel.
- Loops **proximal** to the **obstruction** soon become **dilated with air** and/or **fluid.**
  - This can occur within a few hours of a complete small bowel obstruction.
- Loops **distal** to an **obstruction** will eventually become **decompressed** or **airless** as their contents are evacuated.
- In a mechanical obstruction, **the loop(s) that will become the most dilated** will either be the **loop of bowel with the largest resting diameter before the onset of the obstruction** (e.g., the **cecum** in the large bowel) or the **loop(s) of bowel just proximal to the obstruction.** The **transition point,** the site of obstruction and the location where the bowel changes in caliber from dilated to collapsed, will frequently be visible on CT.
- Most patients with a mechanical obstruction will present with some form of **abdominal pain, abdominal distension,** and **constipation.** Patients may present with **vomiting early** in the course of a **proximal small bowel obstruction** and **later** in the course of the illness with **a distal small bowel obstruction.**
- **Prolonged obstruction** with persistently elevated intraluminal pressures can lead to vascular compromise, **necrosis,** and **perforation** in the affected loop of bowel. At this point, bowel sounds may become hypoactive or absent.
- Let's look at each of the four abnormal bowel gas patterns in detail (Table 16-1).
  - For each of the four abnormalities, we will look at their **pathophysiology, causes, key imaging features,** and **diagnostic pitfalls.**

TABLE 16-1   ABNORMAL GAS PATTERNS—SUMMARY

	Air in Rectum or Sigmoid	Air in Small Bowel	Air in Large Bowel
Normal	Yes	Yes—1-2 loops	Rectum and/or sigmoid
Localized ileus	Yes	2-3 dilated loops	Rectum and/or sigmoid
Generalized ileus	Yes	Multiple dilated loops	Yes—dilated
SBO	No	Multiple dilated loops	No
LBO	No	None—unless ileocecal valve incompetent	Yes—dilated

LBO, Large bowel obstruction; *SBO,* small bowel obstruction.

TABLE 16-2   CAUSES OF A LOCALIZED ILEUS

Site of Dilated Loops	Cause(s)
Right upper quadrant	Cholecystitis
Left upper quadrant	Pancreatitis
Right lower quadrant	Appendicitis
Left lower quadrant	Diverticulitis
Midabdomen	Ulcer or kidney/ureteral calculus

FUNCTIONAL ILEUS: LOCALIZED SENTINEL LOOPS

- **Pathophysiology**
  - ◆ **Focal irritation** of a loop or loops of bowel **occurs most often from inflammation of an adjacent visceral organ;** for example, pancreatitis may affect bowel loops in the left upper quadrant, diverticulitis in the left lower quadrant.
  - ◆ The **loop(s) affected** are almost always loops of **small bowel,** and because they **herald the presence of underlying pathology,** they are called *sentinel loops.*
  - ◆ The **irritation causes these loops to lose their normal function** and **become aperistaltic,** which in turn, leads to **dilatation** of these loops.
  - ◆ Because a **functional ileus does not produce the degree of obstruction that a mechanical obstruction does,** some gas continues to pass through the defunctionalized bowel past the point of the localized ileus.
  - ◆ **Air** usually reaches and **is visible in the rectum** or **sigmoid** on conventional radiographs.
- **Causes of a localized ileus**
  - ◆ The **dilated loops of bowel tend to occur** in the **same anatomic area** as the **inflammatory** or **irritative process** of the adjacent abdominal organ, although this may not always be the case.
  - ◆ On conventional radiographs, the exact cause of the functional ileus can only be inferred by the location of the ileus. The cause of the localized ileus is frequently visible on CT scans of the abdomen.
  - ◆ Table 16-2 summarizes **sites of a localized ileus** and their **most common cause.**

- **Key imaging features of a localized ileus**
  - ◆ On conventional radiographs, there are **one** or **two** *persistently dilated* loops of small bowel.
    - • *Persistently* means that these **same loops remain dilated** on **multiple views** of the abdomen (supine, prone, upright abdomen) or on **serial studies** done over the course of time.
    - • *Dilated* means the small bowel loops are **persistently larger than 2.5 cm.** Small bowel loops involved in a functional ileus **usually do not dilate as greatly as those that are mechanically obstructed.**
    - • Infrequently, the sentinel loop may be **large bowel,** rather than small bowel. This can especially occur in the **cecum,** with diseases such as appendicitis.
  - ◆ There are **frequently air–fluid levels** seen in **sentinel loops.**
  - ◆ There is usually **gas in the rectum** or **sigmoid** in a localized ileus (Fig. 16-1).

 Pitfalls—Differentiating a localized ileus from an early SBO

  - ◆ A **localized ileus may resemble an** *early* **mechanical SBO;** that is, there may be a few dilated loops of small bowel with air in the colon in both. *Early* means the **patient has had symptoms for a day or two.** Patients who have had obstructive symptoms for a week or more usually no longer demonstrate imaging findings of an "early" obstruction.
  - ◆ **Solution.** A combination of the clinical and laboratory findings and CT scanning of the abdomen, which demonstrates the underlying pathology, should differentiate localized ileus from small bowel obstruction (Fig. 16-2).

FUNCTIONAL ILEUS: GENERALIZED ADYNAMIC ILEUS

- **Pathophysiology**
  - ◆ In a generalized adynamic ileus, the **entire bowel is aperistaltic or hypoperistaltic.** Swallowed **air dilates,** and **fluid fills most loops of both small and large bowel.**
  - ◆ A generalized adynamic ileus is almost always the **result of abdominal or pelvic surgery,** in which the bowel is manipulated during the surgery.
- **Causes of a generalized adynamic ileus** are summarized in Table 16-3.
- **Key imaging features of a generalized adynamic ileus**
  - ◆ The **entire bowel is usually air-containing and dilated;** this includes both large and small bowel. The stomach may be dilated as well.
  - ◆ The absence of peristalsis and the continued production of intestinal secretions usually produces **many long air–fluid levels in the bowel.**
  - ◆ Since this is not a mechanical obstruction, there should be **gas in the rectum** or **sigmoid.** There is **no transition point** identified on CT of the abdomen.
  - ◆ **Bowel sounds are frequently absent** or **hypoactive** (Fig. 16-3).

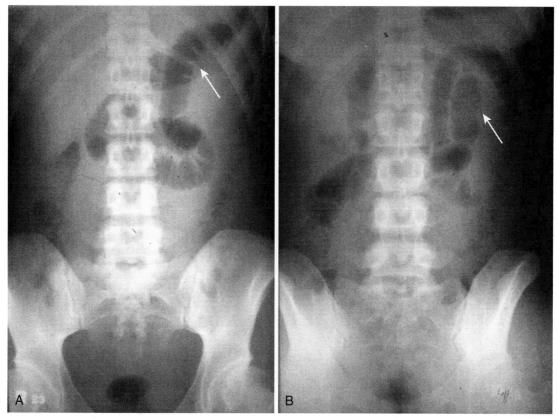

FIGURE 16-1 Sentinel loops from pancreatitis. A single, persistently dilated loop of small bowel is seen in the left upper quadrant *(white arrows)* on both the supine **(A)** and prone **(B)** radiographs of the abdomen representing a **sentinel loop** or localized ileus. A localized ileus is called a *sentinel loop* because it often signals the presence of an adjacent irritative or inflammatory process. This patient had acute pancreatitis.

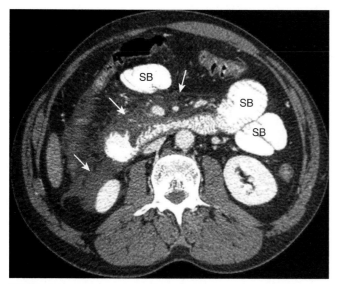

FIGURE 16-2 Pancreatitis producing focal bowel dilatation. Although the cause of a sentinel loop can usually only be inferred from conventional radiographs, computed tomography (CT) scans can depict the underlying abnormality producing the bowel irritation. In this contrast-enhanced, axial CT scan of the upper abdomen with oral contrast, the pancreas is inflamed, enlarged, and edematous *(white arrows)*, and there is infiltration of the peripancreatic fat. This can affect peristalsis in adjacent loops of small bowel (SB) and lead to dilatation of the loops.

 Recognizing a generalized adynamic ileus

◆ Patients do not present to the emergency department with a generalized adynamic ileus unless they are 1 or 2

TABLE 16-3 CAUSES OF A GENERALIZED ADYNAMIC ILEUS

Cause	Remarks
Postoperative	Usually abdominal surgery
Electrolyte imbalance	Especially diabetics in ketoacidosis

days postoperative (abdominal or gynecologic surgery) or they have a severe electrolyte imbalance (e.g., hypokalemia).

◆ Many patients who have either **intestinal pseudoobstruction** (see the end of this chapter) or **aerophagia** can be mistakenly identified as having a generalized ileus on abdominal radiographs.

MECHANICAL OBSTRUCTION: SMALL BOWEL OBSTRUCTION

■ **Pathophysiology**
   ◆ A lesion, either inside or outside of the small bowel, obstructs the lumen.

 Over time, **from the point of obstruction backward,** the **small bowel dilates** from continuously swallowed air and from intestinal fluid, which is still produced by the stomach, pancreatic, biliary systems, and small bowel.

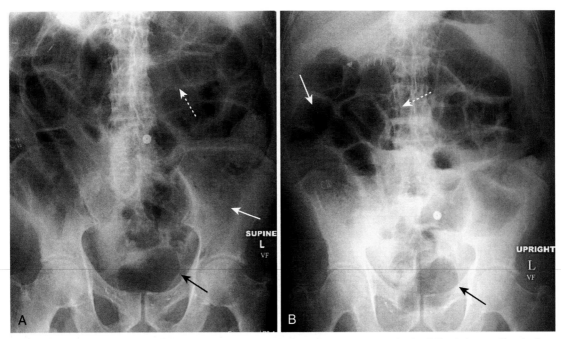

FIGURE 16-3 Generalized adynamic ileus, supine (A) and upright abdomen (B). There are dilated loops of large *(solid white arrows)* and small *(dotted white arrows)* bowel with gas seen down to and including the rectum *(solid black arrows)*. The patient had absent bowel sounds and had undergone colon surgery the day before.

TABLE 16-4 CAUSES OF A MECHANICAL SMALL BOWEL OBSTRUCTION

Cause	Remarks
Postsurgical adhesions	Most common cause of a small bowel obstruction; most frequent following appendectomy, colorectal surgery, and pelvic surgery; **transition point** on small bowel CT without other identifying cause
Malignancy	Primary malignancies of the small bowel are rare; secondary tumors, such as gastric and colonic carcinomas and ovarian cancers, may compromise the lumen of small bowel
Hernia	An inguinal hernia may be visible on conventional radiographs if air-containing loops of bowel are seen over the obturator foramen; easily seen on CT (Fig. 16-4)
Gallstone ileus	May be visible on conventional radiographs or CT if air is seen in the biliary tree and (rarely) a gallstone in RLQ (Chapter 17)
Intussusception	Ileocolic intussusception is the most common form and produces SBO
Inflammatory bowel disease	Thickening of the bowel wall may occur with compromise of the lumen in patients with Crohn disease; this is most likely to occur in the terminal ileum

CT, Computed tomography; *RLQ,* right lower quadrant; *SBO,* small bowel obstruction.

♦ **Peristalsis continues** and **may increase** in an effort to overcome the obstruction.
  • This can lead to **high-pitched, hyperactive bowel sounds.**
♦ As time passes, the peristaltic waves **empty the small bowel along with the colon** of their contents **from the point of obstruction forward.**
  ♦ If the obstruction is **complete** and if enough time has elapsed since the onset of symptoms, **there is usually no air in the rectum** or **sigmoid.**
■ **Causes of a mechanical small bowel obstruction** are summarized in Table 16-4.
■ **Key imaging features of mechanical small bowel obstruction**

♦ On conventional radiographs **there are multiple dilated loops of small bowel** proximal to the point of the obstruction (>2.5 cm).
  • As they begin to dilate, **small bowel loops stack up on one another** forming a *step-ladder* **appearance,** usually beginning in the left upper quadrant, and proceeding—depending on how distal the small bowel obstruction is—to the right lower quadrant (Fig. 16-5).
  • Generally speaking, **the more proximal the small bowel obstruction** (e.g., proximal jejunum), **the fewer the dilated loops** there will be; the **more distal the obstruction** (e.g., at the ileocecal valve), the **greater the number of dilated small bowel loops.**

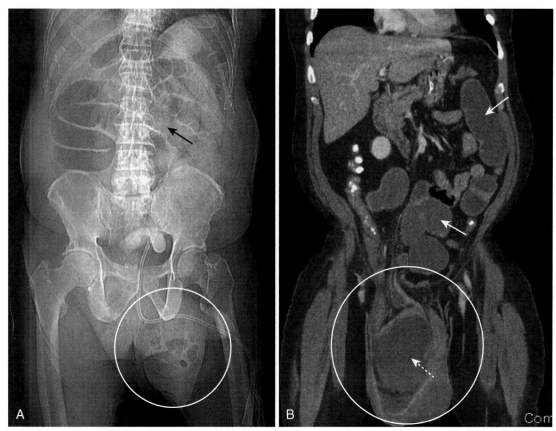

FIGURE 16-4 Small bowel obstruction from inguinal hernia. A, The scout image from a computed tomography (CT) scan of the abdomen reveals dilated loops of small bowel *(solid black arrow)* caused by a left inguinal hernia *(white circle)*. Loops of bowel should normally not be present in the scrotum. **B,** Coronal-reformatted CT scan on another patient shows multiple fluid-filled and dilated loops of small bowel *(solid white arrows)* from an inguinal hernia *(white circle)* containing another dilated loop of small bowel *(dotted white arrow)*.

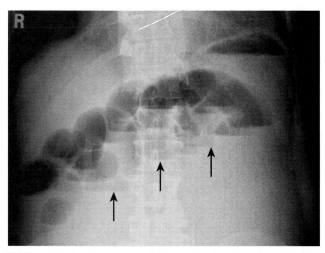

FIGURE 16-5 Step-ladder appearance of obstructed small bowel. As they begin to dilate, small bowel loops stack up, forming a **step-ladder** appearance usually beginning in the left upper quadrant and proceeding—depending on how distal the small bowel obstruction is—to the right lower quadrant *(black arrows)*. The more proximal the small bowel obstruction (e.g. proximal jejunum), the fewer the dilated loops there will be; the more distal the obstruction (e.g., at the ileocecal valve), the greater the number of dilated small bowel loops. This was a distal small bowel obstruction caused by a carcinoma of the colon, which obstructed the ileocecal valve.

♦ On upright or decubitus radiographs, there will usually be **numerous air–fluid levels in the small bowel** proximal to the obstruction.

♦ If enough time has elapsed to decompress and empty the bowel distal to the point of obstruction, **there will be little** or **no gas in the colon, especially in the rectum.**

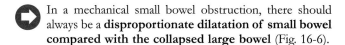 In a mechanical small bowel obstruction, there should always be a **disproportionate dilatation of small bowel compared with the collapsed large bowel** (Fig. 16-6).

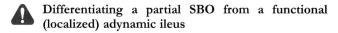 **Differentiating a partial SBO from a functional (localized) adynamic ileus**

♦ An **intermittent** (also known as a *partial* or *incomplete*) **mechanical small bowel obstruction** is one that sometimes **allows some gas to pass the point of obstruction** and can lead to a confusing picture because gas may pass into the colon long after the large bowel would be expected to be devoid of such gas. **Partial** or **incomplete small bowel obstruction** occurs **more often in patients** in whom **adhesions** are the etiologic factors (Fig. 16-7).

♦ CT with or without oral contrast should be able to demonstrate a partial small bowel obstruction or identify the abnormality producing the sentinel loops (Fig. 16-8).

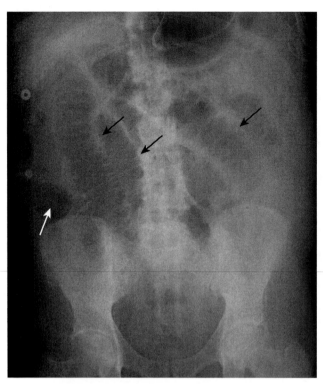

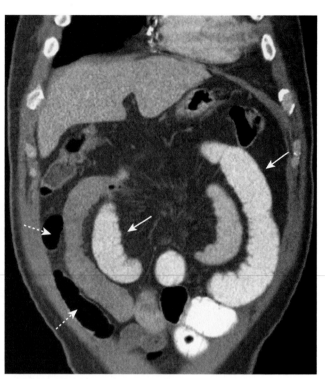

FIGURE 16-6 Mechanical small bowel obstruction. Even though there is still a small amount of air in the right colon *(white arrow)*, the overall bowel gas pattern is one of disproportionate dilation of multiple loops of small bowel *(black arrows)* consistent with a mechanical small bowel obstruction. The obstruction was secondary to adhesions.

FIGURE 16-8 Partial small bowel obstruction. Coronal-reformatted computed tomography scan with oral contrast shows dilated and contrast-containing loops of small bowel *(solid white arrows)*. Although there is still air in the collapsed colon *(dotted white arrows)*, the disproportionate dilatation of small bowel identifies this as a small bowel obstruction.

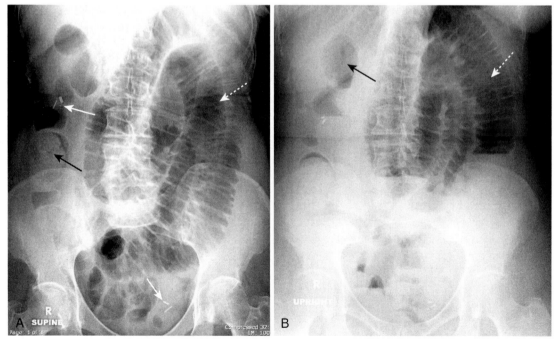

FIGURE 16-7 Partial small bowel obstruction, supine (A) and upright (B). A partial or incomplete mechanical small bowel obstruction is one that allows some gas to pass the point of obstruction, possibly on an intermittent basis. This can lead to a confusing picture because gas may pass into the colon *(solid black arrows)* and be visible long after the large bowel would be expected to be devoid of gas. The important observation is that the small bowel is disproportionately dilated *(dotted white arrows)* compared with the large bowel, a finding suggestive of small bowel obstruction. Partial or incomplete small bowel obstructions occur more often in patients in whom adhesions are the etiologic factors. Notice the clips *(solid white arrows)* attesting to prior surgery.

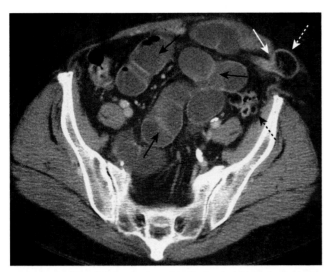

FIGURE 16-9 Small bowel obstruction due to Spigelian hernia. A Spigelian hernia is one that occurs at the lateral edge of the rectus abdominis muscle at the semilunar line. This patient has a transition point *(solid white arrow)* as the small bowel enters the hernia *(dotted white arrow)*. More proximally, there are multiple dilated loops of small bowel *(solid black arrows)*, indicating obstruction. The colon is beyond the point of obstruction and is collapsed *(dotted black arrow)*.

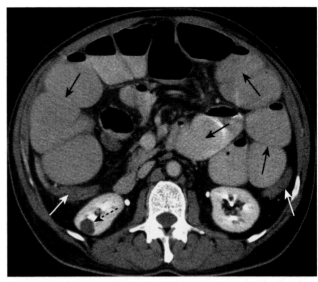

FIGURE 16-10 Small bowel obstruction, computed tomography with oral and intravenous (IV) contrast. There are multiple fluid- and contrast-filled, dilated loops of small bowel *(solid black arrows)* although the colon is collapsed *(white arrows)*, indicating a small bowel obstruction. Bowel wall enhancement, or lack thereof, may be obscured by oral contrast, a drawback to the use of oral contrast. Incidentally noted is a right renal cyst *(dotted black arrow)*.

 CT is the most sensitive study for diagnosing the site and cause of a mechanical small bowel obstruction.

- CT scans for bowel obstruction can be performed with or without oral contrast, utilizing the fluid already present in the bowel as contrast. Orally administered contrast (either barium or iodinated contrast) may help in **identifying dilated loops** of bowel and in finding the **transition point** between the **proximal dilated** bowel and the **distal collapsed** bowel, but the oral contrast might obscure important findings displayed by the use of intravenous contrast.
- **Intravenous contrast** is used for detecting **complications** of bowel obstruction such as **ischemia** and **strangulation.**
- The CT findings of a small bowel obstruction:
  - **Fluid-filled** and **dilated loops of small bowel** (>2.5 cm in diameter) proximal to the point of obstruction.
  - Identification of a **transition point,** which is where the **bowel changes caliber** from **dilated to normal,** indicating the site of the obstruction. In the absence of identifying a mass or other obstructive cause at the transition point, the cause is almost certainly adhesions (Fig. 16-9).
  - **Collapsed small bowel** and/or **colon distal to the point of obstruction** (Fig. 16-10).
  - **Small bowel feces sign.** Proximal to the transition point of a small bowel obstruction, intestinal debris and fluid may accumulate, producing the appearance of fecal material in the small bowel. This is a sign of SBO (Fig. 16-11).
  - **Closed-loop obstruction** occurs when two points of the same loop of bowel are obstructed at a single location. The closed-loop usually remains dilated and may form a U- or C-shaped structure. Most closed-loop

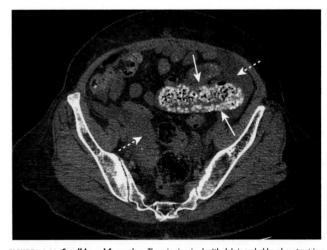

FIGURE 16-11 Small bowel feces sign. There is air mixed with debris and old oral contrast in a dilated loop of small bowel *(solid white arrows)*. There are proximal, fluid-containing, dilated loops of small bowel *(dotted white arrows)*. The patient had a computed tomography scan with oral contrast several days earlier for abdominal pain and returned for this noncontrast scan when symptoms persisted. Intestinal debris and fluid may accumulate in the loop, usually just proximal to a small bowel obstruction, and present with this finding, which resembles fecal material in the colon.

obstructions are caused by adhesions. In the small bowel, a closed-loop obstruction carries a higher risk of strangulation of the bowel. In the large bowel, a closed-loop obstruction is called a *volvulus* (Fig. 16-12).
  - **Strangulation.** Vascular compromise can be identified by circumferential thickening of the wall of the bowel with (frequently) absence of normal wall enhancement following intravenous contrast administration. There may be associated edema of the mesentery and ascites (Fig. 16-13).

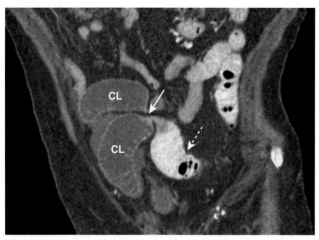

FIGURE 16-12 Closed-loop obstruction, computed tomography. A loop of small bowel is obstructed twice at the same point of twist *(solid white arrow)* producing a closed loop (CL). No oral contrast enters the closed loop but is present in a more proximal loop of small bowel *(dotted white arrow)*. Closed-loop obstructions are important because of their higher incidence of bowel necrosis from strangulation of the bowel.

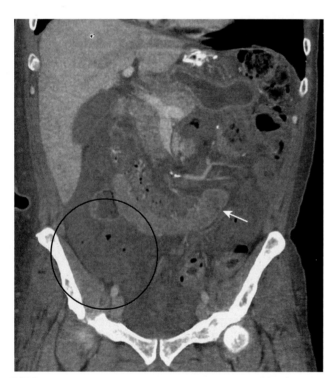

FIGURE 16-13 Bowel necrosis, contrast-enhanced computed tomography (CT). A dilated loop of small bowel demonstrates normal enhancement of the wall *(white arrow)* on this coronal reformat of a contrast-enhanced CT, whereas more distal, dilated loops of small bowel show no wall enhancement *(black circle)*. This is an indication of vascular compromise of the distal loops with bowel necrosis.

MECHANICAL OBSTRUCTION: LARGE BOWEL OBSTRUCTION (LBO)

- **Pathophysiology**
  - A lesion, either inside or outside the colon, causes obstruction to the lumen.
  - Over time, from the point of obstruction **backward**, the **large bowel dilates** with the **cecum frequently**

TABLE 16-5 CAUSES OF A MECHANICAL LARGE BOWEL OBSTRUCTION

Cause	Remarks
Tumor (carcinoma)	Most common cause of LBO; more frequently obstructs when it involves the left colon
Hernia	May be visible on conventional radiographs if air is seen over the obturator foramen
Volvulus	Either the sigmoid (more commonly) or cecum may twist on its axis and obstruct the colon and/or small bowel (see Box 16-2)
Diverticulitis	Uncommon cause of colonic obstruction
Intussusception	Colocolic intussusception usually occurs because of a tumor acting as a lead point

LBO, Large bowel obstruction.

attaining the greatest diameter, even if the obstruction is as far away as the sigmoid colon.

- The large bowel normally functions to reabsorb water, so there are **usually few or no air–fluid levels** in the obstructed colon.
- As time passes, continuing peristaltic waves from the point of obstruction **forward empty the colon distal to the obstruction.**
- There is **usually little** or **no air in the rectum in a mechanical large bowel obstruction.**

■ **Causes of a mechanical large bowel obstruction** are summarized in Table 16-5.

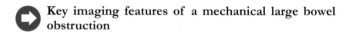 **Key imaging features of a mechanical large bowel obstruction**

- The **colon is dilated to the point of obstruction.**
  - Because there are a limited number of large bowel loops, they tend not to overlap each other (as do the loops of small bowel), so **it is sometimes possible to identify the site of obstruction** as the **last air-containing** segment of the colon (Fig. 16-14).
  - Regardless of the point of obstruction, the **cecum is often the most dilated** segment of the colon. **When the cecum reaches a diameter above 12 to 15 cm, there is danger of cecal rupture.**
- The **small bowel is not dilated** (unless the ileocecal valve becomes **incompetent**). (See the following information.)
- Because it is distal to the point of obstruction, **the rectum contains little** or **no air.**
- Because the large bowel functions to reabsorb water, **there are usually no or very few air–fluid levels in the large bowel.**

⚠ **How an LBO can mimic an SBO**

- **As long as the ileocecal valve** prevents gas from reentering the small bowel in a retrograde direction (such an ileocecal valve is called **competent**), the colon will continue to dilate between the ileocecal valve and the point of colonic obstruction.
  - The **small bowel is not dilated.**

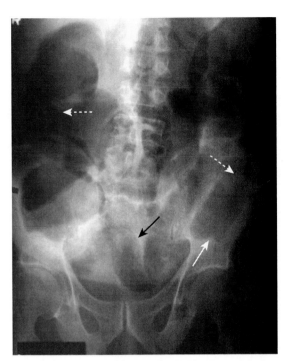

FIGURE 16-14 Mechanical large bowel obstruction. The entire colon is dilated *(dotted white arrows)* to a cut-off point in the distal descending colon *(solid white arrow)*, the site of this patient's obstructing carcinoma of the colon. Some gas has passed backward through an incompetent ileocecal valve and outlines a dilated ileum *(solid black arrow)*. Notice that the large bowel is disproportionately dilated compared with the small bowel, a finding of large bowel obstruction.

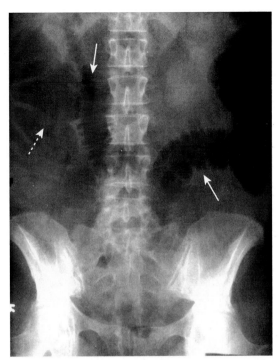

FIGURE 16-15 Large bowel obstruction masquerading as a small bowel obstruction. There are air-filled and dilated loops of small bowel *(solid white arrows)* in this patient who actually had a mechanical large bowel obstruction from a carcinoma of the middescending colon. The pressure in the colon was sufficient to open the ileocecal valve, which then allowed much of the gas in the colon to decompress backward into the small bowel. The cecum still contains air *(dotted white arrow)* and is dilated, a clue that this is really a large bowel obstruction. Abdominal computed tomography can resolve the question of whether the large or small bowel is obstructed.

- ◆ But if the intracolonic pressure rises high enough and the **ileocecal valve opens** (such a valve is called *incompetent*), **gas from the dilated large bowel decompresses backward into the small bowel,** much like the air escaping from a balloon.
- ◆ This can produce images in which there is **disproportionate dilatation of the small bowel** compared with the **decompressed large bowel.** This could lead to a confusing picture that **mimics a mechanical small bowel obstruction** (Fig. 16-15).
- ◆ Solution
  - • Ask for a CT scan of the abdomen. It should show the site of obstruction in the colon rather than the small bowel.
  - • **Barium is not administered by mouth in a patient with a suspected large bowel obstruction** because water will be absorbed from the barium when it reaches the obstructed colon, increasing the viscosity of the barium and possibly leading to impaction.
- ■ **Recognizing a large bowel obstruction on CT**
  - ◆ CT is obtained to identify the cause of the obstruction, assess for free intraperitoneal air, and identify associated lesions, such as metastases to the liver or lymph nodes, if the obstruction is produced by a malignancy.
  - ◆ The large bowel is **dilated to the point of obstruction,** then normal in caliber distal to the obstructing lesion.
  - ◆ The point of obstruction, frequently a carcinoma, can usually be located on CT as a **soft tissue mass.** Hernias

containing large bowel are also easy to identify on CT (Fig. 16-16).

VOLVULUS OF THE COLON

- ■ Volvulus of the colon is a particular kind of large bowel obstruction that produces a striking and characteristic picture that is summarized in Box 16-2.

INTESTINAL PSEUDOOBSTRUCTION (OGILVIE SYNDROME)

- ■ **Ogilvie syndrome (acute intestinal pseudoobstruction)** may occur in older individuals who are usually already hospitalized or at chronic bed rest.
  - ◆ Drugs with anticholinergic effects, such as **antidepressants, phenothiazines, antiparkinsonian agents, and narcotics,** may cause or exacerbate the condition.
- ■ The syndrome is characterized by a **loss of peristalsis,** resulting in sometimes **massive dilatation of the right colon** or **the entire colon and resembling a large bowel obstruction** (Fig. 16-18).
  - ◆ Unlike a mechanical obstruction, **no obstructing lesion can be demonstrated** on CT or with barium enema. Unlike a generalized ileus, **patients have more marked abdominal distension** and bowel sounds may be **normal** or **hyperactive** in almost half of patients with Ogilvie syndrome.

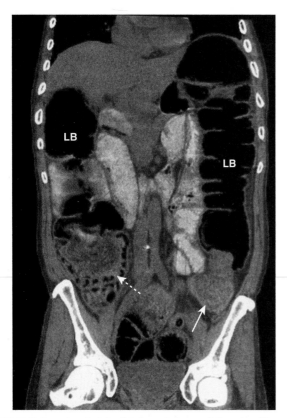

FIGURE 16-16 Large bowel obstruction from carcinoma of the colon. This coronal-reformatted CT scan of the abdomen and pelvis shows dilated cecum containing stool *(dotted white arrow)* and large bowel (LB) to the level of the distal descending colon where a large soft tissue mass is identified *(solid white arrow)*. This mass was an adenocarcinoma of the colon and was surgically removed.

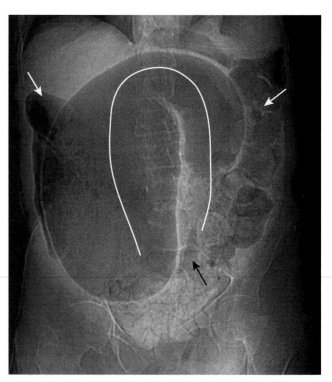

FIGURE 16-17 Sigmoid volvulus, supine abdomen. There is a massively dilated sigmoid colon *(solid white line)* that is twisted upon itself in the pelvis *(black arrow)*. The dilated sigmoid has a **coffee-bean** shape. Since the point of obstruction is in the distal colon, there is air and stool in the more proximal portion of the colon *(white arrows)*. Volvulus can produce massively dilated loops of sigmoid colon.

BOX 16-2 VOLVULUS—A CAUSE OF MECHANICAL LARGE BOWEL OBSTRUCTION

Either the cecum or the sigmoid colon can twist upon itself, producing a mechanical obstruction known as a *volvulus.*

Sigmoid volvulus is more common and tends to occur in older men.

The volvulated sigmoid assumes a massive size, rising up from the pelvis with the wall between the twisted loops of sigmoid forming a line that points from the left lower to the right upper quadrant.

The appearance of the dilated sigmoid has been likened to a **coffee bean** (Fig. 16-17).

When the cecum volvulates, it usually moves across the midline into the left upper quadrant, producing loops of bowel forming a line that characteristically points from the right lower to the left upper quadrant.

A contrast enema can be both diagnostic (the obstructed sigmoid produces a **beak sign**) and therapeutic because the hydrostatic pressure of the enema can sometimes decompress the volvulus.

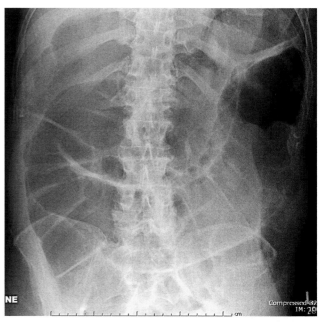

FIGURE 16-18 Ogilvie syndrome. Ogilvie syndrome (acute intestinal pseudoobstruction) may occur in older adults who are usually already hospitalized or on chronic bed rest. Drugs with anticholinergic effects may cause or exacerbate the condition. The syndrome is characterized by a loss of peristalsis, resulting in sometimes massive dilatation of the entire colon resembling a large bowel obstruction, as in this patient. Treatment is pharmacologic stimulation of the bowel.

TAKE-HOME POINTS

RECOGNIZING BOWEL OBSTRUCTION AND ILEUS

The abnormal bowel gas patterns can be divided into two main groups: functional ileus and mechanical obstruction.

There are two varieties of functional ileus: localized ileus (sentinel loops) and generalized adynamic ileus. There are two varieties of mechanical obstruction: small bowel obstruction (SBO) and large bowel obstruction (LBO).

In mechanical obstruction, the gut reacts in predictable ways: loops proximal to the obstruction become dilated, peristalsis attempts to propel intestinal contents through the bowel, and loops distal to the obstruction eventually are evacuated; the loop(s) that will become the most dilated will either be the loop of bowel with the largest resting diameter or the loop(s) of bowel just proximal to the obstruction.

The key findings in a localized ileus are two to three dilated loops of small bowel **(sentinel loops)** with air in the rectosigmoid and an underlying irritative process that frequently is adjacent to the dilated loops.

Some causes of sentinel loops include pancreatitis (left upper quadrant), cholecystitis (right upper quadrant), diverticulitis (left lower quadrant), and appendicitis (right lower quadrant). All can be readily identified using ultrasound or computed tomography (CT).

The key findings in a generalized adynamic ileus are dilated loops of large and small bowel with gas in the rectosigmoid and long air–fluid levels. Postoperative patients develop generalized adynamic ileus.

The key imaging findings in a mechanical small bowel obstruction are disproportionately dilated and fluid-filled loops of small bowel with little or no gas in the rectosigmoid. CT is best at identifying the cause and site of obstruction or its complications.

The most common cause of a SBO is adhesions; other causes include hernias, intussusception, gallstone ileus, malignancy, and inflammatory bowel disease, such as Crohn disease.

A closed-loop obstruction is one in which two points of the bowel are obstructed in the same location, producing the closed loop. In the small bowel, a closed-loop obstruction carries a higher risk of strangulation of the bowel. In the large bowel, a closed-loop obstruction is called a *volvulus*.

The key imaging findings in mechanical LBO include dilatation of the colon to the point of the obstruction and absence of gas in the rectum with no dilatation of the small bowel as long as the ileocecal valve remains competent. CT will often demonstrate the cause of the obstruction.

Causes of mechanical LBO include malignancy, hernia, diverticulitis, and intussusception.

Ogilvie syndrome is characterized by a loss of peristalsis, resulting in sometimes massive dilatation of the entire colon, resembling a large bowel obstruction but without a demonstrable point of obstruction; it can sometimes be confused with a generalized adynamic ileus.

 The supine abdominal radiograph shows **marked bowel dilatation, almost always confined to the colon.**

■ Management is **pharmacologic stimulation of colonic contractions,** usually with drugs such as neostigmine.

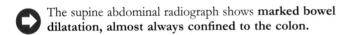 **WEBLINK**

Visit StudentConsult.Inkling.com for more information and a quiz on Abnormal Bowel Gas Patterns.

For your convenience, the following QR Code may be used to link to **StudentConsult.com**. You must register this title using the PIN code on the inside front cover of the text to access online content.

CHAPTER 17
Recognizing Extraluminal Gas in the Abdomen

- Recognition of extraluminal gas is an important finding that can have an immediate effect on the course of treatment. Air is normally not present in the peritoneal or extraperitoneal spaces, bowel wall, or biliary system. **Air outside of the bowel** is called *extraluminal air.*
- **The four most common locations of extraluminal air are:**
  - **Intraperitoneal (pneumoperitoneum)** (frequently called *free air)*
  - **Retroperitoneal air**
  - **Air in the bowel wall (pneumatosis intestinalis)**
  - **Air in the biliary system (pneumobilia)**

SIGNS OF FREE INTRAPERITONEAL AIR

 There are **three major signs of free intraperitoneal air,** arranged below in the order in which they are most commonly seen:

- **Air beneath the diaphragm**
- **Visualization of both sides of the bowel wall**
- **Visualization of the falciform ligament**

AIR BENEATH THE DIAPHRAGM

- Air will rise to the highest part of the abdomen. In the upright position, **free air** will usually reveal itself **under the diaphragm** as a **crescentic lucency** that parallels the undersurface of the diaphragm (Fig. 17-1).
  - The **size of the crescent** will be **roughly proportional to the amount of free air.** The smaller the amount of free air, the thinner the crescent; the larger the amount of free air, the larger the crescent (Fig. 17-2).
- **Although free air is best demonstrated on computed tomography (CT) scans of the abdomen** because of its greater sensitivity in detecting very small amounts of free air (Fig. 17-3), most surveys of the abdomen begin with conventional radiographs. Conventional radiographs **serve as an important screening tool** on which many previously unsuspected cases of free air are discovered.
- On conventional radiographs, **free air is best demonstrated with the x-ray beam directed parallel to the floor** (i.e., a horizontal beam) (see Figs. 14-14 and 14-15).

 Small amounts of free air will not be visible on radiographs in which the x-ray beam is directed vertically downward, such as **supine** or **prone views.**

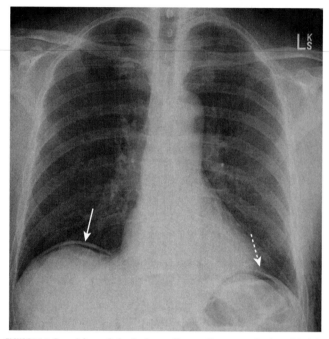

FIGURE 17-1 Free air beneath the diaphragm. There are thin crescents of air beneath both the right *(solid white arrow)* and left *(dotted white arrow)* hemidiaphragms representing free intraperitoneal air. The patient had undergone abdominal surgery 3 days earlier. Free air can remain for up to 7 days after surgery in an adult, but serial studies should demonstrate a progressively decreasing amount of air.

- Free air is **easier to recognize under the right hemidiaphragm** because there is usually only the soft tissue density of the liver in that location. Free air is **more difficult to recognize under the left hemidiaphragm** because air-containing structures like the fundus of the stomach and the splenic flexure already reside in that location and may be mistaken for free air (Fig. 17-4).
- If the patient is unable to stand or sit upright, then a view of the abdomen with the patient lying on the left side (the right side is pointing up) taken with a horizontal x-ray beam may show free air rising above the right edge of the liver. This is the **left lateral decubitus view** of the abdomen (Fig. 17-5).

 Pitfall: Chilaiditi syndrome

- Occasionally, the colon may be interposed between the dome of the liver and the right hemidiaphragm, and unless a **careful search** is made **for the presence of haustral folds** characteristic of the colon, it may be mistaken for free air (Fig. 17-6).

♦ **Solution**
  • If in doubt, obtain a left lateral decubitus view of the abdomen, or if necessary, a CT scan of the abdomen.

VISUALIZATION OF BOTH SIDES OF THE BOWEL WALL

■ In the **normal** abdominal radiograph, we visualize only the air **inside** the lumen of the bowel, **not the wall of the bowel itself.** This is seen because the wall is soft tissue density and is surrounded by tissue of the same density.

■ Introduction of **air into the peritoneal cavity enables us to visualize the wall of the bowel itself** since the wall is now surrounded on both inside and outside by air.

■ The **ability to see both sides of the bowel wall is a sign of free intraperitoneal air** called *Rigler sign* (Fig. 17-7).

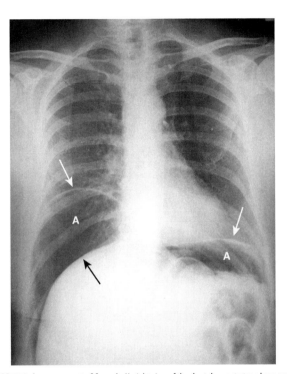

FIGURE 17-2 Large amount of free air. Upright view of the chest demonstrates a large amount of free air (A) beneath each hemidiaphragm *(white arrows)*. The top of the liver *(black arrow)* is made visible by the air above it. The patient had a perforated gastric ulcer.

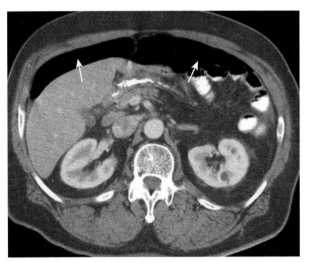

FIGURE 17-3 Free air seen on CT scan of the abdomen. Axial CT scan of the upper abdomen performed with the patient supine shows free air anteriorly *(white arrows)*. The air is not contained within any bowel. Free intraperitoneal air will normally rise to the highest point of the abdomen, which in the supine position is beneath the anterior abdominal wall.

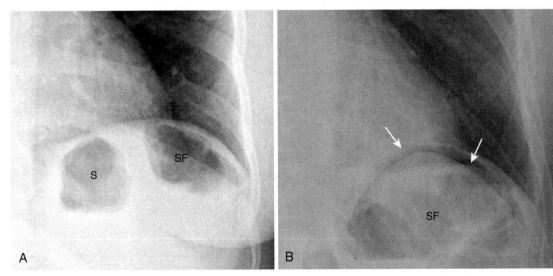

FIGURE 17-4 Normal left hemidiaphragm (A) and free air under hemidiaphragm (B). Close-up views of the left upper quadrant demonstrate the difficulty in recognizing free air beneath the left hemidiaphragm because of the normal location of gas-containing structures such as the stomach (S) and splenic flexure (SF). There is no free air in **(A)**, but the other patient **(B)** does have a crescent of free air *(white arrows)*. It is easier to recognize free air beneath the right hemidiaphragm because there is usually no air above the liver on the right side.

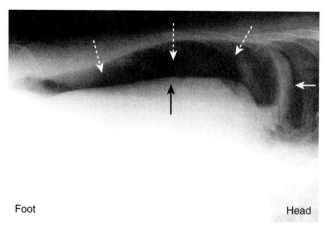

Foot   Head

FIGURE 17-5 **Left lateral decubitus view showing free air.** Close-up of the right upper quadrant in a patient lying on the left side in the left lateral decubitus position shows a crescent of air *(dotted white arrows)* above the outer edge of the liver *(black arrow)*, beneath the right hemidiaphragm *(solid white arrow)*. The head/foot orientation of the patient is indicated. If the patient is unable to stand or sit up for an upright view of the abdomen, a left lateral decubitus view with a horizontal beam can substitute.

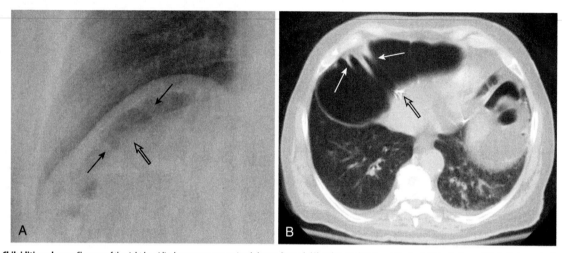

FIGURE 17-6 **Chilaiditi syndrome.** Close-up of the right hemidiaphragm on a conventional chest radiograph **(A)** and an axial CT scan at the level of the diaphragm **(B),** both demonstrating air beneath the diaphragm that could be confused for free air *(open black arrows* in **[A]** and **[B]**). Careful evaluation of this air demonstrates several haustral folds *(solid black arrows* in **[A]** *and solid white arrows* in **[B]**), which indicate this is a loop of colon interposed between the liver and the diaphragm (Chilaiditi syndrome) rather than free air.

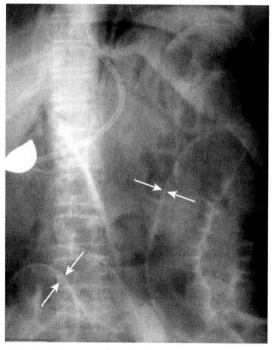

FIGURE 17-7 **Rigler sign.** When air fills the peritoneal cavity, both sides of the bowel wall will be outlined by air *(white arrows)* making the wall of the bowel visible as a discrete line. This is known as ***Rigler sign*** and indicates the presence of a pneumoperitoneum. This patient had a perforated gastric ulcer.

It can be seen on supine, upright, or prone films of the abdomen as long as there is a relatively large amount of free air present.

 Pitfall

♦ When dilated loops of small bowel overlap each other, it may occasionally produce the mistaken impression that you are seeing both sides of the bowel wall (Fig. 17-8).

♦ **Solution**

• Confirm the presence of free air with an upright view, left lateral decubitus view, or CT scan of the abdomen.

VISUALIZATION OF THE FALCIFORM LIGAMENT

■ The **falciform ligament** courses over the **free edge of the liver anteriorly** just to the **right of the upper lumbar spine.** It contains a remnant of the obliterated umbilical artery. **It is normally invisible,** composed of soft tissue and surrounded by tissue of similar density.

■ When a (usually) large amount of free air is present and the **patient is in the supine position, free air** may rise over the anterior surface of the liver, **surround the falciform ligament,** and **render it visible.** Visualization of the falciform ligament is aptly called the *falciform ligament sign* (Fig. 17-9).

■ The curvilinear appearance of the falciform ligament combined with the oval-shaped collection of air that collects beneath and distends the abdominal wall has been likened to the appearance of a football with its laces and is called the *football sign.*

■ Table 17-1 summarizes the **three major signs of free air.**

CAUSES OF FREE AIR

■ The **most common cause of free intraperitoneal air** is **rupture of an air-containing loop of bowel,** either stomach, small bowel, or large bowel.

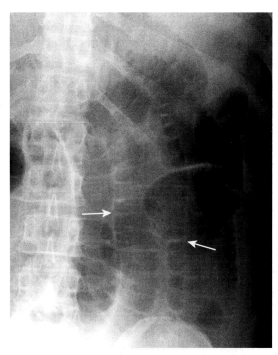

FIGURE 17-8 Overlapping loops mimicking free air. Do not let overlapping loops of dilated small bowel *(white arrows)* fool you into thinking you are seeing both sides of the bowel wall due to free air. Notice that where the loops do not overlap, both sides of the bowel wall are not seen. If there is doubt about the presence of free air, confirmation may be obtained through an upright or left lateral decubitus view of the abdomen or a computed tomography scan of the abdomen.

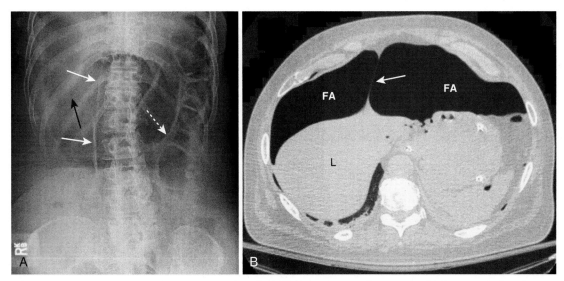

FIGURE 17-9 Falciform ligament sign. A, Free intraperitoneal air may surround the normally invisible falciform ligament on the anterior edge of the liver causing that thin, soft tissue structure to become visible *(solid white arrows)* just to the right of the upper lumbar spine. Notice also that both sides of the stomach wall are visible (Rigler sign) *(dotted white arrow)*, and there is increased lucency to the right upper quadrant *(solid black arrow)* in this patient with a large pneumoperitoneum from a perforated gastric ulcer. **B,** The falciform ligament *(white arrow)* is outlined by free air (FA) on either side of it, anterior to the liver (L).

TABLE 17-1 THREE SIGNS OF FREE AIR

Sign	Remarks
Air beneath diaphragm	Requires patient to be in the upright or left lateral decubitus position and a horizontal x-ray beam unless massive in amount
Visualization of both sides of the bowel wall	Usually requires large amount of free air; will be visible in any position
Visualization of the falciform ligament	Usually requires large amount of free air; patient is usually supine

 Perforated peptic ulcer is the most common cause of a perforated stomach or duodenum and is still the most common cause of free air.

- ♦ **Trauma,** whether accidental or iatrogenic, can also produce free air. **Free air following penetrating trauma** usually **implies a perforation of the bowel,** not free air generated simply by penetration of the abdominal wall itself.
- ♦ **For several days following abdominal surgery (about 5 to 7 days),** whether the surgery had been performed on the bowel or not, **it is normal to see free air** on postoperative studies. The **amount** of free air following surgery **should diminish with each successive study.** A complication of the surgery or of the original disease should be considered **if free air persists for longer than a week** or if the amount **increases** on successive studies.
- ♦ **Perforated diverticulitis and perforated appendicitis** usually produce walled-off abscess collections around the site of the perforation and rarely lead to significant amounts of free air.
- ♦ **Perforation of a carcinoma,** usually of the colon, is unusual but can also lead to free air.

SIGNS OF EXTRAPERITONEAL AIR (RETROPERITONEAL AIR)

- ■ Unlike the collections of free intraperitoneal air that outline loops of bowel and usually move freely in the abdomen, **extraperitoneal air can be recognized** by:
  - ♦ **Streaky, linear appearance outlining extraperitoneal structures**
  - ♦ **Mottled, blotchy appearance** (especially the anterior pararenal space)
  - ♦ **Relatively fixed position, moving little** if at all **with changes in patient positioning**
- ■ Extraperitoneal air may outline extraperitoneal structures such as the following:
  - ♦ **Psoas muscles**
  - ♦ **Kidneys, ureters,** or **urinary bladder**
  - ♦ **Aorta** or **inferior vena cava** (Fig. 17-10)
  - ♦ Inferior border of the diaphragm by collecting in the **subphrenic tissues**
- ■ **Extraperitoneal air may extend through a diaphragmatic hiatus into the mediastinum** (and produce *pneumomediastinum*), or **may extend to the peritoneal**

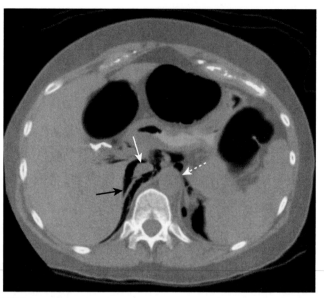

FIGURE 17-10 Extraperitoneal air seen on CT. Air is seen in the retroperitoneum *(solid black arrow)* on this axial CT scan of the upper abdomen. Air outlines the inferior vena cava *(solid white arrow)* and the aorta *(dotted white arrow)*. Unlike free air, extraperitoneal air is streaky, relatively fixed in position and outlines extraperitoneal structures like the vena cava, aorta, psoas muscles, and kidneys.

BOX 17-1 SIGNS OF EXTRAPERITONEAL AIR

Streaky, linear collections of air that outline extraperitoneal structures

Mottled, blotchy collections of air that remain in a fixed position

---

cavity through openings in the peritoneum (and produce *pneumoperitoneum*).
- ■ Box 17-1 summarizes the signs of extraperitoneal air.

CAUSES OF EXTRAPERITONEAL AIR

- ■ **Extraperitoneal air is most frequently the result of bowel perforation** secondary to either:
  - ♦ **Inflammatory disease (e.g., ruptured appendix),** or
  - ♦ **Ulcerative disease (e.g., Crohn Disease** of the ileum or colon)
- ■ **Other causes of extraperitoneal air** include:
  - ♦ **Blunt or penetrating trauma**
  - ♦ **Iatrogenic manipulation** (e.g., perforation of the bowel during **sigmoidoscopy**)
  - ♦ **Foreign body** (e.g., perforation of extraperitoneal ascending colon by an ingested foreign body)
  - ♦ **Gas-producing infection** originating in extraperitoneal organs (such as **perforated diverticulitis**)

SIGNS OF AIR IN THE BOWEL WALL

- ■ Air in the bowel wall is called *pneumatosis intestinalis.*
- ■ Air in the bowel wall is **most easily recognized** on abdominal radiographs **when it is seen in profile** producing a **linear radiolucency (black line) whose contour exactly parallels the bowel lumen** (Fig. 17-11).

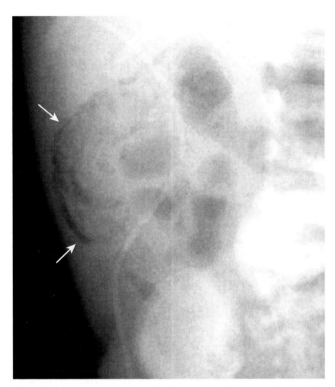

FIGURE 17-11 Pneumatosis seen in profile. Close-up of the right lower quadrant in an infant demonstrates a thin curvilinear lucency that parallels the lumen of the adjacent bowel *(white arrows)*, an appearance characteristic of gas in the bowel wall seen in profile. In infants the most common cause for this finding is **necrotizing enterocolitis (see Chapter 28)**, a disease found mostly in premature infants in which the terminal ileum is most affected. Pneumatosis intestinalis is pathognomonic for necrotizing enterocolitis in infants.

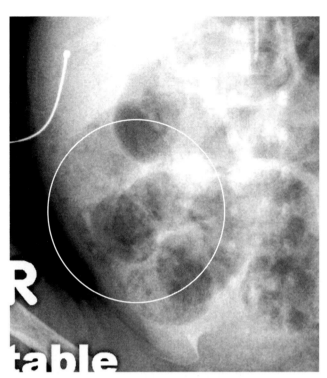

FIGURE 17-12 Pneumatosis seen *en face*. Close-up of the right lower quadrant in another infant shows multiple faint, mottled lucencies in the right lower quadrant, which is the appearance of pneumatosis intestinales when seen *en face*. The density has the same appearance as air mixed with stool, but can be distinguished from stool because it occurs in areas stool might not be expected, and it does not change over time. This infant also had necrotizing enterocolitis.

■ **Air** in the bowel wall **seen *en face*** is **more difficult to recognize** but frequently has a **mottled appearance that resembles gas mixed with fecal material** (Fig. 17-12).
  ♦ Clues to help differentiate pneumatosis from fecal material include:
    • **Presence of such mottled gas in an area of the abdomen unlikely to contain colon**
    • **Lack of change** in the appearance of the mottled gas pattern over several images in **differing positions**
■ Table 17-2 summarizes the signs of air in the bowel wall.

CAUSES AND SIGNIFICANCE OF AIR IN THE BOWEL WALL

■ Pneumatosis intestinalis can be divided into two major categories:
  ♦ **A rare primary form called *pneumatosis cystoides intestinalis,*** which usually **affects the left colon** producing **cystlike collections of air in the submucosa or serosa** (Fig. 17-13)
  ♦ **A more common secondary form**, which can occur in the following:
    • **Chronic obstructive pulmonary disease—** presumably secondary to air from ruptured blebs dissecting through the mediastinum to the abdomen

TABLE 17-2 SIGNS OF AIR IN THE BOWEL WALL

Sign	Remarks
Linear radiolucency paralleling the contour of air in the adjacent bowel lumen	When seen in profile
Mottled appearance that resembles air mixed with fecal material	May occur in an area of the abdomen not expected for colon; does not change over time
Globular, cystlike collections of air that parallel the contour of the bowel	Unusual, benign condition affecting colon, usually left colon

  • Diseases in which there is **necrosis of the bowel wall** such as:
    ▪ **Necrotizing enterocolitis** in infants
    ▪ **Ischemic bowel disease** in adults (Fig. 17-14)
  • **Obstructing lesions of the bowel** that raise intraluminal pressure, such as:
    ▪ **Hirschsprung disease** or **pyloric stenosis** in children
    ▪ **Obstructing carcinomas** in adults

 Pneumatosis intestinalis **associated with diseases that produce necrosis of the bowel** is usually a **more ominous prognostic sign** than pneumatosis associated with **obstructing lesions of the bowel** or chronic obstructive pulmonary disease.

- **Complications of pneumatosis intestinalis** can include:
  - ◆ **Rupture into the peritoneal cavity,** leading to intraperitoneal free air (**pneumoperitoneum**)
  - ◆ Dissection of **air into the portal venous system** (Fig. 17-15)

SIGNS OF AIR IN THE BILIARY SYSTEM

- Air in the biliary system presents as **one** or **two tubelike, branching lucencies in the right upper quadrant overlying the central portion of the liver, which conform to the location** and **appearance of the major bile ducts:** the common duct, cystic duct, and/or the hepatic ducts (Fig. 17-16).
- Box 17-2 summarizes the signs of **air in the biliary system.**

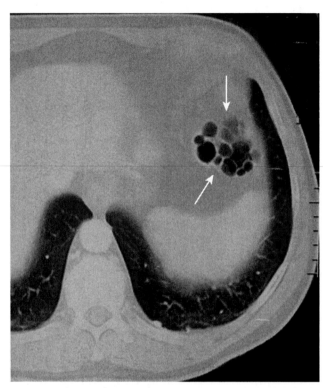

FIGURE 17-13 Pneumatosis cystoides intestinalis. Axial CT scan of the upper abdomen windowed for lung technique shows a cluster of air-containing cysts *(solid white arrows)* associated with the left colon, characteristic of **pneumatosis cystoides intestinales,** a rare but benign condition in which air-containing cysts form in the submucosa or serosa of the bowel.

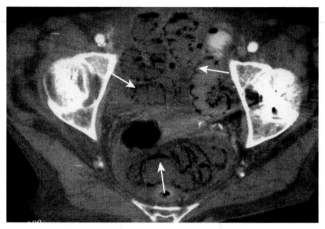

FIGURE 17-14 Necrosis of bowel from mesenteric ischemia. Axial computed tomography image of the pelvis demonstrates multiple loops of bowel with punctate collections of air throughout their walls consistent with pneumatosis *(white arrows).* The patient had widespread ischemia of the bowel from mesenteric vascular disease. Pneumatosis that results from bowel necrosis is an ominous sign.

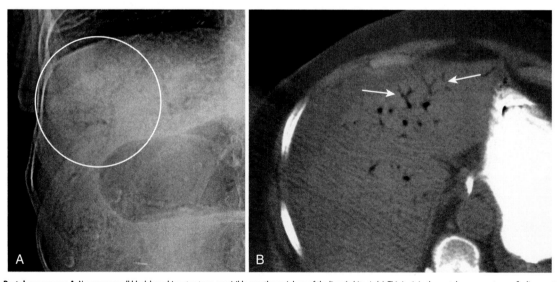

FIGURE 17-15 Portal venous gas. A, Numerous small black branching structures are visible over the periphery of the liver *(white circle).* This is air in the portal venous system, a finding most often associated with necrotizing enterocolitis in infants, but which can also be seen in adults, usually with bowel necrosis. Unlike air in the biliary system, this air is peripheral rather than central and has numerous branching structures instead of the few tubular extructures seen with pneumobilia. **B,** Close-up of axial CT scan through the liver shows air in the portal venous system *(white arrows)* in a patient with mesenteric vascular disease.

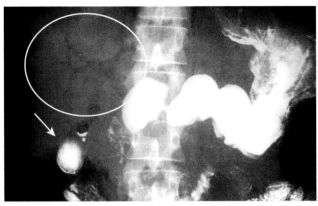

FIGURE 17-16 **Air in the biliary tree.** Frontal view of the upper abdomen from an upper gastrointestinal series demonstrates several air-containing tubular structures over the central portion of the liver consistent with air in the biliary system *(white circle)*. There is also barium in the gallbladder *(white arrow)*. This patient had a history of a prior sphincterotomy for gallstones so that reflux of air and barium into the biliary system would be expected.

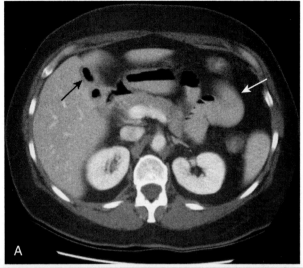

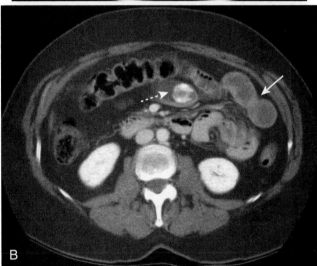

FIGURE 17-17 **Gallstone ileus.** The three key findings of gallstone ileus are present on this study. **A,** Axial CT scan of the upper abdomen shows air in the lumen of the gallbladder *(black arrow)* and dilated small bowel *(white arrow)* consistent with a mechanical small bowel obstruction. **B,** At a lower level, another axial CT scan of the abdomen shows a large calcified gallstone inside the small bowel *(dotted white arrow)* and additional proximal, dilated loops of small bowel *(solid white arrow)*. The gallstone had eroded through the wall of the gallbladder into the duodenum and then began a journey down the small bowel before becoming impacted and producing obstruction.

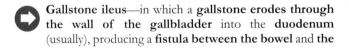

BOX 17-2   SIGNS OF AIR IN THE BILIARY TRACT
Tubelike, branching lucencies in the right upper quadrant overlying the liver
Tubular structures are central in location and few in number compared with portal venous air, which is peripheral in location and fills innumerable vessels
Gas in the lumen of the gallbladder

CAUSES OF AIR IN THE BILIARY SYSTEM

■ Gas in the biliary system **may be a "normal" finding** if the sphincter of Oddi, which guards the entrance of the common bile duct as it enters the duodenum, is open (said to be *incompetent*).

■ Prior **sphincterotomy** such as might be done to allow gallstones to exit from the ductal system into the bowel

■ **Prior surgery** that results in the **reimplantation of the common bile duct** into another part of the bowel (i.e., choledochoenterostomy) is frequently accompanied by gas in the biliary ductal system.

■ Pathologic conditions that can produce pneumobilia include **uncommon causes** such as:

➡ **Gallstone ileus**—in which a **gallstone erodes through the wall of the gallbladder** into the **duodenum** (usually), producing a **fistula between the bowel and the** **biliary system**. The **gallstone impacts in the small bowel**, usually in the narrower terminal ileum, and produces a mechanical obstruction (here called an *ileus*) (Fig. 17-17).

♦ **Gas-forming pyogenic cholangitis**, particularly from *Escherichia coli*

TAKE-HOME POINTS

RECOGNIZING EXTRALUMINAL AIR IN THE ABDOMEN

Gas in the abdomen outside of the normal confines of the bowel is called *extraluminal air.*

The four most common locations for extraluminal air are intraperitoneal (**pneumoperitoneum**—frequently called ***free air***), retroperitoneal air, air in the bowel wall (pneumatosis), and air in the biliary system (pneumobilia).

The three key signs of free air are air beneath the diaphragm, visualization of both sides of the bowel wall (**Rigler sign**), and visualization of the falciform ligament.

The most common causes of free air are perforated peptic ulcer, trauma (whether accidental or iatrogenic), perforated diverticulitis, perforated appendicitis, and perforation of a carcinoma, usually of the colon.

The key signs of extraperitoneal (retroperitoneal) air are a streaky, linear appearance or a mottled, blotchy appearance outlining extraperitoneal structures and its relatively fixed position, moving very little or not at all with changes in patient positioning.

Extraperitoneal air outlines extraperitoneal structures such as the psoas muscles, kidneys, aorta, and inferior vena cava.

Causes of extraperitoneal air include bowel perforation secondary to either inflammatory or ulcerative disease, blunt or penetrating trauma, iatrogenic manipulation, and foreign body ingestion.

The key signs of air in the bowel wall include linear radiolucencies paralleling the contour of air in the adjacent bowel lumen, a mottled appearance that resembles air mixed with fecal material, or uncommonly, globular, cystlike collections of air that parallel the contour of the bowel.

Causes of air in the bowel wall (pneumatosis intestinalis) include a rare primary form called ***pneumatosis cystoides intestinales*** and a more common secondary form that includes diseases in which there is necrosis of the bowel wall, such as necrotizing enterocolitis in infants, ischemic bowel disease in adults, and obstructing lesions of the bowel that raise intraluminal pressure, such as Hirschsprung disease in children and obstructing carcinomas in adults.

Pneumatosis intestinales associated with diseases that produce necrosis of bowel is usually a more ominous prognostic sign than pneumatosis associated with obstructing lesions of the bowel or chronic obstructive pulmonary disease.

Signs of air in the biliary system include tubelike, branching lucencies in the right upper quadrant overlying the liver, which are central in location and few in number, and gas in the lumen of the gallbladder.

Causes of pneumobilia include incompetence of the sphincter of Oddi, prior sphincterotomy, prior surgery that results in the reimplantation of the common bile duct into another part of the bowel, and gallstone ileus.

The triad of findings in gallstone ileus are air in the biliary system, small bowel obstruction, and visualization of the gallstone itself.

 **WEBLINK**

Visit StudentConsult.Inkling.com for more information and a quiz on Recognizing Free Air.

For your convenience, the following QR Code may be used to link to **StudentConsult.com**. You must register this title using the PIN code on the inside front cover of the text to access online content.

CHAPTER 18

Recognizing Abnormal Calcifications and Their Causes

■ Soft tissue calcifications lend themselves to an intuitive approach that ties together a diverse group of diseases. Although this chapter focuses primarily on abdominal calcifications, the same principles and approach apply to dystrophic calcification found anywhere in the body.

■ Most soft tissue calcification occurs in tissue that is already abnormal. Such calcification is called *dystrophic calcification.*

■ The **nature of most calcifications can be determined by examining two of their characteristics:**
  ♦ Their **pattern of calcification**
  ♦ Their **anatomic location**

PATTERNS OF CALCIFICATION

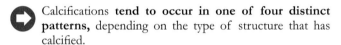 Calcifications **tend to occur in one of four distinct patterns,** depending on the type of structure that has calcified.

■ The patterns are:
  ♦ **Rimlike**
  ♦ **Linear** or **tracklike**
  ♦ **Lamellar (or laminar)**
  ♦ **Cloudlike, amorphous, or popcorn**

RIMLIKE CALCIFICATION

■ Rimlike calcifications imply **calcification that has occurred in the wall of a hollow viscus.** A "hollow viscus" in this context means a structure containing fluid, fat, or air and enclosed by an outer wall.

■ Examples of structures that manifest rimlike calcifications include:
  ♦ **Cysts**—calcification in any one of these is relatively uncommon.
    • **Renal cysts**
    • **Splenic cysts**
    • **Extraabdominal sites** such as:
      ▪ Mediastinal cysts, such as pericardial and bronchial cysts (Fig. 18-1)
      ▪ Popliteal cysts
  ♦ **Aneurysms**
    • **Aortic aneurysm**
      ▪ Aortic aneurysm can sometimes be recognized on a lateral radiograph of the lumbar spine.
      ▪ The abdominal aorta should normally measure <3 cm in diameter, a measurement that requires both apposing walls be visible (Fig. 18-2).
    • **Splenic artery or renal artery** aneurysms

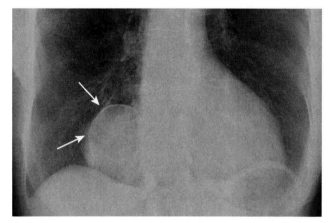

FIGURE 18-1 Calcified pericardial cyst. There is a rimlike calcification *(white arrows)* that identifies the structure containing the calcification as cystic or saccular. The calcification is in the right cardiophrenic angle, an ideal location for pericardial cysts. Pericardial cysts almost always occur on the right side and are most common at the cardiophrenic angle, as in this case. They are usually asymptomatic and discovered when a chest radiograph is obtained for another reason.

• **Extraabdominal sites** such as:
  ▪ Femoral artery aneurysms
  ▪ Cerebral aneurysms
♦ **Saccular organs, such as the gallbladder** or **urinary bladder**
  • **Porcelain gallbladder**
    ▪ An uncommon entity (named after the gross appearance of the gallbladder, which resembles porcelain) that occurs with chronic inflammation and stasis and is associated with gallstones and an increased incidence of carcinoma of the gallbladder (Fig. 18-3)
  • **Urinary bladder**
    ▪ Uncommon occurrence in diseases such as **schistosomiasis, bladder cancer,** and **tuberculosis**

■ Table 18-1 highlights key facts about rimlike calcifications.

LINEAR OR TRACKLIKE CALCIFICATION

■ Linear or tracklike calcification implies **calcification that has occurred in the walls of tubular structures** (Fig. 18-4).

■ Examples include calcifications in the walls of:
  ♦ **Arteries**
    • **Common in atherosclerosis** and seen anywhere in the body

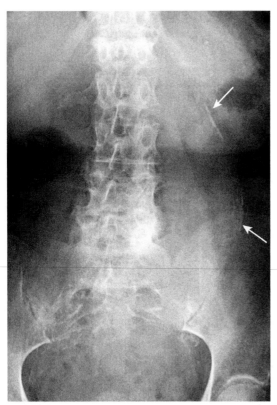

FIGURE 18-2 Calcified aortic aneurysm. Calcification in the wall of the abdominal aorta is a common finding in atherosclerosis, especially in those with diabetes mellitus. In this patient, the aorta demonstrates a rimlike calcification *(white arrows)*. The opposite wall is also calcified, but overlaps the spine, which is why calcifications in the aorta are usually easier to identify on lateral abdominal radiographs. When the diameter of the abdominal aorta exceeds its normal diameter by more than 50%, an aneurysm is present.

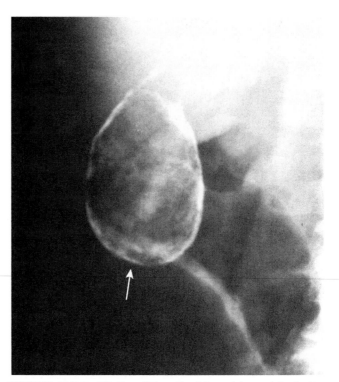

FIGURE 18-3 Calcified gallbladder wall. The rimlike calcification *(white arrow)* identifies this as one that occurs in the wall of a cyst or saccular organ. The calcification is in the right upper quadrant, the location of the gallbladder. This is a **porcelain gallbladder,** an uncommon entity that occurs with chronic inflammation and stasis and is associated with gallstones in over 90% of cases and an increased incidence of carcinoma of the gallbladder in about 20% of cases.

TABLE 18-1 RIMLIKE CALCIFICATIONS

Organ of Origin	Remarks
Renal cyst	Thick and irregular calcifications, though uncommon, may indicate the presence of renal cell carcinoma
Splenic cysts	May be a manifestation of hydatid cyst, old trauma, or prior infection
Aortic aneurysms	Occurs more often in diabetics with advanced atherosclerosis
Gallbladder	Associated with chronic stasis; called *porcelain gallbladder* for its gross appearance; higher incidence of carcinoma of the gallbladder

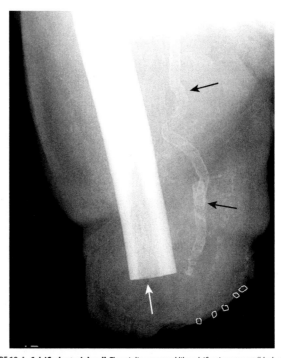

FIGURE 18-4 Calcified arterial wall. There is linear or tracklike calcification present *(black arrows)*, which implies calcification that has occurred in the walls of tubular structures. In the leg, this is calcification in the femoral artery. Such wall calcification occurs in arteries, not veins, and is usually secondary to atherosclerosis, frequently associated with diabetes, or in patients with chronic renal disease. This patient obviously suffered one of the complications of diabetes and has had an above-the-knee amputation *(white arrow)* of a previously gangrenous leg.

- ♦ **Walls of veins do not calcify.**
  - In veins, **long, linear thrombi** or **small, focal thrombi** (the latter called *phleboliths*) may calcify (see Fig. 14-18).
- ♦ **Tubular structures**
  - **Fallopian tubes** and **vas deferens**
    - ▪ Seen more often in diabetics (Fig. 18-5)
  - **Ureter**
    - ▪ Uncommon finding seen in schistosomiasis and even more rarely in tuberculosis

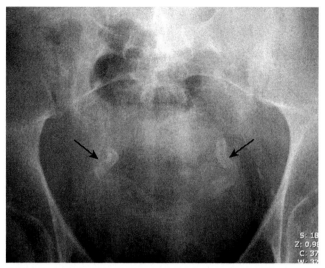

FIGURE 18-5 **Calcification of the vas deferens.** This patient manifests two tracklike calcifications *(black arrows)* symmetrically on each side of the urinary bladder that end in the urethra. The type of calcification identifies it as occurring in the wall of a tubular structure. The location identifies it as calcification in the walls of the vas deferens, which occurs more commonly and earlier in diabetics than as a natural, degenerative process.

TABLE 18-2 LINEAR OR TRACKLIKE CALCIFICATIONS

Organ of Origin	Remarks
Walls of smaller arteries	Mostly seen in atherosclerosis accelerated by diabetes and renal disease
Fallopian tubes or vas deferens	Usually accelerated by diabetes
Ureters	Uncommon occurrence described with schistosomiasis and, rarely, tuberculosis

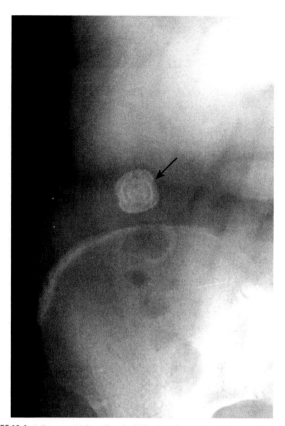

FIGURE 18-6 **Gallstone with lamellated calcification.** There is a lamellated calcification in the right upper quadrant *(black arrow)*. The alternating bands of calcium and less dense material identify it as a calculus that has formed in a viscus in this region. The anatomic location places it in the gallbladder. Most gallstones are not visible on conventional radiographs. Ultrasound is the most accurate imaging modality to detect gallstones.

- Table 18-2 highlights key facts about linear or tracklike calcifications.

LAMELLAR OR LAMINAR CALCIFICATION

- **Lamellar** (or laminar) **calcification implies calcification that forms around a nidus inside a hollow lumen.** A "hollow lumen" refers to a structure such as the gallbladder or urinary bladder that contains a fluid.

➡ **Calcification in concentric layers** begins with a central nidus around which alternating layers of calcified and noncalcified material form as a result of the prolonged movement of the stone within the hollow viscus (Fig. 18-6).

- Lamellar or laminated calcifications are usually called *stones* or *calculi* (singular: calculus) and include:
  - ♦ **Renal calculi**
    - Computed tomography (CT) is the study of choice for the detection of renal and ureteral calculi.
    - Conventional radiographs are only about 50% to 60% sensitive for displaying renal calculi despite the

fact that about 90% of renal calculi contain calcium (Fig. 18-7).
  - ♦ **Gallstones**
    - Ultrasound is the study of choice for the detection of gallstones.
    - Only about 10% to 15% of gallstones contain enough calcification to be visible on conventional radiographs (Fig. 18-8).
  - ♦ **Bladder stones**
    - Bladder stones usually develop secondary to chronic bladder outlet obstruction; they are very prone to develop lamination (Fig. 18-9).
- Table 18-3 highlights key facts about laminated or lamellar calcifications.

CLOUDLIKE, AMORPHOUS, OR POPCORN CALCIFICATION

- **Cloudlike, amorphous,** or **popcorn calcification is calcification that has formed inside of a solid organ or tumor,** and examples include:
  - ♦ **Body of the pancreas**
    - Pathognomonic for chronic pancreatitis (Fig. 18-10)
  - ♦ **Leiomyomas of uterus**
    - Uterine fibroids or leiomyomas very commonly degenerate and calcify over time (Fig. 18-11).

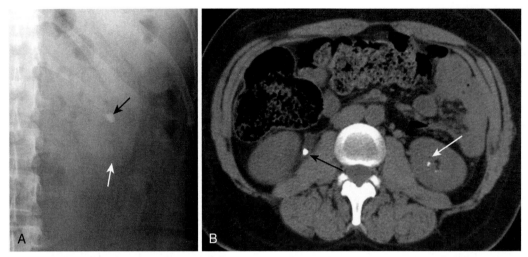

FIGURE 18-7 **Renal calculus, conventional radiograph (A), and axial computed tomography (CT) scan (B). A,** There is a calcification *(black arrow)* that overlies the shadow of the left kidney *(white arrow)*. Although it is too small to recognize lamination, its location suggests a renal calculus. **B,** In a different patient, an image from an unenhanced axial CT called a *stone search* reveals a large calcification in the proximal right ureter *(black arrow)* and several smaller calcifications in the left intrarenal collecting system *(white arrow)*. Because of its greater sensitivity, a CT stone search done without intravenous contrast has primarily replaced conventional radiography for the identification of renal and ureteral calculi.

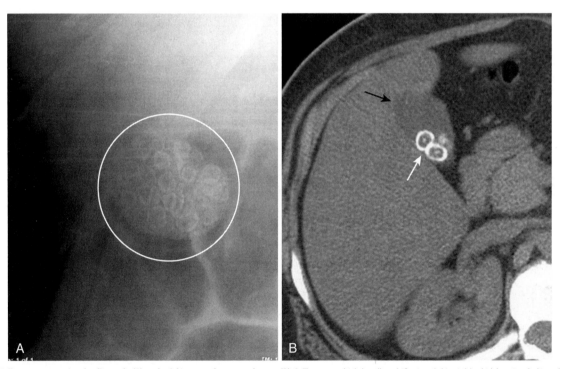

FIGURE 18-8 **Gallstones, conventional radiographs (A), and axial computed tomography scan (B). A,** There are multiple lamellar calcifications *(white circle)*, which have interlocking edges, suggesting that they all formed in a hollow viscus in proximity to each other. These are called ***faceted stones*** for their characteristic shapes. **B,** In a different patient, a close-up view of an unenhanced axial CT scan of the right upper quadrant shows several gallstones *(white arrow)*, two of which clearly have a central nidus surrounded by laminated, concentric rings of noncalcified and calcified material. The gallbladder *(black arrow)* contains bile fats and is less dense than the adjacent liver.

TABLE 18-3 LAMINAR OR LAMELLATED CALCIFICATIONS

Organ of Origin	Remarks
Kidney	Most calcified renal stones are composed of calcium oxalate crystals; most form due to stasis, infection
Gallbladder	Most calcified gallstones are calcium bilirubinate; form due to chronic infection and stasis
Urinary bladder	Most bladder calculi contain urate crystals; form most often from outlet obstruction

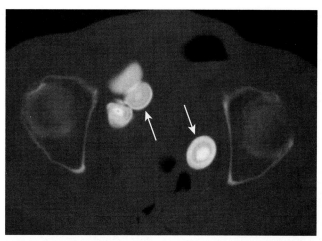

FIGURE 18-9 Urinary bladder stones. There are lamellar calcifications *(white arrows)* seen on this axial computed tomography scan through the level of the pelvis and windowed to show the laminations better. The laminations imply that these calcifications have formed inside of a hollow viscus. The anatomic location of these calculi places them in the urinary bladder.

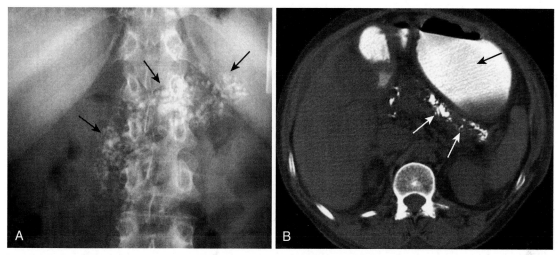

FIGURE 18-10 Chronic calcific pancreatitis, conventional radiograph (A), and axial computed tomography scan (B). A, A close-up view of the left upper quadrant of a conventional radiograph of the abdomen shows amorphous calcifications *(black arrows)*, implying calcification in a solid organ or tumor. The anatomic distribution of the calcification corresponds to the pancreas. **B,** In another patient, a nonenhanced image of the upper abdomen shows calcifications distributed along the course of the body and tail of the pancreas *(white arrows)*. There is oral contrast in the stomach *(black arrow)*. These calcifications are pathognomonic of chronic pancreatitis, an irreversible disease occurring mostly secondary to alcoholism that leads to atrophy of the gland, and diabetes.

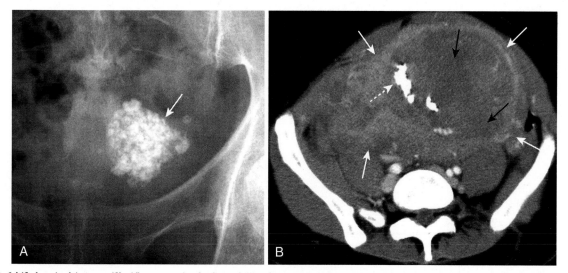

FIGURE 18-11 Calcified uterine leiomyoma (fibroid) on conventional radiograph (A) and computed tomography (B). A, There is an amorphous or popcorn calcification *(white arrow)* in the pelvis of this 48-year-old female. The type of calcification suggests formation in a solid organ or tumor. This is the anatomic location and the classical appearance of calcified uterine leiomyomas (fibroids). **B,** Another patient has large uterine fibroids *(solid white arrows)*, portions of which have necrosed *(solid black arrows)* and calcified *(dotted white arrow)*. Ultrasound is the study of choice in diagnosing uterine fibroids.

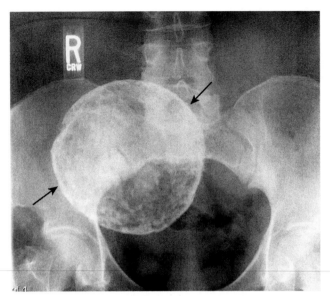

FIGURE 18-12 Calcified rim in uterine leiomyoma. This is a rimlike calcification *(black arrows)* in the pelvis of a female patient; it would be appropriate to consider that this calcification formed in the wall of a hollow viscus or saccular structure. In fact, this is a characteristic pattern of calcification in the outer wall of a degenerated uterine leiomyoma (fibroid). It is a rimlike calcification, but the structure in which it originally formed was solid. Cystic lesions of the ovary might produce this appearance, and ultrasound of the pelvis would be the study of choice to identify the organ of origin.

  Pitfall

- Sometimes a **solid tumor will outgrow its blood supply,** and **the center of the tumor will undergo necrosis,** leaving only a viable "outer shell."
- The subsequent calcification will be **more rimlike than amorphous.**
- Uterine fibroids are especially prone to this appearance (Fig. 18-12).

♦ **Lymph nodes**
- Lymph nodes can calcify anywhere in the body, mostly the result of prior granulomatous infection such as old tuberculosis (see Fig. 12-14).

♦ **Kidneys**
- **Medullary nephrocalcinosis** is macroscopic calcification deposited in the pyramids of the renal medulla, usually secondary to a metabolic derangement such as hyperparathyroidism (Fig. 18-13).

♦ **Mucin-producing adenocarcinomas** of the stomach, ovary, and colon (Fig. 18-14) may calcify both their primary tumors and any metastatic deposits.

♦ And outside of the abdomen in structures such as:
- **Meningiomas**

■ Table 18-4 highlights key facts about amorphous, cloudlike, or popcorn calcifications.

■ Table 18-5 summarizes the key findings of the four patterns of abnormal calcification.

  No matter what its cause, the presence of calcification implies a process that is subacute or chronic.

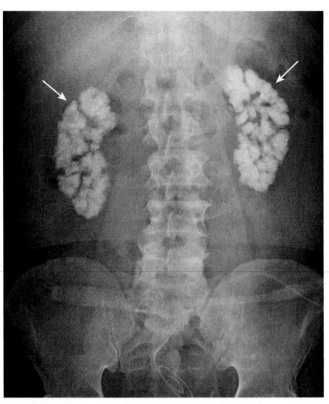

FIGURE 18-13 Medullary nephrocalcinosis. There are cloudlike calcifications seen bilaterally *(white arrows),* suggesting these calcifications have formed within a solid organ or tumor. The calcifications conform to the distribution of the renal collecting systems. This is medullary nephrocalcinosis, a condition not synonymous with renal calculi because nephrocalcinosis signifies a metabolic derangement. This patient had primary hyperparathyroidism.

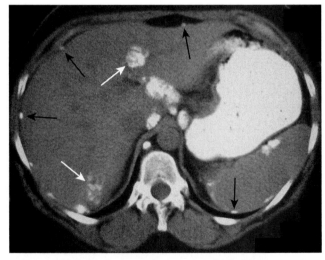

FIGURE 18-14 Calcified ovarian metastases. An unenhanced axial computed tomography scan of the upper abdomen shows multiple amorphous calcifications, some within the liver *(white arrows)* and others that stud the peritoneal surface of the abdomen *(black arrows).* This patient had a mucin-producing adenocarcinoma of the ovary, which metastasized to the peritoneum and liver. Mucin-producing tumors of the stomach and colon can also produce calcified metastases, but ovarian malignancy would be the most common to metastasize to the peritoneum.

TABLE 18-4 AMORPHOUS, CLOUDLIKE, OR POPCORN CALCIFICATIONS

Organ of Origin	Remarks
Pancreas	Chronic pancreatitis, frequently secondary to alcoholism
Uterine fibroids (leiomyomas)	Degenerating fibroids calcify
Mucin-producing tumors	Mucin-producing tumors of the ovary, stomach, or colon may calcify, as can their metastases
Meningioma	Benign, extraaxial brain tumor of older individuals that calcifies about 20% of the time

TABLE 18-5 IDENTIFYING THE FOUR TYPES OF ABNORMAL CALCIFICATION

Type of Calcification	Implies	Examples
Rimlike	Formed in wall of hollow viscus	Cysts, aneurysms, gallbladder
Linear or tracklike	Formed in walls of tubular structures	Ureters, arteries
Lamellar or laminar	Formed in stones	Renal, gallbladder, bladder calculi
Amorphous, cloudlike, popcorn	Forms in a solid organ or tumor	Uterine fibroids, some mucin-producing tumors

TABLE 18-6 LOCATION, LOCATION, LOCATION

Anatomic Quadrant in the Abdomen	Pattern of Calcification	Possible Organ of Origin	Cause
RUQ	Rimlike	Gallbladder wall	Chronic infection
	Tracklike	Hepatic artery	Atherosclerosis
	Laminated	Gallbladder	Gallstones
	Amorphous	Head of pancreas	Chronic pancreatitis
LUQ	Rimlike	Splenic cyst	Amebic infection
	Tracklike	Splenic artery	Atherosclerosis
	Laminated	Kidney	Renal stone
	Amorphous	Tail of pancreas	Chronic pancreatitis
RLQ	Rimlike	Iliac artery	Iliac artery aneurysm
	Tracklike	Iliac artery	Atherosclerosis
	Laminated	Appendix	Appendicolith
	Amorphous	Uterus	Fibroids
LLQ	Rimlike	Iliac artery	Iliac artery aneurysm
	Tracklike	Iliac artery	Atherosclerosis
	Laminated		
	Amorphous	Uterus or ovaries	Ovarian tumor

LLQ, Left lower quadrant; *LUQ,* left upper quadrant; *RLQ,* right lower quadrant; *RUQ,* right upper quadrant.

LOCATION OF CALCIFICATION

- Identifying the **pattern** of calcification helps in identifying its **type.**
- Identifying the anatomic **location** of the calcification helps to identify its **organ** or **tissue of origin.**
  - ♦ Combining the **type** of calcification with its **anatomic location** should provide the key to the **cause** of most pathologic calcifications.
- Table 18-6 summarizes some of the possibilities in the abdomen.

 TAKE-HOME POINTS

RECOGNIZING ABNORMAL CALCIFICATIONS AND THEIR CAUSES

Calcifications can be characterized by the pattern of their calcification and their anatomic location.

There are four distinct patterns: (1) rimlike, (2) linear or tracklike, (3) lamellar (or laminar), and (4) cloudlike, amorphous, or popcorn.

Rimlike calcifications imply calcification that has occurred in the wall of a hollow viscus, that is, a saccular structure containing fluid.

Examples of rimlike calcifications include the walls of cysts, aneurysms, or saccular organs like the gallbladder.

Linear or tracklike calcifications imply calcification that has occurred in the walls of tubular structures.

Examples of tracklike calcifications include the walls of arteries and tubular structures, such as the ureters, fallopian tubes, and vas deferens.

Lamellar (or laminar) calcifications imply calcification that forms around a nidus inside a hollow (usually fluid-containing) lumen.

Examples of lamellar calcifications include renal calculi, gallstones, and bladder stones.

Cloudlike, amorphous, or popcorn calcification is calcification that has formed inside of a solid organ or tumor.

Examples of amorphous or popcorn calcifications include the pancreas, leiomyomas of the uterus, lymph nodes, and mucin-producing adenocarcinomas.

Combining the type of calcification with its anatomic location should provide the keys to the causes of most pathologic calcifications.

 WEBLINK

Visit StudentConsult.Inkling.com for more information and a quiz on Abdominal Calcifications.

For your convenience, the following QR Code may be used to link to **StudentConsult.com**. You must register this title using the PIN code on the inside front cover of the text to access online content.

CHAPTER 19
Recognizing the Imaging Findings of Trauma

Trauma is the leading cause of death, hospitalization, and disability in Americans from the age of 1 year through age 45 years. The major imaging findings of most organ system's trauma will be discussed as a group in this chapter. Table 19-1 summarizes some of the traumatic injuries that are discussed in other chapters.

- Trauma-related injuries are divided into the two major mechanisms that produce them.
  - **Blunt trauma** is usually the result of motor vehicle accidents and is the more common of the two. Most of this chapter deals with the sequelae of blunt trauma.
  - **Penetrating trauma** is usually the result of accidental or criminal stabbings or gunshot wounds. Table 19-2 is an overview of the initial imaging work for **penetrating** injuries to various body parts.

TABLE 19-1   OTHER MANIFESTATIONS OF TRAUMA

Injury	Discussed in
Pleural effusion/hemothorax	Chapter 8
Aspiration	Chapter 9
Pneumothorax, pneumomediastinum, and pneumopericardium	Chapter 10
Fractures and dislocations	Chapter 24
Head trauma	Chapter 27
Spinal Trauma	Chapter 26

TABLE 19-2   INITIAL IMAGING WORKUP OF PENETRATING TRAUMA

Body Part	Modality of Choice
Head	Computed tomography (CT) scan of the head without and then possibly with contrast
Neck	Plain films are frequently obtained first, followed by CT-angiography if damage to the neck vessels is suspected
Chest	Chest radiographs are obtained first; then a CT scan of the chest as a follow-up in almost all patients, except those who are hemodynamically unstable
Abdomen	Abdominal CT scanning is study of choice in identifying and assessing the severity of liver or spleen injury

CHEST TRAUMA

- Chest injuries in trauma patients are **very common** and are responsible for one out of four of the trauma-related deaths. The overwhelming majority of chest trauma is the result of motor vehicle accidents.

RIB FRACTURES

- **Rib fractures are common** sequelae of blunt chest trauma, and the associated morbidity and mortality from such trauma increases as the number of rib fractures increases. The severity of underlying visceral injury is usually more important than the rib fractures themselves, but their presence might provide clues to unsuspected pathology.

➡ Fractures of the **first three ribs** are relatively **uncommon,** and if they occur following blunt trauma, they **indicate a sufficient amount of force** to produce other internal injuries (Fig. 19-1).

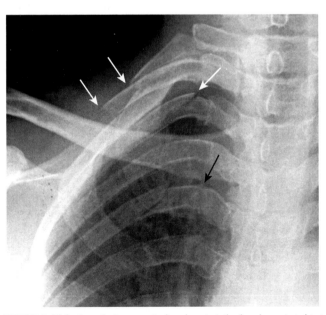

FIGURE 19-1   Rib fractures. Fractures present as linear lucencies in the ribs and are easier to detect when the fracture ends are displaced *(white arrows)*. Fractures of the first three ribs are relatively uncommon, and following blunt trauma, their presence is a clue that the force to the chest may have been sufficient to have produced other internal injuries. Do not mistake the normal costovertebral junction *(black arrow)* for a fracture.

- Fractures of **ribs 4 to 9 are common** and important if they are displaced (pneumothorax) or if there are two fractures in each of three or more contiguous ribs **(flail chest).**
  - ◆ **Flail chest** is almost always accompanied by a pulmonary contusion (see below). Because of the severity of the injuries with which it is usually associated, a flail chest has a significant mortality (Fig. 19-2).

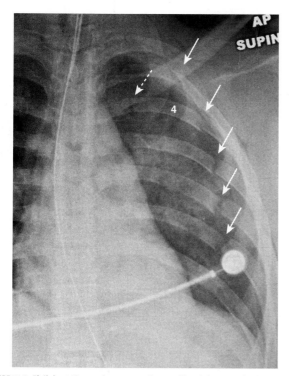

FIGURE 19-2 Flail chest. There were two or more fractures *(dotted white arrow demonstrates second fracture in rib 4)* present in more than three contiguous ribs *(solid white arrows)* in this patient struck by an automobile. The airspace disease in the left lung is an underlying pulmonary contusion, which almost always accompanies a flail chest.

- Fractures of **ribs 10 to 12** may indicate the presence of **underlying trauma** to the **liver** (right side) or the **spleen** (left side), especially if they are displaced (Video 19-1).
- In cases of minor trauma, it is **not unusual for rib fractures to be undetectable on the initial examination** but to become visible in several weeks after callus begins to form.

PULMONARY CONTUSIONS

- Pulmonary contusions are the most frequent complications of blunt chest trauma. They **represent hemorrhage into the lung,** usually at the point of impact.
- **Recognizing a pulmonary contusion:**
  - ◆ The **history of trauma** is of **paramount importance;** contusions present as **airspace disease** that is **indistinguishable from other airspace diseases** such as pneumonia or aspiration.
  - ◆ Contusions tend to be **peripherally placed** and **frequently occur at the point of maximum impact. Air bronchograms** are usually **not present** because blood fills the bronchi and the airspaces (Fig. 19-3).
- Classically, they **appear within 6 hours after the trauma** and because blood in the airspaces tends to be reabsorbed quickly, **disappear within 72 hours,** frequently sooner.

➡ Airspace disease that **lingers more than 72 hours** should **raise suspicions** of another process such as **aspiration, pneumonia,** or a **pulmonary laceration.**

PULMONARY LACERATIONS (HEMATOMA OR TRAUMATIC PNEUMATOCELE)

- Pulmonary hematomas **result from a laceration of the lung parenchyma** and as such, may accompany more severe **blunt trauma** or **penetrating chest trauma.**
- A pulmonary laceration is also called a *traumatic pneumatocele,* or *hematoma.*
- They are sometimes masked by the airspace disease from a surrounding pulmonary contusion, at least for the first few days until the contusion resolves.

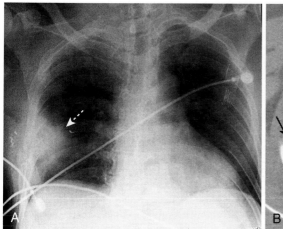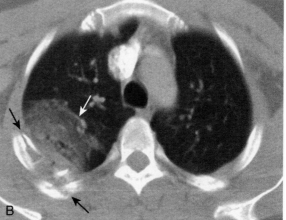

FIGURE 19-3 Pulmonary contusions, chest radiograph, and computed tomography. A, Pulmonary contusions tend to be peripherally placed and frequently at the point of maximum impact *(dotted white arrow).* Air bronchograms are usually not present because blood fills the bronchi and the airspaces. **B,** A second patient, who was an unrestrained passenger in an automobile accident, also has a large contusion *(solid white arrow)* associated with multiple rib fractures *(solid black arrows).*

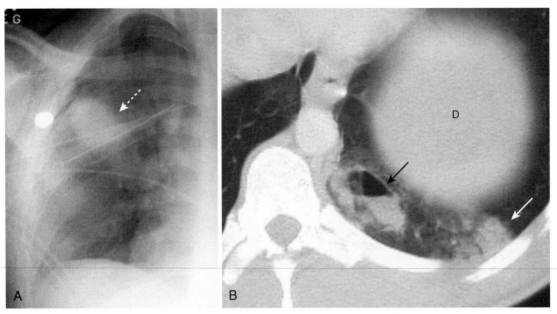

FIGURE 19-4 Pulmonary lacerations, conventional radiograph, and computed tomography. Lacerations are sometimes masked, at least for the first few days, by the airspace disease in a surrounding pulmonary contusion. **A,** If lacerations are completely filled with blood, they will appear as an ovoid mass *(dotted white arrow)*. **B,** If lacerations are partially filled with blood and partially filled with air, they may contain a visible air-fluid level *(black arrow)*. Unlike the neighboring pulmonary contusion *(solid white arrow)*, pulmonary lacerations, especially if they are blood filled, may take weeks or months to completely clear. The top of the left hemidiaphragm (D) is seen in this image.

- ■ **Recognizing a pulmonary laceration:**
  - ◆ Their **appearance will depend on whether they contain blood** and, **if so, how much** blood fills the laceration.
    - • If they are **completely filled** with **blood,** they will appear as a solid, usually **ovoid mass.**
    - • If they are **partially filled with blood** and **partially filled with air,** they may contain a visible **air-fluid level** or demonstrate a **crescent sign** as the blood begins to form a clot and pull away from the wall of the laceration.
    - • If they are **completely filled with air,** they will appear as an **air-containing, cystlike structure** in the lung (Fig. 19-4).
- ■ Unlike pulmonary contusions that clear rapidly, **pulmonary lacerations,** especially if they are blood filled, **may take weeks** or **months to completely clear.**

AORTIC TRAUMA

- ■ Trauma to the aorta is most frequently the result of **deceleration injuries in motor vehicle accidents.** Although the survival rates are improving, **most patients** with rupture of the thoracic aorta **die before reaching the hospital,** and of those who survive an aortic injury, the likelihood of death increases the longer the abnormality remains untreated. Only those with **incomplete** tears, in which the **adventitial lining prevents exsanguination (producing a pseudoaneurysm),** usually survive to be imaged.

➡ The most common site of injury is the **aortic isthmus,** which is the portion of the **aorta just distal to the origin of the left subclavian artery.** Seat-belt injuries may involve the abdominal aorta, but such injuries are far less common than the deceleration injuries to the thoracic aorta.

- ■ Only emergency surgery will prevent approximately 50% of patients with blunt aortic injuries from dying within the first 24 hours if left untreated.
- ■ **Recognizing aortic trauma:**
  - ◆ **Findings seen on conventional radiographs of the chest** are the same as those discussed under "Aortic Dissection" in Chapter 13. A completely **normal chest radiograph** has a relatively **high negative predictive value** for aortic injury, but an **abnormal chest x-ray** has a relatively **low positive predictive value (78%).**
    - • "Widening of the mediastinum" is usually a **poor means of establishing the diagnosis** because it is difficult to assess on a supine, portable chest radiograph, and it is commonly overinterpreted. There may also be **loss of the normal shadow of the aortic knob, a left apical pleural cap** of fluid or blood, a **left pleural effusion,** or **deviation of the trachea or esophagus to the right** (Fig. 19-5).
  - ◆ Under most circumstances, suspected aortic injuries are now studied using computed tomography (**CT**), which allows for rapid image acquisition in one breath-hold and appropriately timed contrast delivery to the aorta (**CT-angiography**). Most experts agree that a negative CT-angiogram obviates the need for angiography. The findings are **frequently subtle** and require experience to recognize, because those patients with the more obvious findings may not have survived to be imaged.

♦ **Findings on contrast-enhanced CT scans of the chest** (Fig. 19-6):

• **Aortic intimal flap.** A lucent defect in the contrast column of the aorta arising from a **tear** in the intima and media.

• **Contour** or **caliber abnormalities.** Abrupt change in the smooth contour or size of the aorta at the point of injury.

• **Periaortic hematoma.** Delineation of a **contrast-filled collection outside of the normal confines of the aorta** representing pseudoaneurysm or extravasation.

• **Mediastinal hematoma.** Increased attenuation in the mediastinum from an admixture of blood and normal fat. May be present in the absence of an aortic injury, presumably due to small vessel trauma.

• **Hemopericardium.** Fluid of high-attenuation (i.e., blood) in the pericardial sac indicates a significant injury to the aorta or heart itself.

♦ Patients with **equivocal** CT findings may go on to a catheter study of the aorta **(aortography).**

ABDOMINAL TRAUMA

■ The role of advanced imaging techniques deserves special mention in abdominal trauma. Radiology has made a significant impact on the lives of traumatized patients by distinguishing those patients who can be managed conservatively from those who need surgical or other interventions and by helping to direct the most appropriate intervention for those who need it.

➡ **CT is the study of choice in abdominal trauma.**

♦ **Intravenous contrast is always used** (unless contraindicated) to identify devascularized areas, hematomas, active extravasation of blood, or extraluminal urine (after contrast passes through the kidneys).

• If a **head CT is also to be done,** it should have **first priority,** before contrast is injected for the abdomen.

♦ **Oral contrast** is usually not administered. **Rectal contrast** is occasionally administered in penetrating trauma, to search for a bowel laceration.

♦ It is always best to consult with the radiologist to tailor the best study to fit the patient's needs.

♦ In some emergency settings, a quick abdominal ultrasound is used in unstable trauma patients to evaluate for hemoperitoneum (Box 19-1).

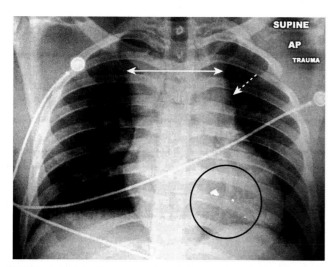

FIGURE 19-5 Mediastinal hematoma. There is "widening of the mediastinum" *(double white arrow)*, an inexact finding on an anteroposterior supine and portable chest radiograph. More important, the shadow of the aortic knob is obscured by something of soft tissue density *(dotted white arrow)*. Since the patient had been shot *(bullet fragments in the black circle)*, these findings did lead to a request for a CT-angiogram, which demonstrated a large mediastinal hematoma.

BOX 19-1 FOCUSED ABDOMINAL SONOGRAM FOR TRAUMA (FAST)

A portable ultrasound utilized on unstable trauma patients solely to identify free peritoneal fluid

Used primarily in place of the diagnostic peritoneal lavage

False negatives occur with abdominal injuries in which there is no hemoperitoneum

---

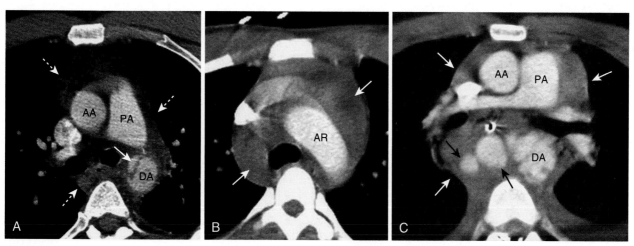

FIGURE 19-6 Aortic trauma, three different patients. A, There is a tear at the level of the aortic isthmus represented by the linear defect in the wall of the descending aorta *(solid white arrow)*. A mediastinal hematoma is also present *(dotted white arrows)*. **B,** There is a large mediastinal hematoma *(solid white arrows)*. **C,** There are periaortic hematomas containing extravasated blood *(black arrows)* and a large mediastinal hematoma *(solid white arrows)*. *AA,* Ascending aorta; *AR,* aortic arch; *DA,* descending aorta; *PA,* pulmonary artery.

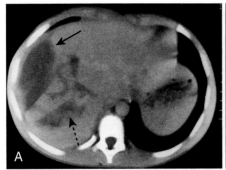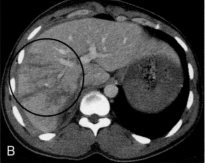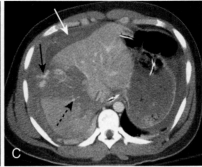

FIGURE 19-7 Hepatic trauma, three different patients. A, There is a lenticular fluid collection involving the lateral portion of the right lobe of the liver that represents a subcapsular hematoma *(solid black arrow)*. There is also a laceration of the right lobe *(dotted black arrow)*. **B,** There are multiple lacerations of the right lobe of the liver *(black circle)*. **C,** There is active extravasation of contrast-enhanced blood *(solid black arrow)* from a large intrahepatic laceration with hematoma *(dotted black arrow)*, as well as both subcapsular blood and hemoperitoneum *(solid white arrow)*.

■ The most commonly affected solid organs in **blunt** abdominal trauma (in order of decreasing frequency) are the **spleen, liver, kidney,** and **urinary bladder.** Traumatic injuries to each of them will be discussed under each organ.

Liver

➡ The liver is discussed first because it is actually the **most frequently injured organ** if both penetrating and blunt trauma are included together. The liver is the largest intraabdominal organ and is fixed in position, making it especially susceptible to injury. Injuries to the liver account for the **majority of deaths from abdominal trauma.**

■ The posterior aspect of the right lobe is injured most frequently. Most hepatic injuries are associated with blood in the peritoneal cavity **(hemoperitoneum).**

■ **Contrast-enhanced CT is the study of choice,** and because of its ability to demonstrate both the nature and extent of the trauma, the overwhelming majority of patients with liver trauma are now managed conservatively and do not require surgery.

■ **CT findings in hepatic trauma** (Fig. 19-7):
  ♦ **Subcapsular hematoma.** Lenticular fluid collections that conform to the shape of the outer contour of the liver, but which frequently flatten the adjacent liver parenchyma. Most occur anterolaterally over the right hepatic lobe.
  ♦ **Lacerations.** Most common finding. Irregularly margined, low attenuation, linear or branching defects, usually at the periphery. *Fracture* is a term that has been used to describe a laceration that avulses a section of the liver.
  ♦ **Intrahepatic hematomas.** Focal, high-attenuation lesions first caused by blood, hematomas may progress to low-attenuation, masslike lesions filled with serous fluid.
  ♦ **Wedge-shaped defects.** Devascularized sections of liver parenchyma, which do not enhance with contrast.
  ♦ **Contusions.** A term used to describe an area of minimal parenchymal hemorrhage; they are lower in attenuation than the surrounding liver and have indistinct margins.
  ♦ **Pseudoaneurysms** and **acute hemorrhages.** Irregular collections of high-attenuation, extravasated contrast

that often require angiography with embolization and/or surgery.

Spleen

■ Splenic trauma is usually caused by **deceleration injuries** in unrestrained occupants of motor vehicle collisions, by a fall from a height, or by being struck by a motor vehicle as a pedestrian.

■ Because the **spleen is the most highly vascular organ,** hemorrhage represents the most serious complication of trauma. Despite its vascular nature and the delayed presentation of many splenic injuries, **most splenic trauma is treated conservatively** (nonsurgical).

■ **CT is the study of choice for evaluating splenic trauma.** Findings include (Fig. 19-8):
  ♦ **Contusion.** Alterations in the normal homogeneous appearance of the spleen, including mottled areas of low attenuation.
  ♦ **Subcapsular hematoma.** Low-attenuation, crescent-shaped collection of fluid in the subcapsular space, which frequently compresses the normal splenic parenchyma.
  ♦ **Laceration.** Irregular, low-attenuation defect that typically transects the spleen.
  ♦ **Intraparenchymal hematoma.** Lacerations filled with blood; they are intrasplenic, rounded areas of low attenuation, which may have a mass effect and enlarge the spleen.
  ♦ **Intraperitoneal fluid** or **blood.** Hemoperitoneum occurs with almost all splenic injuries, including small amounts of blood in the pelvis. Its presence does not necessarily indicate active hemorrhage.

Kidneys

■ Motor vehicle accidents are the most common cause of blunt abdominal trauma to the kidneys in the United States. Almost all patients with renal trauma will have **hematuria.**

■ **Contrast-enhanced CT is the study of first choice** and has almost completely replaced the intravenous urogram and standard cystogram.

■ **CT findings in renal trauma** (Fig. 19-9):
  ♦ **Contusion.** Ill-defined, patchy, low-attenuation areas in the contrast-enhanced kidney.
  ♦ **Subcapsular hematoma.** Crescentic or elliptical densities that compress the denser underlying renal parenchyma.

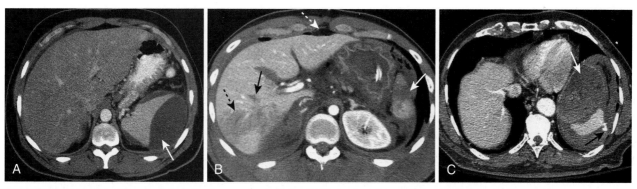

FIGURE 19-8 **Splenic trauma, three different patients. A,** There is crescent-shaped collection of fluid in the subcapsular space, which compresses the normal splenic parenchyma, representing subcapsular hematoma *(solid white arrow)*. **B,** This patient has a splenic *(solid white arrow)* and hepatic *(solid black arrow)* laceration and a large hepatic contusion *(dotted black arrow)*. There is also pneumoperitoneum *(dotted white arrow)*. **C,** This patient has active extravasation of contrast-enhanced blood *(solid black arrow)* and a large intrasplenic hematoma *(solid white arrow)*.

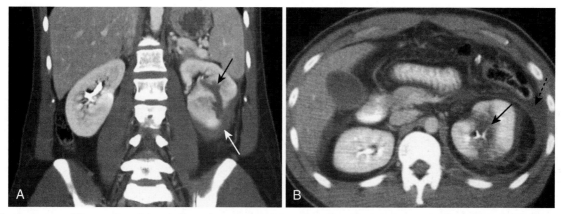

FIGURE 19-9 **Renal trauma, two different patients. A,** Coronal-reformatted contrast-enhanced CT scan shows a low-attenuation linear defect, representing a renal laceration *(black arrow)* and a subcapsular hematoma *(white arrow)*. **B,** Axial CT scan on another patient also shows a renal laceration *(solid black arrow)*, and a perinephric hematoma *(dotted black arrow)*.

♦ **Perinephric hematoma.** Ill-defined fluid collection surrounding the kidney confined by Gerota fascia.
♦ **Laceration.** Low-attenuation, linear, or branching defects in the renal parenchyma. More severe lacerations may extend through the renal hilum into the collecting system, renal artery, or vein. *Fracture* is a term that may be used when the laceration connects the hilum with the cortex.
♦ **Vascular injuries.** If arterial, there may be no flow to the kidney and, hence, no contrast-enhancement. They may also produce wedge-shaped defects in the kidney.
■ **Injuries to the collecting system.** Extraluminal contrast arising from the renal pelvis or ureter (Fig. 19-10).

Shock Bowel

■ Shock bowel usually occurs with **blunt abdominal trauma** in which there is **severe hypovolemia** and **profound hypotension,** with complete reversibility of these findings following resuscitation.
■ **Recognizing shock bowel on CT:**
♦ **Diffuse thickening** of the **small bowel** wall with **increased enhancement**
♦ **Fluid-filled** and **dilated loops of bowel**
♦ Other findings include a **small inferior vena cava** (<1 cm) and **aorta** (<6 mm) and decreased perfusion of the spleen (Fig. 19-11).

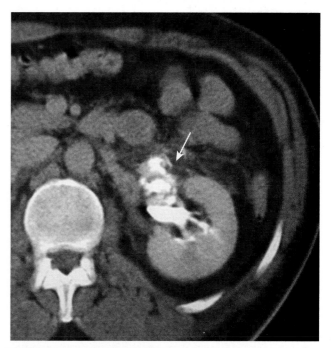

FIGURE 19-10 **Tear of proximal ureter.** There is a tear of the ureter at the level of the left ureteropelvic junction demonstrated by extraluminal contrast *(white arrow)*, representing contrast-containing urine that is leaking from the collecting system. The patient was an unrestrained driver in a motor vehicle accident.

PELVIC TRAUMA

Rupture of the Urinary Bladder

■ About **70% of bladder ruptures occur with pelvic fractures,** and about **10% of patients with pelvic fractures have an associated rupture of the bladder.**

■ They are best demonstrated by a **CT cystogram,** in which contrast is infused under gravity through a Foley catheter into the bladder, but they can also be well-demonstrated by **antegrade filling of the bladder** from renal excretion of intravenously injected contrast.

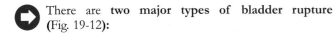

There are **two major types of bladder rupture** (Fig. 19-12):

◆ **Extraperitoneal** bladder rupture is more common (80%) and usually the **result of a pelvic fracture** with **direct puncture** of the bladder. Extraluminal contrast **remains around the bladder,** especially the retropubic space.

◆ **Intraperitoneal** bladder rupture is less common, and usually the **result of a forceful blow to the pelvis with a distended bladder,** especially in children. Rupture usually occurs at the **dome** of the bladder

adjacent to the peritoneal cavity. Contrast runs freely through the **peritoneal cavity, surrounds bowel,** and **extends into the paracolic gutters.**

Urethral Injuries

■ Urethral injuries are associated with significant pelvic trauma in **males,** most often from **blunt trauma.**

■ Urethral injuries should be investigated when there are **straddle fractures** of the pelvis or **penetrating injuries** in the region of the urethra. **Hematuria, blood at the urethral meatus,** and **inability to void** are suggestive clinical findings.

■ Imaging is done most often using **retrograde urethrography (RUG),** in which contrast is instilled retrograde at the urethral meatus and there is retrograde filling of the urethra. This is done before insertion of a Foley catheter into the bladder.

■ The **most common injury** is a rupture of the **posterior urethra** through the urogenital diaphragm into the proximal bulbous urethra. Extraluminal contrast can be seen outside of the urethra in the pelvis and perineum (Fig. 19-13).

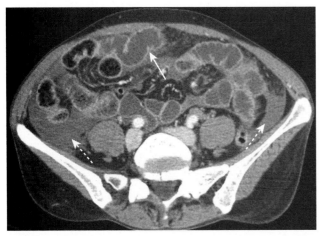

FIGURE 19-11 Shock bowel. There is marked enhancement of the bowel wall with multiple dilated and fluid-filled loops *(solid white arrow)*. There is also retroperitoneal fluid present *(dotted white arrows)*. Shock bowel usually occurs with severe hypovolemia and profound hypotension.

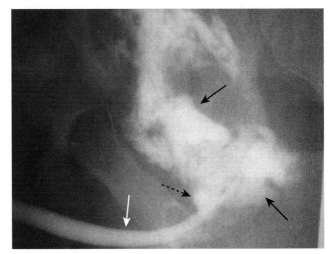

FIGURE 19-13 Urethral trauma. Contrast instilled retrograde through the penile urethra *(white arrow)* is seen to leak from the posterior urethra secondary to a perforation *(dotted black arrow)* and collects outside of the urinary system in the perineum and extraperitoneal bladder spaces *(solid black arrows)*. The patient had pelvic fractures secondary to a fall.

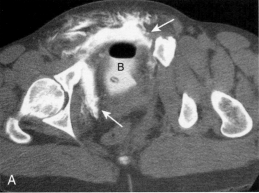
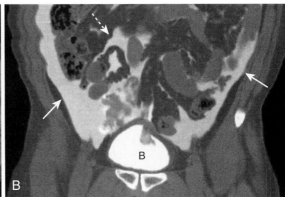

FIGURE 19-12 Bladder ruptures, extraperitoneal and intraperitoneal. A, Contrast-containing urine *(white arrows)* has leaked into the extraperitoneal spaces from a perforated bladder following pelvic fractures. Contrast, the tip of a Foley catheter, and air are seen inside of the partially filled urinary bladder (B). **B,** Intraperitoneal bladder ruptures are less common and may occur with blunt trauma. The contrast flows freely away from the bladder (B) up the paracolic gutters *(solid white arrows)* and outlines loops of bowel *(dotted white arrow)*.

TAKE-HOME POINTS

RECOGNIZING THE IMAGING FINDINGS OF TRAUMA

Trauma is generally divided into blunt and penetrating trauma. Most trauma-related injuries are due to blunt trauma, with motor vehicle accidents contributing the majority.

CT has had a profound impact in traumatized patients by distinguishing those patients who can be managed conservatively from those who need surgical or other interventions.

Rib fractures may herald more serious internal injuries, such as lacerations of the liver, spleen, or pneumothoraces. Most occur in ribs 4 to 9.

Pulmonary contusions are the most common manifestation of blunt chest trauma and represent hemorrhage into the lung, usually at the point of impact. They classically clear in a few days.

Pulmonary lacerations are tears in the lung parenchyma that may be fluid- or air-containing. Their presence may be hidden by a surrounding contusion, and they typically take longer than a contusion to clear.

Aortic injuries usually occur at the isthmus, require rapid recognition for optimum survival, and may take the form on contrast-enhanced CT of intimal flaps, contour abnormalities, or hematomas.

The most commonly affected solid organs in blunt abdominal trauma (in order of decreasing frequency) are the spleen, liver, kidney, and urinary bladder.

The liver is commonly injured in both blunt and penetrating trauma, and its injuries account for the majority of the deaths from abdominal trauma. It may demonstrate lacerations, hematomas, wedge-shaped defects, pseudoaneurysms, and acute hemorrhage.

Because the spleen is highly vascular, hemorrhage is the most serious sequela of splenic trauma; findings include hematoma, laceration, and contusions.

Patients who have had renal trauma almost all have hematuria and may show contusions, lacerations, hematomas, or vascular pedicle injuries on CT. They may also demonstrate extraluminal contrast from an injury to the renal pelvis or ureter.

Shock bowel is a consequence of profound hypotension and shows diffuse small bowel wall thickening with enhancement of dilated and fluid-filled loops on CT.

Bladder ruptures may be either extraperitoneal (more common) or intraperitoneal, the former demonstrating extraluminal contrast surrounding the bladder and the latter showing contrast that flows freely in the peritoneal cavity.

Urethral injuries occur almost exclusively in males, are frequently associated with pelvic fractures, and usually involve the posterior urethra, where extraluminal contrast may be seen in the perineum or extraperitoneally in the pelvis.

 WEBLINK

Visit StudentConsult.Inkling.com for videos and more information.

For your convenience, the following QR Code may be used to link to **StudentConsult.com**. You must register this title using the PIN code on the inside front cover of the text to access online content.

CHAPTER 20
Recognizing Gastrointestinal, Hepatic, and Urinary Tract Abnormalities

- In this chapter, you will learn how to recognize some of the most common abnormalities of the gastrointestinal (GI) tract from the esophagus to the rectum. We will also discuss selected hepatic abnormalities. Chapter 21 on ultrasound will describe some of the more common biliary and pelvic abnormalities.

- Computed tomography (CT), ultrasound, and magnetic resonance imaging (MRI) have essentially replaced conventional radiography, and in some instances, barium studies for the evaluation of the GI tract and the visceral abdominal organs.
- To review some of the terminology used for fluoroscopic studies of the GI tract, see Table 20-1.

TABLE 20-1 TERMINOLOGY

Term	Definition
Fluoroscopy	Utilization by the radiologist of special x-ray producing equipment to observe in real time the dynamic movement of the bowel and to optimally position the patient so as to obtain diagnostic images frequently referred to as *spot films*; in this chapter, the term is used in reference to utilizing x-rays to image the gastrointestinal (GI) tract.
Barium	Barium sulfate in suspension is an inert, radioopaque material prepared in liquid form to study the intraluminal anatomy of the GI tract.
Single-contrast/ double-contrast/ biphasic examination	A single-contrast (also called **full-column**) study usually refers to a GI imaging procedure in which only barium is used as the contrast agent; double contrast (sometimes called **air contrast**) usually refers to a study of the GI tract using both thicker barium and air; a **biphasic examination** is used to study the upper gastrointestinal tract and utilizes an initial double-contrast study followed by a single contrast agent to optimize the study.
Filling defect	A lesion, usually of soft tissue density, that protrudes into the lumen and displaces the intraluminal contrast (e.g., a polyp is a filling defect).
Ulcer	Refers to a persistent collection of contrast that projects outward from the contrast-filled lumen and originates either through a break in the mucosal lining (as in gastric ulcer) or in a GI mass (as in an ulcerating malignancy).
Diverticulum	Refers to a persistent collection of contrast that projects outward from the contrast-filled lumen of the GI tract like an ulcer; unlike an ulcer, the mucosa of a diverticulum is intact; false diverticula represent outpouchings of mucosa and submucosa through the muscularis.
Spot films and overhead films	*Spot films* usually refer to static images obtained by the radiologist, who utilizes fluoroscopy to position the patient for the optimum image; *overhead films* is a term that refers to additional images obtained by the radiologic technologist to complement fluoroscopic spot films using an x-ray tube mounted on the ceiling of the radiographic room (thus, the term *overhead*).
Intraluminal, intramural, extrinsic	Intraluminal (sometimes shortened to luminal) lesions generally arise from the mucosa, such as polyps and carcinomas; intramural (sometimes shortened to mural) lesions arise from the wall (in this chapter from the GI tract), such as leiomyomas and lipomas; extrinsic lesions arise outside of the GI tract (e.g., serosal metastases or endometriosis implants).
En face and in profile	When you look at a lesion directly "head-on," you are seeing it *en face*; a lesion seen tangentially (from the side) is seen in profile; except for those that are perfect spheres, lesions will have a different shape when viewed *en face* and in profile.
Fully distended vs. collapsed	Only loops that are fully distended by contrast can be accurately evaluated, no matter what part of the GI tract is being studied; evaluating certain criteria (such as wall thickness) using collapsed loops may introduce errors of diagnosis.
Change and distensibility	Over time (usually measured in seconds), the walls of all GI luminal structures from esophagus to rectum change in contour, distending and ballooning outward with increasing volumes of barium and/or air. Change and distensibility are normal.
Rigid, stiff, fixed, nondistensible	If the wall of bowel is infiltrated by tumor, blood, edema, or fibrous tissue, for example, the bowel may lose its ability to change and distend; this lack of distensibility is variously called *rigidity*, *stiffening*, *fixed*, or *nondistensible*. This is abnormal.
Irregularity	Except for the normal marginal indentations caused by the folds in the stomach, small bowel, and colon, the walls of the entire GI tract appear relatively smooth and regular; diseases can produce ulceration, infiltration, and nodularity with resultant irregularity of the wall.
Persistence	Almost without exception, an apparent abnormality must be seen on more than one image to be considered a pathologic finding; transient changes in the GI tract caused by peristalsis, ingested food, the presence of stool, or incompletely distended loops of bowel will disappear over time, but true abnormalities will remain constant and persistent.

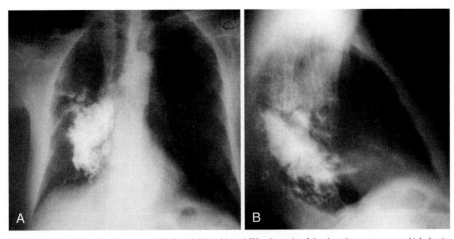

FIGURE 20-1 Aspiration, barium gone wild. Frontal **(A)** and lateral **(B)** radiographs of the chest demonstrate a very high-density material in the right lower lobe. The material is metal density and represents barium that was aspirated into the lung during an upper gastrointestinal series. Barium is inert and did not cause any additional symptoms that the patient was not already having from aspirating his own secretions. It will take some time, but most of this barium will be reabsorbed, most likely leaving only a small amount remaining.

ESOPHAGUS

- Single- and/or double-contrast examinations of the esophagus are performed with the patient drinking liquid barium, either by itself **(single contrast)** or accompanied by a gas-producing agent that provides the "air" in a **double-contrast examination.** Since both the single- and double-contrast techniques have their own strengths, many esophagrams are routinely performed using both techniques, called a *biphasic examination.*

- **Video esophagography (video swallowing function)** is a study of the **swallowing mechanism,** usually performed with fluoroscopy and frequently captured dynamically, either digitally, on videotape, or on film. This is the study of choice for diagnosing and documenting **aspiration,** in which ingested substances pass into the trachea below the level of the vocal cords (Fig. 20-1; Video 20-1).

- **Fluoroscopic observation** of the esophagus can also **reveal abnormalities in esophageal motility.** For example, **tertiary waves** are a common but nonspecific abnormality of esophageal motility, representing disordered and nonpropulsive contractions of the esophagus. They can be observed fluoroscopically and captured on spot films (Fig. 20-2; Video 20-2).

Esophageal Diverticula

- Diverticula of the GI tract are **usually produced when the mucosal** and **submucosal layers herniate through a defect in the muscular layer** of the bowel wall. Wherever they occur in the GI tract, diverticula produce an **outpouching** that projects beyond the borders of the lumen.

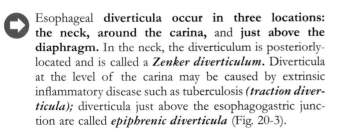 Esophageal **diverticula occur in three locations: the neck, around the carina,** and **just above the diaphragm.** In the neck, the diverticulum is posteriorly-located and is called a *Zenker diverticulum.* Diverticula at the level of the carina may be caused by extrinsic inflammatory disease such as tuberculosis *(traction diverticula);* diverticula just above the esophagogastric junction are called *epiphrenic diverticula* (Fig. 20-3).

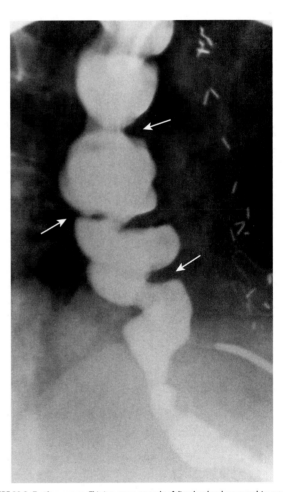

FIGURE 20-2 Tertiary waves. This is a severe example of disordered and nonpropulsive waves of contraction in the esophagus called *tertiary waves (white arrows).* The term *corkscrew esophagus* is sometimes applied to this appearance. Tertiary waves are a nonspecific and very common abnormality that increases with advancing age.

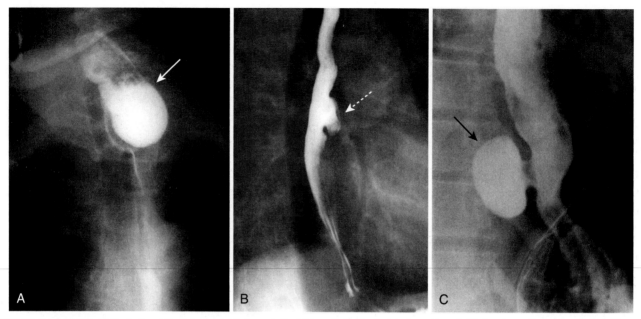

FIGURE 20-3 Esophageal diverticula. Esophageal diverticula characteristically occur in the neck from a localized weakness in the posterior wall of the hypopharynx (**Zenker diverticulum**) *(white arrow)* **(A)**; in the midesophagus from extrinsic disease such as tuberculosis that causes fibrosis, which pulls on the esophagus, forming a **traction diverticulum** *(dotted white arrow)* **(B)**; or just above the diaphragm in the distal esophagus **(epiphrenic diverticulum)** *(black arrow)* **(C)**. Only the traction diverticulum is a **true** diverticulum in that it has all layers of the esophagus involved; the Zenker and epiphrenic are **false** or **pseudodiverticula** because the mucosa and submucosa herniate through a defect in the muscular layer. The Zenker diverticulum is the only one of the three that typically produces symptoms.

Esophageal Carcinoma

- Esophageal carcinoma continues to have a very **poor prognosis** as more than **50% of patients will have metastases upon initial presentation**. The **lack of an esophageal serosa** and a **rich supply of lymphatics** aid in the extension and dissemination of esophageal carcinoma. A combination of long-term alcohol and tobacco use are associated with a higher risk of esophageal carcinoma.
- Esophageal malignancies are either **squamous cell carcinomas** or **adenocarcinomas,** the latter of which is **increasing in prevalence. Adenocarcinomas** arise in esophageal epithelium that has undergone **metaplasia** from **squamous to columnar epithelium (Barrett esophagus),** a process in which **gastroesophageal reflux (GERD) plays a major role.**
- Barium esophagrams are **frequently the initial study** in patients with symptoms such as dysphagia, which suggest this diagnosis.
- Esophageal carcinomas may **appear in one or more of several forms,** including an **annular constricting lesion; polypoid mass; a superficial, infiltrating lesion** or **ulceration;** and **irregularity of the wall.** Most often they present as a mixture of several of these patterns (Fig. 20-4).

Hiatal Hernia and Gastroesophageal Reflux

- Hiatal hernias are divided into the **sliding type** (almost all), in which the **esophagogastric (EG) junction lies above the diaphragm,** or the **paraesophageal type (1%),** in which a **portion of the stomach herniates through the esophageal hiatus, but the EG junction remains below**

the diaphragm. In general, hiatal hernia increases in incidence with age.
- **Most hiatal hernias are asymptomatic,** but there is an association between the presence of some hiatal hernias and clinically significant **gastroesophageal reflux.**
  - ◆ Gastroesophageal **reflux also occurs in patients without** any visible **hiatal hernia,** usually as a result of dysfunction of the lower esophageal sphincter, which normally acts to prevent gastric acid from repeatedly refluxing into the esophagus.

➡ The radiologic findings of hiatal hernia (Fig. 20-5) include a **bulbous area** of the distal esophagus containing oral contrast at the level of the diaphragm, with **failure of the esophagus to narrow** on multiple images **as it passes through the esophageal hiatus; extension of multiple gastric folds above the diaphragm;** and sometimes visualization of a thin, circumferential filling defect in the distal esophagus called a *Schatzki ring.*

  - ◆ A Schatzki ring **marks the position of the esophagogastric junction** so that its appearance above the diaphragm indicates the presence of a sliding hiatal hernia (Fig. 20-6).
- **Gastroesophageal reflux may be evident during fluoroscopy** when barium is seen to move from the stomach retrograde into the esophagus, but reflux is intermittent so that it may not occur during the course of the examination. The absence of reflux during the study **does not exclude reflux,** and demonstration of reflux **does not necessarily indicate the patient has the complications of GERD; that is, esophagitis, stricture,** and Barrett esophagus.

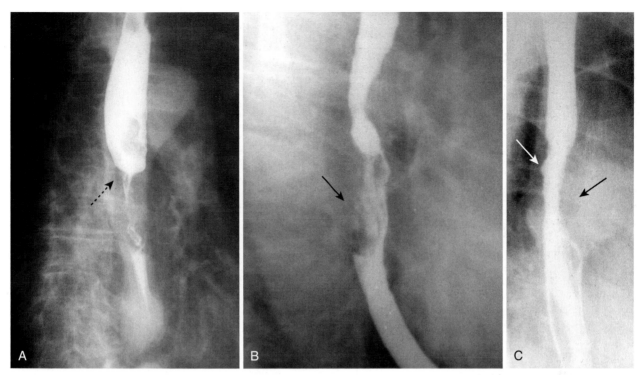

FIGURE 20-4 **Esophageal carcinomas.** Three different patients are shown with different appearances of esophageal carcinoma. **A,** There is an **annular constricting** lesion of the midesophagus *(dotted black arrow)*. The tumor encircles the normal lumen and obstructs it in this case. **B,** There is a **polypoid mass** that arises from the right lateral wall of the esophagus and displaces the barium around it *(black arrow)*. **C,** The wall is irregular and rigid *(black arrow)* and contains a small **ulceration** *(white arrow)*.

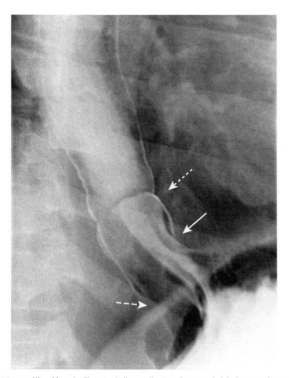

FIGURE 20-5 **Hiatal hernia.** There is a bulbous collection of contrast *(solid white arrow)* representing the stomach herniated above the diaphragm *(dashed white arrow)*. There are gastric folds present in the hernia, identifying it as part of the stomach. Notice the esophagus does not narrow as it normally does when passing through the esophageal hiatus. The narrowing seen above the hernia *(dotted white arrow)* is the esophagogastric junction.

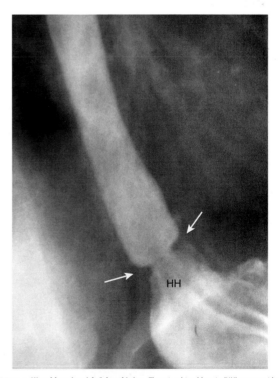

FIGURE 20-6 **Hiatal hernia with Schatzki ring.** There is a hiatal hernia (HH) present, identified by the multiple gastric folds within it and the lack of normal narrowing as the esophagus passes through the diaphragmatic hiatus. Just above the hernia is a thin, weblike filling defect characteristic of a Schatzki ring *(white arrows)*. The Schatzki ring marks the level of the esophagogastric junction.

STOMACH AND DUODENUM

- Today the lumen of the stomach is most often studied by upper endoscopy, and the wall thickness and structures outside of the stomach are studied by CT examination of the abdomen with oral contrast. Nevertheless, biphasic upper gastrointestinal (UGI) examinations, which include a study of the esophagus, stomach, and duodenum, remain a sensitive, cost-effective, readily-available, and noninvasive alternative.

Gastric Ulcers

- In the United States, the **incidence of gastric ulcer disease has been declining**. In adults, infection with *Helicobacter pylori* accounts for almost three out of four cases of gastric ulcer disease. Nonsteroidal antiinflammatory agents account for most of the remainder of cases.

 Most ulcers occur on the **lesser curvature** or **posterior wall** in the region of the **body** or **antrum**. About **95% of all gastric ulcers are benign**. The other **5% will represent ulcerations in gastric malignancies** (Fig. 20-7).

Gastric Carcinoma

- There has been a dramatic **decline in the incidence of gastric carcinoma in the United States**. The **mortality, however, remains quite high** because they are frequently not diagnosed until after they have spread. Most gastric carcinomas (actually, they are adenocarcinomas) **occur in the distal third of the stomach** along the **lesser curvature.**
- Double-contrast UGI images and CT scans of the abdomen can demonstrate gastric carcinomas. CT is utilized for staging the extent of the tumor and degree of spread.
- Gastric carcinomas may be **polypoid, infiltrating** (i.e., *linitis plastica),* or **ulcerative** in form (Fig. 20-8).
- There are other mass lesions that may resemble gastric carcinoma including **leiomyomas,** a benign, wall lesion that characteristically ulcerates; and **lymphoma,** which may produce diffusely thickened folds or multiple masses in the stomach.

DUODENAL ULCER

 Duodenal ulcers are two to three times more common than gastric ulcers. Almost all duodenal ulcers occur in the **duodenal bulb,** the majority on the **anterior wall** of the bulb. They are **overwhelmingly caused by *H. pylori* infection (85% to 95%).**

- Double-contrast UGI series have a sensitivity that exceeds 90% in detecting duodenal ulcers (Fig. 20-9).
- Complications of duodenal ulcers, best demonstrated by CT, include **obstruction, perforation** (into the peritoneal cavity), **penetration** (such as into the pancreas), or **hemorrhage** (Fig. 20-10).

SMALL AND LARGE BOWEL

General Considerations

- Opacification and distension of the bowel lumen is necessary for proper evaluation of the bowel no matter which modality is used.

 Collapsed or **unopacified loops of bowel can introduce errors of diagnosis** related to our inability first to visualize, and then to differentiate, real from artifactual findings or to accurately characterize the abnormality even if recognized. On CT scans of the abdomen and pelvis, unopacified loops of bowel may mimic masses or adenopathy, and wall thickness is difficult to assess if the bowel is not distended.

- Therefore **orally administered contrast,** frequently given in temporally divided doses to allow earlier contrast to reach the colon while later contrast opacifies the stomach, **is routinely utilized for most abdominal CT scans,** except those performed for **trauma,** the **stone search study,** and studies specifically directed toward evaluating vascular structures such as the **aorta.** Oral contrast used for CT examinations is either a dilute solution containing barium or iodinated contrast.

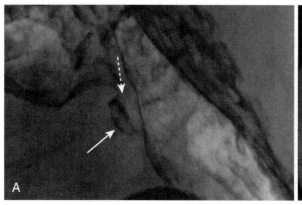 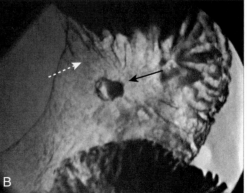

FIGURE 20-7 Benign lesser curvature gastric ulcer. A, Seen in profile there is a large collection of barium that protrudes beyond the expected contour of the normal body of the stomach along the lesser curvature, representing a gastric ulcer *(solid white arrow).* This ulcer collection was present on multiple views (an important characteristic of an ulcer called ***persistence***). There is a mound of edematous tissue that surrounds the ulcer *(dotted white arrow)* called an ***ulcer collar. B,*** Seen *en face* there are numerous gastric folds *(dashed white arrow),* which all radiate to the ulcer margin and a central collection *(black arrow)* representing the ulcer itself. This was a benign gastric ulcer.

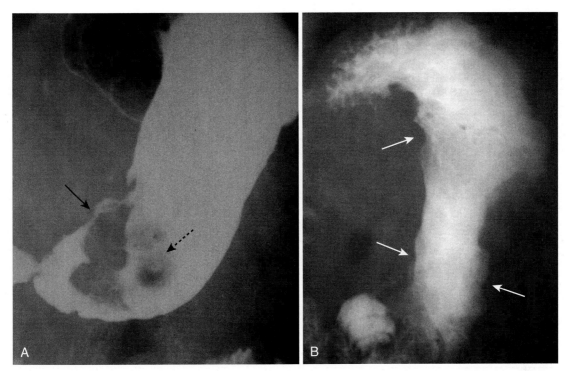

FIGURE 20-8 **Carcinomas of the stomach. A,** There is a large, polypoid filling defect in the antrum of the stomach that displaces the barium around it *(solid black arrow)*. Contained within the mass and seen *en face* is an irregularly shaped collection of barium that represents an ulceration in the mass *(dotted black arrow)*. This was an adenocarcinoma of the stomach. **B,** The entire body of the stomach displays a lack of distensibility, losing the normal motility that every portion of the gastrointestinal tract demonstrates when filled with enough barium and/or air. Instead, the walls of the stomach are concave inward *(white arrows)* and **rigid,** a sign of malignancy. This stomach would display the same appearance on all images. This is the typical appearance for **linitis plastica,** caused by an infiltrating adenocarcinoma of the stomach.

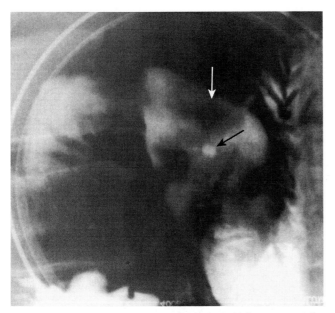

FIGURE 20-9 **Acute duodenal ulcer.** Contained within the duodenal bulb on its anterior wall is a collection of barium *(black arrow)*, shown to be **persistent** on a number of other images, surrounded by a **zone of edema** *(white arrow)* that displaces the barium from around the ulcer. This collection is characteristic of an acute duodenal ulcer. When duodenal ulcers heal, they are likely to do so with scarring that deforms the normal triangular contour of the bulb.

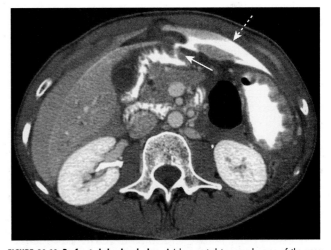

FIGURE 20-10 **Perforated duodenal ulcer.** Axial computed tomography scan of the upper abdomen done with oral and intravenous contrast shows a tract of extraluminal oral contrast from the duodenum *(solid white arrow)* into the peritoneal cavity *(dotted white arrow)*. Obstruction, perforation, and hemorrhage are common complications of ulcer disease. The patient had a perforated duodenal ulcer repaired at surgery.

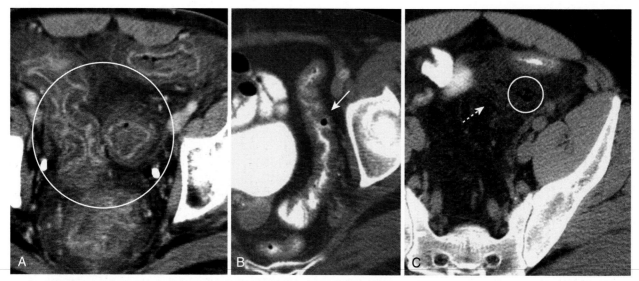

FIGURE 20-11 Key findings on computed tomography (CT) of the gastrointestinal tract. These are findings applicable to any part of the bowel that are key to the diagnosis of bowel abnormalities on CT. **A,** There is thickening and enhancement of the wall of the bowel *(circle)*. When distended, as these loops of large bowel are, the bowel wall is normally very thin. **B,** There is submucosal infiltration of the wall **(thumbprinting)** *(white arrow)*. In this case of ischemic colitis, it most likely represents edema with some hemorrhage. **C,** There is infiltration of the surrounding fat *(dotted white arrow)*, a sentinel finding that usually heralds adjacent inflammation. There is also extraluminal air *(circle)*, a sign of bowel perforation.

There are **several important findings** common to any part of the bowel that are key to the diagnosis of bowel abnormalities by CT:

- **Thickening of the bowel wall.** The normal **small bowel lumen** does not exceed about **2.5 cm** in diameter, and the **wall** is usually **no thicker than 3 mm.** The **colonic wall does not exceed 3 mm** with the lumen distended.
- **Submucosal edema** or **hemorrhage.** Submucosal infiltration produces varying degrees of **thumbprinting,** nodular indentations into the bowel lumen representing focal areas of submucosal infiltration by edema, hemorrhage, inflammatory cells, tumor (lymphoma), or amyloid.
- **Hazy** or **strandlike infiltration of the surrounding fat.** Extension of inflammatory reaction outside of the bowel into the adjacent fat is a sentinel finding that heralds associated disease.
- **Extraluminal contrast** or **extraluminal air.** This indicates the presence of a bowel perforation (Fig. 20-11).

Small Bowel—Crohn Disease

- Crohn disease is a **chronic, relapsing, granulomatous inflammation of the small bowel** and **colon** resulting in ulceration, obstruction, and fistula formation. Crohn disease **typically involves the ileum** and **right colon,** presents with *skip areas* (abnormal bowel interposed between normal bowel), is prone to **fistula formation,** and has a **propensity for recurring** following surgical resection and reanastomosis in whatever loop of bowel becomes the new terminal ileum.
- Crohn disease may be imaged either with a barium **small bowel follow-through** (series) or **CT of the abdomen** and **pelvis.**

Imaging findings in Crohn disease include **narrowing, irregularity,** and **ulceration** of the **terminal ileum** frequently with proximal small bowel dilatation;

separation of the loops of bowel as a result of fatty infiltration of the mesentery surrounding the ileum, making the affected loop(s) stand apart from the surrounding loops of small bowel **(proud loop)**; the **string-sign**—narrowing of the terminal ileum into a near slitlike opening by spasm and fibrosis; and **fistulae**—especially between the ileum and colon, but also to the skin, vagina, and urinary bladder (Fig. 20-12).

LARGE BOWEL

- The intraluminal surface of the colon is most often studied with optical colonoscopy or CT colonography and/or double-contrast barium enema examination. Structures outside of the colon are usually studied by CT examination of the abdomen and pelvis with oral or rectal contrast.

Diverticulosis

- **Colonic diverticula,** like most diverticula of the GI tract, represent herniation of the mucosa and submucosa through a defect in the muscular layer **(false diverticula).**

They occur more frequently with increasing age and may partially be the result of an increase in intraluminal pressure and weakening of the colonic wall. They are **usually multiple (diverticulosis),** are almost always **asymptomatic** (about 90% of the time), but **can become inflamed or bleed. Diverticulosis is the most common cause of massive lower GI bleeding.** When they bleed, the **right-sided** diverticula seem to bleed more than those on the left.

- They occur most often in the sigmoid colon and are readily **identified on either barium enema or CT examination as small spikes or smoothly contoured collections of air and/or contrast attached to the colon** (Fig. 20-13).

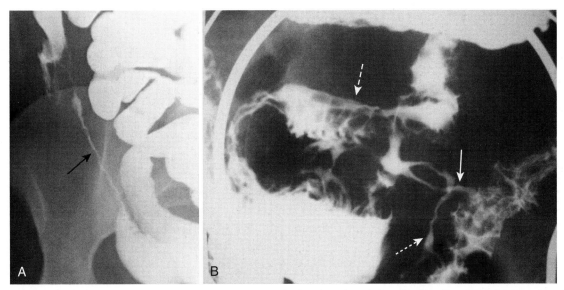

FIGURE 20-12 Crohn disease. A, The terminal ileum *(black arrow)* is markedly narrowed **(string sign)** and stands apart from other loops of small bowel **(proud loop). B,** A close-up image of the right lower quadrant from a small bowel follow-through study in another patient shows multiple streaks of barium *(solid and dotted white arrows),* representing multiple enteric fistulae originating from an abnormal loop of small bowel *(dashed white arrow)* and connecting with each other and the large bowel. Fistula formation is a common complication of this disease.

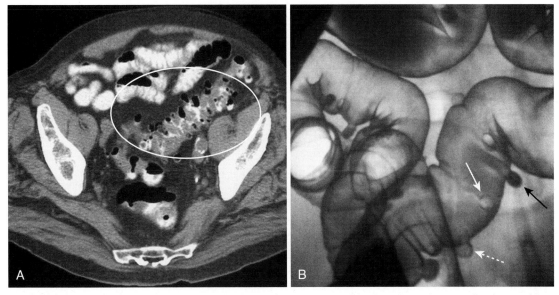

FIGURE 20-13 Diverticulosis. A, In this computed tomography scan of the pelvis, diverticula contain air and appear as small, usually round outpouchings, especially in the region of the sigmoid colon *(white oval). B,* There are numerous outpouchings containing barium seen in the sigmoid colon of this air–contrast barium enema examination. Some diverticula are filled with barium *(solid black arrow),* whereas others contain air and are outlined with barium *(dotted white arrow).* Where a diverticulum is seen *en face,* it produces a circular density *(solid white arrow),* which can mimic the appearance of a polyp.

Diverticulitis

- Diverticula can become inflamed and perforate **(diverticulitis),** most often secondary to mechanical irritation and/or obstruction. **CT is the modality of choice** for the **diagnosis of diverticulitis** since the pericolonic soft tissues can be visualized using CT, which is impossible with either barium enema or optical endoscopy.

- **CT** findings of diverticulitis start with the **presence of diverticula** and include **thickening** of the adjacent colonic wall (>4 mm); **pericolonic inflammation**—hazy areas of increased attenuation and/or streaky and disorganized linear and amorphous densities in the pericolonic fat; **abscess formation**—multiple small bubbles of air or pockets of fluid contained within a pericolonic soft tissue, masslike density; and **perforation of the colon**—extraluminal air or contrast either around the site of the perforation, or less likely, free air in the peritoneal cavity (Fig. 20-14).

Colonic Polyps

- The **incidence of polyps increases with age,** and the **incidence of malignancy increases with the size** of the polyp. Patients with polyposis syndromes, in which multiple adenomatous polyps are present, have a much higher risk of developing a colonic malignancy.

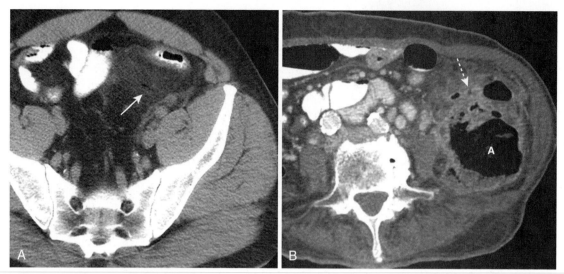

FIGURE 20-14 Diverticulitis, computed tomography (CT). A, Infiltration of the pericolonic fat is demonstrated by a hazy increase in attenuation *(white arrow)* of the normal fat. Focal infiltration of fat is a common characteristic of inflammatory disease. **B,** There is a large abscess cavity (A) in the left lower quadrant in this close-up of a CT scan of the lower abdomen. There are adjacent small bubbles of gas that are not contained within bowel and infiltration of the normal fat *(dotted white arrow)*. These findings are secondary to a confined perforation with abscess formation from diverticulitis.

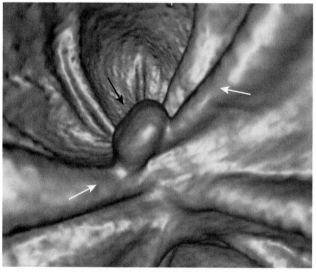

FIGURE 20-15 Polyp on computed tomography (CT) colonography. CT colonography utilizes CT scanning of the abdomen to allow for the three-dimensional reconstruction of the appearance of the inside of the bowel lumen without the use of an endoscope. A polyp in the descending colon *(black arrow)* is seen as a distinct mass and the normal haustral folds *(white arrows)* are ridgelike structures present throughout the large bowel.

Most colonic polyps are **hyperplastic polyps** that have **no malignant potential. Adenomatous polyps carry** a low **potential for malignancy that increases with the size of the polyp** so that those **>1.5 cm in size** have about a **10% chance of being malignant.** Therefore the early detection and removal of adenomatous polyps will decrease the chances of malignant transformation.

■ Colonic polyps can be visualized using either barium enema examination, CT colonography, or optical colonoscopy.

■ *CT colonography* is a technique made possible by newer, faster CT and MRI scanners and complex computer algorithms that allow for the three-dimensional reconstruction of the appearance of the inside of the bowel lumen, including time-of-flight (motion) displays without the use of an endoscope. CT colonography also allows for visualization of the other abdominal structures outside of the colon (Fig. 20-15; Video 20-3).

■ Polyps may be **sessile** (attach directly to the wall) or **pedunculated** (attach to the wall by a stalk) (Fig. 20-16). A polyp **may contain numerous fronds** that produce an irregular, wormlike surface with numerous crypts that may collect barium **(villous polyp).** Villous polyps tend to be larger and have more of a malignant potential than other adenomatous polyps (Fig. 20-17).

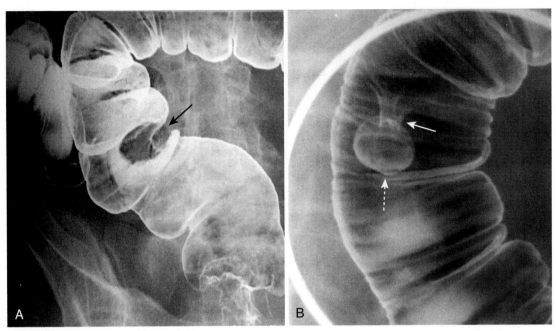

FIGURE 20-16 Sessile and pedunculated polyps of the colon. Polyps can be recognized as persistent filling defects in the colon; barium is displaced by the polyp. **A,** There is a **sessile** filling defect in the pool of barium along the medial wall of the sigmoid colon *(black arrow)*. The size of the lesion should raise concern for malignancy. **B,** Another patient with a filling defect in the sigmoid colon outlined by barium *(dotted white arrow)*. In this patient, the polyp is attached to the wall of the colon by a stalk *(solid white arrow)*. Polyps on a stalk are called ***pedunculated polyps***.

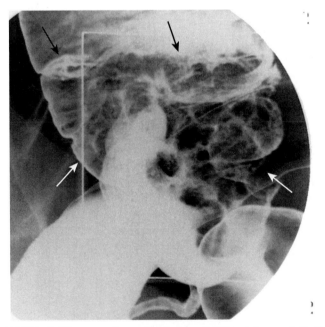

FIGURE 20-17 Villous tumor of cecum. There is a large, polypoid mass in the cecum *(outlined by both black and white solid arrows)*. Contained within the mass is an interlacing network of white lines representing barium that is trapped within the interstices of the frondlike projections of this tumor. This is a characteristic appearance for a villous adenomatous tumor.

■ Occasionally, a polyp may serve as a **lead point** for an **intussusception** in which the polyp drags and prolapses one part of the bowel into the lumen of the bowel immediately ahead of it. The bowel proximal to the intussusception is usually obstructed and dilated. Intussusception may produce a characteristic **coiled-spring** appearance on barium enema or CT (Fig. 20-18).

Colonic Carcinoma

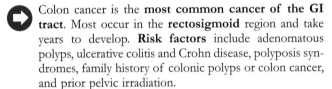 Colon cancer is the **most common cancer of the GI tract.** Most occur in the **rectosigmoid** region and take years to develop. **Risk factors** include adenomatous polyps, ulcerative colitis and Crohn disease, polyposis syndromes, family history of colonic polyps or colon cancer, and prior pelvic irradiation.

■ The imaging findings of carcinoma of the colon include the presence of a persistent, large, polypoid filling defect; annular constriction of the colonic lumen, producing an **apple-core lesion** (Fig. 20-19); and **frank** or **microperforation** manifest by infiltration of the pericolonic fat with streaky or hazy densities of increased attenuation, with or without the presence of extraluminal air.

■ Other findings of carcinoma of the colon can include large bowel obstruction (Fig. 20-20), either antegrade obstruction and/or retrograde obstruction demonstrated by the inability of rectally administered barium to pass the point of the colon cancer and **metastases**, especially to the liver and the lungs.

Colitis

■ Colitis is **inflammation of the large bowel.** There are numerous **causes** of colitis, including infectious, ulcerative and granulomatous, ischemic, radiation-induced, and antibiotic-associated causes. Because many forms of colitis appear similar radiographically, clinical history is of paramount importance.

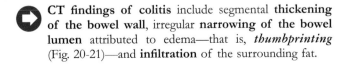 **CT findings of colitis** include segmental **thickening of the bowel wall,** irregular **narrowing of the bowel lumen** attributed to edema—that is, *thumbprinting* (Fig. 20-21)—and **infiltration** of the surrounding fat.

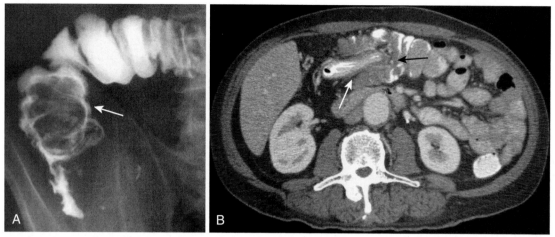

FIGURE 20-18 Intussusception, barium enema, and computed tomography scan. A, When one loop of bowel prolapses inside the loop immediately distal to it, the resultant obstruction produces a **coiled-spring** appearance on barium enema examination because two loops of bowel are superimposed on one another *(white arrow)*. **B,** Another patient with an intussusception. A loop of large bowel *(white arrow)* is seen prolapsing into the loop distal to it *(black arrow)*, producing a filling defect and obstructing the lumen.

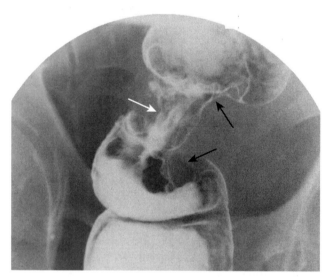

FIGURE 20-19 Annular constricting carcinoma of the rectum. There is a characteristic **apple-core** lesion of the rectum caused by circumferential growth of a colonic carcinoma. The margins of the lesion *(black arrows)* demonstrate what is called an ***overhanging edge,*** where tumor tissue projects into and overhangs the normal lumen, typical of this type of lesion. The "core" of the "apple" *(white arrow)* is composed of tumor tissue—all of the normal colonic mucosa has been replaced. Identification of such a lesion is pathognomonic for carcinoma.

Appendicitis

- Pathophysiologically, the development of appendicitis is invariably preceded by obstruction of the appendiceal lumen. CT is now the modality of choice in diagnosing appendicitis. Appendicitis is also diagnosed using ultrasound and MRI.
- An **appendicolith is a calcified concretion found in the appendix** of about 15% of all people. It will manifest, especially on CT, as a calcification in the lumen of the appendix. The combination of **abdominal pain** and the **presence of an appendicolith** is associated with **appendicitis about 90% of the time** and indicates a higher probability for **perforation.**

 The key CT findings of acute appendicitis are identification of a **dilated appendix** (>6 mm) (Fig. 20-22), which does **not** fill with oral contrast; **periappendiceal**

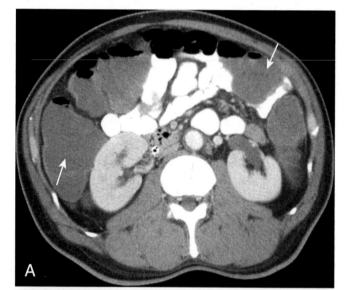

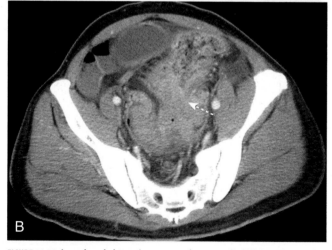

FIGURE 20-20 Large bowel obstruction, computed tomography (CT). Two CT images from the same patient demonstrate **(A)** dilated and fluid-filled loops of large bowel *(white arrows)*. The orally administered contrast had not yet reached the large bowel at the time of this study, but it fills the small bowel. A large soft tissue mass in the sigmoid **(B)** *(dotted white arrow)* represents the patient's obstructing carcinoma of the sigmoid colon.

inflammation is evidenced by streaky, disorganized, linear, high-attenuation densities in the surrounding fat (see Fig. 20-22, *A*) and **increased contrast enhancement of the wall of the appendix** as a result of inflammation.

■ **Perforation** occurs in up to 30% of cases and is **recognized by small quantities of periappendiceal extraluminal air** or **a periappendiceal abscess.** Since obstruction of the appendiceal lumen is a prerequisite for appendicitis, the presence of a large amount of free intraperitoneal air should point to another diagnosis (see Fig. 20-22, *B*).

■ The use of ultrasound in appendicitis will be discussed in Chapter 21.

■ The **appendix** of this book lists the **studies of first choice** for many **different clinical scenarios** relating to **abdominal pain.**

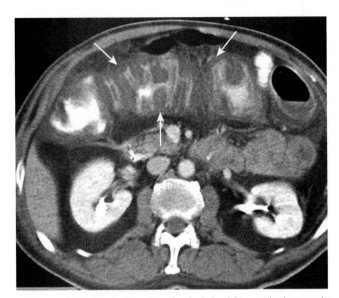

FIGURE 20-21 Colitis. The colon demonstrates **thumbprinting** *(white arrows)* and a pattern that is called the ***accordion sign***. This patient had *Clostridium difficile* colitis, formerly called *pseudomembranous colitis* and now known to be caused almost exclusively by toxins produced by *C. difficile*. It frequently follows antibiotic therapy. The diagnosis is usually made clinically by visualization of the pseudomembrane on endoscopy. The **accordion sign** represents contrast that is trapped between enlarged folds and indicates the presence of marked edema or inflammation, but it is not specific for *C. difficile* colitis.

PANCREAS

Pancreatitis

■ The two **most common causes of pancreatitis** are **alcoholism** and **gallstones.** Inflammation of pancreatic tissue leading to disruption of the ducts and spillage of pancreatic juices occurs readily because of the lack of a capsule surrounding the pancreas.

➡ Pancreatitis is a **clinical diagnosis,** with CT serving to document either a **cause** (e.g., gallstones) or **complication** (e.g., pseudocyst formation).

■ Recognizing acute pancreatitis on CT:
   ◆ **Enlargement** of all or part of the pancreas (normal measurements for the pancreas are **3 cm for the head, 2.5 cm for the body,** and **2 cm for the tail**) (Fig. 20-23).
   ◆ **Peripancreatic stranding** or **fluid collections.**
   ◆ **Low-attenuation** lesions in the pancreas from **necrosis.** Areas of nonviable pancreas usually develop **early** in the course of the disease. To be visible, intravenous (IV) contrast administration is used for predicting the prognosis.

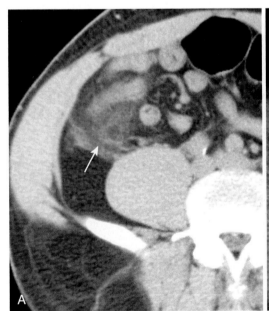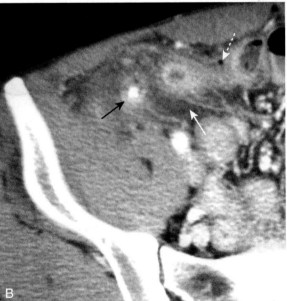

FIGURE 20-22 Appendicitis, computed tomography. A, There is infiltration of the periappendiceal fat in the right lower quadrant manifest by the increased attenuation in the mesenteric fat *(white arrow)*. Focal infiltration of fat is a common characteristic of inflammatory diseases and helps in their localization. **B,** Contained within the lumen of the appendix is a small calcification *(black arrow)*, or **appendicolith.** There is inflammatory infiltration of the surrounding fat producing high attenuation *(solid white arrow)*. A very small amount of air is present outside of the appendiceal lumen from a confined perforation *(dotted white arrow)*. An appendicolith can be found in about one quarter of the cases of acute appendicitis. The combination of an appendicolith and acute appendicitis is a strong predictor of a perforated appendix.

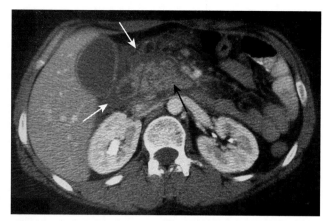

FIGURE 20-23 Acute pancreatitis. The body of the pancreas is enlarged *(black arrow)*. There is infiltration of the peripancreatic fat *(white arrows)*. These findings are consistent with acute pancreatitis in the proper clinical setting. This patient had a markedly elevated amylase and lipase.

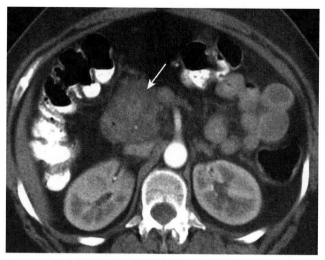

FIGURE 20-25 Pancreatic adenocarcinoma. The head of the pancreas is enlarged by a mass *(white arrow)*. Normally, the head of the pancreas should be roughly the same size as the width of the lumbar vertebral body nearest to it. Most pancreatic adenocarcinomas are located in the head (75%), and jaundice is a common presenting sign.

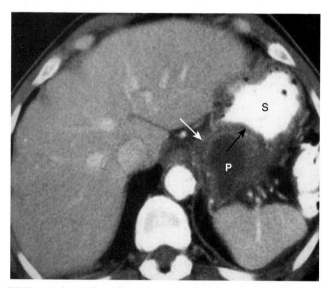

FIGURE 20-24 Pancreatic pseudocyst. A pseudocyst (P) of the pancreas occurs when fibrous tissue encapsulates a walled-off collection of pancreatic juices released from the inflamed pancreas. Pseudocysts may have an enhancing wall *(white arrow)*. The cyst is indenting a loop of adjacent bowel, in this case the posterior wall of the stomach (S). The indentation on a loop of bowel by an extrinsic mass is called a ***pad sign*** *(black arrow)*.

♦ **Pseudocyst formation.** Fibrous tissue encapsulates a walled-off collection of pancreatic juices released from the inflamed pancreas. The **wall of a pseudocyst is usually visible by CT** and **may enhance with contrast** (Fig. 20-24).

Chronic Pancreatitis

■ Chronic pancreatitis is a continuous and irreversible disease of the pancreas most often secondary to **alcohol abuse,** leading to **fibrosis, atrophy of the gland, ductal dilatation,** and frequently **diabetes.**

The **hallmarks** of the disease are **multiple, amorphous calcifications** that form within the **dilated ducts** of the **atrophied gland** (see Fig. 18-10).

Pancreatic Adenocarcinoma

■ Risk factors for pancreatic adenocarcinoma include **alcoholism, cigarette smoking, chronic pancreatitis,** and **diabetes.** Pancreatic adenocarcinoma has an **exceedingly poor prognosis;** most tumors are unresectable and incurable at the time of diagnosis. **Most of the time** (75%), the tumor is located in the **head** of the pancreas; about 10% occur in the body and 5% in the tail. About half of the patients present with **jaundice,** and most of the time **there is associated pain.**

♦ **Ultrasound is the study of choice in the initial workup of the jaundiced patient** (see Chapter 21).

Recognizing pancreatic adenocarcinoma on CT

♦ **Focal pancreatic mass,** usually **hypodense** to the remainder the gland (Fig. 20-25).
♦ **Ductal dilatation,** usually involving **both** the **pancreatic** and **biliary ducts.** The normal **pancreatic duct** measures **less than 4 mm** in the head and tapers to the tail; the **common duct** should be **less than 7 mm** in diameter.
♦ Other findings of pancreatic carcinoma include spread to contiguous organs, enlarged lymph nodes and/or ascites.

HEPATOBILIARY ABNORMALITIES
Liver—General Considerations

■ CT evaluation of **liver masses** is usually done with a **combination** of scans obtained **before** and **after intravenous contrast** injection. **Postcontrast** scans are obtained in **two phases:** one is done quickly **(hepatic arterial phase),** and then a second is done about a minute later **(portal venous phase),** the combination helping to best define and characterize liver masses. This combination of three separate scans done **without contrast,** then during the **arterial phase,**

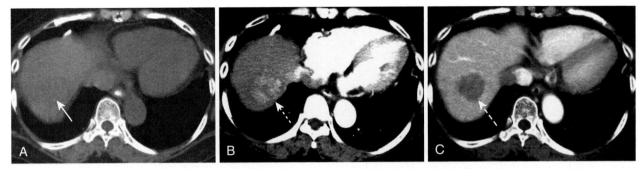

FIGURE 20-26 Triple-phase computed tomography scan of the liver, hepatocellular carcinoma. Evaluation of liver masses is usually done with a combination of scans, including an unenhanced scan **(A)** and then two postcontrast scans: one obtained quickly (hepatic-arterial phase) **(B)** and a second (portal-venous phase) slightly delayed **(C)**. The combination of three scans is called a *triple-phase scan.* This case shows the typical findings of a focal hepatocellular carcinoma. Most are low density (hypodense) or the same density as normal liver (isodense) without contrast *(white arrow in A)*, enhance on the arterial phase with intravenous contrast (hyperdense) *(dotted arrow in B)*, and then return to hypodense or isodense on the venous phase *(dashed arrow in C)*.

followed by the **venous phase,** is called a *triple-phase scan* (Fig. 20-26).

- **MRI** is increasingly utilized as the modality of choice in the evaluation of both focal and diffuse liver disease. MRI can often definitively characterize hepatic lesions described as indeterminate on CT scan such as cysts, benign hepatic neoplasms such as hemangioma and focal nodular hyperplasia, hepatocellular carcinoma, and metastatic disease. It is **particularly useful in the evaluation of small (10 mm or less) lesions** compared with CT.

- For routine MRI imaging of the liver, an intravenous contrast agent called *gadolinium* is typically administered, and multiple postgadolinium enhanced images are obtained. Gadolinium will be discussed in Chapter 22.

Fatty Infiltration

- The term *nonalcoholic fatty liver disease* is used now to refer to a spectrum of diseases of the liver ranging from **hepatic steatosis** (fatty infiltration of the liver) to **NASH** (nonalcoholic steatohepatitis) to **cirrhosis.**

➡️ Fatty infiltration (also known as *hepatic steatosis*) of the liver is a **very common** abnormality in which there is fat accumulation in the hepatocytes in such conditions as obesity, diabetes, hepatitis, or cirrhosis. Most patients with a fatty liver are **asymptomatic.** The **fatty infiltration** may be **diffuse** or **focal,** and focal lesions may be **solitary** or **multiple.**

- When diffuse, the **liver is usually slightly enlarged.** The **blood vessels stand out prominently** but are usually **neither obstructed** nor **displaced.**

- **Normally on noncontrast CT scans, the liver is always denser than or equal to the density of the spleen.** With **fatty infiltration of the liver, the spleen is denser than the liver** without intravenous contrast (Fig. 20-27).

⚠️ **Focal** fatty infiltration can produce an appearance that **mimics tumors,** but fatty infiltration usually produces **no mass effect,** and fatty infiltration has the ability to **appear** and **disappear** in a matter of weeks, quite unlike tumor masses.

- **MRI is the most accurate modality in the evaluation of a fatty liver,** using a phenomenon called *chemical shift*

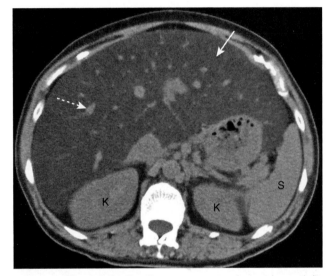

FIGURE 20-27 Diffuse fatty liver. Normally, on noncontrast computed tomography scans, the liver is always denser than or equal to the density of the spleen. In this patient with diffuse fatty infiltration of the liver, the spleen (S) is denser than the liver *(solid white arrow)*. Notice how the vessels stand out in the fatty liver *(dotted white arrow)*. *K, Kidney.*

imaging to detect the presence of microscopic, intracellular lipid present in such a liver. Chemical shift relates to the way that lipid and water protons behave in the magnetic field (Fig. 20-28).

Cirrhosis

- Cirrhosis is a chronic, irreversible disease of the liver, which features destruction of normal liver cells and diffuse fibrosis and appears to be the final common pathway of many abnormalities, including **hepatitis B** and **C, alcoholism, nonalcoholic fatty infiltration of the liver,** and miscellaneous diseases such as **hemochromatosis** and **Wilson disease.** Complications of cirrhosis include **portal hypertension, ascites, renal dysfunction, hepatocellular carcinoma, hepatic failure,** and **death.**

➡️ Recognizing cirrhosis of the liver on CT:

♦ **Early** in the disease, the liver may demonstrate **diffuse fatty infiltration.** As the disease progresses, the liver **contour**

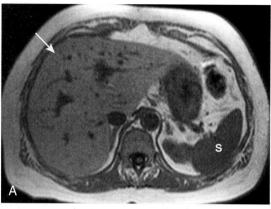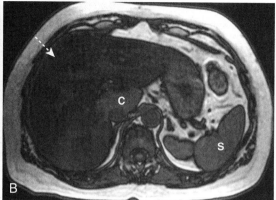

FIGURE 20-28 **Fatty liver, MRI.** Using a phenomenon called *chemical shift imaging* to detect the presence of microscopic, intracellular lipid in a fatty liver, MRI is the most accurate imaging modality in identifying a fatty liver. Chemical shift relates to the way that lipid and water protons behave in the magnetic field. **A,** The liver *(white arrow)* appears normal, brighter than the spleen (S). **B,** This is called an *opposed-phase image,* and it demonstrates marked signal loss (signal dropout) throughout the liver *(dotted white arrow),* indicating a fatty liver. Most of the liver is now darker than the spleen (S), except for the caudate lobe (C), which is normal.

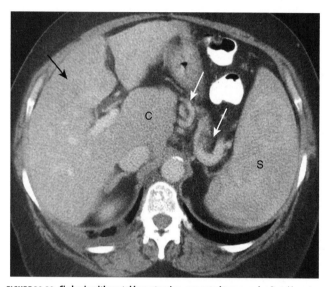

FIGURE 20-29 **Cirrhosis with portal hypertension, computed tomography.** Portal hypertension can lead to dilated vessels around the stomach, splenic hilum *(white arrows),* and esophagus, representing varices. Splenomegaly may develop (S). There is characteristic enlargement of the caudate lobe (C) relative to the right lobe of the liver *(black arrow),* especially in alcoholic cirrhosis.

BOX 20-1   DIFFERENTIATING ASCITES FROM PLEURAL EFFUSION

Patients may have a combination of ascites and pleural effusion for a number of reasons, including cirrhosis, ovarian tumors, metastatic disease, hypoproteinemia, and congestive heart failure.

Differentiation between ascitic fluid and pleural fluid at the lung base on computed tomography examinations may be difficult because both may appear posterior to the liver.

Ascitic fluid will appear **anterior to the hemidiaphragm** in the axial plane (see Fig. 20-30, *A*).

Pleural effusion will be located **posterior to the hemidiaphragm.**

Ascitic fluid **will not spread to the "bare" area of the liver posteriorly** where there is no peritoneal lining (see Fig. 15-02).

Therefore, **fluid that appears posterior to the bare area is pleural in location** (see Fig. 20-30, *B*).

SPACE-OCCUPYING LESIONS OF THE LIVER

- One of the primary aims of imaging studies, no matter what part of the body is being studied, is to accurately differentiate between benign and malignant processes using techniques that do not place the patient in danger or subject them to unnecessary pain. This is the central goal in evaluating liver masses as well.
- CT studies are best at demonstrating liver masses when performed **both with** and **without contrast** because either study alone may fail to reveal an **isodense** mass, which is one that has the identical attenuation as the surrounding normal tissue.
- **MRI is useful in characterizing liver masses,** particularly small lesions 10 mm or less.

Metastases

 Metastases are the **most common malignant hepatic masses.** Although **most are multiple,** metastases also represent the **most common cause** of a **solitary malignant mass** in the liver.

becomes lobulated. The **liver shrinks in volume,** with the **right lobe** characteristically becoming **smaller** whereas the **caudate lobe** and **left lobe** become **disproportionately larger,** especially in alcoholic cirrhosis (Fig. 20-29).

- There is a **mottled, inhomogeneous** appearance to the liver parenchyma following intravenous contrast enhancement; this is attributed to a mixture of regenerating nodules, focal fatty infiltration, and fibrosis.
- **Portal hypertension** may develop, which can lead to **dilated vessels** around the stomach, splenic hilum, and esophagus, representing **varices. Splenomegaly** may result.

Ascites may be present. Sometimes it can be difficult to differentiate between ascites and pleural effusion on CT examinations (Box 20-1; Fig. 20-30).

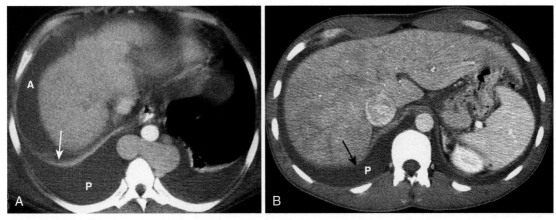

FIGURE 20-30 **Differentiating pleural effusion from ascites. A,** Patients may have a combination of ascites (A) and pleural effusions (P) for a number of reasons, cirrhosis being one of them. Ascitic fluid will appear anterior to the hemidiaphragm *(white arrow)* in the axial plane. Pleural effusion will be located posterior to the hemidiaphragm. **B,** Ascites will never completely encircle the liver because of the "bare area" *(black arrow)*, not covered by peritoneum (see Fig. 15-2). Fluid posterior to the bare area must be in the pleural space (P).

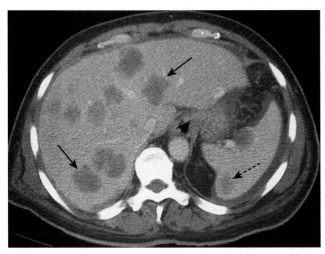

FIGURE 20-31 **Metastases to the liver and spleen.** Metastases usually appear as multiple, low-attenuation masses *(solid black arrows)*. There are also low-attenuation lesions in the spleen *(dotted black arrow)*. The patient had a primary adenocarcinoma of the colon.

■ **Most liver metastases originate in the gastrointestinal tract,** in particular the **colon,** and almost all reach the liver via the bloodstream. Other primary sites of metastatic spread to the liver include stomach, pancreas, esophagus, lung, melanoma, and breast.

■ **Recognizing liver metastases on CT** and **MRI:**
  ♦ Metastases are usually **multiple, low-attenuation masses** (Fig. 20-31). **Larger metastases** may demonstrate areas of necrosis that can be recognized as mottled areas of low attenuation within the mass. **Mucin-producing carcinomas,** such as might originate in the stomach, colon, or ovary, can **calcify** in both the primary tumor and the metastases (see Fig. 18-14).
  ♦ **MRI** is **as sensitive** as CT in detecting liver metastases, but it is usually reserved for a **problem-solving role.** In general, MRI is more expensive than CT scanning and may suffer from more motion artifact than CT.

Hepatocellular Carcinoma (Hepatoma)

■ Hepatocellular carcinoma (HCC) is the **most common primary malignancy** of the liver. Virtually all arise in livers

with preexisting abnormalities such as **cirrhosis** and/or **hepatitis. Most are solitary**, but up to one out of five can be multiple, mimicking metastases. **Vascular invasion is common,** particularly of the portal system.

■ **Recognizing hepatocellular carcinoma on CT** and **MRI:**
  ♦ There are **three patterns** of presentation for hepatocellular carcinoma: **solitary mass** (see Fig. 20-26), large **multiple nodules,** and **diffuse infiltration** throughout a segment, lobe or the entire liver.
  ♦ On CT, most HCCs are **low density (hypodense)** or the same density as normal liver **(isodense) without contrast;** then they enhance on the **arterial phase** with IV contrast **(hyperdense)** and return to **hypodense** or **isodense** on the **venous phase** (Fig. 20-32).
  ♦ Low-attenuation areas from **necrosis** are common. **Calcification** may be present.
  ♦ **MRI** can demonstrate certain features that are fairly specific for **hepatocellular carcinoma** and detect intrahepatic metastases and venous invasion. Unlike benign lesions such as hemangiomas of the liver, which tend to retain intravenously administered gadolinium contrast, hepatocellular carcinomas **show washout** of the contrast material

Cavernous Hemangiomas

⮕ Cavernous hemangiomas are the **most common primary liver tumor** and **second in frequency to metastases for localized liver masses.** They are more common in **women,** are usually **solitary,** and are almost always **asymptomatic.** They are complex structures composed of multiple, large, vascular channels lined by a single layer of endothelial cells.

■ **Recognizing cavernous hemangiomas of the liver on CT** and **MRI:**
  ♦ **Cavernous hemangiomas are usually hypodense lesions on unenhanced CT scans.** They have a characteristic nodular enhancement **from the periphery** inward following injection of intravenous contrast and **become isodense** in the venous phase.
  ♦ Contrast tends to be **retained** within the numerous vascular spaces of the lesion so that they

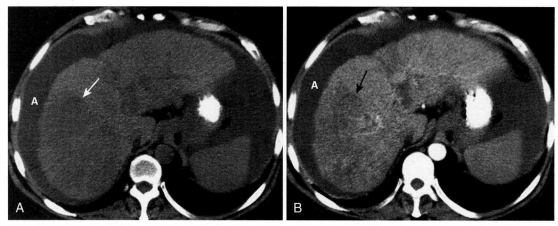

FIGURE 20-32 Diffuse hepatocellular carcinoma of the liver, computed tomography. There are three patterns of appearance for hepatocellular carcinoma: solitary mass (see Fig. 20-26), multiple nodules, and diffuse infiltration throughout a segment, lobe (as in this case), or the entire liver. **A,** A typical low-attenuation lesion is seen in the right lobe of the liver on the nonenhanced scan *(white arrow)*. **B,** The arterial phase demonstrates patchy enhancement *(black arrow)*, indicating the probability of tumor necrosis in the low-attenuation areas. There is ascites present (A). The overall volume of the liver is decreased, and the contour is lobulated from underlying cirrhosis.

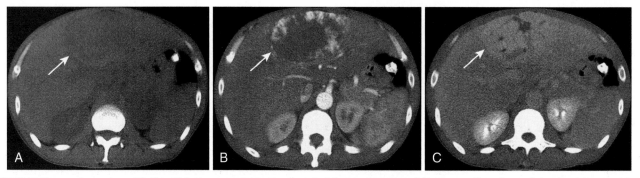

FIGURE 20-33 Cavernous hemangioma of the liver, triple-phase computed tomography study. A, Cavernous hemangiomas *(white arrows on all images)* are usually hypodense lesions on unenhanced scans. **B,** They characteristically enhance from the periphery inward following injection of intravenous contrast during the arterial phase and eventually become isodense. **C,** Contrast then tends to be retained within the numerous vascular spaces of the lesion so that it characteristically appears denser than the rest of the liver on delayed scans.

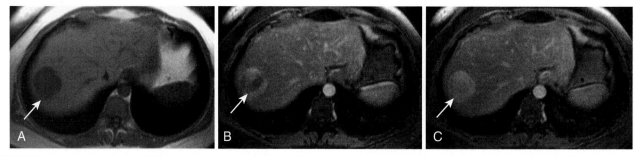

FIGURE 20-34 Cavernous hemangioma of the liver, magnetic resonance imaging. A, This image (an axial T1-weighted image) demonstrates a well-circumscribed, slightly lobular dark mass in the right hepatic lobe *(white arrow in all images)*. **B,** Subsequent images following the administration of intravenous contrast (gadolinium) show peripheral-to-central enhancement, until the entire mass homogeneously enhances on a delayed 10-minute image. **C,** The combination of this enhancement pattern and the signal characteristics of the lesion allows an unequivocal diagnosis of hemangioma.

characteristically appear **denser than the rest of the liver on delayed** (10 minute) **scans** (Fig. 20-33).

- ◆ **MRI is frequently the preferred modality** in the evaluation of hemangiomas, because it is **more sensitive** than a nuclear medicine–tagged red blood cell scan and more specific than a multiphase CT scan.
- ◆ Similar to CT, **hemangiomas on MRI** usually have a characteristic nodular enhancement from the periphery inward following injection of intravenous contrast.

Contrast tends to be retained within the numerous vascular spaces of hemangiomas so that they typically appear brighter than the rest of the liver on delayed (10-minute) scans (Fig. 20-34; Video 20-4).

Hepatic Cysts

- ■ Believed to be **congenital** in origin, they are easily identified as **sharply marginated**, spherical lesions of **low attenuation (fluid density)** compared with the remainder of the

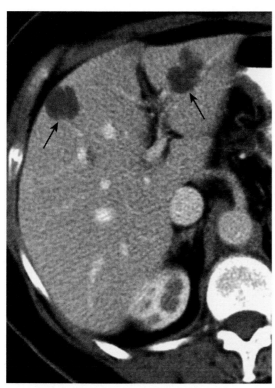

FIGURE 20-35 Hepatic cysts, computed tomography. Believed to be congenital in origin, hepatic cysts are easily identified as sharply marginated, spherical lesions of low attenuation (fluid density) compared with the remainder of the liver on both unenhanced and enhanced scans *(arrows)*. They are homogeneous in density.

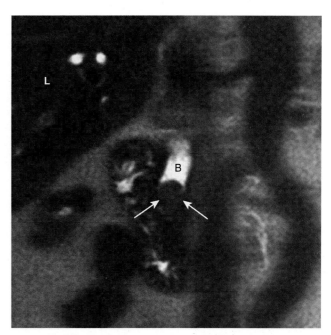

FIGURE 20-36 Choledocholithiasis and biliary ductal dilatation on magnetic resonance cholangiopancreatography (MRCP). This is a coronal close-up of the right upper quadrant utilizing magnetic resonance imaging. There is a large obstructing gallstone *(arrows)* in the distal common bile duct (B), which is dilated to 13 mm. Because of the signal characteristics of bile, an MRCP can be done without the need for the injection of contrast. *L,* Liver.

liver on both unenhanced and enhanced CT scans. MRI is much better than CT at characterizing cysts. They are usually **solitary** and are **homogeneous** in density (Fig. 20-35).

BILIARY SYSTEM
Magnetic Resonance Cholangiopancreatography

- Magnetic resonance cholangiopancreatography (MRCP) is a **noninvasive** way to image the biliary tree **without requiring injection of contrast material.** MRCP utilizes MRI imaging sequences that make fluid-filled structures such as the bile ducts, pancreatic ducts, and gallbladder appear extremely bright, whereas everything else looks dark. Patients are imaged during a single breath-hold.
- MRCP is excellent at depicting **biliary** or **ductal strictures, ductal dilatation, stones in the bile ducts (choledocholithiasis)** (Fig. 20-36), **gallstones, adenomyomatosis of the gallbladder, choledochal cysts, and pancreas divisum.**
- If there is a concern for malignancy, such as pancreatic adenocarcinoma or cholangiocarcinoma, as the cause of pancreaticobiliary ductal dilatation, then additional pulse sequences following the administration of gadolinium can be obtained. Contrast administration allows better detection of malignancy.

URINARY TRACT
Kidneys—General Considerations

- The kidneys are retroperitoneal organs encircled by varying amounts of fat and enclosed within a fibrous capsule. As long as they are functioning properly, the **kidneys are the route** through which intravenously injected, iodinated **contrast** and **gadolinium-based MRI contrast agents are excreted from the body.** Following injection of iodinated contrast agents, the kidneys will therefore **enhance.**
- Ultrasound is utilized in the characterization of renal masses, particularly in assessing their cystic or solid nature and involvement of the collecting system or surrounding vessels.

Space-Occupying Lesions
Renal cysts

- Simple renal cysts are a **very common** finding on CT and ultrasonography (US) scans of the abdomen occurring in more than half of the population over 55 years of age.
- Simple cysts are **benign, fluid-filled** structures, which are frequently **multiple** and **bilateral.** On CT scans they tend to have a **sharp margin** where they meet the normal renal parenchyma (Fig. 20-37). They have density measurements (Hounsfield numbers) of **water density** (−10 to +20). They **do not enhance with contrast** (see Fig. 20-37, *A*).
- On **ultrasound examinations,** simple cysts are **echo-free** (anechoic) masses with **strong through transmission** of the ultrasound signal; they have **sharp borders** where they meet the renal parenchyma and **round** or **oval shapes.** Thickening of the wall or dense internal echoes raise suspicion for a malignant lesion (see Fig. 20-37, *B*).

Renal cell carcinoma (hypernephroma)

- Renal cell carcinoma is the **most common primary renal malignancy** in adults. Solid masses in the kidneys of adults are usually renal cell carcinomas. They have a propensity for

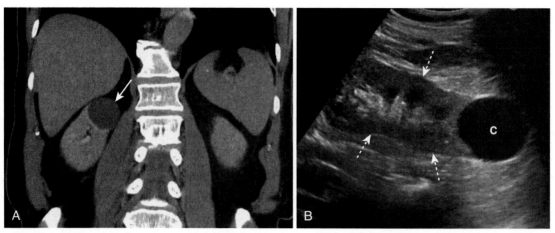

FIGURE 20-37 **Renal cysts, computed tomography (CT) urogram, and ultrasonography. A,** This is an image from a CT urogram, which demonstrates a low-attenuation mass *(white arrow)* in the upper pole of the right kidney, which is homogeneous in density and sharply marginated. These findings are characteristic of a simple cyst. **B,** A sagittal ultrasound of the kidney *(dotted white arrows)* in another patient demonstrates an anechoic mass (C) with features consistent with a simple cyst of the lower pole.

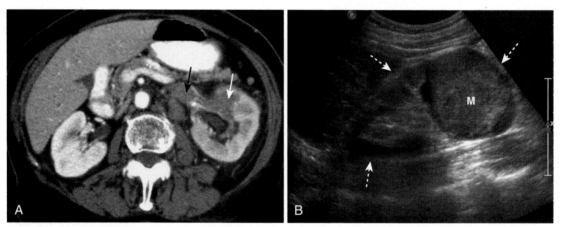

FIGURE 20-38 **Renal cell carcinomas, computed tomography and ultrasonography. A,** There is a low-density mass involving the anterior portion of the left kidney *(white arrow)*. The tumor is seen to extend directly into the left renal vein *(black arrow)*, a feature of renal cell carcinomas. **B,** Sagittal ultrasound on another patient with renal cell carcinoma shows an echogenic mass (M) occupying the midportion of the kidney *(dotted white arrows)*.

extending into the renal veins, into the **inferior vena cava,** and producing **nodules in the lung.**

■ When they metastasize to bone, they are **purely lytic** and **often expansile.**

 Recognizing renal cell carcinoma on CT (Fig 20-38):

♦ A dedicated CT scan for renal cell carcinoma usually consists of images obtained before and after intravenous contrast administration.

♦ Ranging from **completely solid** to **completely cystic,** renal cell carcinomas are **usually solid lesions, which may contain low-attenuation areas of necrosis.** Even though renal cell carcinomas **enhance with intravenous contrast,** they still tend to remain **lower in density** than the surrounding normal kidney.

♦ **Renal vein invasion occurs in up to one in three cases** and may produce filling defects in the lumen of the renal veins (see Fig. 20-38, *A*).

■ On **ultrasound,** smaller renal cell carcinomas are **usually hyperechoic;** as the lesion increases in size and undergoes necrosis, it may be hypoechoic. Its wall, however, should be thicker and more irregular than a simple cyst (see Fig. 20-38, *B*).

■ The primary role of **MRI of the kidneys** is in the evaluation of **small masses** less than 1.5 cm or any renal mass characterized as indeterminate for malignancy by either CT or US.

PELVIS

General Considerations

■ **Ultrasound is the study of first choice in evaluation of suspected abnormalities of the female pelvis** (see Chapter 21).

■ **MRI** (see Chapter 22) has assumed an increasingly important role in defining the anatomy of the uterus and ovaries and in clarifying questions in patients in whom US findings are

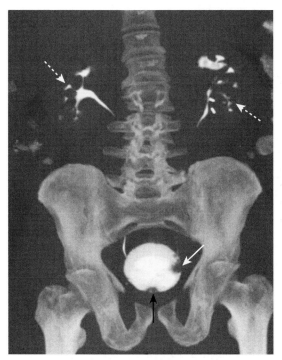

FIGURE 20-39 **Transitional cell carcinoma of the bladder, computed tomography urogram.** There is a filling defect in the left lateral wall of the contrast-filled bladder *(solid white arrow)*, representing a tumor. The defect at the base of the bladder *(black arrow)* is caused by the prostate gland. The calyceal collecting systems *(dotted white arrows)* are normal.

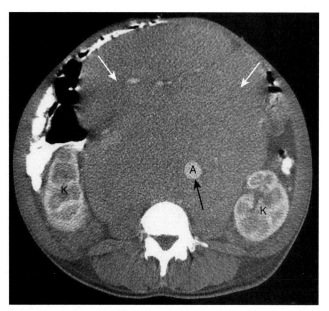

FIGURE 20-40 **Massive lymphadenopathy, lymphoma.** There is massive abdominal lymphadenopathy occupying most of the abdomen *(white arrows)* and displacing the kidneys (K) laterally and the aorta (A) anteriorly far from its normal location next to the spine *(black arrow)*. The patient had non-Hodgkin lymphoma.

confusing. MRI is also used in staging and surgical planning.

■ MRI can be particularly useful in evaluating ovarian dermoid cysts, endometriosis, and hydrosalpinges (fluid-filled fallopian tubes) and in determining whether an ovarian cystic lesion is simple (benign) or contains a solid component (often malignant).

URINARY BLADDER

Bladder Tumors

■ Most malignant bladder tumors are **transitional cell tumors.** Transitional tumors may occur simultaneously anywhere along the uroepithelium from the bladder to the ureter to the kidney. The **primary tumor** appears as **focal thickening of the bladder wall** and/or produces a **filling defect in the contrast-filled bladder** (Fig. 20-39).

ADENOPATHY

Lymphoma

■ Lymphoma can involve any part of the GI or GU tract. Although thoracic lymphoma is almost always caused by Hodgkin disease, **the stomach is the most common extranodal site of gastrointestinal non-Hodgkin lymphoma.**

■ Extranodal involvement and noncontiguous spread to other organs or nodes are features of non-Hodgkin lymphoma.

 Recognizing the CT findings of lymphoma:

◆ **Multiple enlarged lymph nodes.** Pelvic lymph nodes are considered pathologically enlarged if they exceed 1 cm in their shortest dimension.

◆ **Conglomerate masses of coalesced nodes.** Bulky tumor masses that can encase and obstruct vessels.

◆ Lymphadenopathy will classically **displace the aorta** and/or **vena cava anteriorly** (Fig. 20-40).

■ Other malignancies can produce abdominal and/or pelvic adenopathy besides lymphoma, and even benign disease such as sarcoid can produce abdominal adenopathy.

 TAKE-HOME POINTS

RECOGNIZING GASTROINTESTINAL, HEPATIC, AND URINARY TRACT ABNORMALITIES

Computed tomography (CT), ultrasound (US), and magnetic resonance imaging (MRI) have essentially replaced conventional radiography and, in some instances, barium studies for the evaluation of the gastrointestinal (GI) tract.

Some of the terminology used in describing fluoroscopic studies of the GI tract is defined.

Esophageal diverticula occur in the neck (Zenker), around the carina (traction), and just above the diaphragm (epiphrenic); only the Zenker diverticulum tends to produce symptoms.

Esophageal carcinoma has a poor prognosis with an increasing incidence of adenocarcinomas forming in Barrett esophagus, a condition in which gastroesophageal reflux (GERD) plays a major

Continued

role in stimulating metaplasia of the squamous to columnar epithelium.

Esophageal carcinomas appear in one or more of several forms including an annular-constricting lesion, a polypoid mass, and a superficial, infiltrating type lesion.

Hiatal hernias are a common abnormality that may be associated with GERD, although GERD can occur even in the absence of a demonstrable hernia; they are usually of the sliding variety in which the esophagogastric junction lies above the diaphragm.

The radiologic findings of **gastric ulcer** include a persistent collection of barium that extends outward from the lumen beyond the normal contours of the stomach, usually along the lesser curvature or posterior wall in the region of the body or antrum; the ulcer may have radiating folds that extend to the ulcer margin and a surrounding margin of edema.

The key finding in **gastric carcinoma** is a mass that protrudes into the lumen and produces a filling defect, displacing barium; gastric carcinomas may be associated with rigidity of the wall, nondistensibility of the lumen, and irregular ulceration or thickening of the gastric folds (>1 cm), especially localized to one area of the stomach.

The radiologic findings of **duodenal ulcers** include a persistent collection of contrast, more often seen *en face* with surrounding spasm and edema. Healing of duodenal ulcers produces scarring and deformity of the bulb.

Any imaging evaluation of the bowel should ideally be carried out with the bowel distended with air or contrast, since collapsed and unopacified loops of bowel can introduce artifactual errors of diagnosis.

Key abnormal findings of bowel disease on CT are thickening of the bowel wall, submucosal edema, hemorrhage, hazy infiltration of fat, and extraluminal air or contrast.

Crohn disease is a chronic, relapsing, granulomatous inflammation of the small bowel and colon, usually involving the terminal ileum, resulting in ulceration, obstruction, and fistula formation; it may have "skip areas", is prone to fistula formation, and has a propensity for recurring following surgical removal of an involved segment.

Colonic **diverticulosis** increases in incidence with increasing age, most often involves the sigmoid colon, and is almost always asymptomatic, although it can lead to diverticulitis or massive GI bleeding, especially from right-sided diverticula.

CT is the study of choice for imaging **diverticulitis,** and the findings include pericolonic inflammation, thickening of the adjacent colonic wall (>4 mm), abscess formation, and/or confined perforation of the colon.

Most **colonic polyps** are hyperplastic and have no malignant potential; adenomatous polyps carry a malignant potential that is related in part to their size. Colonic polyps can be visualized using barium enema examination, CT colonography, or optical colonoscopy.

Imaging signs of colonic polyps include a persistent filling defect in the colon with or without a stalk; some larger, villous adenomatous polyps have a higher malignant potential, and may contain barium within the interstices of their fronds.

The imaging findings of **colonic carcinoma** are a persistent, large, polypoid or annular constricting filling defect of the colon, which may have frank or microperforation, or large bowel obstruction and metastases especially to the liver and the lungs.

Colitis of any etiology can cause thickening of the bowel wall, narrowing of the lumen, and infiltration of the surrounding fat.

CT is the study of choice in diagnosing **appendicitis.** Findings include a dilated appendix (>6 mm), which does not fill with oral contrast, periappendiceal inflammation, increased enhancement of the wall of the appendix with intravenous (IV) contrast, and sometimes identification of an appendicolith (fecalith).

The two most common causes of **pancreatitis** are gallstones and alcoholism; pancreatitis is a clinical diagnosis with CT serving to document a cause or a complication of the disease. CT findings include enlargement of the pancreas, peripancreatic stranding, pancreatic necrosis, and pseudocyst formation.

Pancreatic adenocarcinoma has a very unfavorable prognosis, occurs most often in the pancreatic head, and usually manifests as a focal hypodense mass, which may be associated with dilatation of the pancreatic and/or biliary ducts.

Fatty infiltration of the liver is very common and can produce focal or diffuse areas of decreased attenuation that characteristically do not displace or obstruct the hepatic vessels; the liver appears less dense than the spleen.

In its later stages, **cirrhosis** produces a small liver (especially the right lobe) with a lobulated contour, inhomogeneous appearance of the parenchyma, prominent left and caudate lobes, splenomegaly, varices, and ascites.

Evaluation of liver masses is frequently done utilizing a **triple-phase CT scan** that includes a precontrast scan and two postcontrast scans, one in the hepatic-arterial phase and then another in the portal-venous phase.

Metastases are the most common malignant hepatic masses, and they mostly originate in the GI tract; they appear as multiple, low-density masses that may necrose as they become larger.

Hepatocellular carcinoma is the most common primary hepatic malignancy; these lesions are usually solitary and typically enhance with IV contrast on CT.

Cavernous hemangiomas are usually solitary, more common in females, and typically produce no symptoms; they have a characteristic centripetal pattern of enhancement and frequently retain contrast longer than the remainder of the liver.

Magnetic resonance cholangiopancreatography (MRCP) is a noninvasive way to image the biliary tree without requiring injection of contrast material; it can be used to demonstrate biliary strictures, gallstones, and congenital anomalies.

Renal cysts are a very common finding, are frequently multiple and bilateral, do not enhance, and typically have sharp margins where they meet the normal renal parenchyma. On US, simple cysts are well-defined anechoic masses.

Renal cell carcinoma is the most common primary renal malignancy and shows a propensity for extension into the renal vein and for metastasizing to lung and bone; on CT, it is usually a solid mass that enhances with IV contrast but remains less dense than the normal kidney. On US, it is frequently an echogenic mass.

Ultrasound is the imaging study of first choice in evaluating the female pelvis.

Abdominal and/or pelvic adenopathy may be caused by lymphoma, other malignancies, or benign diseases such as sarcoid; multiple enlarged nodes or conglomerate masses of nodes may be seen on CT.

 **WEBLINK**

Visit StudentConsult.Inkling.com for additional figures, videos, and quizzes on Gastrointestinal Diseases and Computed Tomography of Abdominal Abnormalities.

For your convenience, the following QR Code may be used to link to **StudentConsult.com**. You must register this title using the PIN code on the inside front cover of the text to access online content.

CHAPTER 21
Ultrasonography: Understanding the Principles and Recognizing Normal and Abnormal Findings

Ultrasound (US) makes use of **probes** with an acoustical frequency that is hundreds of times greater than humans can hear. *Ultrasonography* is the medical imaging modality that uses that acoustical energy to localize and characterize human tissues.

HOW IT WORKS

- The creation of a sonographic image **(sonogram)** depends on three major components: the production of a high-frequency sound wave, the reception of a reflected wave or echo, and the conversion of that echo into the actual image.
- The sound wave is produced by a **probe** or **transducer** that sends out extremely short bursts of acoustical energy at a given frequency.
  - For most imaging, the probe is placed **externally on the skin surface** and swept back and forth over the area to be scanned while the images so obtained are viewed in real time on a monitor. To produce the best contact between the probe and the skin, a coupling gel is applied to the skin surface first (Video 1-3).
  - On occasion, more detailed images can be obtained by inserting the probe into the body, such as is done with **transvaginal, transrectal,** and **transesophageal** sonography.
- Like all sound waves, the pulses produced by the transducer travel at different speeds, depending on the **density** of the medium through which they are traveling.
- Where the wave strikes an **interface** between tissues of **differing densities**, some of the sound will be **transmitted forward** and some will be **reflected back** to the transducer.

> How much sound is transmitted versus how much is reflected is a property of the tissues that make up the interface and is called the *acoustical impedance.* **Large differences** in acoustical impedance will result in **greater sound reflection; small differences** in acoustical impedance will result in **greater transmission.**

  - If the pulse encounters **fluid,** most of the acoustical energy is **transmitted.** If the pulse encounters **gas** or **bone,** so much of the acoustical energy is **reflected** back that it is usually not possible to define deeper structures.
- When the echo arrives back at the transducer (in a matter of microseconds), it is converted from sound into electrical pulses that are then sent to the scanner itself.
- Using an onboard computer, the scanner determines the length of **time** it took for the echo to be received, the frequency of the reflected echo, and the magnitude or **amplitude** of the signal. With this information, a sonographic image of the scanned body part can be generated by the computer and recorded digitally, either as static images or as a **cine.**
- A tissue that reflects many echoes is said to be **echogenic (hyperechoic)** and is usually depicted as **bright** or **white** on the sonogram; a tissue that has few or no echoes is said to be **sonolucent (hypoechoic** or **anechoic)** and is usually depicted as being **dark** or **black.**
- Images can be produced in any plane by adjusting the direction of the probe. By convention, two **common imaging planes** are used. One of these planes is **along the long axis** of the body or body part being scanned and is called the *sagittal* or *longitudinal plane.* The other plane is **perpendicular to the long axis** of the body or body part being scanned and is called the *transverse plane.*

> Also by convention, sonographic images are viewed with the patient's **head** to your **left** and the patient's **feet** toward your **right; anterior** is **up** and **posterior** is **down.**

- There are several types of US used in medical imaging. They are described in Table 21-1.

TABLE 21-1   TYPES OF ULTRASOUND

A-mode	Simplest; spikes along a line represent the signal amplitude at a certain depth; used mainly in ophthalmology.
B-mode	Mode most often used in diagnostic imaging; each echo is depicted as a dot, and the sonogram is made up of thousands of these dots; can depict real-time motion.
M-mode	Used to show moving structures, such as blood flow or motion of the heart valves.
Doppler	Uses the Doppler effect to assess blood flow; used for vascular ultrasound. *Pulsed Doppler* devices emit short bursts of energy that allow for an accurate localization of the echo source.
Duplex ultrasonography	Used in vascular studies; refers to the simultaneous use of both (1) grayscale or color Doppler to visualize the structure of, and flow within, a vessel and (2) spectral (waveform) Doppler to quantitate flow.

DOPPLER ULTRASONOGRAPHY

■ You are probably familiar with the common examples of a passing train whistle or police siren as illustrations of the *Doppler effect,* which generally states that sound changes in frequency as the object producing the sound approaches or recedes from your ear (Video 21-1).

■ Sonography makes use of the Doppler effect to determine if an object, usually blood, is **moving toward** or **away from** the transducer and at what **velocity** it is moving.
- ◆ The transducer sends out a signal of known frequency; the frequency of the echo returned is compared with the known frequency of the original signal.
- ◆ If the returned echo has a **lower frequency** than the original, then the object is **moving away** from the transducer. If the returned echo has a **higher frequency** than the original, then the object is **moving toward** the transducer.

➡ The **direction of flow** of blood is represented sonographically by the colors **red** and **blue**. By convention, **red indicates flow toward** and **blue indicates flow away** from the transducer.

ADVERSE EFFECTS OR SAFETY ISSUES

■ US procedures are **well tolerated.** Scans can be obtained relatively quickly, done at the bedside if necessary, and for the most part, require no patient preparation other than abstinence from food before abdominal studies.

■ US has the short-term potential of causing **minor elevation of heat** in the area being scanned, though not at levels used in diagnosis.

➡ There are **no known long-term side effects** that have been scientifically demonstrated to be caused by the use of medical US in humans. Nevertheless, like all medical procedures, US should be used only when medically necessary. The United States Food and Drug Administration warns against the use of US during pregnancy for the purpose of producing "keepsake photos or videos" (Table 21-2).

TABLE 21-2 ADVANTAGES AND DISADVANTAGES OF ULTRASONOGRAPHY

Advantages	Disadvantages
No ionizing radiation	Difficulty penetrating through bone
No known long-term side effects	Gas-filled structures reduce its utility
"Real-time" images	Penetration may be difficult in obese patients
Produces little or no patient discomfort	Dependent on the skills of the operator doing the scanning
Small, portable, inexpensive, ubiquitous	

MEDICAL USES OF ULTRASONOGRAPHY

■ We will look at several common abnormalities in which US plays a primary imaging role.

Biliary System

■ **US is the study of first choice for abnormalities of the biliary system.** Patients who present with the relatively common complaint of right upper quadrant pain usually undergo an US examination first. Computed tomography (CT) may be helpful in patients with difficult or unusual anatomy, for detecting masses, or in determining the extent of disease already diagnosed, but **CT is less sensitive than US in detecting gallstones.**

Normal ultrasound anatomy

■ The **gallbladder** is an elliptical sac that lies in the interlobar fissure between the right and left lobes of the liver. Although different layers of its wall have different echogenic properties, the gallbladder overall consists of a fluid-filled **sonolucent lumen** surrounded by an **echogenic wall.** In the fasting patient, the gallbladder is about **4 × 10 cm in size,** and the **wall is normally no thicker than 3 mm** (Fig. 21-1).

Gallstones and acute cholecystitis

■ **Cholelithiasis** is estimated to affect more than 20 million Americans. In almost all cases, acute cholecystitis starts with a gallstone **impacted in the neck of the gallbladder** or **cystic duct.** The presence of gallstones does not, by itself, mean that the patient's pain is emanating from the gallbladder, because asymptomatic gallstones are common. Cholecystitis can also occur less commonly in the absence of stones **(acalculous cholecystitis).**

■ **Gallstones usually fall to the most dependent part of gallbladder,** which will depend on the patient's position at the time of the scan. This helps to differentiate gallstones

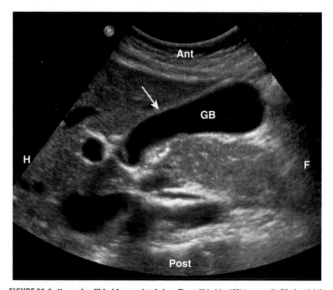

FIGURE 21-1 Normal gallbladder, sagittal view. The gallbladder (GB) is normally filled with bile and is sonolucent. The wall of the gallbladder is less than 3 mm in size and slightly echogenic *(white arrow)*. By convention on a sagittal view, the patient's head is to your left *(H)*, feet to your right *(F)*, anterior *(Ant)* is up, and posterior *(Post)* is down.

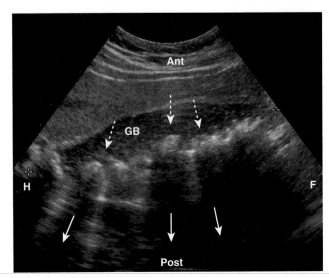

FIGURE 21-2 Cholelithiasis, sagittal view. There are numerous echogenic stones *(dotted white arrows)* in the gallbladder (GB). The stones cast acoustical shadows as they reflect most of the sound waves *(solid white arrows)*. *Ant,* Anterior; *F,* feet; *H,* head; *Post,* posterior.

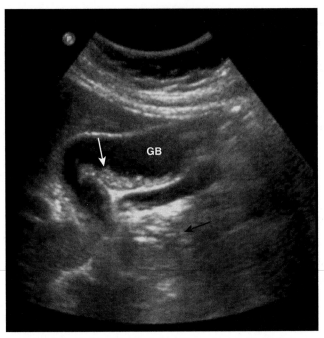

FIGURE 21-4 Sludge in the gallbladder. Sludge *(white arrow)* in the gallbladder (GB) is associated with biliary stasis. Although it may be echogenic, sludge does not produce acoustical shadowing as gallstones do. (The **absence** of shadowing is shown by the *black arrow*).

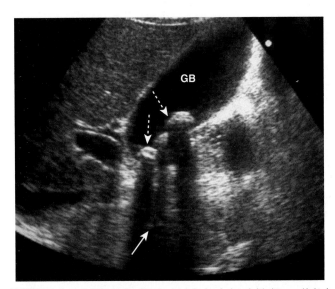

FIGURE 21-3 Acoustical shadowing. There is a band of reduced echoes *(solid white arrow)* behind echogenic gallstones *(dotted white arrows)* that reflect most, but not all, of the sound waves. The presence of acoustical shadowing can have diagnostic value in identifying the presence of calculi in the gallbladder (GB).

from polyps or tumors, which may be attached to a nondependent surface. Gallstones are characteristically **echogenic** and produce **acoustical shadowing** because they reflect most of the signal (Fig. 21-2).

 Acoustical shadowing describes a band of reduced echoes behind an echo-dense object (e.g., a gallstone) that reflects most, but not all, of the sound waves. Although acoustical shadowing reduces the diagnostic effectiveness of US through such tissues as bone and bowel gas, its presence can have diagnostic value in identifying the presence of calculi, such as in the gallbladder and kidney (Fig. 21-3).

Biliary sludge can be found in the lumen of the gallbladder and is an aggregation that may contain cholesterol crystals, bilirubin, and glycoproteins. It is often associated with biliary stasis. Although it may be echogenic, **sludge does not produce acoustical shadowing as gallstones do** (Fig. 21-4).

■ **Recognizing acute cholecystitis on US:**
   ♦ The presence of **gallstones,** possibly impacted in the neck of the gallbladder or cystic duct) (Fig. 21-5)
   ♦ **Thickening of the gallbladder wall** (>3 mm) (see Fig. 21-5, *A*)
   ♦ **Pericholecystic fluid** (fluid around the gallbladder) (see Fig. 21-5, *B*)
   ♦ A positive sonographic **Murphy sign** (A positive *Murphy sign* in this case is pain that is elicited by compression of the gallbladder with the US probe.)
■ In the presence of gallstones and gallbladder wall thickening, US has a positive predictive value for acute cholecystitis as high as 94%.
■ Radionuclide scans **(hepatoiminodiacetic acid [HIDA] scans) are also used in the diagnosis of acute cholecystitis.**
   ♦ Hepatoiminodiacetic acid (the "HIDA" in HIDA scans) is tagged with a radioactive tracer (technetium-99m), injected intravenously, and imaged with a special camera after it has been excreted by the liver into the bile and emptied into the small intestine.
   ♦ In patients with **obstruction of the cystic duct,** the tracer will not appear in the **gallbladder;** in patients with **obstruction of the common bile duct,** the tracer will **not appear in the small intestine.** Either finding is usually caused by an obstructing gallstone (Fig. 21-6).

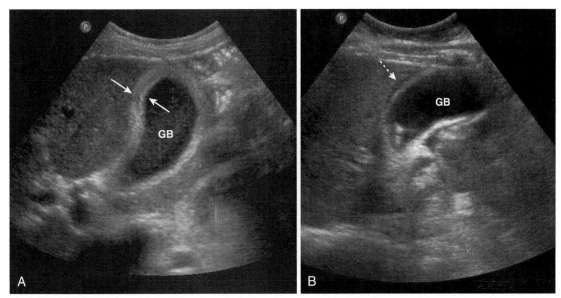

FIGURE 21-5 Acute cholecystitis, sagittal views, two patients. A, Thickening of the gallbladder (GB) wall *(white arrows)*. The wall should be 3 mm or less. This wall is markedly thickened at 6 mm. **B,** There is an echo-free crescent *(dotted white arrow)* surrounding a thickened gallbladder (GB) wall, representing pericholecystic fluid. The patient had a positive sonographic Murphy sign.

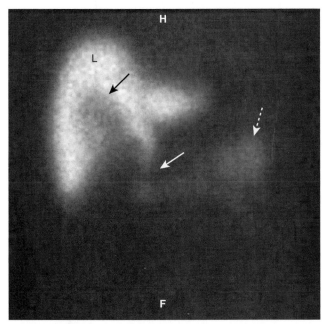

FIGURE 21-6 HIDA scan in cystic duct obstruction. HIDA, a radioactive labeled isotope, concentrates in the liver (L) and is then excreted into the biliary ducts. On this delayed image, the common bile duct *(solid white arrow)* and small bowel *(dotted white arrow)* fill normally because the common bile duct is patent, but the cystic duct and gallbladder do not fill because the cystic duct is obstructed, by a stone in this case. There is a tracer-free (photopenic) area in the gallbladder fossa because the nuclide cannot fill the gallbladder lumen *(solid black arrow)*. *F,* Feet; *H,* head.

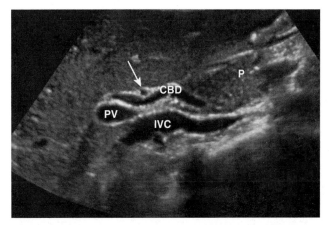

FIGURE 21-7 Normal common bile duct, portal vein, and hepatic artery, sagittal view. The common bile duct (CBD) measures 3 mm (normal <6 mm). The *white arrow* points to the hepatic artery, seen on end. The portal vein (PV) is posterior to the common duct, and the inferior vena cava (IVC) is seen posterior to the portal vein. The pancreas (P) is anterior to the CBD.

Bile ducts

■ **US plays a key role in evaluation of the intrahepatic and extrahepatic bile ducts and the pancreatic duct.** The intrahepatic biliary radicals drain into the left and right hepatic ducts, which join to form the common hepatic duct (CHD). Where the cystic duct from the gallbladder joins the CHD marks the origin of the common bile duct (CBD), which drains either within or adjacent to the head of the pancreas via the ampulla of Vater into the second portion of the duodenum. **The CBD lies anterior to the portal vein and lateral to the hepatic artery in the porta hepatis** (Fig. 21-7).

➡ The CHD and proximal CBD can be visualized normally on virtually all US studies of the right upper quadrant. **The CHD measures no more than 4 mm** (inner wall to inner wall) in diameter, and the **CBD measures no more than 6 mm** in diameter. The **pancreatic duct measures less than 2 mm.**

■ Normal intrahepatic bile ducts are not visible. When the CBD is obstructed, the **extrahepatic ducts dilate before the intrahepatic ducts.** Over time, both the intrahepatic and extrahepatic ducts will be dilated (Fig. 21-8).

■ **Causes of bile duct obstruction** include gallstones, pancreatic carcinoma, strictures, sclerosing cholangitis, cholangiocarcinoma, and metastatic disease.

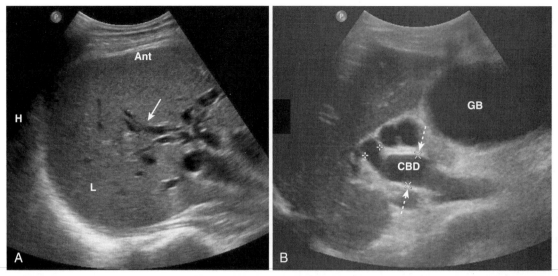

FIGURE 21-8 Dilated intrahepatic and extrahepatic ducts in two different patients, sagittal images. A, The intrahepatic biliary ducts are normally not visible by ultrasound. In this case, they are dilated *(white arrow)* from obstruction by a pancreatic carcinoma (not shown). **B,** The common bile duct (CBD) is dilated at 15 mm *(dotted white arrows)*, and the gallbladder (GB) is distended from an obstructing stone (not shown). *H,* Head; *L,* liver.

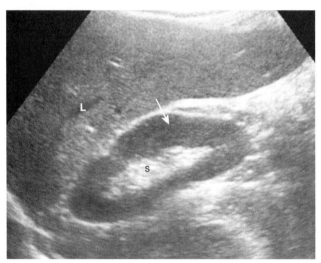

FIGURE 21-9 Normal right kidney, longitudinal. The renal sinus (S) is home to the renal pelvis and the major branches of the renal artery and vein. Because the renal sinus contains fat, it normally appears brightly echogenic. The normal renal pelvis is not visible in the renal sinus. The renal parenchyma *(white arrow)* has uniformly low echogenicity and is usually less echogenic than the adjacent liver (L) or spleen.

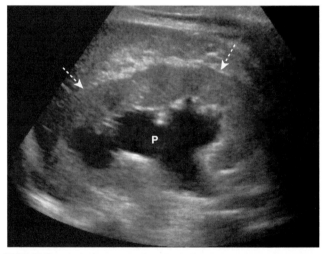

FIGURE 21-10 Hydronephrosis, sagittal view, right kidney. The renal sinus contains a markedly dilated, fluid-filled, and therefore anechoic renal pelvis (P) in this patient with hydronephrosis. The renal parenchyma remains normal in size and echogenicity *(dotted white arrows)*. The patient had an obstructing stone at the ureteropelvic junction.

Urinary Tract
Normal ultrasound anatomy

- The kidneys normally measure **9 to 12 cm in length, 4 to 5 cm in width, and are 3 to 4 cm thick.** The **renal sinus** is home to the renal pelvis and the major branches of the renal artery and vein. Because the renal sinus contains fat, it normally appears **brightly echogenic. The calyces are normally not visible.** The **medullary pyramids are hypoechoic.** The renal parenchyma has **uniformly low echogenicity,** which is **usually less** than that of the adjacent **liver** and **spleen** (Fig. 21-9).
- Renal masses are discussed in Chapter 20.

Hydronephrosis

- **Hydronephrosis** is defined as dilatation of the renal pelvis and calyces.
- In patients experiencing renal colic, US is used primarily to **evaluate for the presence of hydronephrosis** because the ureters are difficult to visualize with US. The **stone search** itself is almost always carried out by CT scanning (see Chapter 18).
- The typical appearance of obstructive uropathy is a **dilated calyceal system.** The echogenic **renal sinus** contains a **dilated, fluid-filled, and therefore anechoic renal pelvis.** The ureter may be dilated to the level of the obstructing stone. Severe hydronephrosis may distort the appearance of the kidney (Fig. 21-10).

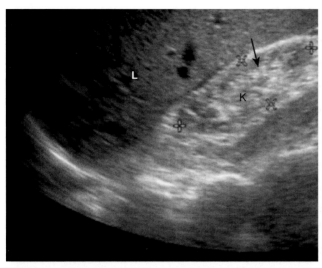

FIGURE 21-11 Chronic medical renal disease, sagittal view. The right kidney (K) is small, measuring 6 × 3 cm (cursor marks). The renal parenchyma *(black arrow)* is more echogenic (brighter) than the adjacent liver (L), the reverse of the normal echo pattern. This patient had chronic glomerulonephritis caused by long-standing diabetes.

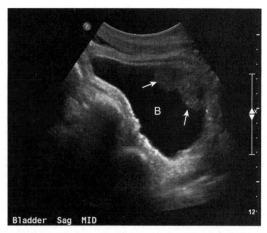

FIGURE 21-12 Bladder tumor. This sagittal ultrasound image of the urinary bladder (B) shows a large, polypoid mass arising from the anterior bladder wall *(white arrows).* A bladder stone would be unlikely to rest on the nondependent surface of the bladder. On biopsy, this was found to be a transitional cell carcinoma, a type that accounts for 90% of bladder tumors.

Medical renal disease

■ **Medical renal disease** refers to a host of diseases that primarily affect the **renal parenchyma.** They include such diseases as glomerulonephritis, diseases that cause the nephrotic syndrome, and renal involvement in collagen vascular diseases.

■ In the early stages of medical renal disease, the kidneys may appear normal. Later, changes in the echo architecture occur, but these are usually nonspecific as to the actual etiology. The **renal parenchyma becomes more echogenic (brighter) than the liver** and **spleen,** the reverse of the normal echo pattern. **Renal size** is also an important reflection of the chronicity of a disease, with the renal parenchyma almost always **decreasing in size with chronic disease.** A biopsy using US guidance may be performed to determine the etiology of the disease (Fig. 21-11).

Urinary bladder

■ As a fluid-filled structure, the urinary bladder lends itself well to US evaluation. US can be used to assess the thickness of the bladder wall and the presence of bladder tumors, stones, and diverticulae and can also be used to estimate the amount of postvoid residual urine. Measurement of residual urine is helpful in the assessment of urinary incontinence, bladder outlet obstruction, and in the neurogenic bladder.

■ A full bladder is also used as an acoustic window to assess the prostate in men and pelvic structures in women (Fig. 21-12).

Scrotal ultrasound

■ Differential diagnosis of **acute scrotal pain** includes epididymitis, orchitis, testicular torsion, testicular trauma, and herniation of abdominal contents into the scrotum. Prompt intervention is required in cases of testicular torsion, trauma, and incarcerated hernias to salvage the affected testis.

■ **Testicular torsion,** which usually occurs in adolescents, is an especially important diagnosis because timely detorsion is

curative. Color Doppler US (which can reveal scrotal blood flow) is used in the assessment of acute scrotal symptoms (pain or swelling). There is usually no flow to the affected testis in cases of torsion (Fig. 21-13).

Abdominal Aortic Aneurysms

■ An aneurysm is defined as a localized dilation of an artery by **50% or more above its normal size.** Most aortic aneurysms occur in the abdominal aorta **inferior to the origin of the renal arteries** and **frequently extend into one** or **both iliac arteries.**

♦ Most aortic aneurysms are either **fusiform** in shape or produce **uniform dilatation** of the entire vessel.

➡ The **abdominal aorta normally measures no more than 3 cm in diameter (outer wall to outer wall).**

♦ The **size** of an aneurysm is directly **related to its risk of rupture.** For aneurysms **less than 4 cm** in diameter, there is a **less than a 10%** chance of rupture; for aneurysms **4 to 5 cm** in diameter, the risk of rupture increases to **almost 25%.**

■ **Recognizing an abdominal aortic aneurysm:**

♦ **US is the screening study of choice** when an asymptomatic, pulsatile abdominal mass is palpated.

♦ Because it is moving, blood within the lumen of the aorta will appear anechoic; thrombus in the wall of the aneurysm will appear echogenic (Fig. 21-14).

♦ Unenhanced CT has the advantage of depicting the absolute size of the aneurysm, but **to define the extent of mural thrombus and the presence of dissection, intravenous contrast must be used.**

Female Pelvic Organs

➡ **US is the imaging study of choice in evaluating a pelvic mass** or **pelvic pain in the female.** Leiomyomas confined to the myometrium are the most common tumors of the uterus. Endometrial carcinomas are usually confined to the uterus at the time of discovery. The most

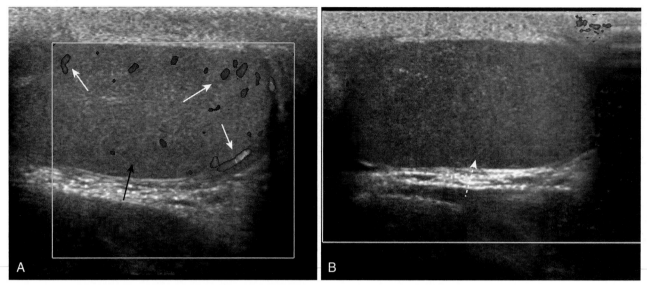

FIGURE 21-13 Scrotal ultrasound, torsion of the left testicle. Color Doppler images of both testicles are shown. **A,** The right testicle *(black arrow)* demonstrates normal blood flow *(white arrows).* **B,** The torsed left testicle is enlarged and demonstrates no blood flow because the vessels leading into the testicle are occluded by the torsion *(dotted arrow).* Testicular torsion is a urologic emergency.

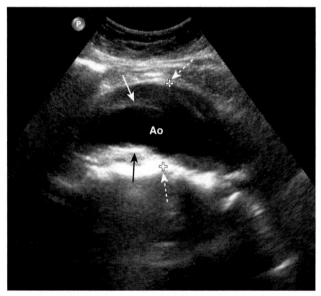

FIGURE 21-14 Abdominal aortic aneurysm, sagittal view. There is fusiform dilatation of the abdominal aorta (Ao) measuring 4.9 cm between marks *(dotted white arrows).* There is hypoechoic thrombus in the aneurysm *(solid black and white arrows).* Ultrasonography is the screening study of choice when an asymptomatic, pulsatile abdominal mass is palpated. For aneurysms 4 to 5 cm in diameter, the risk of rupture increases to almost 25%.

common mass in the ovary is a functional cyst. Generally, uterine masses are solid, and ovarian masses are cystic.

Normal US anatomy of the uterus

- The uterus is made up of a thick muscular layer **(myometrium)** and a mucous surface **(endometrium).** It is divided into a **body** (with **cornu** to receive the fallopian tubes) and the **cervix.** Anterior to the uterus is the peritoneal space called the *anterior cul-de-sac.* The **posterior cul-de-sac** is also called the *rectouterine recess.*
- The uterus is usually **anteverted** (axis of the cervix relative to the vagina) and **anteflexed** (position of the uterine body

relative to the cervix). In the adult, the uterus has a **pear shape** with maximum dimensions of approximately **8 cm in length, 5 cm in width,** and **4 cm in the anteroposterior dimension.** The size of the normal uterus increases with multiparity; with aging, the uterus decreases in size (Fig. 21-15).

- The normal **endometrial cavity** is collapsed, forming a **thin echogenic stripe** or **line** between the apposing surfaces of the endometrium. By convention, the endometrium is measured as a double thickness on a sagittal view of the miduterus. The appearance of both the endometrium and the ovaries varies, depending on the phase of the menstrual cycle.
- Normal fallopian tubes are not usually visualized on US.

➡ The standard **transabdominal** (through the abdominal wall) study of the uterus is done with a **full urinary bladder.** The full bladder provides an **acoustical window** to the uterus by pushing bowel loops upward out of the pelvis and also helps to delineate the bladder itself.

- **Transvaginal** studies are done with higher-frequency probes and thus provide **better-resolution** images. They are done with the **bladder empty.** Transabdominal and transvaginal studies are considered complementary techniques.
- **Sonohysterography** is a procedure in which saline is instilled in the uterine cavity in conjunction with images obtained using transvaginal US. It provides excellent visualization of the uterine cavity and can be used to detect small uterine polyps, submucosal myomas, and adhesions.

Uterine leiomyomas (fibroids)

- Leiomyomas are **benign smooth muscle tumors** of the uterus that occur in up to 50% of women over the age of 30. Although most women with fibroids are **asymptomatic,** fibroids can cause **pain, infertility, menorrhagia,** and urinary or bowel symptoms if they grow large enough.
- **US (transabdominal** and **transvaginal) is the imaging study of choice** in evaluating uterine fibroids. Magnetic resonance imaging (MRI) is used mainly to evaluate

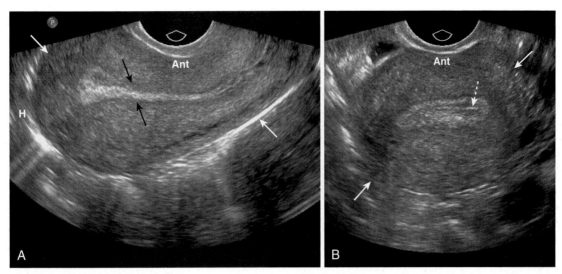

FIGURE 21-15 Normal uterus, longitudinal and transverse. The uterus *(solid white arrows)* has a pear shape with maximum dimensions of approximately 8 cm in length, 5 cm in width, and 4 cm in the anteroposterior dimension. **A,** The endometrium lines the uterine cavity and is measured in the sagittal view *(black arrows).* **B,** The collapsed uterine cavity is within the center of the endometrium and is visualized as a thin, echogenic line *(dotted white arrow). Ant,* Anterior; *H,* head.

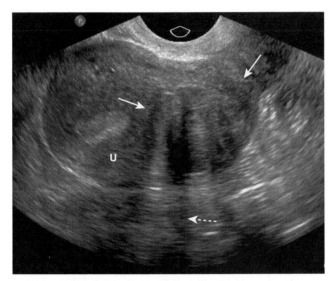

FIGURE 21-16 Leiomyoma of the uterus, sagittal view. A heterogeneously hypoechoic mass *(solid white arrows)* in the uterus (U) is shown. Uterine leiomyomas may display some areas that contain many echoes and others that demonstrate few echoes. They frequently absorb enough sound to produce acoustical shadowing *(dotted white arrow).* Where they are necrotic, they may be cystic. If they contain calcium, the calcium will produce acoustical shadowing.

complicated cases or in surgical planning. Fibroids are also frequently visualized on CT scans of the pelvis, usually performed for other reasons.

- **Recognizing uterine leiomyomas on US:**
  - Uterine leiomyomas are **heterogeneously hypoechoic, solid masses,** meaning they may display some areas that contain many echoes and others that demonstrate few echoes. When degenerated, they may become calcified and absorb enough sound to produce **acoustical shadowing** (Fig. 21-16). They may become necrotic with cystic components or echogenic with fat.
  - It is important to characterize the location of fibroids as **submucosal** (which has implications for bleeding and infertility), **myometrial** (most common),

or **subserosal** (pedunculated, which can grow very large).
- **Recognizing uterine leiomyomas on CT scans**
  - Characteristically, they are **lobulated soft tissue masses** that **frequently calcify** with **amorphous or popcorn calcification.** When they grow large, they frequently **undergo central necrosis,** which manifests as **low-attenuation** areas (see Fig. 18-11).
  - The **myometrium** is very vascular and **markedly enhances** on contrast-enhanced CT scans of the pelvis. The viable portions of uterine myomas also enhance dramatically following injection of intravenous contrast.

Adenomyosis

- "Adenomyosis" refers to **ectopic endometrial tissue within the myometrium.** It usually occurs in women 35 to 50 years old and can produce **dysmenorrhea** and **menorrhagia.**
- Adenomyosis is usually imaged with transvaginal US. On US, the uterus may be **enlarged** and contain myometrial **cystic spaces,** especially involving a **widened posterior uterine wall** and **decreased uterine echogenicity,** but without the lobulations seen with leiomyomas (Fig. 21-17).
- Unlike US, MRI is not limited by the concomitant presence of calcified uterine fibroids and is better able to differentiate adenomyosis from multiple, small uterine fibroids.

Ovarian cysts, tumors, and pelvic inflammatory disease

- US is the **imaging study of choice** for evaluating the ovaries.

Normal ovarian anatomy and physiology

- In premenopausal women, the ovaries are approximately **2 × 3 × 4 cm in size,** frequently containing cystic follicles. The ovaries atrophy after menopause (Fig. 21-18).

➡ The normal ovary changes in appearance not only with age but also with the phase of each menstrual cycle.

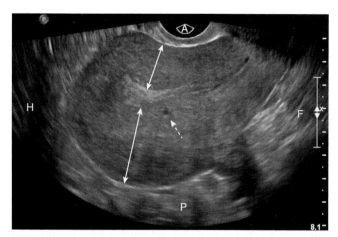

FIGURE 21-17 Adenomyosis of the uterus. On this transvaginal ultrasound, the uterus is enlarged. The myometrium is markedly thickened *(double white arrows)*. There is a cyst visible in the myometrium *(dotted white arrow)*. Adenomyosis represents ectopic endometrial tissue within the myometrium and usually occurs in women 35 to 50 years of age and produces dysmenorrhea and menorrhagia. *A,* Anterior; *F,* foot; *H,* head; *P,* posterior.

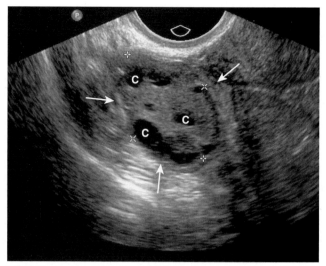

FIGURE 21-18 Normal right ovary, sagittal view. In premenopausal women, the ovaries are approximately 2 × 3 × 4 cm in size *(white arrows)*, usually containing small cystic follicles (C) as shown here.

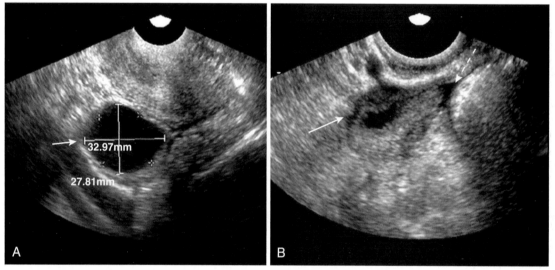

FIGURE 21-19 Dominant follicle ruptures during scan. Under normal hormonal stimulation, one egg-containing follicle becomes dominant and attains a size of about 2.5 cm at the time of ovulation. **A,** The dominant follicle is 3.3 × 2.8 cm at the start of the study *(white arrow)*. **B,** A few minutes later, during the study, the follicle ruptures (ovulation occurs), and the follicle shrinks dramatically in size *(solid white arrow)*. A small, physiologic amount of free fluid appears in the peritoneal cavity *(dashed white arrow)*.

Under normal hormonal stimulation, one egg-containing follicle becomes dominant and attains a size of about 2.5 cm at the time of ovulation (Fig. 21-19).

Ovarian cysts

- If one of the nondominant follicles fills with fluid and does not rupture, a **follicular cyst** is formed. Follicular cysts are **unilateral, usually asymptomatic,** and usually **involute** during the next menstrual cycle or two. If they persist or enlarge, they are pathologic.
- A *corpus luteum* is the structure that forms after expulsion of an egg from the dominant ovarian follicle. If the corpus luteum fills with fluid, a **corpus luteal cyst** will develop. Though **less common** than follicular cysts, corpus luteal cysts are **more symptomatic (pain)** and **can be larger.** Still, **most** usually **involute** in 6 to 8 weeks.

- **Follicular cysts** and **corpus luteum cysts** are called *functional cysts* of the ovary because they occur as a result of the hormonal stimuli associated with ovulation.
- Both are characteristically **well-defined, thin-walled, anechoic structures with homogeneous internal fluid density.** They may contain echogenic material if hemorrhage occurs into the "cyst" (Fig. 21-20). These cysts are physiologic and do not require follow-up or other imaging.
- **Nonfunctional cystic lesions** of the ovary include **dermoid cysts, endometriomas,** and **polycystic ovaries.** A specific diagnosis can usually be made with US.
  - **Dermoid cysts** are **mature teratomas** composed of cells from all three germ layers—ectoderm, mesoderm, and endoderm—although the ectodermal elements (hair, bone) predominate. They are most commonly found in women of reproductive age, are bilateral in up

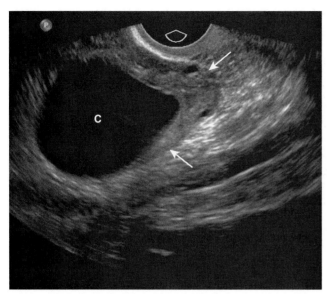

FIGURE 21-20 Simple ovarian cyst, sagittal view. Contained within the left ovary *(white arrows)* is a well-defined, thin-walled, anechoic structure representing a simple ovarian cyst (C). Ovarian cysts may contain echogenic material if hemorrhage into the cyst occurs.

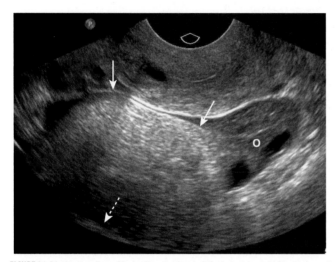

FIGURE 21-21 Ovarian dermoid cyst. There is a large, solid echogenic mass *(solid white arrows)* in the right ovary (O) with acoustical shadowing *(dotted white arrow)*. Dermoid cysts of the ovary are most commonly found in women of reproductive age and are bilateral in up to 25% of cases.

to 25% of cases, and may serve as a lead point for ovarian torsion (Fig. 21-21).

- **Endometriomas,** part of the disease *endometriosis,* are cystic ovarian lesions, sometimes called *chocolate cysts* because they are filled with brownish-red blood. Endometriomas can become large and multilocular.
- **Polycystic ovarian disease** is an endocrine abnormality that allows for numerous ovarian follicles (>12/ovary) to develop in various stages of hormonal growth and atresia. When associated with oligomenorrhea, hirsutism, and obesity, the constellation is called the *Stein-Leventhal syndrome* (Fig. 21-22).

Ovarian tumors

- Most ovarian tumors arise from the **surface epithelium** that covers the ovary, with **serous tumors** being more **common**

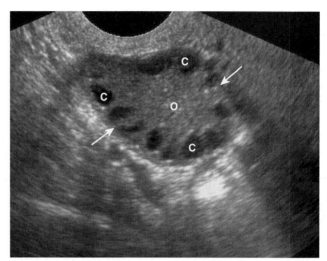

FIGURE 21-22 Polycystic ovarian disease. Polycystic ovarian disease is an endocrine abnormality that allows for numerous ovarian follicles to develop in various stages of hormonal growth and atresia. This ovary (O) is enlarged *(white arrows)* and contains multiple peripheral cysts, some of which are labeled (C). When associated with oligomenorrhea, hirsutism, and obesity, polycystic ovarian disease is called the *Stein-Leventhal syndrome.*

(serous cystadenomas, adenocarcinomas) than mucinous tumors (mucinous cystadenomas, adenocarcinomas). Most serous **tumors** and the overwhelming number of mucinous tumors are **benign**.

 On US, primary tumors of the ovary are **usually cystic in nature.** Keys to differentiating ovarian cysts from tumors include **thick** and **irregular walls** and **papillary projections** seen in tumors (Fig. 21-23).

- Staging of ovarian cancer is best done with CT or MRI.

Pelvic inflammatory disease

- *Pelvic inflammatory disease (PID)* is a term used to describe a **group of infectious diseases affecting the uterus, fallopian tubes,** and **ovaries.** Most cases of PID **begin as a transient endometritis** and ascend to infection of the tubes and ovaries. Patients can have pain, vaginal discharge, adnexal tenderness, and elevated white blood cell counts. Complications include **infertility, chronic pain,** or **ectopic pregnancy.**
- **Recognizing PID on US** (Fig 21-24):
  - **Enlarged ovaries** with multiple cysts and periovarian inflammation
  - Fluid-filled and dilated fallopian tube **(pyosalpinx)** (see Fig. 21-24, *A*)
  - Fusion of the dilated fallopian tube and ovary **(tubo-ovarian complex)** (see Fig. 21-24, *B*)
  - Multiloculated mass with septations **(tubo-ovarian abscess)**
- CT can be used for cases of complicated PID or for patients whose history may not suggest that diagnosis.

Ascites

There are normally only a few milliliters of fluid in the peritoneal cavity. **Ascites** is the **excessive accumulation of fluid** in that cavity. In the recumbent position, ascitic fluid flows up the **right paracolic gutter** to the right

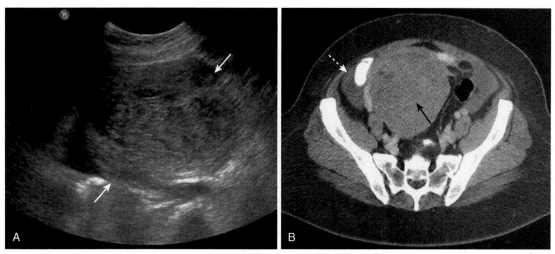

FIGURE 21-23 Ovarian tumor, US and CT. A, A transverse scan through the right adnexa demonstrates a large, solid ovarian mass *(white arrows).* **B,** CT of the same patient shows a large heterogeneous mass arising from the right adnexa *(black arrow).* A small amount of ascites is seen *(dotted white arrow).* This patient had a fibroma of the ovary, a tumor sometimes associated with ascites and pleural effusion **(Meigs syndrome).**

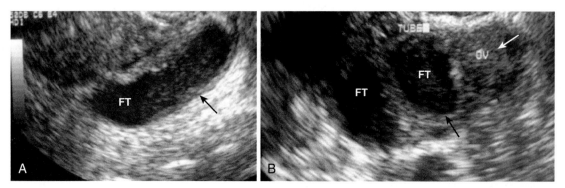

FIGURE 21-24 Pelvic inflammatory disease, longitudinal and transverse views, US. A, A fluid-filled and dilated fallopian tube (FT) containing pus and debris *(black arrow)* representing a **pyosalpinx** is shown. **B,** Because of progressive inflammation, there is fusion of a dilated and tortuous fallopian tube (FT; *solid black arrow)* and the adjacent ovary (OV; *white arrow),* producing a **tubo-ovarian complex.**

subphrenic space, so ascites is generally easier to detect by US in the right upper quadrant between the liver and the diaphragm.

- Ascitic fluid that is a **transudate** is primarily **sonolucent.** Fluid collections that are **exudates** or contain hemorrhage or pus may contain **echoes** (Fig. 21-25).
- US is frequently used to identify the best **location** and to provide guidance to perform a **paracentesis** to remove ascitic fluid.

Appendicitis

- The normal appendix may not be visualized on US. The diameter of the appendix is usually **less than 6 mm.** When visible, the **normal appendix compresses** when pressure is applied with the transducer.
- The pathophysiology of acute appendicitis begins with obstruction of the appendiceal lumen, followed by progressive distension of the obstructed appendix until perforation occurs and a periappendiceal abscess forms.

In **acute appendicitis,** the appendix may be recognized on US as a blind-ending, aperistaltic tube with a **diameter of 6 mm or more.** The inflamed appendix is **noncompressible** (using a technique called *graded compression).*

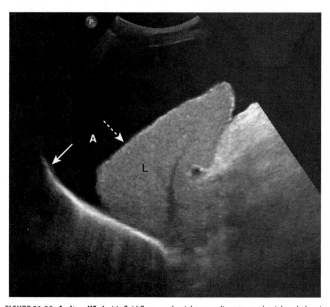

FIGURE 21-25 Ascites, US. Ascitic fluid flows up the right paracolic gutter to the right subphrenic space, so ascites (A) is generally easier to detect by ultrasound in the right upper quadrant between the liver (L) and the diaphragm *(solid white arrow).* The liver is contracted and has a nodular margin *(dotted white arrow),* both features of cirrhosis. If the fluid is a transudate, as in this case of ascites caused by cirrhosis, the fluid will be anechoic. Exudates will produce internal echoes.

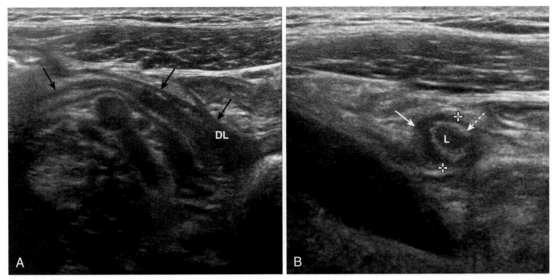

FIGURE 21-26 **Acute appendicitis, longitudinal and transverse views. A,** A blind-ending tubular structure *(black arrows)* that has a thick wall and a distended lumen (DL), representing an inflamed and distended appendix, is shown. **B,** In this transverse view, the appendix *(solid white arrow)* measures 7 mm between marks (normal <6 mm). The lumen (L) contains fluid, and there is an echogenic ring that represents the mucosa *(dotted white arrow)* surrounding it. The appendix was tender when palpated with the probe.

It may be **tender when palpated** with the probe. In about one third of appendicitis cases, a **fecalith** will be present (Fig. 21-26).

- US can be used for other gastrointestinal and abdominal wall abnormalities, including hernias, intussusception, bowel wall thickening, and masses. Endoscopic US allows for further characterization of the bowel wall and helps in guidance for biopsies.

Pregnancy

- US has provided a **safe** and **reliable** means of visualizing the fetus in utero and the ability to do so repeatedly during the course of a pregnancy, if necessary. The overwhelming majority of mothers in North America and Western Europe undergo at least one US evaluation sometime during pregnancy.
- Even before pregnancy, US can be used for assessing the time of ovulation to aid in the process of successful fertilization.
- The uses of US during pregnancy are outlined in Box 21-1.
- The goals of sonography during pregnancy may differ, depending on the timing of the scan. During the **first trimester,** goals are to exclude an ectopic pregnancy, estimate the age of the pregnancy, determine viability, and determine the number of embryos (type of twins) present. During the **second** and **third trimesters,** goals may include estimates of amniotic fluid volume, detection of fetal anomalies, determination of placental and fetal positioning, or guidance for invasive studies to determine the likelihood of fetal viability in the event of a premature birth (Video 21-2).
- We'll look at three important uses of US during pregnancy: ectopic pregnancy, fetal abnormalities, and molar pregnancy.

Ectopic pregnancy

- Most ectopic pregnancies are **tubal in location** and occur near the **fimbriated (ovarian) end.** The classical clinical findings of **pain, abnormal vaginal bleeding,** and a

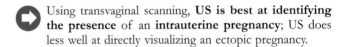

BOX 21-1   USES OF ULTRASOUND DURING PREGNANCY

Fetal presence and gestational age
Fetal abnormalities and viability
The presence of multiple pregnancies
Placental localization
Amniotic fluid volume
Intrauterine growth retardation
Helping to guide invasive studies such as amniocentesis, chorionic villus sampling, and intrauterine transfusions

palpable adnexal mass occur in only about half of cases. The incidence of ectopic pregnancies is increasing, most likely because of increasing risk factors, but the mortality rate has declined, in part because of their early diagnosis by US.

Using transvaginal scanning, **US is best at identifying the presence** of an **intrauterine pregnancy;** US does less well at directly visualizing an ectopic pregnancy.

- If a **gestational sac** (the earliest sonographic finding in pregnancy, appearing at about 4-5 weeks gestational age) or a **yolk sac** (the first structure to be seen normally in gestational sac) or **embryo** is identified **in the uterine cavity,** an **ectopic pregnancy** is effectively **excluded.** Endovaginal examinations are usually performed to find the gestational sac (Fig. 21-27).

Although it is possible to have **simultaneous** intrauterine and extrauterine pregnancies, such *heterotopic pregnancies* are **extremely rare,** other than in patients undergoing fertility treatment, so that identification of an intrauterine pregnancy effectively excludes an ectopic pregnancy.

- Conversely, the demonstration of a **live embryo outside the uterus is diagnostic of an ectopic pregnancy.** This is not a common occurrence with most ectopic pregnancies (Fig. 21-28).

- Most often, an ectopic location is diagnosed by a combination of findings that includes **absence of an identifiable intrauterine pregnancy,** often with an **extrauterine, extraovarian complex cystic mass,** and in which the quantitative **serum human chorionic gonadotropin hormone (β-HCG)** rises above the discriminatory level, at which point a normal intrauterine pregnancy **should almost always be identified.** If those criteria are met, **an ectopic pregnancy is presumed present.**
  - ♦ The **beta (β) subunit of HCG** is specific for the hormone produced by placental tissue shortly after the implantation of a fertilized ovum in the uterus. The levels of HCG roughly double every 2 to 3 days in a normal pregnancy. At a β-HCG level **more than 3,000 milli-international units per milliliter (mIU/mL), a normal intrauterine pregnancy should be visible** with transvaginal US. If not, the current recommendation calls for a repeat β-HCG test and US before undertaking treatment for ectopic pregnancy.

 Serial β-HCG determinations may help in differentiating an ectopic pregnancy from an *early abortion,* both of which may display similar sonographic findings. Patients with an early abortion will display **falling β-HCG levels** on serial serum studies, whereas these levels will rise in **ectopic pregnancies,** though usually **more slowly** than in normal intrauterine pregnancies.

- An ectopic pregnancy is also presumed present when there are **large amounts of free fluid (blood) inside the abdominal cavity.** Small amounts of free fluid can develop for other reasons, such as spontaneous abortion, ruptured ovarian cysts, and normal ovulation.
- Ectopic pregnancies are **managed** either surgically (usually laparoscopic surgery) or medically with an abortifacient such as methotrexate. Some spontaneously resolve.

Fetal abnormalities

- US is used extensively to monitor normal fetal growth and development.
- Certain fetal anomalies that are known to be universally fatal after birth, such as anencephaly or complete ectopia cordis, can be recognized by US in utero. The accurate interpretation of sonograms by someone trained and accredited in obstetric US is critical in detecting such anomalies (Fig. 21-29).
- Some of the many fetal anomalies that can be diagnosed by US in utero are described in Table 21-3.
- In detecting fetal anomalies, US has played a major role in obstetric management of pregnancy. As more reliable markers for chromosome abnormalities that can be quickly assessed without prolonged waiting periods become available, the role of US should increase.

Molar pregnancy

- **Molar pregnancy** is the most common of a group of disorders of the placenta, which also includes **invasive mole** and **choriocarcinoma.** Pathologically, molar pregnancies feature cystic (grapelike or hydatidiform) degeneration of chorionic villi and proliferation of the placental trophoblast.
- A molar pregnancy is suggested by **uterine size that is disproportionately large** for the dates of gestation, β-HCG levels in excess of 100,000 mIU/mL (normal pregnancies are <60,000 mIU/mL), vomiting, vaginal bleeding, and toxemia.

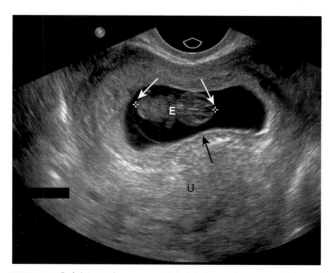

FIGURE 21-27 Early intrauterine pregnancy. A single live intrauterine pregnancy *(white arrows)* is contained within the gestational sac *(black arrow)* inside the uterus (U). Using a measurement called the *crown-rump length* (between the white arrows), the embryo (E) was estimated at 9 weeks of age. Sonographic gestational age begins on the first day of the last normal menstrual cycle because, for most individuals, that is a more certain date than the date of ovulation.

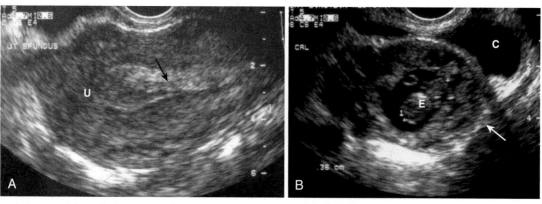

FIGURE 21-28 Ectopic pregnancy. A, A normal endometrial stripe *(black arrow)* with no evidence of a pregnancy in the uterus (U) is shown. **B,** An adnexal mass *(solid white arrow)* containing an embryo (E) can be visualized. There is fluid in the cul-de-sac (C). The demonstration of an embryo outside the uterus is diagnostic of an ectopic pregnancy.

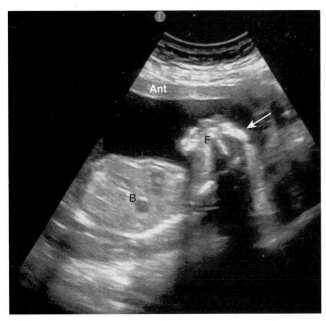

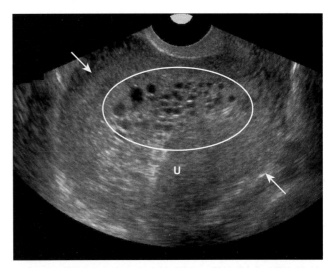

FIGURE 21-30 **Molar pregnancy, sagittal view.** The uterus (U) is enlarged and filled with echogenic tissue *(white arrows)*. There are innumerable, relatively uniform-sized cystic spaces that represent hydropic villi *(oval)* in this complete molar pregnancy. In a complete molar pregnancy, there is no fetus.

FIGURE 21-29 **Anencephaly, sagittal view.** This image shows a single live intrauterine pregnancy. The body (B) and face (F) are present, but the cranium and all of the cerebrum and cerebellum are absent *(white arrow)*. Anencephaly involves a failure of closure of the neural tube during the 3rd to 4th weeks of development. It virtually always leads to fetal demise, stillbirth, or neonatal death. *Ant,* Anterior.

TABLE 21-3 IN UTERO ABNORMALITIES DIAGNOSABLE BY ULTRASOUND

Organ System	Abnormalities
Central nervous system	Hydrocephalus, abnormalities of the prosencephalon, agenesis of the corpus callosum, intrauterine infections, cysts, meningomyelocele, anencephaly (Fig. 21-29)
Skeletal anomalies	Dwarfism, skeletal dysplasias, achondroplasia, osteogenesis imperfecta, asphyxiating thoracic dysplasia, limb anomalies
Gastrointestinal abnormalities	Esophageal atresia and tracheoesophageal fistula, duodenal atresia, small- and large-bowel obstruction, abdominal wall defects, congenital diaphragmatic hernias, choledochal cyst
Genitourinary tract abnormalities	Renal agenesis, congenital ureteropelvic junction and ureterovesical junction obstruction, bladder outlet obstruction, multicystic dysplastic kidney, polycystic kidney disease
Cardiac anomalies	Hypoplastic left heart syndrome, tricuspid atresia, endocardial cushion defects, Ebstein anomaly, tetralogy of Fallot, transposition of the great vessels, coarctation of the aorta, cardiac arrhythmias

- **Sonographic findings of molar pregnancy:**
  - ◆ **Enlarged uterus** filled with echogenic tissue that enlarges the endometrial cavity
  - ◆ **Innumerable,** uniform-sized **cystic spaces** that represent hydropic villi (Fig. 21-30)
  - ◆ **Enlarged** and **cyst-filled ovaries**
  - ◆ In a complete molar pregnancy, there is no fetus.
- Therapy is uterine evacuation. About 20% of those with complete molar pregnancies may harbor persistent

trophoblastic tissue, so follow-up is carried out with serial HCG determinations.

Vascular Ultrasound

- Vascular US studies combine morphologic images of the vessels with the simultaneous recording of the velocity of flow displayed by the **Doppler spectral waveform.** The combination of the two is called *duplex sonography,* and its use helps ensure that the sample being measured accurately represents the area of interest (see Video 21-3).
- The **Doppler spectral waveform** is a graphic representation of the **velocity of flow over time** within a focused area. It is depicted along an *x* (time) and *y* (velocity) axis. Flow toward the transducer is displayed above the baseline; flow away is displayed below the baseline. Different arteries have distinctive spectral waveforms, depending, in part, on whether there is normally a high or low resistance to flow within them.
- **Color flow Doppler imaging** adds the dimension of superimposing moving blood (shown in color) over a grayscale image of the anatomic structure, enabling a more rapid identification of potential abnormalities. The Doppler spectral waveform quantitates the flow.
- **Carotid ultrasonography** has become the study of choice for the noninvasive assessment of extracranial atherosclerotic disease. Extracranial carotid occlusive disease accounts for more than one half of strokes. Carotid US is also used to evaluate bruits, as preoperative screening before other major vascular surgery, and to assess the patency of the vessel after endarterectomy (Fig. 21-31).

Carotid artery stenosis

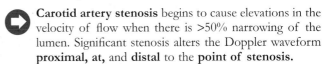 **Carotid artery stenosis** begins to cause elevations in the velocity of flow when there is >50% narrowing of the lumen. Significant stenosis alters the Doppler waveform **proximal, at,** and **distal** to the **point of stenosis.**

- US is used to assess the **thickness of the vessel wall** (it gets thicker with atherosclerosis) and the **presence** and

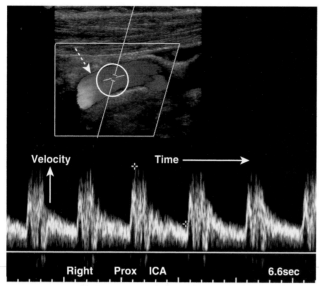

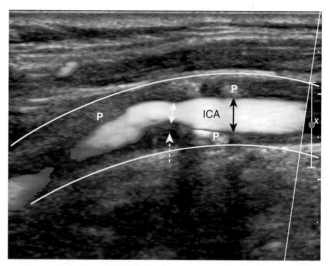

FIGURE 21-31 Normal proximal right internal carotid artery (ICA), duplex sonogram. The upper display *(dashed white arrow)* depicts flow in the artery. (The original is displayed in color online.) Within the *white circle* is the Doppler sampler volume, the measurements for which are shown on the lower graphical display called the *Doppler (spectral) waveform*. The x-axis represents time, and the height of the waveforms represents the velocity of blood flow within the Doppler sampler volume. The velocity normally increases with each systole and decreases with diastole. *Prox,* Proximal.

FIGURE 21-32 Stenotic right internal carotid artery (ICA), sagittal view. The *curved white lines* indicate the normal site of the wall of the right internal carotid artery (ICA). The lumen is narrowed *(black double-arrow)* by the presence of plaque (P) throughout the wall. At the site of the *dotted white arrow*, the lumen narrows to <50% of its normal diameter *(white double-arrow)* as a result of thicker plaque.

nature of plaque, as well as to analyze the Doppler spectral waveform.

- Any vessel in the body of large enough size that is accessible to the US transducer can be studied. Besides the carotid arteries, the jugular veins, vertebral artery, renal arteries, and peripheral arteries are commonly studied, as well as dialysis access routes and transplants (Video 21-4). CT-angiography may also be used for vascular studies, but it requires ionizing radiation and intravenous contrast dye.

- **Arterial flow in the extremities** produces a **high-resistance waveform** caused by the downstream high resistance of the arterial bed. With significant arterial disease, there is a **focal increase** in the velocity of flow at the point of stenosis. A normal high-resistance flow becomes low resistance when the scan sample is distal to the point of obstruction (Fig. 21-32).

Deep venous thrombosis (DVT)

- The majority of patients with DVT are **asymptomatic.** The most serious complication of DVT is pulmonary embolism.

- The highest-yielding sonographic examination for DVT using US occurs in the **symptomatic** patient who has symptoms **above the knee.** US has a much lower sensitivity in asymptomatic patients.

- The US examination for DVT is performed by examining the leg along the course of certain anatomic landmarks: the common femoral vein, the proximal deep femoral vein, the greater saphenous vein, and the popliteal vein.

➡ Sonographic evaluation for DVT of the leg is based mainly on the principle that **normal venous structures will easily compress** and **completely collapse** by the transducer, whereas veins harboring thrombi will not compress. Sonographic evaluation also seeks to visualize the echogenic thrombus itself (Fig. 21-33).

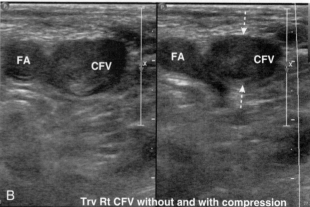

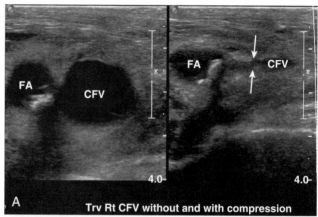

FIGURE 21-33 Normal common femoral vein (CFV) and deep venous thrombosis (DVT). Normal venous structures will compress and collapse because of pressure from the transducer, whereas veins harboring thrombi will not compress. Arteries also will not collapse. These are paired images done both **without** and **with** compression of the CFV in two different patients. **A,** In this normal patient, the CFV collapses normally with compression *(solid white arrows).* The femoral artery (FA) does not. **B,** In this patient with a DVT, the CFV does not collapse with compression *(dotted white arrows),* implying the presence of a clot contained within it.

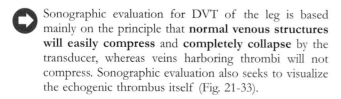

TAKE-HOME POINTS

ULTRASONOGRAPHY: UNDERSTANDING THE PRINCIPLES AND RECOGNIZING NORMAL AND ABNORMAL FINDINGS

Creation of a sonographic image (sonogram) depends on three major components: the production of a high-frequency sound wave, the reception of a reflected wave or echo, and the conversion of that echo into the actual image.

A tissue that reflects many echoes is said to be echogenic (hyperechoic) and is usually depicted as bright or white on the sonogram A tissue that has few or no echoes is said to be sonolucent (hypoechoic or anechoic) and is usually depicted as being dark or black.

Sonography makes use of the Doppler effect to determine if an object, usually blood, is moving toward or away from the transducer and at what velocity it is moving.

There are no known long-term side effects that have been scientifically demonstrated to be caused by the use of medical US in humans.

Gallstones are characteristically echogenic and produce **acoustical shadowing** because they reflect most of the signal.

Biliary sludge can be found in the lumen of the gallbladder and is often associated with biliary stasis. Although it may be echogenic, sludge does not produce acoustical shadowing as gallstones do.

The typical appearance of obstructive uropathy is a dilated calyceal system. The echogenic renal sinus contains a dilated, fluid-filled, anechoic renal pelvis.

In medical renal disease, the renal parenchyma becomes more echogenic (brighter) than the liver and spleen, the reverse of the normal echo pattern.

Scrotal torsion is a medical emergency. It occurs mostly in adolescents and produces acute scrotal pain. US can confirm the diagnosis by demonstrating absence of flow in the torsed testicle.

Ultrasonography is the screening study of choice when an asymptomatic, pulsatile abdominal mass is palpated. The normal abdominal aorta measures <3 cm in diameter.

Leiomyomas confined to the myometrium are the most common tumors of the uterus. The most common mass in the ovary is a functional cyst. Generally, uterine masses are solid, and ovarian masses are cystic.

Adenomyosis is ectopic endometrial tissue within the myometrium. On US, the uterus may be enlarged and contain myometrial cystic spaces, a thickened posterior uterine wall, and decreased uterine echogenicity.

Follicular cysts and corpus luteum cysts are called *functional cysts* of the ovary. Follicular cysts are more common. Functional cysts are characteristically well-defined, thin-walled, anechoic structures with homogeneous internal fluid density. They may contain echogenic material if hemorrhage occurs into the cyst.

Nonfunctional cysts of the ovary include dermoid cysts, endometriomas, and polycystic ovaries.

Tumors of the ovaries most often arise from the surface covering and are either serous or mucinous. Most serous tumors, along with the overwhelming number of mucinous tumors, are benign.

Pelvic inflammatory disease (PID) is a term used to describe a group of infectious diseases affecting the uterus, fallopian tubes, and ovaries, with most beginning as a transient endometritis.

When the patient is in the recumbent position, ascitic fluid flows up the right paracolic gutter to the right subphrenic space, so ascites is generally easier to detect by US in the right upper quadrant between the liver and the diaphragm.

In acute appendicitis, the appendix may be a blind-ending, aperistaltic tube with a diameter of 6 mm or more. It is noncompressible and may be tender when palpated with the probe. In about one third of the cases of appendicitis, a fecalith will be present.

US is a safe and reliable means of visualizing the fetus in utero and provides the ability to do so repeatedly during the course of a pregnancy if necessary.

Most ectopic pregnancies are tubal in location and occur near the fimbriated (ovarian) end. An ectopic pregnancy can be effectively excluded if an intrauterine pregnancy is present and included if an extrauterine pregnancy is seen. Most often, an ectopic pregnancy is diagnosed by a combination of absence of an identifiable intrauterine pregnancy with a β-HCG that rises above a certain level.

A molar pregnancy is suggested by uterine size that is disproportionately large for the dates of gestation and β-HCG levels >100,000 mIU/mL (normal pregnancies are <60,000 mIU/mL).

Vascular US studies combine morphologic images of the vessels with the simultaneous recording of the velocity of flow displayed by the Doppler spectral waveform. Carotid stenosis begins to cause changes in the velocity of flow when there is >50% narrowing of the lumen.

Sonographic evaluation for DVT of the leg is based mainly on the principle that normal venous structures will be easily compressed and collapsed by the transducer, whereas veins harboring thrombi will not compress. Sonographic evaluation can also be used to visualize the echogenic thrombus itself.

 **WEBLINK**

Visit StudentConsult.Inkling.com for videos and more information.

For your convenience, the following QR Code may be used to link to **StudentConsult.com**. You must register this title using the PIN code on the inside front cover of the text to access online content.

CHAPTER 22
Magnetic Resonance Imaging: Understanding the Principles and Recognizing the Basics

Daniel J. Kowal, MD

HOW MAGNETIC RESONANCE IMAGING WORKS

- Because magnetic resonance imaging (MRI) utilizes the **molecular** composition of tissues, especially water, it is particularly sensitive to detecting **soft tissue abnormalities** in much higher detail than computed tomography (CT) and is also well-suited to evaluating **changes in tissue composition over time,** providing a window into the acuity of a disease.
- MRI uses a very strong magnetic field to manipulate the electromagnetic activity of atomic nuclei in a way that releases energy in the form of radiofrequency (RF) signals, which are recorded by the scanner's receiving coils and then computer-processed to form an image.
  - ♦ The functioning of **clinical MRI scanners is based on hydrogen nuclei** (which contain one proton) because of their abundance in the human body.
- Each proton has a **positive electrical charge,** and because protons also have a **spin,** this charge is **constantly moving.** You might remember that a **moving electrical charge** is also an **electrical current,** and because an electrical current induces a **magnetic field,** each proton has its own small magnetic field (called a *magnetic moment*).
- When a patient enters an MRI scanner, the minimagnet protons all align with the more powerful external magnetic field of the MRI magnet. Most of these protons will point **parallel to the field,** and others will point **antiparallel** to the field, but they will all align with the external magnetic field of the MRI.
- Protons do not like to be couch potatoes, so they **precess** (i.e., wobble like a spinning top) along the magnetic field lines of the MRI.
- We will return to our little wobbling protons in a (magnetic) moment.

HARDWARE THAT MAKES UP AN MRI SCANNER

Main Magnet

- The main magnet in an MRI scanner is usually a *superconducting magnet.*
- Superconducting magnets contain a conducting coil that is cooled down to **superconducting temperatures** in order to carry the current (4 K or −269° C). At temperatures that low (close to absolute zero), **resistance to the flow of electricity in the conductor practically disappears.**

 Therefore an electric current sent once through this ultra-cold conducting material will **flow continuously** and create a **permanent magnetic field. The magnet in an MRI scanner is always "on."**

- Most scanners today have a **magnetic field strength** between **0.5 and 3 Tesla (T). Open MRI scanners,** those that do not completely encircle the patient in the scanning circle, have **lower field strengths of 0.1 to 1.0 T.** By comparison, the Earth's magnetic field is only about **50 μT.**

Coils

- The **coils** placed within the magnet are an important part of the MRI scanner. These coils are responsible for either **transmitting** the **RF pulses (transmitter coils)** that excite the protons or **receiving** the signal (or **echo**) given off by these excited protons **(receiver coils).**
- These coils are subjected to strong electrical currents and exist within a strong magnetic field that produces the repetitive "knocking" sound during an MRI scan.

Computer

- A computer dedicated to the MRI scanner processes the RF signals obtained by the receiver coils and converts them into an image.

WHAT HAPPENS ONCE SCANNING BEGINS

- When the patient is placed in the scanner magnet, the **transmitter coils** send a short electromagnetic pulse (measured in milliseconds) called a *radiofrequency pulse.* Remember that at this point the spinning protons in the patient have already **aligned** with the external magnetic field of the magnet.
- This RF pulse is sent at a particular frequency that changes the orientation of the protons.
- When the **RF pulse is turned off,** the displaced protons **relax** and **realign** with the main magnetic field, and **energy** subsequently **released** in the **form of RF signals** (the echo) is **detected** by the **receiver coils.**

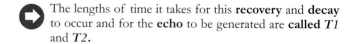 The lengths of time it takes for this **recovery** and **decay** to occur and for the **echo** to be generated are **called *T1* and *T2*.**

- **T1 relaxation** (or **recovery**) is the time it takes for the tissue to recover to its longitudinal state (parallel to the magnetic field); that is, the amount of time after the RF pulse was administered.

- **T2 relaxation** (or **decay**) is the time it takes for the tissue to regain its transverse orientation (perpendicular to the magnetic field) after the RF pulse was administered.
- **In summary,** as soon as the RF pulse stops, relaxation begins and the spinning nuclei release energy that is subsequently detected by the receiver coil and ultimately used to produce an image.

Pulse Sequences

- **Pulse sequences** consist of a set of **imaging parameters** determined in advance by protocols for **specific diseases** and **body parts** and then preselected by the MRI technologist at the computer. A particular **imaging protocol** (e.g., a routine MRI brain protocol) **consists of a series of multiple pulse sequences** that determine the way different tissues will appear on the scan. A pulse sequence can last between 20 seconds and 15 minutes.

➡️ There are **two main pulse sequences: spin echo (SE)** and **gradient recalled echo (GRE).** SE sequences have a higher signal-to-noise ratio, but GRE sequences are faster sequences preferred for rapid imaging techniques. **All of the pulse sequences used in MRI scanning are based on these two pulse sequences.**

Repetition Time and Echo Time

- **Repetition time (TR)** and echo time **(TE)** parameters are set by the MRI operator at the console prior to scanning and determine how the image is "weighted."
- **TR** is the **repetition time between two RF pulses,** and it influences the amount of **T1 weighting.**
  - Pulse sequences that feature a **short TR** (meaning a short amount of time between the RF pulses) will create what is called a *T1-weighted image.*
- **TE** is the **echo time between a pulse** and **its resultant echo,** and it influences the amount of **T2 weighting.**
  - Pulse sequences that feature a **long TE** (meaning a long amount of time between the RF pulse and the echo) will create what is called a *T2-weighted image.*

HOW CAN YOU IDENTIFY A T1-WEIGHTED OR T2-WEIGHTED IMAGE?

- Different tissues have different T1 and T2 values, which is why fat, muscle, and bone, for example, will appear differently not only from each other but also with different pulse sequences.

Bright Versus Dark

- **Tissues** that have a **short T1** will be **bright.**
- **Tissues** with a **long T2** will be **bright.**
- **Bright** translates into **whiter** or having **increased signal intensity** on MRI scans. **Dark** translates into **blacker** or having **decreased signal intensity** on MRI.

➡️ A key point is that **water** will be **dark** on **T1-weighted images** and **bright** on **T2-weighted images. Water is T1-dark and T2-bright.**

- A "bright" way to remember the fact that water is T2-bright is that the number "2" is both in H_2O (water) and T2 weighted.
- Therefore, when looking at any MR image, **first try to find something you know is fluid (water),** such as the **cerebrospinal fluid in the ventricles** and **spinal canal** or **urine** in the **bladder.**
- If the fluid is **dark,** then you are probably looking at a **T1-weighted image** (Fig. 22-1, *A*).
- If the fluid is **bright,** then chances are you are looking at a **T2-weighted image** (see Fig. 22-1, *B*).
- Certain tissues and structures are typically **bright on T1-weighted images.**
  - **Fat:** subcutaneous and intraabdominal fat, fat within yellow bone marrow, fat-containing tumors (see Fig. 22-1, *A*)
  - **Hemorrhage:** varies depending on the age of the hemorrhage (Fig. 22-2)
  - **Proteinaceous fluid:** in renal or hepatic cysts, cystic neoplasms
    - However, a simple cyst containing water will be dark on T1-weighted images (and bright on T2-weighted

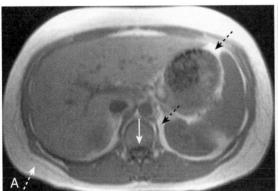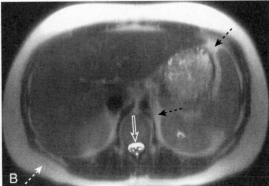

FIGURE 22-1 Normal T1-weighted and T2-weighted axial images of the abdomen. Because cerebrospinal fluid is similar to water in density, it appears dark on the T1-weighted image **(A)** *(solid white arrow)* and bright on the T2-weighted image **(B)** *(open white arrow).* Subcutaneous fat *(dotted white arrows)* and intraabdominal fat *(dotted black arrows)* are bright on both the T1- and T2-weighted images.

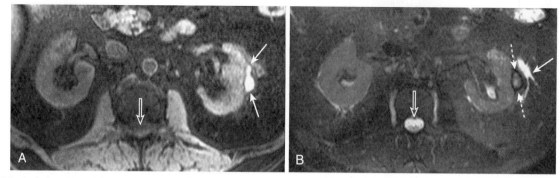

FIGURE 22-2 Subcapsular hematoma of the kidney. A, Axial T1-weighted, fat-suppressed image demonstrates a bright subcapsular hematoma *(solid white arrows)* involving the left kidney laterally. We can tell that this is a T1-weighted image because cerebrospinal fluid (CSF) in the spinal canal is dark *(open white arrow)*. **B,** Axial T2-weighted, fat-suppressed image demonstrates a slightly bright left subcapsular hematoma but also a dark rim of hemosiderin *(dotted white arrows)*, indicating surrounding older blood. There is a small amount of adjacent left perinephric fluid *(solid white arrow)*. The bright signal of the CSF helps us to recognize this image as a T2-weighted image *(open white arrow)*.

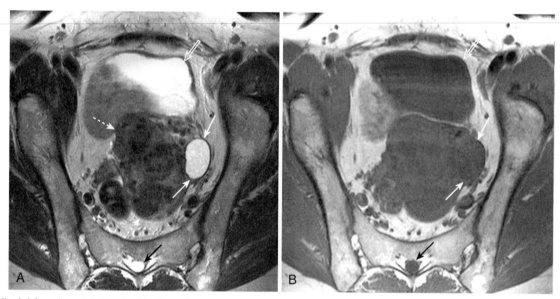

FIGURE 22-3 Simple left ovarian cyst. A, Axial T2-weighted image demonstrates an ovoid, homogeneously bright lesion in the left ovary *(solid white arrows)* adjacent to a fibroid uterus *(dotted white arrow)*. Note that urine in the bladder *(open white arrow)* and cerebrospinal fluid (CSF) in the spinal canal *(solid black arrow)* are both bright, which helps us identify this image as a T2-weighted image. **B,** Axial T1-weighted image shows that this left ovarian lesion is dark *(solid white arrows)* and therefore consistent with simple fluid. Urine in the bladder *(open white arrow)* and CSF in the spinal canal *(closed black arrow)* are also dark.

images), because remember that water is T1-dark (Fig. 22-3).

♦ **Melanin:** for example, melanoma (Fig. 22-4)
♦ **Gadolinium and other paramagnetic substances (manganese, copper)**
■ Certain tissues and structures are typically **bright on T2-weighted images.**
   ♦ **Fat:** subcutaneous and intraabdominal fat, fat within yellow bone marrow, fat-containing tumors (see Fig. 22-1, *B*)
   ♦ **Water, edema** (Figs. 22-5 and 22-6), **inflammation, infection, cysts** (see Fig. 22-3)
   ♦ **Hemorrhage:** varies depending on the age of the hemorrhage (see Fig. 22-2)
■ Notice that **both fat and hemorrhage can be T1-bright and T2-bright.**

Suppression

■ A useful feature of MRI is the ability to **cancel out** or **suppress** the signal from certain tissues selectively, thus making

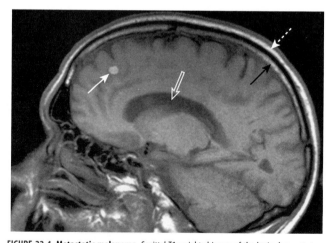

FIGURE 22-4 Metastatic melanoma. Sagittal T1-weighted image of the brain demonstrates a bright mass *(solid white arrow)* in the frontal lobe representing metastatic melanoma. Notice that both the yellow bone marrow within the skull *(solid black arrow)* and the overlying subcutaneous fat of the scalp *(dotted white arrow)* are bright. We can tell that this is a T1-weighted image because the cerebrospinal fluid in the lateral ventricles is dark *(open white arrow)*.

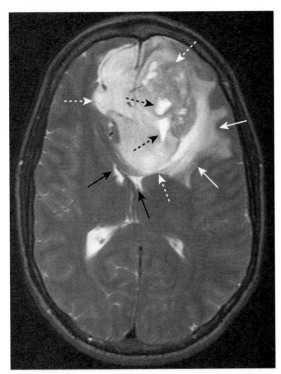

FIGURE 22-5 Glioblastoma multiforme with surrounding edema. Axial T2-weighted image demonstrates bright vasogenic-type edema *(solid white arrows)* surrounding a large, lobulated frontal lobe mass *(dotted white arrows)*, representing glioblastoma multiforme, an aggressive brain tumor. There are a few bright areas of cystic degeneration *(dotted black arrows)* within this mass. The frontal horns of the lateral ventricles are compressed *(solid black arrows)*.

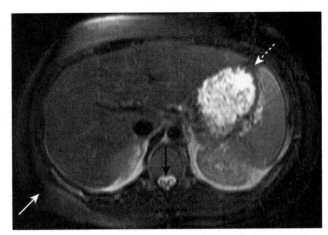

FIGURE 22-7 Normal T2-weighted fat-suppressed axial image of the abdomen. We can tell that this is a T2-weighted image because the cerebrospinal fluid in the spinal canal is bright *(solid black arrow)*. Also, it is a fat-suppressed image because the subcutaneous *(solid white arrow)* and intraabdominal *(dotted white arrow)* fat are dark. Fat is normally bright on a T2-weighted image without fat suppression.

that tissue look **dark** on the image and making other structures and pathologies more conspicuous.

- One body tissue that is **often suppressed** is **fat.**
- **Fat** is normally bright on T1-weighted images but will be **dark on T1 fat-suppressed images** (Fig. 22-7).
- This feature is useful when attempting to identify fat-containing lesions such as **ovarian dermoid cysts, adrenal myelolipomas,** and **liposarcomas** (Fig. 22-8), because they will appear to change from **bright** on the **non–fat-suppressed** images to **dark** on the **fat-suppressed images.**
- Fat suppression is **also essential** for evaluation of tissues **after the administration of gadolinium contrast dye.**

Other Pulse Sequence Types

- There are many other pulse sequences besides T1-weighted and T2-weighted images that typically comprise a particular MRI scan protocol, such as **diffusion-weighted images, proton density-weighted images,** and an entire alphabet soup of images with catchy acronyms, such as *STIR, FLAIR,* and *MRA/TOF* (standing for *short tau inversion recovery, fluid-attenuated inversion recovery,* and *time-of-flight magnetic resonance angiography,* respectively).
- **Functional MRI,** or **fMRI,** correlates the brain's changing blood flow requirements with changes in neural activity and translates them into differences in the MRI signal, particularly on T2-weighted images. fMRI is increasingly being used to map neural activity in the brain or spinal cord.

MRI CONTRAST AGENTS: GENERAL CONSIDERATIONS

 Gadolinium is the most common intravenous contrast agent used in clinical MRI.

♦ Gadolinium is a rare earth, heavy metal ion that is chelated to different compounds to form MRI contrast agents. When chelated to an acid known as *gadolinium diethylenetriaminepentaacetic acid (Gd-DTPA),* it forms gadopentetate dimeglumine, a commonly used contrast agent.

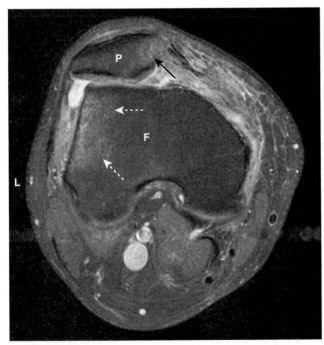

FIGURE 22-6 Bone marrow edema due to transient lateral patellar dislocation. Axial proton-density, fat-saturated image (a T2-like sequence) demonstrates bright bone marrow edema involving the lateral (L) femoral (F) condyle *(dotted white arrows)* and the medial aspect of the patella (P) *(solid black arrow)*. This edema is due to recent lateral patellar dislocation with contusion of the patella as it struck the tibia.

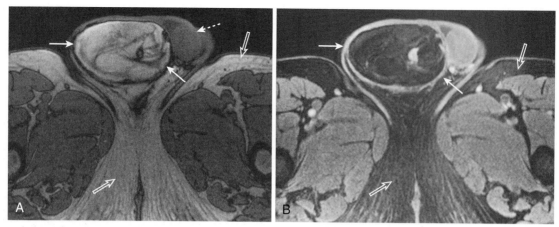

FIGURE 22-8 Fat in a liposarcoma of the right spermatic cord. (A) Axial T1-weighted image of the scrotum demonstrates a bright, heterogeneous right scrotal mass *(solid white arrows)*. The left testis is unremarkable *(dotted white arrow)*, and the right testis is not visible. Note that subcutaneous fat is normally bright *(open white arrows)*. **(B)** Axial T1-weighted, fat-suppressed, gadolinium-enhanced image demonstrates that the bright signal in the right scrotal mass is now dark, consistent with fat *(solid white arrows)*. The subcutaneous fat of the thighs is also dark *(open white arrows)*. This was an unusual malignancy of the spermatic cord derived from fat cells.

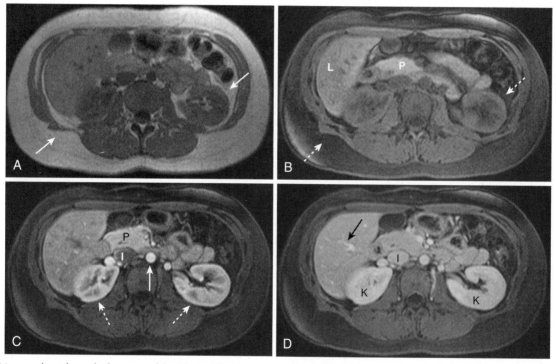

FIGURE 22-9 Fat suppression and normal enhancement of the abdomen following the administration of intravenous gadolinium. A, T1-weighted image of a normal abdomen demonstrates normal bright subcutaneous and intraabdominal fat *(solid white arrows)*. **B,** T1-weighted, fat-suppressed image shows that the signal from the fat has been suppressed *(dotted white arrows)* and is now dark. Intraabdominal organs such as the pancreas (P) and liver (L) now appear brighter relative to the adjacent suppressed fat. **(C)** T1-weighted, fat-suppressed, early-phase postgadolinium image shows the normal enhancement of the aorta *(solid white arrow)*, which enhances earlier than the inferior vena cava (I). The kidneys demonstrate normal corticomedullary phase enhancement *(dotted white arrows)*, and the pancreas (P) enhances maximally during this phase. **(D)** T1-weighted, fat-suppressed, later-phase postgadolinium image shows that the hepatic veins *(solid black arrow)* and inferior vena cava (I) are now well enhanced. The kidneys (K) now demonstrate normal homogeneous enhancement, the optimal phase for detecting renal masses.

- Gadolinium-based contrast agents are **used in the same way that iodinated contrast media are used in CT.** They can be injected either intravascularly or intraarticularly.
- After intravenous injection, Gd-DTPA enters the blood pool, enhances organ parenchyma, and then is **excreted by the kidneys** via glomerular filtration.
  - ◆ Other special types of gadolinium-based contrast agents have a component of biliary excretion.
- **Gadolinium's effect is to shorten the T1 relaxation times** of hydrogen nuclei (and to a lesser extent, to shorten T2).

T1 shortening will cause a brighter signal on T1-weighted images than the same images without gadolinium, and it is for this reason that **images obtained after gadolinium administration are usually T1-weighted** to take advantage of this effect.

- Fat is bright on T1 even before the administration of gadolinium. In order to increase detection of contrast enhancement in fat, the precontrast and postcontrast images are typically fat-suppressed (meaning darkened) to enhance the effect of the gadolinium (Fig. 22-9).

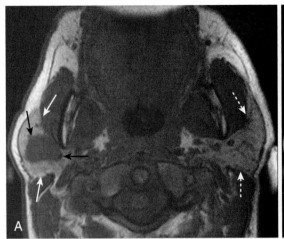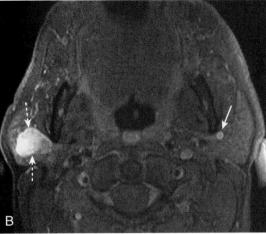

FIGURE 22-10 Right parotid pleomorphic adenoma. A, Axial T1-weighted image of the neck demonstrates a dark, ovoid mass with a slightly lobular contour *(solid black arrows)* located within the right parotid gland *(solid white arrows)*. The left parotid gland is unremarkable *(dotted white arrows)*. **B,** Axial T1-weighted, fat-suppressed, postgadolinium-enhanced image demonstrates that the right parotid mass is very bright due to intense enhancement *(dotted white arrows)*. Surgical resection of this mass yielded a pleomorphic adenoma, the most common benign tumor of the parotid gland. A clue that this image is a gadolinium-enhanced image is the presence of bright, enhancing vascular structures such as the left retromandibular vein *(solid white arrow)*.

 Structures that become **bright on postgadolinium images are typically vascular** (such as tumors) (Fig. 22-10) or **inflamed tissues,** and are described as **enhancing.**

MRI SAFETY ISSUES

Claustrophobia

- Patients can occasionally experience such extreme claustrophobia in the small confines of the MRI scanner that **they will be unable to begin** or **complete** the study. Pretreatment with sedatives may help in appropriate clinical situations.
- Alternatively, the patient can be scanned in an **open magnet,** which is not as confining. The trade-off, however, is that open magnets in general have lower magnetic field strength and poorer spatial resolution.

Ferromagnetic Objects

- Any ferromagnetic object inside the patient **can be moved** by the magnetic field of the MRI scanner and potentially damage adjacent tissues. Such internal ferromagnetic objects also hold the potential to become **heated** and **cause burns** to surrounding tissues.

 Ferromagnetic **objects in a location where motion of the object may be harmful to the patient represent an absolute contraindication to MRI.** These objects include medically inserted items, such as cerebral aneurysm repair clips, vascular clips, and surgical staples. Many vascular clips and staples are now manufactured to be MRI-compatible.

- Some **foreign bodies, such as bullets, shrapnel,** and **metal in the eyes** (as can sometimes be found in metalworkers) can also be ferromagnetic.
  - Conventional orbital radiographs must be examined in patients who may have a history of metallic foreign bodies in the eyes before they undergo MRI, and, if

metal is present, an alternative means of imaging should be used.
- Ferromagnetic objects outside the patient, such as **oxygen tanks, scissors, scalpels,** and **metallic tools,** also pose a risk to the patient as **they could become airborne** once they enter the magnetic field and for this reason are strictly forbidden in the MRI scanning room.

Mechanical or Electrical Devices

- MRI cannot usually be performed in patients who have **pacemakers, pain stimulator implants, insulin pumps, other implantable drug infusion pumps,** and **cochlear implants.** An exception is a newer type of pacemaker approved by the Food and Drug Administration (FDA) that is specifically engineered so that patients can safely undergo MRI.

Can a Patient with an Implanted Medical Device Undergo an MRI Examination?

- The answer is, it depends.
- In 2005, ASTM International, formerly known as the American Society for Testing and Materials, defined terminology, now recognized by the FDA, for medical devices in patients undergoing MRI examinations.
  - *MR Safe* refers to an item that has **no known hazards in all MRI environments.** These products, such as plastic tubing, are nonmetallic, nonconductive, and nonmagnetic.
  - *MR Conditional* refers to an item that has been demonstrated to **pose no known hazards in a particular MRI environment under specific conditions of use.** MRI environments differ as to the strength of the main magnet, the gradient magnetic fields, and the RF fields, to name a few parameters.
  - *MRI Unsafe* refers to any item **known to pose hazards in all MRI environments.** MR Unsafe items might include magnetic items such as a pair of ferromagnetic scissors.

- There are extensive online lists of items with detailed descriptions of each item's degree of MRI safety and descriptions contained in the item manufacturer's package information, as well as on the various manufacturers' websites.

Pregnant Patients

- Although there are no known biological risks associated with MRI in adults, the effects of MRI on the fetus are not definitively known.
- However, the American College of Radiology states that pregnant patients **can** undergo MRI scans at any stage of pregnancy if it is decided that the risk/benefit ratio to the patient weighs in favor of obtaining the study (Fig. 22-11).
- MRI should probably **not** be performed **electively** in early-term pregnancy due to the unknown risk to the fetus.
- The use of **gadolinium contrast agents is not recommended in pregnant patients,** because gadolinium crosses the placenta, is subsequently excreted by the fetal kidneys, and has unknown effects on the fetus.

Nephrogenic Systemic Fibrosis

- In patients with renal insufficiency, gadolinium-based contrast agents have been associated with a rare, painful, debilitating, and sometimes fatal disease called *nephrogenic systemic fibrosis (NSF).*
- NSF produces fibrosis resembling scleroderma of the skin, eyes, joints, and internal organs.
- Patients with **preexisting renal dysfunction,** especially those on dialysis, are at the **greatest risk** for developing NSF due to the use of gadolinium-based contrast agents.
  - ♦ **Caution** must be exercised **when administering gadolinium** to patients who have **moderate renal disease** (particularly with estimated glomerular filtration rate [eGFR] of 30-45 mL/min/1.73 m^2). The use of gadolinium is typically avoided in patients with severe renal disease (eGFR <30 mL/min/1.73 m^2).

DIAGNOSTIC APPLICATIONS OF MRI

- Some of the many clinical uses of MRI are outlined in Table 22-1. The chapters in which some of the diseases are discussed in greater detail are indicated.

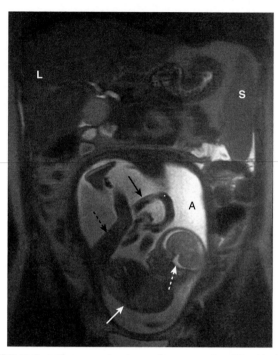

FIGURE 22-11 Magnetic resonance imaging scan in pregnancy. Coronal T2-weighted image demonstrates an intrauterine pregnancy. The maternal liver (L) and spleen (S) are partially imaged. The bright amniotic fluid (A) and fetal cerebrospinal fluid *(dotted white arrow)* help us to recognize this as a T2-weighted image. The fetal body *(solid white arrow)* and leg *(dotted black arrow)* can be seen clearly. The umbilical cord *(solid black arrow)* is partially visible.

TABLE 22-1   DIAGNOSTIC APPLICATIONS OF MAGNETIC RESONANCE IMAGING

System	Organ	Diseases
Musculoskeletal	Evaluate menisci, tendons, muscles	Meniscal tears; ligamentous and tendon injuries
	Bones	Bone marrow contusions; occult or stress fractures
	Osteomyelitis	High negative predictive value if normal
	Spine (Chapter 26)	Disk disease and marrow infiltration; differentiating scarring from prior surgery or from new disease
Neurologic	Brain (Chapter 27)	Ideal for studying brain, especially posterior fossa; tumor, infarction; multiple sclerosis
	Peripheral nerves	Impingement; injury
Gastrointestinal	Liver (Chapter 20)	Characterize liver lesions; detect small lesions; cysts, hemangiomas; hepatocellular carcinoma; focal nodular hyperplasia; hemochromatosis; fatty infiltration
	Biliary system (Chapter 20)	Magnetic resonance cholangiopancreatography for strictures, ductal dilatation
	Small and large bowel	Magnetic resonance enterography; appendicitis in pregnant females
Endocrine/reproductive	Adrenal glands	Adenomas; adrenal hemorrhage
	Female pelvis	Anatomy of uterus and ovaries; leiomyomas; adenomyosis; ovarian dermoid cysts; endometriosis; hydrosalpinx
	Male pelvis	Staging of rectal, bladder, and prostate carcinoma
Genitourinary	Kidneys	Renal masses; cysts versus masses

TAKE-HOME POINTS

MAGNETIC RESONANCE IMAGING: UNDERSTANDING THE PRINCIPLES AND RECOGNIZING THE BASICS

MRI uses a very strong magnetic field to influence the electromagnetic activity of hydrogen nuclei, also called *protons*.

Protons each have a charge and possess a spin. The constant movement of protons generates a small magnetic field, causing the proton to behave like a minimagnet. When the protons are placed in the much more powerful magnetic field of the MRI scanner, they all align with this external magnetic field.

A radiofrequency (RF) pulse, transmitted by a transmitter coil, displaces the protons from their original alignment with the external magnetic field of the scanner.

When the RF pulse is turned off, the displaced protons relax and realign with the main magnetic field, producing a RF signal (the echo) as they do so. Receiver coils receive this signal (or echo) given off by the excited protons. A computer reconstructs the information from the echo to generate an image.

The main magnet in an MRI scanner is usually a superconducting magnet that is cooled to extremely low temperatures in order to carry the electrical current.

Pulse sequences consist of a set of imaging parameters that determine the way a particular tissue will appear. The two main pulse sequences on which all MRI pulse sequences are based are called *spin echo (SE)* and *gradient recalled echo (GRE)*.

T1 and T2 are both time constants. T1 is called the *longitudinal relaxation time,* and T2 is called the *transverse relaxation time.*

TR is the repetition time between two RF pulses. A short TR creates a T1-weighted image.

TE is the echo time between a pulse and its resultant echo. A long TE creates a T2-weighted image.

On T1-weighted images, fat, hemorrhage, proteinaceous fluid, melanin, and gadolinium are typically bright (white).

On T2-weighted images, fat, water, edema, inflammation, infection, cysts, and hemorrhage are typically bright.

Fat is T1-bright and T2-bright. Water is T1-dark and T2-bright.

Suppression is a feature of MRI that will cancel out or eliminate signals from certain tissues and is most often used for fat. Although normally T1-bright, fat appears dark on T1-weighted, fat-suppressed images. Fat suppression is particularly useful for tissue characterization after administration of gadolinium.

Gadolinium is the most common intravenous contrast agent used in clinical MRI, and its effect is to shorten the T1 relaxation time of hydrogen nuclei, yielding a brighter signal. Vascular structures such as tumors and areas of inflammation enhance after gadolinium administration and become more conspicuous.

Ferromagnetic objects must be kept outside the MRI scanning room, as they could become airborne when exposed to the magnetic field. For patients who may have metallic foreign bodies in their eyes, conventional orbital radiographs must be obtained to determine if metal is present.

In pregnancy, MRI is preferred in the second and third trimesters, and gadolinium is contraindicated.

Nephrogenic systemic fibrosis is a debilitating fibrotic disease that can occur in patients with renal insufficiency who receive intravenous gadolinium-based contrast agents. Therefore, gadolinium is typically avoided in patients with severe renal disease.

 **WEBLINK**

Visit StudentConsult.Inkling.com for more information.

For your convenience, the following QR Code may be used to link to **StudentConsult.com**. You must register this title using the PIN code on the inside front cover of the text to access online content.

CHAPTER 23

Recognizing Abnormalities of Bone Density

NORMAL BONE ANATOMY

Conventional Radiography

- On conventional radiographs, bones consist of a dense **cortex** of **compact bone** that completely envelops a less dense **medullary cavity** containing **cancellous bone** arranged as **trabeculae,** separated primarily by blood vessels, hematopoietic cells, and fat. The proportion of cortical versus trabecular bone varies in different skeletal sites and even at different locations in the same bone; that is, the **cortex is naturally thicker in some places than in others.**
- When viewed in **tangent** on conventional radiographs, the **cortex** produces a **smooth, dense white shell** of varying thickness that appears as a dense white band along the outer margins of the bone.
- The **medullary cavity** on conventional radiographs appears as a core of less-dense, **grayish density inside the cortical shell, interlaced with a fine network of bony trabecular markings.** The *corticomedullary junction* is the edge between the inner margin of the cortex and the medullary cavity (Fig. 23-1).

- It is important to remember that **the cortex completely surrounds the entire bone** but on conventional radiographs is best seen when it is viewed in profile (i.e., where the x-ray beam passes tangentially to the bone).

- ➡️ Almost all **examinations of bone start with conventional radiographs** obtained with at least **two views exposed at a 90° angle to each other** (called *orthogonal views*) to localize abnormalities better and to visualize as much of the circumference of the bone as possible.

- Still, conventional radiographs cannot visualize the entire circumference of a tubular bone, and they are not particularly sensitive for demonstrating musculoskeletal soft tissue abnormalities other than soft tissue swelling.

Computed Tomography and Magnetic Resonance Imaging

- Computed tomography (CT) and magnetic resonance imaging (MRI) are able to demonstrate the entire circumference and internal matrix of bone, including—especially with

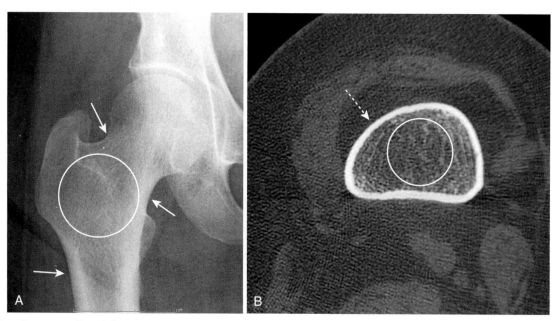

FIGURE 23-1 Normal appearance of bone. A, This is an anteroposterior view of the hip. When viewed in tangent, the cortex can be seen as a white line, varying in thickness in different parts of the bone *(white arrows).* In the medullary cavity, cancellous bone can be seen to contain an interlacing network of trabeculae *(white circle).* **B,** This image was produced by axial CT through the proximal femur. The entire 360° circumference of the cortex *(dotted white arrow)* can be seen surrounding the less dense medullary cavity, which contains both bony trabeculae and fat *(white circle).* The image is optimized to display bone so that the muscles and subcutaneous fat are less well seen.

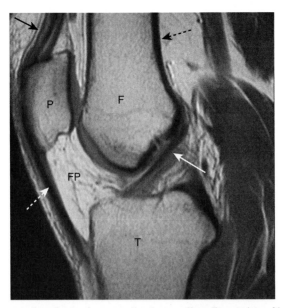

FIGURE 23-2 Normal magnetic resonance imaging scan of knee. This sagittal view of the knee demonstrates the superior display of the internal matrix of bone and the surrounding soft tissues. There is fatty marrow in the distal femur (F), proximal tibia (T), and patella (P). The quadriceps *(solid black arrow)* and patellar *(dotted white arrow)* tendons are shown. The anterior cruciate ligament *(solid white arrow)* is visible. There is high-signal fat in the infrapatellar fat pad (FP). Notice how the cortex of bone has a very weak signal *(dotted black arrow).*

TABLE 23-1   CHANGES IN BONE DENSITY

Density	Extent	Examples Used in This Chapter
Increased density	Diffuse	Diffuse osteoblastic metastases
	Focal	Localized osteoblastic metastases
		Avascular necrosis of bone
		Paget disease
Decreased density	Diffuse	Osteoporosis
		Hyperparathyroidism
	Focal	Localized osteolytic metastases
		Multiple myeloma
		Osteomyelitis

- Both osteoclastic and osteoblastic activity depend on the presence of a **viable blood supply** to bring those cells to the bone.
- Bones also **respond to mechanical forces**—for example, the contractions of muscles and tendons, the process of bearing weight, and constant use or prolonged disuse—which help to form and/or maintain the shape as well as the content of each bone.
- In this chapter, abnormalities of bone density are arbitrarily divided into two major categories based primarily on their appearance on conventional radiographs—those that produce a pattern of either **increased** or **decreased** bone density. Those two patterns are further subdivided by **extent of disease**—**focal** versus **diffuse** (or **generalized**) changes (Table 23-1).
- If **MRI** were used as the basis for categorization of bone (marrow) disorders, these disorders would be divided into four categories:
  - ◆ **Reconversion** refers to the reversal of the normal conversion of marrow cells such that red marrow repopulates bone from which it had been replaced by yellow marrow. Chronic anemias such as sickle cell disease are examples that fall in this category.
  - ◆ **Marrow replacement** involves replacement by, for example, metastatic cells, multiple myeloma, or leukemia.
  - ◆ **Myeloid depletion** involves the loss of red marrow by agents such as radiation, chemotherapy, or aplastic anemias.
  - ◆ **Myelofibrosis** involves the replacement of marrow by fibrous tissue such as might result from chemotherapy or radiation treatments.
- Fractures and dislocations, arthritis, and spinal diseases are discussed in subsequent chapters.

MRI—surrounding soft tissues not visible on conventional radiographs. This is accomplished by computer-aided reformatting and a superior capability to display more subtle differences in tissue densities (Fig. 23-2).

- In addition to bony trabeculae, the marrow cavity contains **red** and **yellow bone marrow. The red marrow** produces the precursors of blood cells. **Yellow marrow** contains fat. The composition of marrow is different for different bones of the body and changes throughout life to become **less hematopoietically active with age,** so that, by around 30 years of age, most of the appendicular skeleton contains only yellow marrow, whereas most of the red marrow resides in the axial skeleton. Unlike conventional radiography, **MRI is an excellent means of studying** the components of marrow, a fact that makes it so useful in the study of **marrow pathology.**
- Whereas the cortex is the part of the bone most easily visualized on conventional radiographs, **cortical bone has a very low signal intensity on conventional MRI** sequences.

THE EFFECT OF BONE PHYSIOLOGY ON BONE ANATOMY

- Bones reflect the general metabolic status of the individual. Their composition requires a **protein-containing, collagenous matrix (osteoid)** upon which **bone minerals,** principally **calcium phosphate,** are transformed into cartilage and bone.
- Bones are continuously undergoing a remodeling processes that includes **resorption of old** or **diseased bone by osteoclasts** and **formation of new bone by osteoblasts.** Whereas osteoblasts are responsible for bone matrix production, osteoclasts resorb both the matrix and minerals.

RECOGNIZING A GENERALIZED INCREASE IN BONE DENSITY

- On conventional radiographs and CT scans, there is an **overall whiteness (sclerosis)** to all or most of the bones.
- This leads to a **diffuse loss of visualization of the normal network of bony trabeculae in the medullary cavity** because of replacement of the normal intertrabecular fatty marrow by bone-producing elements.
- There is also a **loss of visualization of the normal corticomedullary junction** because of the abnormally increased density of the medullary cavity relative to the cortex (Fig. 23-3).

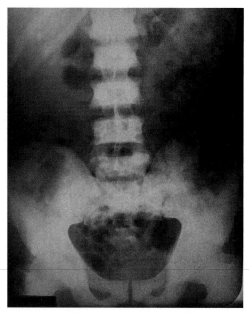

FIGURE 23-3 Diffuse metastatic disease from carcinoma of the prostate. The bones shown here are diffusely sclerotic. You can no longer see the normal trabeculae or the junction between the medullary cavity and the cortex, because the medullary cavities have been filled in with osteoblastic metastatic disease that obscures these normal boundaries and increases the overall bone density. Contrast this picture with that of Paget disease of the pelvis shown in Figure 23-11.

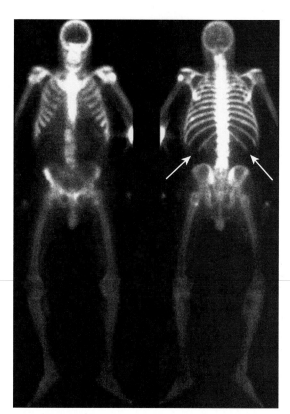

FIGURE 23-4 Radionuclide bone superscan. Anterior *(left)* and posterior *(right)* views of the axial and appendicular skeleton show the distribution of bone radiotracer uptake throughout the skeleton. This picture shows the so-called **superscan** produced by osteoblastic metastatic disease involving every bone, leading to high uptake throughout the skeleton, with poor or absent renal excretion of the radiotracer. The *white arrows* point to the absence of excretion by the kidneys, another characteristic of a superscan.

Osteoblastic Metastatic Disease

■ Diffuse, blood-borne, metastatic disease from **carcinoma of the prostate** is the **prototype for generalized increase in bone density.** Osteoblastic activity occurs beyond the control of normal physiologic constraints.

■ Metastatic disease to bone **occurs in over 80% of autopsied patients with carcinoma of the prostate. Multiple bone metastases** from carcinoma of the prostate occur much **more frequently** than do **solitary bone lesions,** which are discussed later in this chapter.

■ With diffuse bone metastases, a so-called **superscan** may be seen on radionuclide bone scan. The superscan demonstrates high radiotracer uptake throughout the skeleton, with poor or absent renal excretion of the radiotracer (Fig. 23-4).

RECOGNIZING A FOCAL INCREASE IN BONE DENSITY

➡ **Focal sclerotic lesions** can affect the cortex and/or medullary cavity. Those that affect the **cortex** will usually produce **periosteal new-bone formation (periosteal reaction),** which leads to an appearance of **thickening of the cortex.** Those that affect the **medullary cavity** will result in **punctate, amorphous sclerotic lesions** surrounded by the normal medullary cavity (Fig. 23-5).

■ Examples of diseases that cause a **focal increase in bone density** are shown in Box 23-1.

Carcinoma of the Prostate

■ Osteoblastic metastatic disease, like that from the prostate, can produce a focal increase in bone density, as well as a generalized increase.

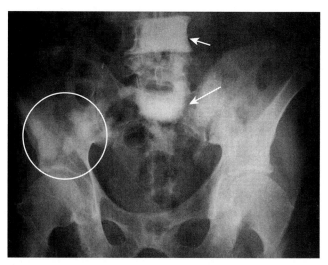

FIGURE 23-5 Focal sclerotic metastases from carcinoma of the prostate. Sclerotic lesions can be seen in the L4 and S1 vertebral bodies *(white arrows)*. It is no longer possible to distinguish the junction between the cortex and the medullary cavity in either of those vertebral bodies. Also present are multiple sclerotic lesions in the right ilium *(white circle)* and scattered throughout the pelvis. Sclerotic lesions in bone are a common finding in metastatic carcinoma of the prostate.

BOX 23-1 FOCAL INCREASE IN BONE DENSITY

Carcinoma of the prostate (which can also cause a diffuse increase in bone density)

Avascular necrosis of bone

Paget disease

BOX 23-2 FINDING METASTASES TO BONE: BONE SCAN

Intravenous administration of a minute amount of technetium-99M methylene diphosphonate (MDP), a radioactively tagged tracer that affixes to the surface of bone.

Technetium-99M is the radionuclide used to tag MDP, the portion that directs the tracer to bone.

Activity in bone depends, in part, on its blood supply and rate of bone turnover. Processes with extremely high or extremely low bone turnover may produce false-negative scans.

Osteoblastic lesions almost always show increased activity (uptake of radiotracer). Even osteolytic metastases usually show increased uptake because of the repair that occurs in most, but not all, osteolytic processes.

The bone scan is much less sensitive in detecting multiple myeloma, so conventional radiographic surveys of the skeleton are the initial study of choice when searching for myeloma lesions.

Bone scans are highly sensitive, but not very specific; a positive scan almost always requires another imaging procedure (conventional radiography, CT, or MRI) to rule out nonmalignant causes of a positive bone scan (e.g., fractures or infection).

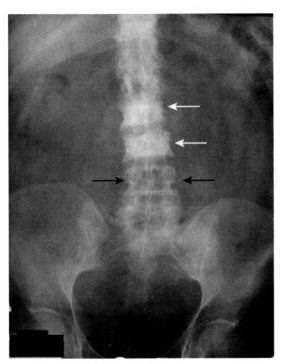

FIGURE 23-6 Focal increase in bone density from carcinoma of the breast. A frontal view of the lumbar spine and pelvis demonstrates abnormally dense vertebral bodies, most marked at L2 and L3 *(white arrows)*. Notice how the pedicles are obscured by the abnormally increased density of the vertebral body compared with the normal pedicles at L4 *(black arrows)*. Dense, white vertebrae are called ***ivory vertebrae.*** Osteoblastic carcinoma of the breast and prostate are two causes of an ivory vertebra.

TABLE 23-2 SOME CAUSES OF AVASCULAR NECROSIS OF BONE

Location	Example of Disease
Intravascular	Sickle cell disease Polycythemia vera
Vascular	Vasculitis (lupus- and radiation-induced)
Extravascular	Trauma (fractures)
Idiopathic	Exogenous steroids and Cushing disease Legg-Calve-Perthes disease

- A substance secreted by tumor cells from **metastatic carcinoma of the prostate may stimulate osteoblastic activity** and produce focal areas of localized increased density (i.e., **sclerotic bone lesions**). These lesions are **most often seen in the vertebrae, ribs, pelvis, humeri, and femora** (Fig. 23-6).

- The **radionuclide bone scan is currently the study of choice for detecting skeletal metastases,** regardless of the suspected primary (Box 23-2).

Avascular Necrosis of Bone

- Avascular necrosis (AVN) of bone (also called ***ischemic necrosis, aseptic necrosis,*** or ***osteonecrosis***) results from cellular death and leads to collapse of the affected bone. It usually involves those bones that have a relatively poor collateral blood supply (e.g., scaphoid in the wrist, femoral head) and tends to affect the hematopoietic elements of marrow earliest, so **MRI is the most sensitive modality for detecting AVN.**

- There are a myriad of causes of avascular necrosis. Some of the more common ones are listed in Table 23-2.

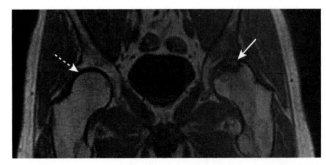

FIGURE 23-7 Magnetic resonance imaging study showing avascular necrosis. This T1-weighted coronal view of both hips demonstrates a normal high signal from the fatty marrow in the right femur *(dotted white arrow)* but decreased signal in the left femoral head extending to the subchondral bone of the left hip joint *(solid white arrow)*. The joint space is preserved. Magnetic resonance imaging is the most sensitive method of detecting avascular necrosis of the hip.

- On conventional radiographs, the **region of avascular necrosis appears denser than the surrounding bone.** On **MRI,** there is usually a **decrease from the normal high signal** produced by fatty marrow (Fig. 23-7).

- The **devascularized bone becomes denser** and therefore appears more sclerotic than the remainder of the bone. This

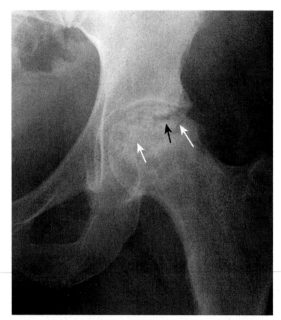

FIGURE 23-8 Avascular necrosis of the left femoral head in a patient on long-term steroids for systemic lupus erythematosus. This close-up view of the left femoral head shows a zone of increased sclerosis in the superior aspect of the femoral head *(white arrows)*, a characteristic finding of avascular necrosis of the head. The linear, subcortical lucency *(black arrow)* represents subchondral fractures seen with this disease, called the ***crescent sign.*** Notice that the disease is isolated to the femoral head and involves neither the joint space nor the acetabulum; that is, this is not an arthritis.

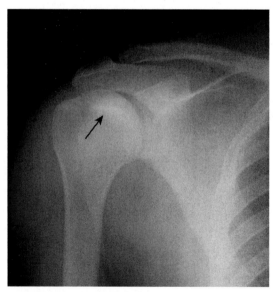

FIGURE 23-9 Avascular necrosis of humeral head. Increased density can be seen at the very top of the humeral head *(black arrow)* in this patient with sickle cell disease. Because the white cap on the bone looks like snow on a mountaintop, this sign of avascular necrosis has been called ***snowcapping.*** Avascular necrosis in sickle cell disease does not usually manifest until young adulthood. With the use of magnetic resonance imaging, it has been found to be more prevalent than originally thought when plain films alone were used.

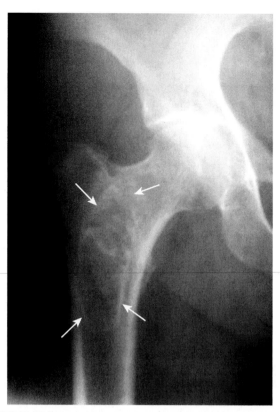

FIGURE 23-10 Old medullary bone infarct. This image shows an amorphous calcification in the medullary cavity of the proximal femur *(white arrows)*. In general, the differential diagnosis for such an intramedullary calcification includes bone infarct or enchondroma. The characteristic thin sclerotic membrane surrounding this lesion identifies it as a bone infarct. This patient had been on long-term steroids for asthma.

Paget Disease

- Paget disease is a **chronic disease of bone** that most often occurs in older men. It is now believed to be caused by chronic paramyxoviral infection. It is **characterized by varying degrees of increased bone resorption** and **increased bone formation,** with the latter predominating in those cases seen in more progressive forms of the disease.
- The end result is almost always a **denser bone** that, despite its density, is **mechanically inferior** to normal bone and thus **susceptible to pathologic fractures** or bone-softening deformities such as **bowing.** The **pelvis is most frequently involved,** followed by the **lumbar spine, thoracic spine, proximal femur,** and **calvarium.**

 Paget disease is usually diagnosed using conventional radiography. The following are the **imaging hallmarks of Paget disease:**

- **Thickening of the cortex**
    - To recognize thickening of the cortex, compare the thickness of the cortex of the suspicious area with another part of the same bone or, if visible, the same bone on the opposite side of the body.
- **Accentuation of the trabecular pattern**
    - There is coarsening and thickening of the trabeculae (Fig. 23-11).
- **Increase in the size of the bone involved**
    - The "classical" history for Paget disease, rendered less useful as fashions have changed, was a gradual increase

occurs especially in the **femoral head** (Fig. 23-8) and the **humeral head** (Fig. 23-9).
- On conventional radiographs, old **medullary bone infarcts** are recognized as dense, amorphous deposits of bone within the medullary cavities of long bones and are frequently marginated by a thin, sclerotic membrane (Fig. 23-10).

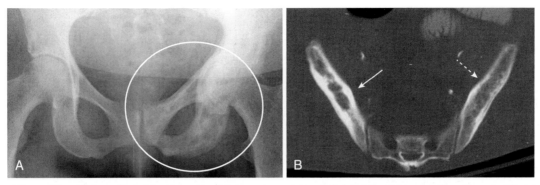

FIGURE 23-11   Paget disease of the pelvis, two patients. A, Frontal view of the pelvis shows an increase in bone density in the left hemipelvis, accentuation and coarsening of the trabeculae, and thickening of the cortex *(white circle),* the hallmarks of Paget disease of bone. Compare the left hemipelvis with the normal right side. **B,** Axial computed tomographic scan of the pelvis of another patient with Paget disease shows thickening of the cortex and accentuation of the trabeculae in the right ilium *(solid white arrow).* Compare it with the normal left side *(dotted white arrow).*

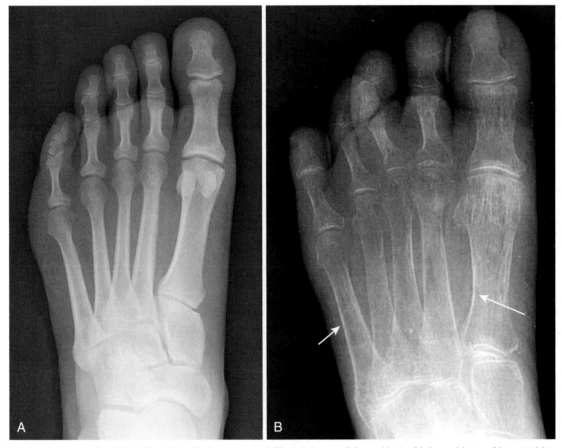

FIGURE 23-12   Normal and osteoporotic foot. A, Normal frontal view of the foot to contrast with **(B),** which shows overall decreased density of the bone and thinning of the cortices *(white arrows)* secondary to disuse because of osteoporosis in this patient. Conventional radiographs are insensitive in the diagnosis of osteoporosis, and they are subject to technical variations that can mimic the disease, even in a healthy individual. More sensitive methods, such as dual-energy x-ray absorptiometry, should be used to confirm the diagnosis.

in a man's hat size as the calvarium increased in size because of this disease.

RECOGNIZING A GENERALIZED DECREASE IN BONE DENSITY

- The following are findings associated with a generalized decrease in bone density:
  - ◆ There may be a **diffuse loss of the normal network of bony trabeculae in the medullary cavity** because of the decrease in, and thinning of, many of the smaller trabecular structures.
  - ◆ **Accentuation of the normal corticomedullary junction** may be present in which the **cortex,** although thinner than normal, **stands out more strikingly** because of the decreased density of the medullary cavity (Fig. 23-12).
  - ◆ **Compression of vertebral bodies**
  - ◆ **Pathologic fractures** in the hip, pelvis, or vertebral bodies

BOX 23-3   DIFFUSE DECREASE IN BONE DENSITY

Osteoporosis
Hyperparathyroidism

■ Examples of diseases that cause a **diffuse decrease in bone density** are shown in Box 23-3.

Osteoporosis

■ *Osteoporosis* is defined as a systemic skeletal disorder **characterized by low bone mineral density** (BMD) and is generally divided into **postmenopausal** and **age-related bone loss**.
  ♦ **Postmenopausal osteoporosis** is characterized by **increased bone resorption** caused by osteoclastic activity. **Age-related bone loss** begins around ages 45 to 55 and is characterized by a **loss of total bone mass**.
■ Additional factors that **increase the risk of osteoporosis** include exogenous steroid administration, Cushing disease, estrogen deficiency, inadequate physical activity, and alcoholism.
■ Osteoporosis **predisposes individuals to pathologic fractures** in bones such as the femoral neck, compression fractures of the vertebral bodies, and fractures of the distal radius (Colles fractures).
■ **Conventional radiographs** are relatively **insensitive for detecting osteoporosis.** Almost **50% of the bone mass must be lost** before it is **recognizable** on conventional radiographs. Findings on conventional radiographs include **overall decreased density of bone, thinning of the cortex,** and **decrease in the visible number of trabeculae in the medullary cavity.**
■ Currently, **dual-energy x-ray absorptiometry (DEXA) is** the most accurate and **widely recommended method** for BMD measurements.
  ♦ DEXA scans are obtained by using a filtered x-ray source that produces two distinct energies that are differentially absorbed by bone and soft tissue, respectively, allowing for more accurate calculation of bone density by subtracting out the error introduced by varying amounts of overlying soft tissue.
  ♦ The x-ray dose is very low, and the spine or hip is generally used for density measurements.

Hyperparathyroidism

■ *Hyperparathyroidism* is a condition caused by excessive secretion of **parathormone (PTH)** by the parathyroid glands. PTH exerts its effects on bones, the kidneys, and the gastrointestinal tract. Its **effect on bones is to increase resorption** by stimulating osteoclastic activity. Calcium is removed from the bone and deposited in the bloodstream.
■ There are **three forms** of hyperparathyroidism: primary, secondary, and tertiary (Table 23-3).
■ The **diagnosis of hyperparathyroidism is based on clinical and laboratory findings;** however, there are numerous findings of the disease on conventional radiographs, and other imaging studies are also evaluated to direct surgery on the glands if indicated. Imaging studies of the parathyroid

TABLE 23-3   FORMS OF HYPERPARATHYROIDISM

Type	Remarks
Primary	Usually caused by a single adenoma in most patients (80% to 90%) and almost always results in hypercalcemia
Secondary	Results from hyperplasia of the glands secondary to imbalances in calcium and phosphorus levels, seen mostly with chronic renal disease
Tertiary	Occurs in patients with long-standing secondary hyperparathyroidism in whom autonomous hypersecretion of parathyroid hormone develops, leading to hypercalcemia

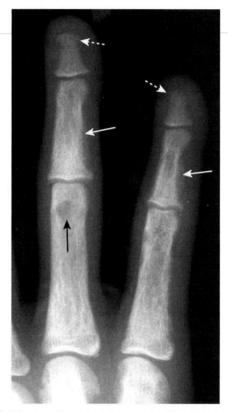

FIGURE 23-13 Subperiosteal resorption in hyperparathyroidism. The radiologic hallmark of hyperparathyroidism is subperiosteal bone resorption, seen especially well on the radial aspect of the middle phalanges of the index and middle fingers *(solid white arrows)*. Here the cortex appears shaggy and irregular compared with the cortex on the opposite side of the same bone, which is well defined. This patient also displays two other findings of hyperparathyroidism: a small brown tumor *(black arrow)* and resorption of the terminal phalanges *(dotted white arrows)*.

glands themselves may include ultrasound, nuclear medicine parathyroid scans, and MRI scans.

 Some of the **findings of hyperparathyroidism on conventional radiographs:**

■ Overall **decrease in bony density**
■ **Subperiosteal bone resorption,** especially on the radial side of the middle phalanges of the index and middle fingers (Fig. 23-13)
■ **Erosion of the distal clavicles** (Fig. 23-14)

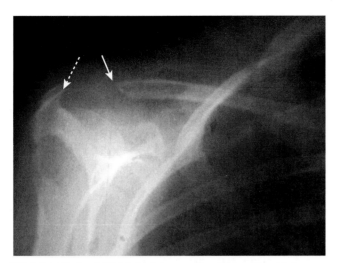

FIGURE 23-14 **Erosion of distal clavicle in hyperparathyroidism.** Another relatively common site of bone resorption in hyperparathyroidism is the distal end of the clavicle. Here the distal clavicle *(solid white arrow)*, which should articulate with the acromion *(dotted white arrow)* has been resorbed, increasing the distance between it and the acromion. Other potential sites of bone resorption are the terminal phalanges, as shown in Figure 23-13 **(acro-osteolysis)**, the lamina dura of the teeth, the medial aspect of the tibia, the humerus, and the femur.

- **Well-circumscribed lytic lesions** in the long bones, called **Brown tumors**, and a **salt-and-pepper appearance** of the skull (Fig. 23-15).

RECOGNIZING A FOCAL DECREASE IN BONE DENSITY

- These lesions are most often produced by **focal infiltration of bone by cells other than osteocytes.**
- Examples of diseases that cause a **focal decrease in bone density** are shown in Box 23-4.

Osteolytic Metastatic Disease

- **Osteolytic metastatic disease** can produce focal destruction of bone (Box 23-5).
- The **medullary cavity is almost always involved,** from which the disease may erode into and destroy the cortex as well. When the medullary cavity alone is involved, there must be almost a 50% reduction in mass in order for the lesion to be recognizable when viewed en face on conventional radiographs.
- MRI, on the other hand, is excellent at demonstrating the status of the medullary cavity and so is much more sensitive than conventional radiography to the presence of metastatic disease (Fig. 23-16).
- In **some cases, only the cortex is involved.** Cortical metastases may be easier to visualize on conventional radiographs because relatively less cortical destruction is needed for them to become apparent, especially if the lesions happen to be viewed in tangent.

 Classical findings of osteolytic metastases on conventional radiographs:

- ◆ **Irregularly shaped, lucent bone lesions,** which can be single or multiple
  - These lytic lesions are frequently characterized as belonging to one (or sometimes more) of three

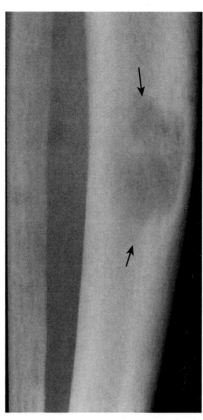

FIGURE 23-15 **Brown tumor.** A geographic, lytic lesion in the midshaft of the tibia is shown *(black arrows)*. Brown tumors (also called **osteoclastomas**) are benign lesions that represent the osteoclastic resorption of a localized area of (usually) cortical bone and its replacement with fibrous tissue and blood. Their high hemosiderin content gives them a characteristic brown color; they were not named after a "Dr. Brown." The lesions can look like osteolytic metastases or multiple myeloma, so that the clinical history of hyperparathyroidism is key. They can be seen with both primary and secondary hyperparathyroidism.

BOX 23-4 FOCAL DECREASE IN BONE DENSITY

Metastatic disease to bone

Multiple myeloma

Osteomyelitis

BOX 23-5 METASTATIC DISEASE TO BONE

Metastases to bone are far more common than primary bone tumors.

Metastases to bone fall into two major categories: those that stimulate the production of new bone are called *osteoblastic,* and those that destroy bone are called *osteolytic;* some metastases include lesions in which osteoblastic and osteolytic changes are both present.

Metastatic bone lesions from any source are very uncommon distal to the elbow or the knee; when present in these locations, they are usually widespread and caused by lung or breast cancer.

The radionuclide bone scan is currently the study of choice for detecting skeletal metastases.

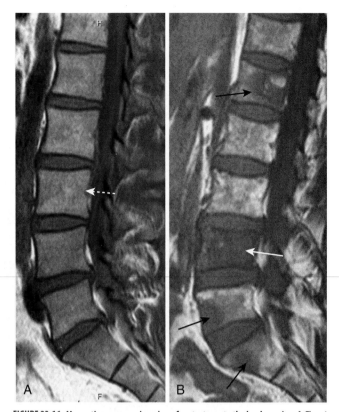

FIGURE 23-16 Magnetic resonance imaging of metastases to the lumbar spine. A, There is a normal signal in the lumbar vertebral bodies on this T1-weighted sagittal view of the lumbar spine *(dotted white arrow)*. **B,** In this patient with a primary breast carcinoma, there are multiple metastatic deposits replacing the normal marrow in the lumbar spine and sacrum *(black arrows)*. The body of L4 is completely replaced by tumor *(white arrow)*.

patterns: **permeative, mottled,** or **geographic** in order of increasing size of the smallest and most discrete lesion visible (Fig. 23-17).

♦ Osteolytic metastases typically **incite little** or **no reactive bone formation** around them. They can be **expansile** and **soap-bubbly** (i.e., contain bony septations), especially in renal and thyroid carcinomas (Fig. 23-18).

♦ In the spine, they may preferentially **destroy the pedicles,** because of their blood supply (the **pedicle sign**), which can help to differentiate metastases from multiple myeloma (see below), which tends to spare the pedicle early in the disease (Fig. 23-19).

■ The most common causes of osteoblastic and osteolytic bone metastases are listed in Table 23-4.

Multiple Myeloma

■ **Multiple myeloma,** the most common primary malignancy of bone in adults, can occur in a **solitary form,** often seen as a soap-bubbly, expansile lesion in the spine or pelvis (called a *solitary plasmacytoma*) or in a **disseminated form** with multiple, **punched-out** lytic lesions throughout the axial and proximal appendicular skeleton.

> **Findings of multiple myeloma on conventional radiographs:**

♦ The most common early manifestation is **diffuse** and **usually severe osteoporosis.**

♦ **Plasmacytomas** appear as **expansile, septated lesions, frequently with associated soft tissue masses** (Fig. 23-20).

♦ Later, in its disseminated form, **multiple, small, sharply circumscribed** (described as **punched-out**)

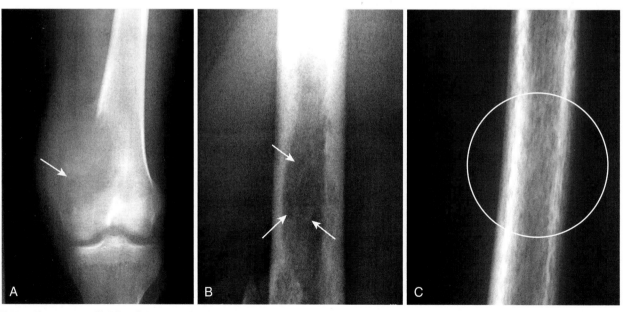

FIGURE 23-17 Three patterns of lytic bone lesions. A, A solitary bone lesion is shown with a gradual zone of transition between it and the normal bone and complete destruction of the cortex *(white arrow)*, called a *geographic lesion*. **B,** Several ill-defined lytic lesions *(white arrows)* with indistinct margins are shown, implying a more aggressive malignancy. This is called a *moth-eaten pattern*. **C,** A close-up view of the femur shows innumerable, small, irregular holes in the bone *(white circle)*, called a *permeative pattern*. Permeative lesions are called *round-cell lesions* because of the shape of the cells that produce them. Such lesions include Ewing sarcoma, myeloma, and leukemia.

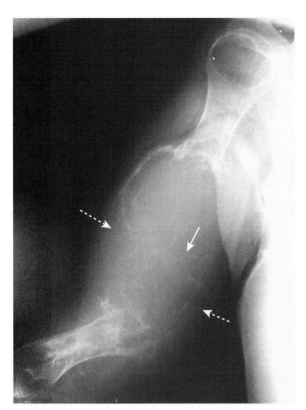

FIGURE 23-18 Expansile renal cell carcinoma metastasis. This image shows a very aggressive and expansile osteolytic metastasis in the humerus from a primary renal cell carcinoma. Notice that the cortex has been destroyed in several areas *(dotted white arrows)* and that the lesion has a characteristic *soap-bubbly* appearance produced by fine septae *(solid white arrow)*. Thyroid carcinoma and a solitary plasmacytoma could also produce these findings.

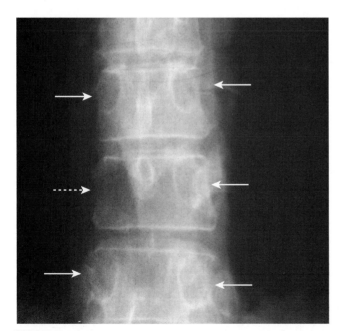

FIGURE 23-19 Pedicle sign. Destruction of the right pedicle of T10 can be seen *(dotted white arrow)*. Each vertebral body should normally have two oval-shaped pedicles, one on each side, visible on the frontal radiograph of the spine *(solid white arrows)*. In the spine, osteolytic metastases may preferentially destroy the pedicles because of their blood supply, producing the **pedicle sign**, although most metastatic lesions to the spine also involve the body. In multiple myeloma, the pedicle tends to be spared early in the disease.

TABLE 23-4 CAUSES OF OSTEOBLASTIC AND OSTEOLYTIC BONE METASTASES

Osteoblastic	Osteolytic
Prostate carcinoma (most common in older men)	Lung cancer (most common osteolytic lesion in men)
Breast carcinoma is usually osteolytic, but can be osteoblastic, especially if treated	Breast cancer (most common osteolytic lesion in women)
Lymphoma	Renal cell carcinoma
Carcinoid tumors (rare)	Thyroid carcinoma

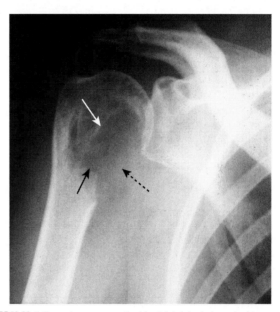

FIGURE 23-20 Solitary plasmacytoma, shoulder. A lytic lesion in the proximal humerus *(solid black arrow)* is shown, destroying the cortex *(dotted black arrow)* and containing multiple septations *(solid white arrow)*. This is the so-called *soap-bubbly* appearance that can be seen with solitary plasmacytomas, a precursor to the more disseminated form of multiple myeloma. Renal cell and thyroid carcinoma can also produce such a picture.

lytic lesions of approximately the same size are present, usually without any accompanying sclerotic reaction around them (Fig. 23-21).

■ Classically, **in detecting the lesions of multiple myeloma, conventional radiographs are more sensitive than radionuclide bone scans,** which tend to underestimate the number and extent of lesions because of the absence of reactive bone formation.

Osteomyelitis

■ *Osteomyelitis* refers to the **focal destruction of bone,** most often by a blood-borne **infectious agent,** the most common of which is *Staphylococcus aureus.*

■ **In children,** the osteolytic lesion **tends to occur at the metaphysis** because of its rich blood supply.

⇨ Findings of acute osteomyelitis on conventional radiographs:

♦ **Focal cortical bone destruction**
♦ **Periosteal new bone formation** (Fig. 23-22)

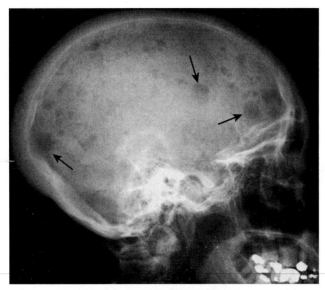

FIGURE 23-21 **Multiple myeloma.** Innumerable lytic lesions *(black arrows)* can be seen in the skull. They are small, uniform in size, and have sharply marginated edges, the so-called **punched-out** lytic defects seen in multiple myeloma. Metastases to the skull can produce a similar picture, but tend to be fewer in number and less well-defined.

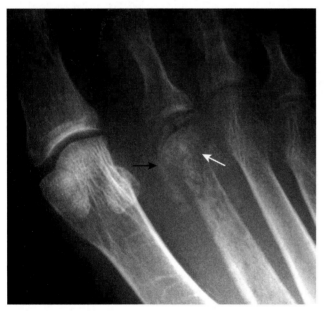

FIGURE 23-22 **Acute osteomyelitis of the second metatarsal.** Close-up view of the second toe in a 59-year-old woman with diabetes shows the hallmarks of bone destruction *(white arrow)* and periosteal new bone formation *(black arrow)* in osteomyelitis. Most cases of osteomyelitis of the foot are caused by a diabetic vasculopathy.

♦ Inflammatory changes accompanying the infection may produce **soft tissue swelling** and **focal osteoporosis** caused by hyperemia.

■ In **adults,** the **infection tends to involve the joint space more** often than in children, producing not only osteomyelitis but also **septic arthritis** (see Chapter 25).

■ **Conventional radiographs can take up to 10 days to display the first findings of osteomyelitis,** so other imaging modalities, such as MRI and nuclear medicine, are frequently used for earlier diagnosis.

BOX 23-6   INSUFFICIENCY FRACTURES

Insufficiency fractures are a type of pathologic fracture in which mechanically weakened bone fractures as a result of a normal or physiologic stress.

Insufficiency fractures are most common secondary to osteoporosis in postmenopausal women.

Common sites include the pelvis, thoracic spine, sacrum, tibia, and calcaneus.

Unlike other fractures that manifest by a lucency in the bone, most insufficiency fractures display a sclerotic band (representing healing) on conventional radiographs.

CT, MRI, or nuclear medicine bone scans are more sensitive than conventional radiographs in detecting insufficiency fractures (Fig. 23-23).

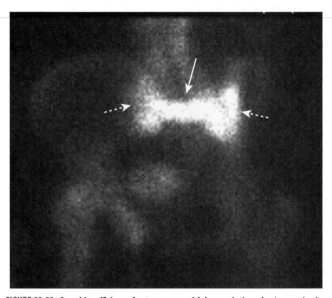

FIGURE 23-23 **Sacral insufficiency fractures seen with bone scintigraphy.** Increased radiotracer uptake can be seen in vertical fractures through the sacral ala *(dotted white lines)* and a horizontal fracture through the body of the sacrum *(solid white line).* This presentation has been called the *Honda sign* because it resembles the carmaker's insignia. By definition, insufficiency fractures occur in abnormal bones that undergo normal stress. The sacrum is a common site for such fractures in osteoporosis.

■ There are a variety of radionuclide bone scans that can demonstrate osteomyelitis, the most specific of which is currently a **tagged white-cell scan,** in which a sample of the patient's white blood cells is removed, tagged with a radioactive isotope (frequently indium), and injected back into the patient, who is then imaged with a camera specific for nuclear studies to detect a site of abnormally increased radioactive tracer uptake.

PATHOLOGIC FRACTURES

■ Pathologic fractures are those that **occur in bone with a preexisting abnormality.** Pathologic fractures tend to occur with **minimal** or **no trauma.**

♦ **Insufficiency fractures** are a type of pathologic fracture (Box 23-6).

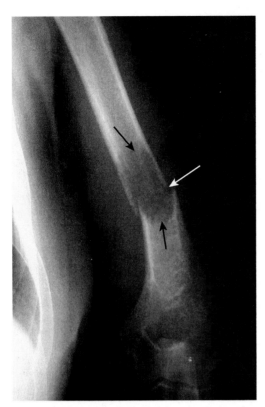

FIGURE 23-24 Pathologic fracture. Pathologic fractures are those that occur with minimal stress in bones with preexisting abnormalities. In this patient with metastatic renal cell carcinoma to the humerus, there is a geographic, lytic lesion seen in the distal humerus *(black arrows)* through which a transverse fracture has occurred *(white arrow).*

- Diseases that produce either an **increase** in bone density or a **decrease** in bone density tend to **weaken the normal architecture** of bone and predispose patients to pathologic fractures.
- Diseases that predispose patients to pathologic fractures may be **local** (e.g., metastases) or **diffuse** (e.g., osteoporosis). In general, pathologic fractures occur more often in the **ribs, spine,** and **proximal appendicular skeleton** (humeri and femurs).
- **Recognizing pathologic fractures** (Fig. 23-24)
  - First, there has to be a fracture present.
  - The bone surrounding the fracture will demonstrate abnormal density or architecture.
  - **Delayed healing** is **common** in pathologic fractures.
- Attempts to predict "impending" fractures in diseased bone have, by and large, proven to be unreliable.
- The treatment of a pathologic fracture depends in part on successful treatment of the underlying condition that produced it.

 **WEBLINK**

Visit StudentConsult.Inkling.com for more information and a quiz on Abnormalities of Bone Density.

For your convenience, the following QR Code may be used to link to **StudentConsult.com**. You must register this title

TAKE-HOME POINTS

RECOGNIZING ABNORMALITIES OF BONE DENSITY

Bones consist of a **cortex** of compact bone surrounding a **medullary cavity** containing cancellous bone arranged as trabeculae, separated by blood vessels, hematopoietic cells, and fat.

On conventional radiographs, the cortex is best seen in tangent. On CT scans, the entire cortex is visualized. MRI is particularly sensitive to assessment of the marrow. Both CT and MRI scans are superior to conventional radiographs for evaluation of soft tissues.

Bone undergoes continuous change caused by a combination of biochemical and mechanical forces.

Increased osteoclastic activity can produce focal or generalized decreases in bone density, and increased osteoblastic activity can produce focal or diffuse increased bone density. Osteoblastic metastases, especially from carcinoma of the prostate and breast, can produce focal or generalized increases in bony density.

Other diseases that can increase bone density include avascular necrosis of bone and Paget disease.

Hallmarks of **Paget disease** include thickening of the cortex, accentuation of the trabecular pattern, and enlargement and increased density of the affected bone.

Osteolytic metastases, especially from lung, renal, thyroid, and breast cancer, can produce focal areas of decreased bone density, as can solitary plasmacytomas. Plasmacytomas are considered to be precursors to multiple myeloma, the most common primary tumor of bone.

Examples of diseases that can cause a generalized decrease in bone density include osteoporosis and hyperparathyroidism.

Osteoporosis is characterized by low bone mineral density and is most often either postmenopausal or age related. Osteoporosis predisposes persons to pathologic fractures.

Pathologic fractures are those that occur with minimal or no trauma in bones that had a preexisting abnormality.

Examples of diseases that can cause a focal decrease in bone density include metastases, multiple myeloma, and osteomyelitis.

The radionuclide bone scan is the modality of choice in screening for skeletal metastases. MRI is used primarily to solve specific questions related to a lesion's composition and extent.

There must be almost a 50% reduction in the mass of bone in order for a difference in density to be perceived on conventional radiographs. MRI is much more sensitive to the presence of medullary metastatic disease.

using the PIN code on the inside front cover of the text to access online content.

CHAPTER 24
Recognizing Fractures and Dislocations

RECOGNIZING AN ACUTE FRACTURE

- Radiographs of broken bones attract a lot of attention. Fractures are a favorite amongst those learning radiology, perhaps because of how common and seemingly straightforward they are. In this chapter we'll tell you how to recognize a fracture, describe it, name it, and avoid missing it.
- A fracture is described as a **disruption in the continuity of all** or **part of the cortex of a bone.**
    - If the **cortex** is **broken through** and **through**, the fracture is called *complete*.
    - If **only a part of the cortex** is fractured, it is called *incomplete*. Incomplete fractures tend to occur in bones that are "softer" than normal, such as those in children, or in adults with bone-softening diseases such as **Paget disease** (see Chapter 23).
    - Examples of incomplete fractures in children are the **greenstick fracture**, which involves only one part of the cortex, and the **torus fracture (buckle fracture)**, which represents compression of the cortex (Fig. 24-1).

 Radiologic features of acute fractures (Box 24-1; Fig. 24-2)

- **Fracture lines** when viewed in the correct plane **tend to be "blacker" (more lucent)** than other lines normally found in bones, such as **nutrient canals** (see Fig. 24-2, *A*).
- There may be an **abrupt discontinuity of the cortex**, sometimes associated with **acute angulation of the normally smooth contour of bone** (see Fig. 24-2, *B*).

BOX 24-1   CHARACTERISTICS OF AN ACUTE FRACTURE

Abrupt disruption of all or part of the cortex

Acute changes in the smooth contour of a normal bone

Fracture lines are black and linear

Where fracture lines change their course, they tend to be sharply angulated

Fracture fragments are jagged and not corticated

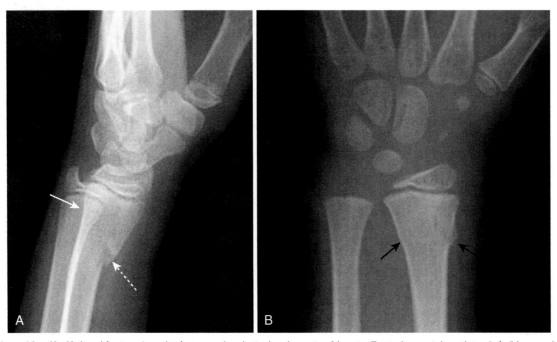

FIGURE 24-1 Greenstick and buckle (torus) fractures. Incomplete fractures are those that involve only a portion of the cortex. They tend to occur in bones that are "softer" than normal, such as those in children (above) or in adults with bone-softening diseases such as Paget disease. **A,** There is a greenstick fracture, which involves only one part of *(dotted white arrow)* rather than the entire cortex *(solid white arrow)*. **B,** This is a buckle fracture, in which there is buckling of the cortex *(black arrows)*.

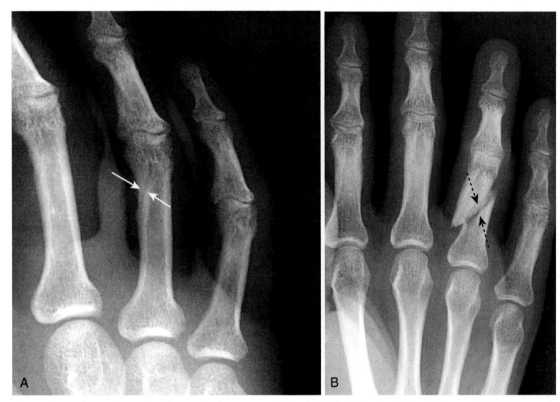

FIGURE 24-2 Nutrient canal versus fracture. Fracture lines, when viewed in the correct orientation, tend to be *"blacker"* (more lucent) than other lines normally found in bones, such as nutrient canals. **A,** This is a nutrient canal *(white arrows)*, whereas a true fracture is seen in another patient in **(B)** *(dotted black arrows)*. Notice how the nutrient canal has a sclerotic (whiter) margin and is confined to the cortex, which is not the case with fracture lines that are darker and traverse the cortex and medullary cavity. The edges of a fracture tend to be jagged and rough.

TABLE 24-1 DIFFERENTIATING FRACTURES, OSSICLES, AND SESAMOIDS

	Acute Fracture	Sesamoids and Accessory Ossicles*
Abrupt disruption of cortex	Yes	No
Bilaterally symmetrical	Almost never	Almost always
"Fracture line"	Not sharp, jagged	Smooth
Bony fragment has a cortex completely around it	No	Yes

*Old, unhealed fractures will not be bilaterally symmetrical.

♦ Fracture lines tend to be **straighter in their course yet more acute in their angulation** than any naturally occurring lines (such as **epiphyseal plates**) (Fig. 24-3).
♦ The **edges of a fracture tend to be jagged** and **rough.**

 Pitfalls—sesamoids, accessory ossicles, and unhealed fractures (Table 24-1; Fig. 24-4)

♦ **Sesamoids** are bones that form in a tendon as it passes over a joint. The patella is the largest and most famous sesamoid bone.
♦ **Accessory ossicles** are accessory epiphyseal or apophyseal ossification centers that do not fuse with the parent bone.

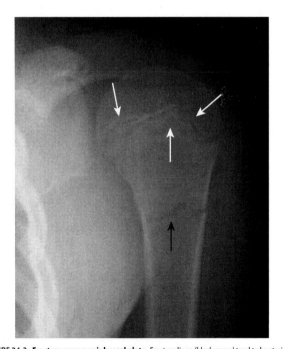

FIGURE 24-3 Fracture versus epiphyseal plate. Fracture lines *(black arrow)* tend to be straighter in their course and more acute in their angulation than any naturally occurring lines, such as the epiphyseal plates in the proximal humerus *(white arrows)*. Because the top of the metaphysis has irregular hills and valleys, the epiphyseal plate has an undulating course that will allow you to see it in tangent, both on the anterior and posterior margins of the humeral head. This gives the mistaken appearance that there is more than one epiphyseal plate.

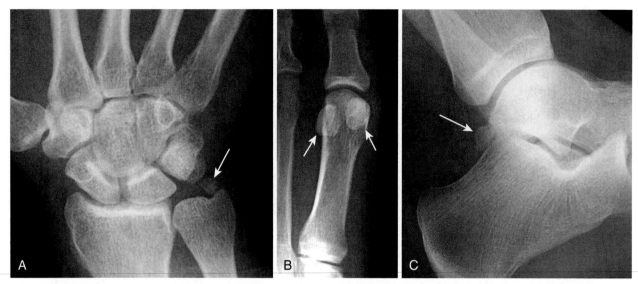

FIGURE 24-4 Pitfalls in fracture diagnosis. A, Old, unhealed fracture fragments *(white arrow).* **B,** Sesamoids (bones that form in a tendon as it passes over a joint) *(white arrows).* **C,** Accessory ossicles (accessory epiphyseal or apophyseal ossification centers that do not fuse with the parent bone, such as this **os trigonum**; *white arrow).* These examples can sometimes mimic acute fractures. Unlike fractures, these small bones are corticated (i.e., there is a white line that completely surrounds the bony fragment), and their edges are usually smooth. Sesamoids and accessory ossicles are usually bilaterally symmetrical.

- **Old, unhealed fracture fragments** can sometimes mimic acute fractures (see Fig. 24-4, *A*).
- Unlike fractures, these small bones are **corticated** (i.e., there is a white line that completely surrounds the bony fragment) and their edges are usually **smooth.**
- In the case of sesamoids and accessory ossicles, they are **usually bilaterally symmetrical** so that a view of the opposite extremity will usually demonstrate the same bone in the same location. They also **occur at anatomically predictable sites.**
  - There are almost always sesamoids present in the thumb, the posterolateral aspect of the knee **(fabella,)** and the great toe (see Fig. 24-4, *B*).
  - Accessory ossicles are most common in the foot (see Fig. 24-4, *C*).

RECOGNIZING DISLOCATIONS AND SUBLUXATIONS

- In a **dislocation,** the bones that originally formed the two components of a joint are no longer in apposition to each other. Dislocations occur only at joints (Fig. 24-5, *A*).
- In a **subluxation,** the bones that originally formed the two components of a joint are in **partial contact** with each other. Subluxations also occur only at joints (see Fig. 24-5, *B*).
- Some characteristics of dislocations of the shoulder and hip are described in Table 24-2.

DESCRIBING FRACTURES

- There is a common lexicon used in describing fractures to facilitate a reproducible description and to ensure reliable and accurate communication.
- **Fractures are usually described using four major parameters:** (Table 24-3)
  - The **number of fragments**
  - The **direction of the fracture line**
  - The **relationship of the fragments** to each other
  - Whether the fracture communicates **with the outside atmosphere**

TABLE 24-2 DISLOCATIONS OF THE SHOULDER AND HIP

Shoulder	Hip
Anterior, subcoracoid most common	Posterior and superior more common
Caused by a combination of abduction, external rotation, and extension	Frequently caused by knee striking dashboard, transmitting force to hip
Associated with fractures of humeral head **(Hill-Sachs deformity)** and glenoid **(Bankart fracture)**	Associated with fractures of posterior rim of the acetabulum

TABLE 24-3 HOW FRACTURES ARE DESCRIBED

Parameter	Terms Used
Number of fracture fragments	Simple or comminuted
Direction of fracture line	Transverse, oblique (diagonal), spiral
Relationship of one fragment to another	Displacement, angulation, shortening and rotation
Open to the atmosphere (outside)	Closed or open (compound)

HOW FRACTURES ARE DESCRIBED—BY THE NUMBER OF FRACTURE FRAGMENTS

- If the fracture produces **two fragments**, it is called a *simple fracture.*
- If the fracture produces **more than two** fragments, it is called a *comminuted fracture.* Some comminuted fractures have special names (Fig. 24-6).
  - A **segmental fracture** is a comminuted fracture in which a portion of the shaft exists as an isolated fragment (see Fig. 24-6, *A*).
  - A **butterfly fragment** is a comminuted fracture in which the central fragment has a triangular shape (see Fig. 24-6, *B*).

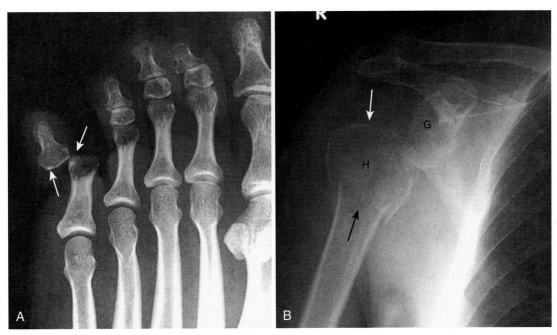

FIGURE 24-5 Dislocation and subluxation. A, In a dislocation, the bones that originally formed the two components of the interphalangeal joint are no longer in apposition to each other *(white arrows)*. The terminal phalanx is dislocated laterally. **B,** In a subluxation, the bones that originally formed the two components of a joint are in partial contact with each other. The humeral head (H) is subluxed inferiorly *(white arrow)* in the glenoid (G) because of a large hematoma in the joint secondary to a fracture of the surgical neck of the humerus *(black arrow)*. The hematoma itself is not visible by conventional radiography.

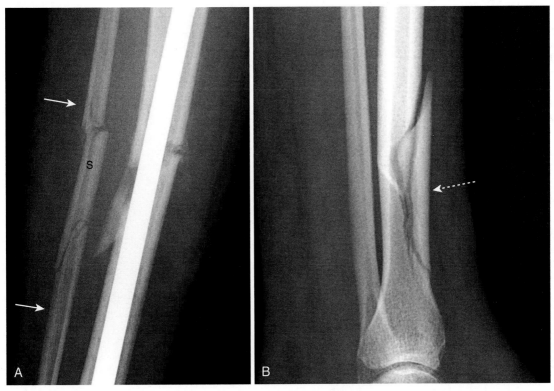

FIGURE 24-6 Segmental fracture and butterfly fractures. These are two comminuted fractures. **A,** There is a segmental fracture in which a portion of the shaft exists as an isolated fragment. Notice how the fibula has a center segment (S) and two additional fragments, one on either side *(white arrows)*. **B,** A butterfly fragment is a comminuted fracture in which the central fragment has a triangular shape *(dotted white arrow)*.

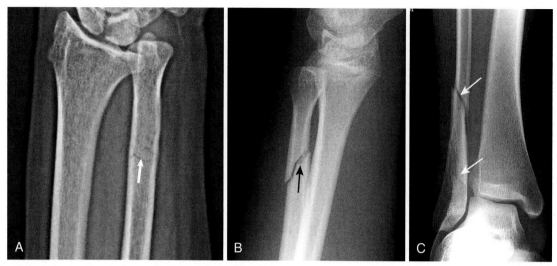

FIGURE 24-7 **Transverse, diagonal, and spiral fracture lines. A,** In a transverse fracture *(white arrow)*, the fracture line is perpendicular to the long axis of the bone. **B,** Diagonal or oblique fractures *(black arrow)* are diagonal in orientation relative to the normal axis of the bone. **C,** Spiral fractures *(white arrows)* are usually caused by twisting or torque injuries.

TABLE 24-4 DIRECTION OF FRACTURE LINE AND MECHANISM OF INJURY

Direction of Fracture Line	Mechanism
Transverse	Force applied perpendicular to long axis of bone; fracture occurs at site of force
Diagonal (also known as oblique)	Force applied along the long axis of bone; fracture occurs somewhere along shaft
Spiral	Twisting or torque injury

HOW FRACTURES ARE DESCRIBED—BY THE DIRECTION OF THE FRACTURE LINE
(Table 24-4)

■ In a **transverse fracture** the fracture line is perpendicular to the long axis of the bone. Transverse fractures are caused by a force directed **perpendicular to the shaft** (Fig. 24-7, *A*).
■ In a **diagonal** or **oblique fracture** the fracture line is diagonal in orientation relative to the long axis of the bone. Diagonal or oblique fractures are caused by a force usually applied along the **same direction as the long axis** of the affected bone (see Fig. 24-7, *B*).
■ With a **spiral fracture** a twisting force or torque produces a fracture similar to those that might be caused by planting the foot in a hole while running. Spiral fractures are usually unstable and often associated with soft tissue injuries, such as tears in ligaments or tendons (see Fig. 24-7, *C*).

HOW FRACTURES ARE DESCRIBED—BY THE RELATIONSHIP OF ONE FRACTURE FRAGMENT TO ANOTHER

⬁ By convention, abnormalities of the position of bone fragments secondary to fractures describe **the relationship of the distal fracture fragment relative to the proximal fragment.** These descriptions are based on the position the distal fragment would have normally assumed had the bone not been fractured.

■ There are four major parameters most commonly used to describe the relationship of fracture fragments. Some fractures display more than one of these abnormalities of position. The four parameters are:
  ♦ **Displacement**
  ♦ **Angulation**
  ♦ **Shortening**
  ♦ **Rotation**
■ **Displacement** describes **the amount by which the distal fragment is offset**, front-to-back and side-to-side, from the proximal fragment (Fig. 24-8). Displacement is most often described either in terms of *percent* (e.g., the distal fragment is displaced by 50% of the width of the shaft) or by *fractions* (e.g., the distal fragment is displaced half the width of the shaft of the proximal fragment) (see Fig. 24-8, *A*).
■ **Angulation** describes **the angle between the distal** and **proximal fragments** as a function of the degree to which the distal fragment is deviated from the position it would have assumed were it in its normal position. Angulation is described in degrees and by position (e.g., the distal fragment is angulated 15° anteriorly relative to the proximal fragment) (see Fig. 24-8, *B*).
■ **Shortening** describes **how much, if any, overlap there is of the ends of the fracture fragments,** which translates into how much shorter the fractured bone is than it would be had it not been fractured (see Fig. 24-8, *C*).
  ♦ The opposite term from shortening is *distraction, which* refers to **the distance the bone fragments are separated from each other** (see Fig. 24-8, *D*).
  ♦ Shortening (overlap) or distraction (lengthening) is usually described by a number of centimeters (e.g., there are 2 cm of shortening of the fracture fragments).
■ **Rotation** is an unusual abnormality in fracture positioning almost always involving the **long bones,** such as the femur or humerus. Rotation describes the orientation of the joint

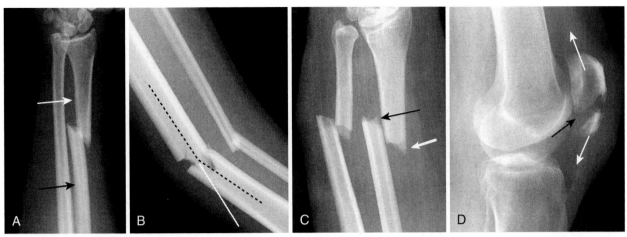

FIGURE 24-8 Fracture parameters. The orientation of fracture fragments is described by using these four parameters. **A,** *Displacement* describes the amount by which the distal fragment *(white arrow)* is offset, front-to-back and side-to-side, from the proximal fragment *(black arrow)*. **B,** *Angulation* describes the angle between the distal and proximal fragments *(dotted black line)* as a function of the degree to which the distal fragment is deviated from its normal position *(solid white line)*. **C,** *Shortening* describes how much, if any, overlap occurs at the ends of the fracture fragments *(white and black arrows)*. The opposite term from shortening is ***distraction* (D),** which refers to the distance the bone fragments are separated from each other *(two white arrows show pull of tendons on fracture fragments of patella; black arrow points to distraction of fracture)*.

at one end of the fractured bone relative to the orientation of the joint at the other end of the same bone.

♦ For example, normally when the hip joint is pointing forward, the knee joint is also pointing forward. If there is **rotation** about a fracture of the femoral shaft, the hip joint could be pointing forward, whereas the knee joint is oriented in another direction (Fig. 24-9). To appreciate rotation, both the joint above and the joint below a fracture must be visualized, preferably on the same radiograph.

HOW FRACTURES ARE DESCRIBED—BY THE RELATIONSHIP OF THE FRACTURE TO THE ATMOSPHERE

■ A **closed fracture** is the **more common** type of fracture, in which there is **no communication** between the fracture fragments and the outside atmosphere.

■ In an **open** or **compound fracture,** there is **communication between the fracture** and **the outside** atmosphere; that is, a fracture fragment penetrates the skin (Fig. 24-10). Compound fractures have implications regarding the way in which they are treated in order to avoid the complication of osteomyelitis. Whether a fracture is open or not is best diagnosed clinically.

AVULSION FRACTURES

■ Avulsion is a **common mechanism** of fracture production in which the fracture fragment **(avulsed fragment)** is pulled from its parent bone by **contraction of a tendon** or **ligament.**

■ Although avulsion fractures can and do occur at any age, they are **particularly common in younger individuals** engaging in athletic endeavors—in fact, they derive many of their names from the type of athletic activity that produces them; for example, **dancer's fracture, skier's fracture, and sprinter's fracture.**

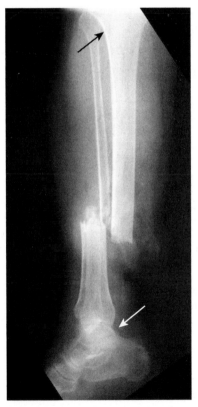

FIGURE 24-9 Rotation. An unusual abnormality in fracture positioning, almost always involving the long bones, which describes the orientation of the joint at one end of the fractured bone relative to the orientation of the joint at the other end of the fractured bone. To appreciate rotation, both the joint above and the joint below a fracture must be visualized, preferably on the same radiograph. In this patient, the proximal tibia *(black arrow)* is oriented in the frontal plane, and the distal tibia and ankle *(white arrow)* are rotated and oriented laterally.

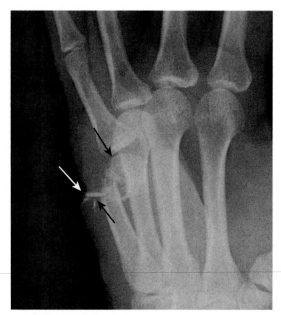

FIGURE 24-10 **Open (compound) fracture, 5ᵗʰ metacarpal.** Most fractures are closed, in which there is no communication between the fracture fragments and the outside atmosphere. Open or compound fractures *(black arrows)* have a communication between the fracture and the outside *(white arrow)*. Whether a fracture is open or not is best evaluated clinically. Treatment of a compound fracture must also consider the higher incidence of infection, which can occur in these injuries.

TABLE 24-5   COMMON AVULSION FRACTURES AROUND THE PELVIS

Avulsed Fragment	Muscle That Inserts on That Fragment
Anterior, superior iliac spine	Sartorius muscle
Anterior, inferior iliac spine	Rectus femoris muscle
Ischial tuberosity	Hamstring muscles
Lesser trochanter of femur	Iliopsoas muscle

- They **occur in anatomically predictable locations** because tendons insert on bones in a known location (Table 24-5), and the avulsed fragment is **typically small** (Fig. 24-11).
- They sometimes **heal with such exuberant callus formation** that they can be mistaken for a bone tumor (Fig. 24-12).

Salter-Harris Fractures—Epiphyseal Plate Fractures in Children

- Salter-Harris fractures are discussed in Chapter 28.

Child Abuse

- Child abuse is discussed in Chapter 28.

STRESS FRACTURES

- Stress fractures **occur as a result of numerous microfractures** in which bone is subjected to repeated stretching and compressive forces.

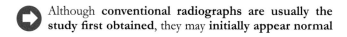 Although **conventional radiographs are usually the study first obtained,** they may **initially appear normal**

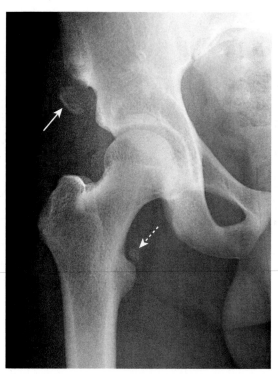

FIGURE 24-11 **Avulsion fractures, ASIS, and lesser trochanter.** Avulsion fractures are common fractures in which the avulsed fragment is pulled from its parent bone by contraction of a tendon or ligament. They are particularly common in younger individuals who engage in athletic endeavors. There is an avulsion of the anterior superior iliac spine (ASIS) *(solid white arrow)*, which is the site of the insertion of the sartorius muscle. There is also an avulsion of a portion of the lesser trochanter, on which the iliopsoas muscle inserts *(dotted white arrow)*. The patient had participated in track and field events a week prior to these injuries.

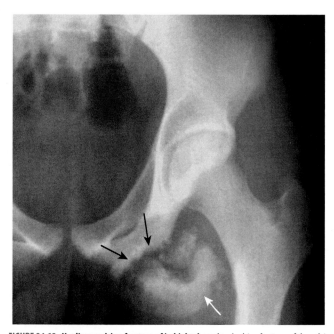

FIGURE 24-12 **Healing avulsion fracture of ischial tuberosity.** Avulsion fractures of the pelvis occur in anatomically predictable locations (tendons insert on bones in known locations), and they are typically small fragments. Sometimes they heal with such exuberant callus formation that they can be mistaken for a bone tumor. There is a healing fracture *(black arrows)* of the ischial tuberosity caused by contraction of the hamstring muscles. There is a great deal of external callus present *(white arrow)*.

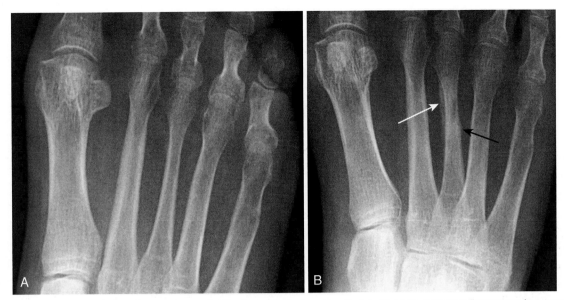

FIGURE 24-13 Stress fracture, two frontal views taken 5 weeks apart. A, Although conventional radiographs are the study of first choice, they may initially appear normal in as many as 85% of cases of stress fractures, so it is common for a patient to complain of pain yet have a normal-appearing radiograph, as seen here one day after pain began. **B,** The fracture may not be diagnosable until after periosteal new bone formation forms *(white arrow)* or, in the case of a healing stress fracture of cancellous bone, the appearance of a thin, dense zone of sclerosis across the medullary cavity *(black arrow).* This radiograph was taken 5 weeks after the first.

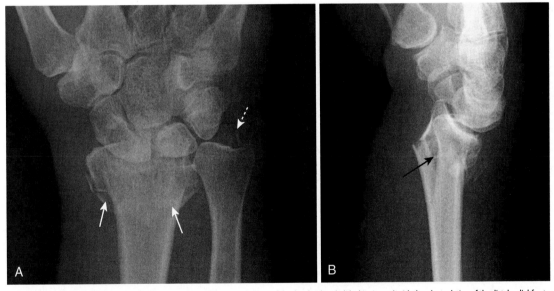

FIGURE 24-14 Colles fracture, frontal (A) and lateral (B) views. A Colles fracture is a fracture of the distal radius *(solid white arrows)* with dorsal angulation of the distal radial fracture fragment *(black arrow)* caused by a fall on the outstretched hand (sometimes abbreviated as *FOOSH*). There is frequently an associated fracture of the ulnar styloid *(dotted white arrow).*

in as many as 85% of stress fractures, so it is common for a patient to complain of pain yet have a normal-appearing radiograph at first.

- The **fracture may not be diagnosable** until after periosteal new bone formation occurs or, in the case of a healing stress fracture of cancellous bone, **the appearance of a thin, dense zone of sclerosis across the medullary cavity** (Fig. 24-13).
- Radionuclide **bone scan will usually be positive much earlier** than conventional radiographs: **within 6 to 72 hours after the injury.**
- Some **common locations for stress fractures** are the **shafts of long bones** such as the proximal femur or proximal tibia, the calcaneus, and the 2nd and 3rd metatarsals **(march fractures).**

COMMON FRACTURE EPONYMS

- There are almost as many fracture eponyms as there are types of fractures.
  - We will concentrate on five of the most commonly used eponyms.
- **Colles fracture** is a **fracture of the distal radius with dorsal angulation** of the distal radial fracture fragment caused by a fall on the outstretched hand (sometimes abbreviated as *FOOSH*). There is frequently an associated fracture of the ulnar styloid (Fig. 24-14).

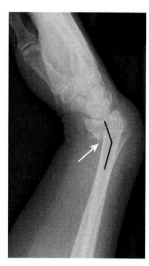

FIGURE 24-15 **Smith fracture.** A Smith fracture is a fracture of the distal radius *(white arrow)* with palmar angulation of the distal radial fracture fragment *(black line angle)*, the reverse of a Colles fracture. It is caused by a fall on the back of the flexed hand.

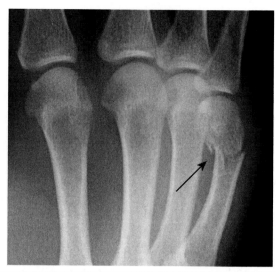

FIGURE 24-17 **Boxer's fracture.** A boxer's fracture is a fracture of the neck of the 5th metacarpal with palmar angulation of the distal fracture fragment *(black arrow)*. It is most often the result of punching a person. Despite its name, it is not a fracture commonly sustained by professional boxers, whose 2nd and 3rd metacarpals and radius bear the brunt of the force.

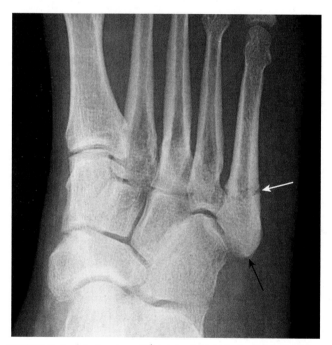

FIGURE 24-16 **Jones fracture, base of 5th metatarsal.** A Jones fracture is a transverse fracture of the base 5th metatarsal *(white arrow)*. It occurs about 1 to 2 cm from the tuberosity of the 5th metatarsal *(black arrow)* and frequently takes longer to heal than an avulsion fracture of the tuberosity. It is caused by plantar flexion of the foot and inversion of the ankle.

fracture fragment. Sometimes the 4th metacarpal may be involved. It is most often the result of punching a person or wall (Fig. 24-17).

■ **March fracture** is a type of **stress fracture** caused by repeated microfractures to the foot from trauma (such as marching) and most often affects the **shafts of the 2nd** and **3rd metatarsals** (see Fig. 24-13).

SOME EASILY MISSED FRACTURES OR DISLOCATIONS

■ Look at these areas carefully when evaluating for a possible fracture; then look a second and third time.
■ **Scaphoid fractures (common)**
   ♦ Scaphoid (navicular) fractures are clinically suspected if there is tenderness in the **anatomic snuff box** after a fall on an outstretched hand. Look for **hairline-thin radiolucencies** on special angled views of the scaphoid (Fig. 24-18). Fractures across the waist of the scaphoid can lead to **avascular necrosis of the proximal pole** of that bone.
   • Because of peculiarities of vascularization, a fracture through the midportion **(waist) of the scaphoid** bone in the wrist **interrupts blood supply to the proximal pole,** whereas the remainder of the bones of the wrist continue to undergo the process of bone turnover. The result is an **apparent relative increase in the density of the devascularized part compared with the remainder of the bone** (Fig. 24-19).
■ **Buckle fractures of radius** and/or **ulna in children (common)**
   ♦ These are common fractures in children. Look for **acute** and **sudden angulation of the cortex,** especially near the wrist (see Fig. 24-1). These are impacted fractures and usually heal quickly with no deformity.

■ **Smith fracture** is a fracture of the distal radius with palmar angulation of the distal radial fracture fragment (a reverse Colles fracture). It is caused by a fall on the back of the flexed hand (Fig. 24-15).
■ **Jones fracture** is a transverse fracture of the 5th metatarsal about 1 to 2 cm from its base caused by plantar flexion of the foot and inversion of the ankle. A Jones fracture may take longer to heal than the more common avulsion fracture of the base of the 5th metatarsal (Fig. 24-16).
■ **Boxer's fracture** is a fracture of the neck of the 5th metacarpal (little finger) with palmar angulation of the distal

■ **Radial head fracture (common)**
  ♦ A fracture of the radial head is the **most common fracture of the elbow in an adult.** Look for a crescentic lucency of fat along the posterior aspect of the distal humerus produced by normally invisible **intracapsular, extrasynovial fat** that is lifted away from the bone by swelling of the joint capsule, resulting from a **traumatic hemarthrosis**—the **positive posterior fat-pad sign** (Fig. 24-20).

■ **Supracondylar fracture of the distal humerus in children (common)**
  ♦ This is the **most common fracture of the elbow in a child.** Most of these fractures produce **posterior displacement** of the distal humerus.
  ♦ On a true lateral film, the **anterior humeral line** (a line drawn tangential to the anterior humeral cortex) should bisect the middle third of the ossification center of capitellum. In a **supracondylar fracture, this line passes anterior to its normal location** (Fig. 24-21).

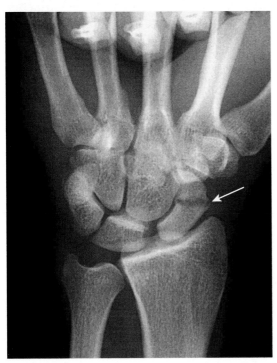

FIGURE 24-18 Scaphoid fracture. Scaphoid fractures are common. They are suspected clinically if there is tenderness in the anatomic snuff box after a fall on an outstretched hand. Look for linear fracture lines on special angled views of the scaphoid *(white arrow)*. Fractures across the waist of the scaphoid can lead to avascular necrosis of proximal pole of that bone.

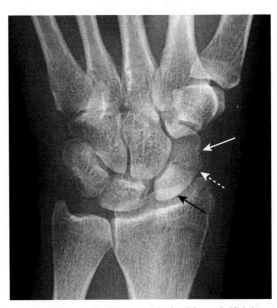

FIGURE 24-19 Avascular necrosis of the proximal pole of the scaphoid. A close-up frontal view of the wrist demonstrates that the proximal pole of the scaphoid *(black arrow)* is denser than the distal pole *(solid white arrow)*. There is a fracture through the waist of the scaphoid *(dotted white arrow)*. Because of the peculiar blood supply of the scaphoid (from distal to proximal), fractures through the wrist may interrupt the proximal blood supply while the remainder of the bones of the wrist, having normal blood supply, become demineralized. This makes the proximal pole of the scaphoid appear denser relative to the other bones of the wrist.

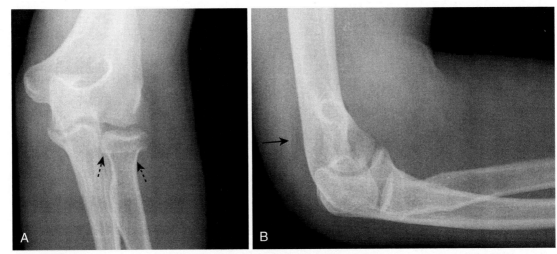

FIGURE 24-20 Fracture of radial head with joint effusion, frontal (A) and lateral (B) views. A, Radial head fractures *(dotted black arrows)* are the most common fractures of the elbow in an adult. **B,** Look for fat density appearing as a crescentic lucency along the dorsal aspect of the distal humerus *(solid black arrow)* caused by intracapsular, extrasynovial fat that has lifted away from the bone by swelling of the joint capsule as a result of traumatic hemarthrosis—the **positive posterior fat-pad sign**. Virtually all studies of bones will include at least two views at 90° angles to each other called *orthogonal views*. Many protocols call for two additional oblique views, which enable you to visualize more of the cortex in profile.

- **Posterior dislocation of the shoulder (uncommon)**
  - The humeral head is fixed in internal rotation and looks like a lightbulb in all views of the shoulder. Look at the axillary or Y-view to see if the head still lies within the glenoid fossa. On the Y-view (an oblique view of the shoulder), the head will lie lateral to the glenoid in a posterior dislocation (Fig. 24-22).

- **Hip fractures in older adults (common)**
  - Hip fractures are common and frequently related to osteoporosis. Conventional radiographs of the femoral neck should be performed with the patient's leg in internal rotation so as to display the neck in profile. Look for **angulation of the cortex** or zones of **increased density indicating impaction** (Fig. 24-23).

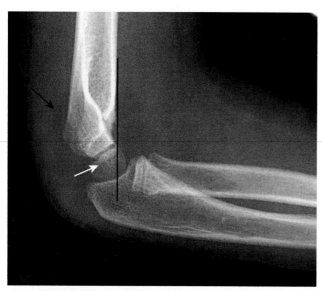

FIGURE 24-21 **Supracondylar fracture.** A supracondylar fracture of the distal humerus is a common fracture in children, and its findings may be subtle. Most of these fractures produce posterior displacement of the capitellum of the distal humerus. On a true lateral film, the anterior humeral line (a line drawn tangential to the anterior humeral cortex and shown here in black) should bisect the middle portion of the capitellum *(white arrow)*. When there is a supracondylar fracture, this line will pass more anteriorly, as it does here. There is a positive posterior fat pad sign present *(black arrow)*.

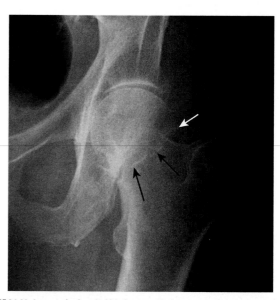

FIGURE 24-23 **Impacted subcapital hip fracture.** Hip fractures are relatively common fractures in older adults and are frequently related to osteoporosis. Look for angulation of the cortex *(white arrow)* and zones of increased density *(black arrows)* indicating impaction. Conventional radiographs of the femoral neck should be obtained with the patient's leg in internal rotation (as shown here) so as to display the neck in profile. Hip fractures can be very subtle and sometimes require additional imaging such as MRI or bone scan for their diagnosis.

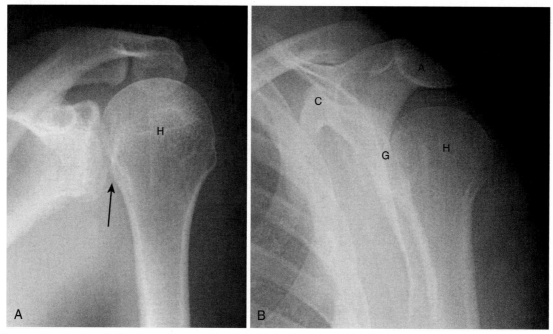

FIGURE 24-22 **Posterior dislocation of the shoulder.** Posterior dislocations of the shoulder are much less common than anterior dislocations, but more difficult to diagnose. On the frontal view **(A)** look for the humeral head (H) to be persistently fixed in internal rotation and resemble a lightbulb, no matter how the patient turns the forearm. There is also an increased distance between the head and the glenoid *(black arrow)*. **B,** On the Y-view, the head (H) will lie under the acromion (A), a posterior structure of the scapula, not the anteriorly located coracoid process. (C). G, Glenoid.

- ◆ Sometimes hip fractures may be very subtle and require an MRI or radionuclide bone scan for diagnosis.
- ■ **Look for indirect signs indicating the possibility of an underlying fracture** (Table 24-6).

TABLE 24-6   INDIRECT SIGNS OF POSSIBLE FRACTURE

Sign	Remarks
Soft tissue swelling	Frequently accompanies a fracture, but if present, does not necessarily mean that a fracture is present.
Disappearance of normal fat stripes	The **pronator quadratus fat stripe** on the volar aspect of the wrist, for example, may be displaced with a fracture of the distal radius (see Fig. 24-24).
Joint effusion	The **positive posterior fat pad** sign seen on the dorsal aspect of the distal humerus from a traumatic joint effusion is an example (see Fig. 24-20, *B*).
Periosteal reaction	Sometimes the healing of a fracture will be the first manifestation that a fracture was present, especially with stress fractures of the foot

FRACTURE HEALING

- ◆ Fracture healing is determined by many factors, including the **age of the patient,** the **fracture site,** the **position of the fracture fragments**, the degree of **immobilization,** and the **blood supply** to the fracture site (Table 24-7).
- ◆ Immediately following a fracture, there is hemorrhage into the fracture site.
- ◆ Over the next several weeks, osteoclasts act to remove the diseased bone. The **fracture line may actually minimally widen** at this time.

TABLE 24-7   FACTORS THAT AFFECT FRACTURE HEALING

Accelerate Fracture Healing	Delay Fracture Healing
Youth	Old age
Early immobilization	Delayed immobilization
Adequate duration of immobilization	Too short a duration of immobilization
Good blood supply	Poor blood supply
Physical activity after adequate immobilization	Steroids
Adequate mineralization	Osteoporosis, osteomalacia

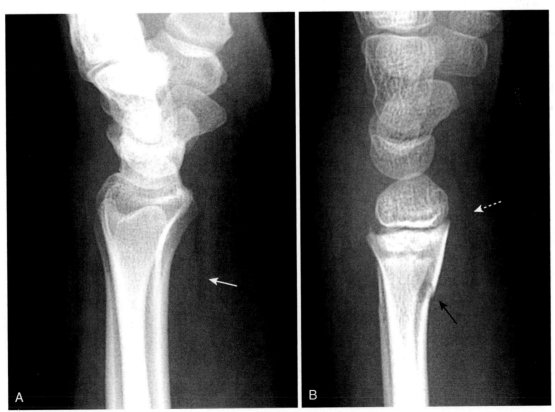

FIGURE 24-24 Normal and abnormal pronator quadratus fat plane. Soft tissue abnormalities can provide clues to the presence of subtle fractures or help confirm the significance of a questionable finding. **A,** Here is an example of a normal fascial plane produced by the pronator quadratus *(white arrow points to lucency)* on the volar aspect of the wrist, compared with the bulging fascial plane *(dotted white arrow)* in **(B),** which has occurred because of soft tissue swelling accompanying a fracture of the distal radius *(black arrow).*

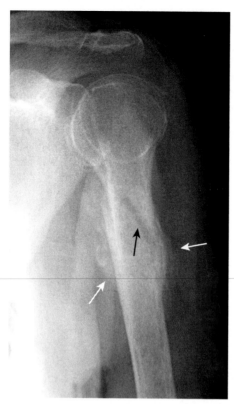

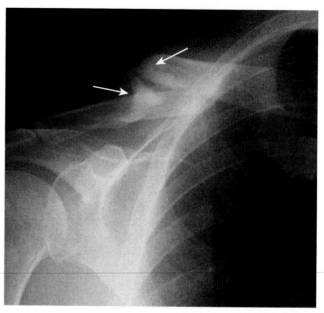

FIGURE 24-26 Nonunion of clavicular fracture. Nonunion is a radiologic diagnosis that implies fracture healing is not likely to occur because the processes leading to the repair of bone have ceased. It is characterized by smooth and sclerotic fracture margins with distraction of the fracture fragments *(white arrows)*. A **pseudarthrosis,** complete with a synovial lining, may form at the fracture site.

FIGURE 24-25 Healing humeral fracture. Immediately following a fracture, there is hemorrhage into the fracture site. Over the next several weeks, new bone (callus) begins to bridge the fracture gap. Internal endosteal healing is manifest by indistinctness of the fracture line *(black arrow)*, eventually leading to obliteration of the fracture line. External, periosteal healing is manifest by external callus formation *(white arrows)*, leading to bridging of the fracture site.

♦ Then over the course of several more weeks, **new bone (callus)** begins to bridge the fracture gap (Fig. 24-25).

 Internal endosteal healing is manifest by **indistinctness of the fracture line,** leading to eventual obliteration of the fracture line.

♦ **External periosteal healing** is manifest by external callus formation, eventually leading to **bridging of the fracture site.**

■ **Remodeling** of bone **begins at about 8 to 12 weeks** post-fracture as mechanical forces in part begin to adjust the bone to its original shape.
  ♦ In **children this occurs much more rapidly** and usually leads to a bone that eventually appears normal.

In **adults this process may take years,** and the healed fracture may never assume a completely normal shape.

■ **Complications of the healing process**
  ♦ **Delayed union. The fracture does not heal in the expected time** for a fracture at that particular site (e.g., 6 to 8 weeks for a fracture of the shaft of the radius). **Most cases of delayed union will eventually progress to complete healing** with further immobilization.
  ♦ **Malunion.** Healing of the fracture fragments occurs in a **mechanically** or **cosmetically unacceptable position.**
  ♦ **Nonunion.** This implies that **fracture healing will never occur.** It is characterized by **smooth** and **sclerotic fracture margins** with **distraction of the fracture fragments** (Fig. 24-26). A pseudarthrosis, complete with a synovial lining, may form at the fracture site.
    • **Motion at the fracture site** may be demonstrated under fluoroscopic manipulation or on stress views.

TAKE-HOME POINTS

RECOGNIZING FRACTURES AND DISLOCATIONS

A **fracture** is described as a disruption in the continuity of all or part of the cortex of a bone.

Complete fractures involve the entire cortex, are more common, and typically occur in adults; **incomplete fractures** involve only a part of the cortex and typically occur in bones that are softer, such as those of children; **torus** and **greenstick fractures** are incomplete fractures.

Fracture lines tend to be blacker, more sharply angled, and more jagged than other lucencies in bones such as nutrient canals or epiphyseal plates.

TAKE-HOME POINTS—cont'd

Sesamoids, accessory ossicles, and unhealed fractures may mimic acute fractures, but all will have smooth and corticated margins.

Dislocation is present when two bones that originally formed a joint are no longer in contact with each other; **subluxation** is present when two bones that originally formed a joint are in partial contact with each other.

Fractures are described in many ways, including the number of fracture fragments, direction of the fracture line, relationship of the fragments to each other, and whether or not they communicate with the outside atmosphere.

Simple fractures have two fragments; **comminuted fractures** have more than two fragments. Segmental and butterfly fractures are two types of comminuted fracture.

The direction of fracture lines is described as **transverse, diagonal, or spiral.**

The relationships of the fragments of a fracture are described by four parameters: **displacement, angulation, shortening, and rotation.**

Closed fractures are those in which there is no communication between the fracture and the outside atmosphere; they are much more common than **open** or **compound fractures,** in which there is a communication with the outside atmosphere.

Avulsion fractures are produced by the forceful contraction of a tendon or ligament; they can occur at any age but are particularly common in younger, athletic individuals.

Stress fractures, such as march fractures in the metatarsals, occur as a result of numerous microfractures and frequently are not visible on conventional radiographs taken when the pain first begins; after some time, bony callus formation, and/or a dense zone of sclerosis becomes visible.

Some common named fractures are ***Colles fracture*** (of the radius), ***Smith fracture*** (of the radius), ***Jones fracture*** (of the base of the 5th metatarsal), ***boxer's fracture*** (of the head of the 5th metacarpal), and ***march fracture.***

Some fractures are more difficult to detect than others; the easily missed fractures (and how common they are) include scaphoid fractures (common), buckle fractures of the radius and ulna (common), radial head fractures (common), supracondylar fractures (common), posterior dislocations of the shoulder (uncommon), and hip fractures (common).

Soft tissue swelling, the disappearance of normal fat stripes and fascial planes, joint effusions, and periosteal reaction are indirect signs that should alert you to the possibility of an underlying fracture.

Fractures heal with a combination of endosteal callus (recognized by a progressive indistinctness of the fracture line) and external callus, which bridges the fracture site; many factors affect fracture healing, including the age of the patient, the degree of mobility of the fracture, and its blood supply.

Delayed union refers to a fracture that is taking longer to heal than is usually required for that site; ***malunion*** means the fracture is healing but in a mechanically or cosmetically unacceptable way; ***nonunion*** is a radiologic diagnosis that implies there is little, if any, likelihood the fracture will heal.

 WEBLINK

Visit StudentConsult.Inkling.com for more information and a quiz: Fracture or No Fracture?

For your convenience, the following QR Code may be used to link to **StudentConsult.com**. You must register this title using the PIN code on the inside front cover of the text to access online content.

CHAPTER 25
Recognizing Joint Disease: An Approach to Arthritis

Imaging studies play a key role in the diagnosis and management of arthritis and are the method by which many arthridities are first diagnosed. Other arthridities are initially diagnosed on clinical and laboratory grounds, and imaging is used to document the severity, extent, and course of the disease (Table 25-1).

- Conventional radiographs demonstrate osseous abnormalities but provide only indirect and usually delayed evidence of abnormalities involving the soft tissues, such as the synovial lining of the joint, the articular cartilage, muscles, ligaments, and tendons surrounding the joint. Magnetic resonance imaging (MRI) is used to demonstrate those soft tissue abnormalities.

ANATOMY OF A JOINT

- Figure 25-1 contains a diagram of a typical synovial joint and compares the structures visualized on conventional radiographs and MRI.
- The **articular cortex** is the thin, white line as seen on conventional radiographs that lies within the joint capsule and is usually capped by **hyaline cartilage,** called the *articular cartilage.* The bone immediately beneath the articular cortex is called *subchondral bone.*

TABLE 25-1 ARTHRITIS—HOW IS IT DIAGNOSED?

Usually Diagnosed Clinically	Frequently Diagnosed Radiologically
Septic (pyogenic) arthritis	Osteoarthritis
Psoriatic arthritis	Early rheumatoid arthritis
Gout	Calcium pyrophosphate deposition disease
Hemophilia	Ankylosing spondylitis
	Septic (tuberculosis)
	Charcot (neuropathic) joint—late

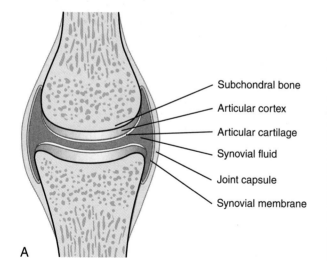

A

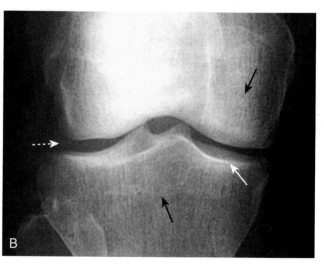

B

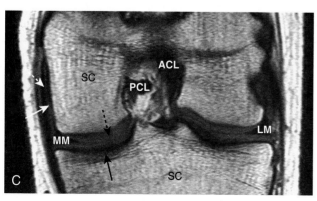

C

FIGURE 25-1 Diagram, radiograph, and MRI of a true joint. A, A representation of a synovial joint shows the **articular cortex** within the **joint capsule** that is usually capped by **articular cartilage.** The bone immediately beneath the articular cortex is called *subchondral bone.* Within the joint capsule is the **synovial membrane** and **synovial fluid. B,** On conventional radiograph, the articular cortex *(solid white arrow)* and subchondral bone *(solid black arrows)* are visible, but the cartilage and synovial fluid are not *(dotted white arrow).* **C,** A T1-weighted coronal MRI of the knee shows the medial (MM) and lateral (LM) menisci, anterior (ACL) and posterior (PCL) cruciate ligaments, articular cartilage *(dotted black arrow),* joint capsule *(dotted white arrow),* synovial fluid *(solid white arrow),* and marrow in the subchondral bone (SC). The cortex of the bone *(solid black arrow)* produces little signal and is dark.

- Lining the **joint capsule** is the **synovial membrane** containing **synovial fluid.** The synovial membrane is frequently the earliest structure involved by an arthritis.
- Conventional radiographs will demonstrate abnormalities of the **articular cortex** and the **subchondral bone** and will provide late, indirect evidence of the integrity of the articular cartilage.
- On conventional radiographs, the synovial membrane, synovial fluid, and the articular cartilage are usually not directly visible. However, all are visible using MRI.
- While MRI is more sensitive in directly visualizing the soft tissues in and around a joint, **conventional radiographs remain the study of first choice** in evaluating for the presence of an arthritis.

CLASSIFICATION OF ARTHRITIS

- First, let's define **what an arthritis is**, and then we'll outline a classification of arthritides. An **arthritis is a disease that affects a joint** and **usually the bones on either side of the joint**, almost always accompanied by **joint space narrowing** (Fig. 25-2).
- We will divide arthritides into **three main categories** (Fig. 25-3; Table 25-2):
    - ♦ **Hypertrophic arthritis** is characterized, in general, by **bone formation** at the site of the involved joint(s). The bone formation may occur within the confines of the parent bone **(subchondral sclerosis)** or protrude from the parent bone **(osteophyte)** (see Fig. 25-3, *A*).
    - ♦ **Erosive arthritis** indicates underlying inflammation and is characterized by tiny, marginal, irregularly shaped lytic lesions in or around the joint surfaces called *erosions* (see Fig. 25-3, *B*).
    - ♦ **Infectious arthritis** is characterized by joint swelling, osteopenia, and **destruction of long, contiguous segments of the articular cortex** (see Fig. 25-3, *C*).
- Within each of these three main categories, we will look at a few of the more common types of arthritis that also have characteristic imaging findings.

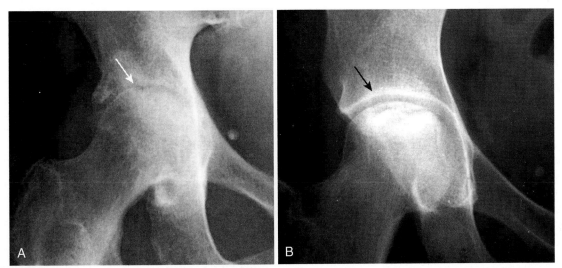

FIGURE 25-2 Arthritis or not? An arthritis is a disease that affects a joint and usually the bones on either side of the joint, almost always with associated joint space narrowing. **A,** This disease meets those specifications. There is narrowing of the hip joint, and both the femoral head and the acetabulum are abnormal *(white arrow)*. This is osteoarthritis of the hip. **B,** There is an abnormality of the femoral head (sclerosis), but the joint space is intact, as is the acetabulum *(black arrow)*. This is avascular necrosis of the femoral head.

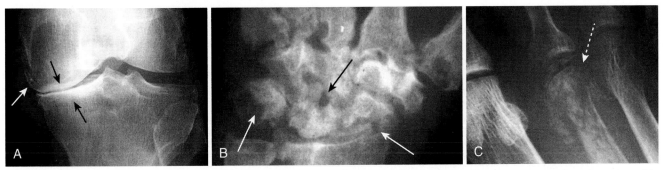

FIGURE 25-3 The imaging hallmarks of the three major categories of arthritis. A, Hypertrophic arthritis features subchondral sclerosis *(black arrows)* and marginal osteophyte production *(white arrow)*. **B,** Erosive (inflammatory) arthritis features characteristic marginal lytic erosions *(white and black arrows)*. **C,** Infectious arthritis features destruction of the articular cortex *(dotted white arrow)*.

TABLE 25-2 CLASSIFICATION OF ARTHRITIS

Category	Hallmarks	Types	Remarks
Hypertrophic arthritis	Bone formation Osteophytes	Primary osteoarthritis	Most common; mechanical stress; hands, hips, and knees most common
		Secondary osteoarthritis	Degenerative joint disease (DJD) secondary to prior trauma or avascular necrosis
		Charcot arthropathy	Fragmentation; joint destruction; sclerosis; most often secondary to diabetes
		Calcium pyrophosphate deposition disease	Chondrocalcinosis; DJD in unusual sites
Erosive arthritis	Erosions	Rheumatoid	Carpals, metacarpal-phalangeal joints, proximal interphalangeal joints of hands; osteoporosis; soft tissue swelling
		Gout	Juxtaarticular erosions with overhanging edges; long latency; metatarsal-phalangeal joint of big toe; no osteoporosis
		Psoriatic	Juxtaarticular erosions of distal interphalangeal joints of hands; pencil-in-cup deformity; enthesophytes
		Hemophilia	Remodeling from hemarthroses and hyperemia; same changes in knee in female—think of juvenile rheumatoid arthritis
		Ankylosing spondylitis	HLA B27+; bilateral sacroiliac (SI) joints; syndesmophytes
		Seronegative spondyloarthropathies	Rheumatoid factor negative; HLA B27+; SI joints; syndesmophytes; reactive, psoriasis
Infectious arthritis	Osteopenia and soft tissue swelling; early and marked destruction of most or all of the articular cortex	Pyogenic	Early destruction of articular cortex; osteoporosis
		Tuberculosis	Gradual and late destruction of articular cortex; marked osteoporosis

HYPERTROPHIC ARTHRITIS

- **Hypertrophic arthritis is typified by bone formation** and includes:
  - ♦ **Osteoarthritis (degenerative arthritis, degenerative joint disease),** which is further divided into:
    - • **Primary osteoarthritis**
    - • **Secondary osteoarthritis**
  - ♦ **Erosive osteoarthritis**
  - ♦ **Charcot arthropathy (neuropathic joint)**
  - ♦ **Calcium pyrophosphate deposition disease (CPPD)**

Primary Osteoarthritis (Also Known as Primary Degenerative Arthritis, Degenerative Joint Disease—DJD)

- This is the **most common form of arthritis,** estimated to affect over 20 million Americans. It results from **intrinsic degeneration of the articular cartilage,** mostly from the mechanical stress of **excessive wear** and **tear** in weight-bearing joints.
- It mostly involves the **hips, knees,** and **hands** and increases in prevalence with increasing age.

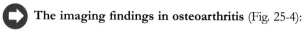 The imaging findings in osteoarthritis (Fig. 25-4):

- ♦ **Marginal osteophyte formation. A hallmark of hypertrophic arthritis,** osseous transformation of cartilaginous excrescences, and metaplasia of synovial lining cells leads to the production these **bony protrusions at or near the joint.**
- ♦ **Subchondral sclerosis.** This is a reaction of the bone to the mechanical stress to which it is subjected when its protective cartilage has been destroyed.
- ♦ **Subchondral cysts.** As a result of chronic impaction, necrosis of bone, and/or imposition of synovial fluid

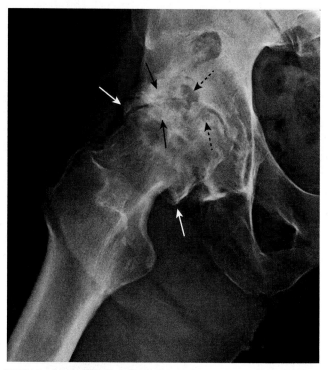

FIGURE 25-4 Osteoarthritis. The hallmarks of osteoarthritis are demonstrated in this patient's right hip. There is marginal osteophyte formation (*solid white arrows*), a process by which there is osseous transformation of cartilaginous excrescences, and metaplasia of synovial lining cells leading to the production of bony protrusions at or near the joint. There is also subchondral sclerosis (*solid black arrows*), representing reaction of the bone to the mechanical stress to which it is subjected when its protective cartilage has been destroyed. There is also subchondral cyst formation (*dotted black arrows*).

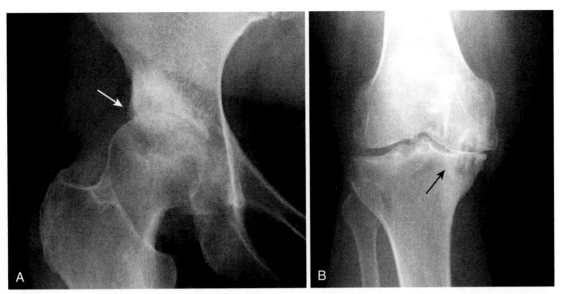

FIGURE 25-5 Osteoarthritis of hip (A) and knee (B). In osteoarthritis destruction of the cartilaginous buffer between the apposing bones of a joint leads to narrowing of the joint space usually on the weight-bearing side of the joint. **A,** In the hip, the superior and lateral surface is weight-bearing and is therefore most affected *(white arrow)*, whereas in the knee **(B)**, the medial compartment is weight-bearing and more affected *(black arrow)*.

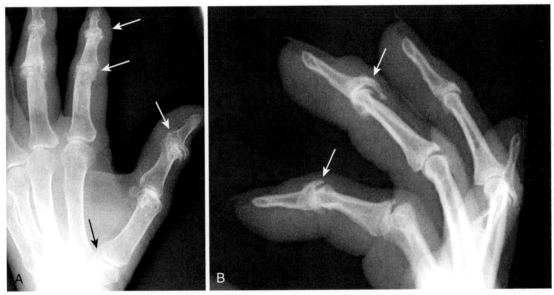

FIGURE 25-6 Osteoarthritis of the hands. A and B, In the hands osteoarthritis affects primarily the distal and then proximal interphalangeal joints. There are osteophytes at the distal and proximal phalangeal joints *(white arrows)*, and the joint spaces are narrowed (both **A** and **B**). There is also subchondral sclerosis present at the carpal-metacarpal joint of the thumb *(black arrow)*. Osteoarthritis of the hands occurs most often in older women.

into the subchondral bone, cysts of varying sizes form in the subchondral bone.

- ♦ **Narrowing of the joint space.** Seen in all forms of arthritis.
- ■ **What joints are involved?**
  - ♦ In osteoarthritis, destruction of the cartilaginous buffer between the apposing bones of a joint leads to narrowing of the joint space most often on the **weight-bearing** side of the joint: **hip (superior** and **lateral)** and **knee (medial)** (Fig. 25-5).
  - ♦ In most patients with osteoarthritis of the interphalangeal joints of the hands, the first carpometacarpal joint **(base of thumb) is also affected** (Fig. 25-6). It is also common for osteoarthritis

to affect the **distal interphalangeal joints,** especially in older females.

Secondary Osteoarthritis (Secondary Degenerative Arthritis)

- ■ Secondary osteoarthritis is a form of degenerative arthritis of synovial joints that occurs because of an underlying, predisposing condition, **most frequently trauma,** that damages or leads to damage of the articular cartilage.
- ■ The radiographic findings of secondary osteoarthritis are the same as those for the primary form, with several special clues to help suggest secondary osteoarthritis:
  - ♦ It **occurs at an atypical age** for primary osteoarthritis (e.g., a 20-year-old with osteoarthritis) (Box 25-1).

BOX 25-1 SOME CAUSES OF SECONDARY OSTEOARTHRITIS

Trauma

Infection

Avascular necrosis

Calcium pyrophosphate deposition disease

Rheumatoid arthritis

BOX 25-2 CAUSES OF CHARCOT JOINTS BY LOCATION

Shoulders—syrinx, spinal tumor, and syphilis

Hips—tertiary syphilis, diabetes

Ankles and feet—diabetes (common) and syphilis (uncommon)

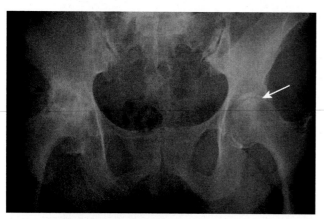

FIGURE 25-7 **Secondary osteoarthritis, right hip.** There is a marked discrepancy between the two hips with severe and advanced osteoarthritis of the right hip *(black arrow)* and a relatively normal left hip *(white arrow)*. This should raise suspicion for a secondary form of osteoarthritis. In the hips, primary arthritis is usually bilateral, and the hips are similarly affected. This patient had a slipped capital femoral epiphysis on the right, and it was never attended to medically.

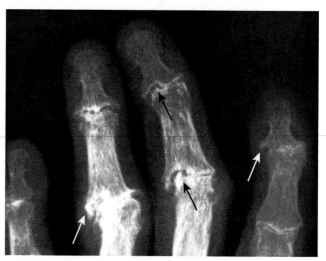

FIGURE 25-8 **Erosive osteoarthritis.** A type of primary osteoarthritis characterized by more severe inflammation and by the development of erosive changes, erosive osteoarthritis may feature bilaterally symmetrical changes like the osteophytes of degenerative joint disease but with marked inflammation (swelling and tenderness). The erosions are typically centrally located within the joint *(black arrows)* and, combined with the small osteophytes associated with the disease *(white arrows)*, produce what has been called the ***gull-wing deformity.***

♦ It has an **atypical appearance for primary osteoarthritis** (e.g., primary osteoarthritis is usually bilateral and often symmetrical; severe osteoarthritic changes of one hip in someone in whom the opposite appears perfectly normal should alert you to the possibility of secondary osteoarthritis).

♦ It may **appear in an unusual location** for primary osteoarthritis (e.g., the elbow joint) (Fig. 25-7).

 Eventually, any arthritis that affects the articular cartilage, no matter what the cause, can lead to changes of secondary osteoarthritis.

Erosive Osteoarthritis

■ Erosive osteoarthritis is a **type of primary osteoarthritis** characterized by more **severe inflammation** and by the development of **erosive changes.** It occurs most often in **perimenopausal females.**

■ Erosive osteoarthritis may feature bilaterally symmetrical changes such as the osteophytes of primary osteoarthritis, but with **marked inflammation (swelling** and **tenderness)** and **erosions** at the affected joints.

♦ The **erosions are typically centrally located within the joint** and, combined with the small osteophytes associated with the disease, may produce the so-called *gull-wing deformity* (Fig. 25-8).

♦ Erosive osteoarthritis most commonly occurs at the **proximal** and **distal interphalangeal joints**

of the fingers, 1st **carpal-metacarpal,** and the **interphalangeal joint of the thumb.**

♦ **Bony ankylosis may occur,** an uncommon finding in primary osteoarthritis.

Charcot Arthropathy (Neuropathic Joint)

■ Charcot arthropathy develops from a **disturbance in sensation,** which leads to **multiple microfractures,** as well as an autonomic imbalance, which leads to hyperemia, **bone resorption, and fragmentation of bone.**

■ Even though the joint lacks sensory feedback, almost three out of four patients with a Charcot joint **complain of some degree of pain,** although it is usually far less than would be expected for the degree of joint destruction present.

■ **Soft tissue swelling is a prominent feature.**

■ The **most common cause of a Charcot joint today is diabetes,** and most Charcot joints are found in the **lower extremities, particularly in the feet** and ankles (Box 25-2).

■ **Radiographic findings in Charcot arthropathy:**

 As a hypertrophic arthritis, a Charcot joint will demonstrate **extensive subchondral sclerosis.**

♦ The **hallmark findings of a Charcot joint,** however, are:

• **Fragmentation** of the bones surrounding the joint, which produces numerous small, bony densities

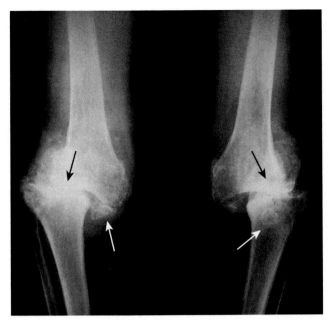

FIGURE 25-9 Charcot arthropathy of knees. As a hypertrophic arthritis, a Charcot joint will demonstrate extensive subchondral sclerosis. The hallmark findings of a Charcot joint, however, are fragmentation of the bones surrounding the joint, which produces numerous small, bony densities within the joint capsule *(white arrows)*, and joint space destruction *(black arrows)*. The most common cause of a Charcot joint of the knee is now diabetes.

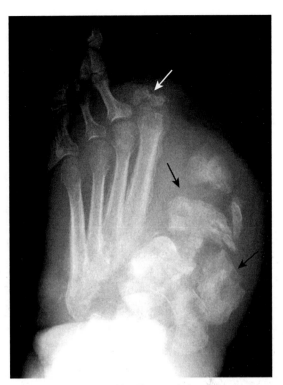

FIGURE 25-10 Charcot arthropathy of foot. This patient had previously undergone an amputation of the phalanges of the second toe *(white arrow)* for diabetic gangrene, but the destruction and marked fragmentation of the great toe and tarsals are manifestations of Charcot neuropathy *(black arrows)*. Charcot neuropathy can produce some of the more dramatic examples of total joint destruction than any other arthritis.

within the joint capsule. Sometimes **many, if not all, of the fragments may be resorbed** and no longer be visible (Fig. 25-9).

- Eventual **destruction of the joint.** Charcot neuropathy is responsible for some of the **most dramatic examples of total joint destruction of any arthritis** (Fig. 25-10).

♦ A Charcot joint shares findings with osteomyelitis, and the two may mimic each other in that they both produce **bone destruction** and **periosteal reaction (from fracture healing).** A radioisotope-tagged white cell bone scan can help to differentiate infection from a Charcot joint.

Calcium Pyrophosphate Deposition Disease (Pyrophosphate Arthropathy)

■ Calcium pyrophosphate deposition disease (CPPD) is an arthropathy resulting from the deposition of **calcium pyrophosphate dihydrate** crystals in and around joints, **mostly in hyaline cartilage** and **fibrocartilage.** This is especially common in the **triangular fibrocartilage of the wrist** and the **menisci of the knee.**

■ The terminology associated with describing this disease can be confusing.

♦ **Chondrocalcinosis** refers only to **calcification** of the articular cartilage or fibrocartilage and is seen in about 50% of adults over the age of 85, most of whom are **asymptomatic.** Chondrocalcinosis can occur in other diseases besides CPPD, such as hyperparathyroidism or hemochromatosis (Fig. 25-11).

♦ **Pseudogout** is a **clinical syndrome** consisting of an **acute, monoarticular arthropathy** characterized by

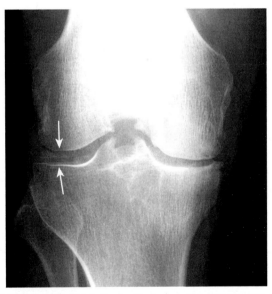

FIGURE 25-11 Chondrocalcinosis. Chondrocalcinosis refers only to calcification of the articular cartilage *(white arrows)* or fibrocartilage and is seen in about 50% of adults over the age of 85, most of whom are asymptomatic. If this patient had acute pain, redness, swelling, and limitation of motion, the combination would be called *pseudogout.*

redness, pain, and **swelling** of the affected joint (most commonly, the knee) from which **calcium pyrophosphate dihydrate** crystals can be aspirated.

♦ **Pyrophosphate arthropathy** is a radiologic diagnosis and the most common form of CPPD.

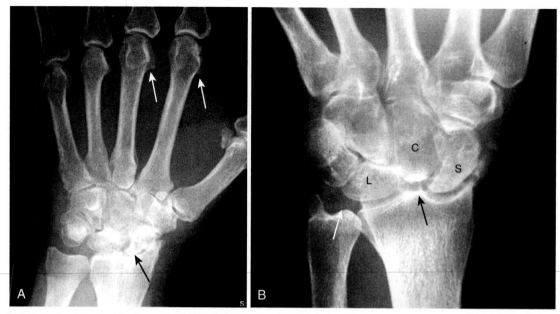

FIGURE 25-12 Calcium pyrophosphate deposition disease. CPPD arthropathy produces changes similar to osteoarthritis but differs from it in that CPPD affects joints not usually affected by primary osteoarthritis. **A,** Hook-shaped bony excrescences along the 2nd and 3rd metacarpal heads are a common finding in CPPD *(white arrows)*. The radiocarpal joint is narrowed *(black arrow)*. **B,** In the wrist, characteristic findings include calcification of the triangular fibrocartilage *(white arrow)*, separation of the scaphoid (S) and the lunate (L) **(scapholunate dissociation)**, and collapse of the capitate (C) toward the radius *(black arrow)*, called **scapholunate advanced collapse (SLAC).**

 Pyrophosphate arthropathy **may be indistinguishable from primary osteoarthritis** but differs from it in several important respects (Fig. 25-12):

♦ CPPD may disproportionately involve joints **not usually affected by osteoarthritis,** such as **the patellofemoral joint space of the knee, the radiocarpal joint, the metacarpal-phalangeal (MCP) joints of the hands,** and the joints of the **wrist.**

♦ **Chondrocalcinosis is usually present** in pyrophosphate arthropathy but is not required for the diagnosis.

♦ **Subchondral cysts** are more common, larger, more numerous, and more widespread than in primary osteoarthritis.

♦ **Hook-shaped bony excrescences** along the second and third metacarpal heads are a common finding in this form of CPPD (see Fig. 25-12, *A*).

♦ **In the wrist,** characteristic findings of CPPD include **calcification of the triangular fibrocartilage,** narrowing of the radiocarpal joint, separation of the scaphoid and the lunate by more than 3 mm **(scapholunate dissociation),** and collapse of the distal carpal row toward the radius **(scapholunate advanced collapse)** (see Fig. 25-12, *B*).

EROSIVE ARTHRITIS

■ **Erosive arthritis** comprises a large number of arthridities, all of which are associated with some degree of **inflammation** and **synovial proliferation (pannus formation),** which participates in the production of **lytic lesions in or near the joint** called *erosions,* especially in the small joints of the hands and feet.

BOX 25-3 SOME CAUSES OF EROSIVE ARTHRITIS

Rheumatoid arthritis
Gout
Psoriatic arthritis
Ankylosing spondylitis (spine)
Rheumatoid variants
Reactive arthritis
Sarcoid
Hemophilia

■ **Pannus** acts like a mass of growing and enlarging synovial tissue, which leads to **marginal erosions** of the **articular cartilage** and **underlying bone.**

■ Box 25-3 lists some of the many causes of erosive arthritis. We shall look at four of the more common.

Rheumatoid Arthritis (RA)

■ Rheumatoid arthritis is **more common in females,** frequently **involving the proximal joints of the hands** and **wrists.** It is **usually bilateral** and **symmetrical.**

■ Conventional radiographs remain the study of first choice in imaging RA.

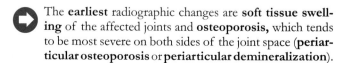 The **earliest** radiographic changes are **soft tissue swelling** of the affected joints and **osteoporosis,** which tends to be most severe on both sides of the joint space **(periarticular osteoporosis** or **periarticular demineralization).**

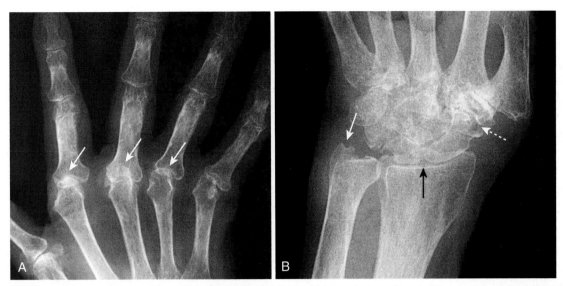

FIGURE 25-13 Rheumatoid arthritis, hand (A) and wrist (B). A, In the hand, the erosions of rheumatoid arthritis tend to involve the proximal joints: that is, the carpal-metacarpal joints, metacarpal-phalangeal (MCP) joints *(white arrows)*, and proximal interphalangeal joints. Late findings in the hands include deformities such as ulnar deviation of the fingers at the MCP joints, subluxation of the MCP joints, and ligamentous laxity leading to deformities of the fingers, which are also present in this hand. **B,** In the wrist, erosions of the carpals *(dotted white arrow)*, ulnar styloid *(solid white arrow)*, and narrowing of the radiocarpal joint space *(solid black arrow)* are commonly seen.

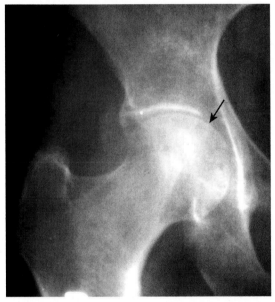

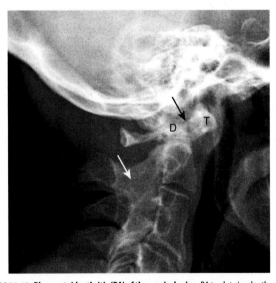

FIGURE 25-14 Rheumatoid arthritis of the hip. The larger joints (hips and knees) usually demonstrate no erosions, but there may be marked uniform narrowing of the joint space with little subchondral sclerosis *(black arrow)*. If this were primary osteoarthritis, you would expect far more subchondral sclerosis and osteophyte production for this degree of joint space narrowing.

FIGURE 25-15 Rheumatoid arthritis (RA) of the cervical spine. RA tends to involve the cervical spine by producing ligamentous laxity, which can lead to forward subluxation of C1 on C2 **(atlanto-axial subluxation).** The distance between the anterior border of the dens (D) and the posterior border of the anterior tubercle of C1 (T) is called the ***predentate space*** and is normally no more than 3 mm. This patient's predentate space measured 8 mm *(black arrow)*. C1 is subluxed forward on C2. RA may also cause fusion of the facet joints *(white arrow).*

- ■ **In the hand,** the **erosions tend to involve the proximal joints:** the carpal-metacarpal joints, metacarpal-phalangeal joints, and proximal interphalangeal joints (Fig. 25-13).
  - ◆ **Late findings in the hands** include deformities such as **ulnar deviation of the fingers at the MCP joints, subluxation of the MCP joints,** and ligamentous laxity, leading to deformities of the fingers (**swan-neck** and **boutonnière deformities**).
- ■ **In the wrist,** erosions of the **carpals, ulnar styloid,** and **narrowing of the radiocarpal joint** space are frequently seen.

- ■ Elsewhere in the body, the **larger joints usually show no erosions,** but there may be **marked uniform narrowing of the joint space with little** or **no subchondral sclerosis** (Fig. 25-14).
- ■ In the spine, RA tends to involve the cervical spine by producing **ligamentous laxity,** which can lead to forward subluxation of C1 on C2 **(atlantoaxial subluxation).** Atlantoaxial subluxation can produce cord compression if severe (Fig. 25-15).
  - ◆ RA may also cause **narrowing, sclerosis,** and **eventual fusion of the facet joints in the cervical spine.**

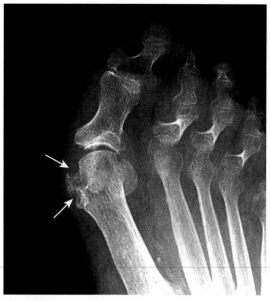

FIGURE 25-16 **Gout.** Gout most commonly affects the metatarsal-phalangeal joint of the great toe, as in this patient. As an erosive arthritis, the hallmark of gout is the sharply marginated erosion, which may have a sclerotic border *(white arrows)*. The erosions in gout tend to be juxtaarticular rather than intraarticular. The overhanging edges of gouty erosions have been called *rat-bites*. The metatarsal-phalangeal joint space is slightly narrowed, and there is no periarticular osteoporosis.

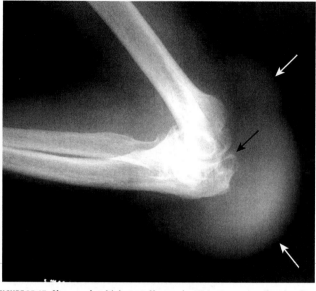

FIGURE 25-17 **Olecranon bursitis in gout.** Olecranon bursitis is a common manifestation of gout *(large soft tissue mass around elbow shown by white arrows)*, and its presence alone should alert you to the possibility of underlying gout. This patient also displays erosions adjacent to the elbow joint *(black arrow)*.

Gout

- Gouty arthritis represents the **inflammatory changes incited by the deposition of calcium urate crystals in the joint.** There is **characteristically an extremely long latent period (5 to 7 years)** between the onset of symptoms and the visualization of bone changes, so **gout is usually a clinical** and **not a radiologic diagnosis** (see Table 25-1).
- Gout tends to be **monoarticular** at its onset and **asymmetric later** in its course.
- It is **more common** in males and **most often affects the metatarsal-phalangeal joint of the great toe** at the time of symptom onset.

 Imaging findings of gout:

- As an erosive arthritis, the **hallmark of gout is the sharply marginated, juxtaarticular erosion, which tends to have a sclerotic border.** The appearance of the overhanging edges of gouty erosions have been called *rat-bites*.
- **Joint space narrowing may be a late finding** in the disease, and there is **characteristically little** or **no periarticular osteoporosis** (Fig. 25-16).
- **Tophi**, collections of urate crystals in the soft tissues, are a late finding in gout and, when present, **rarely calcify.**
- **Olecranon bursitis is common** (Fig. 25-17).

Psoriatic Arthritis

- **Most patients** with psoriatic arthritis **also have skin and nail changes of psoriasis** for many years, although for some the joint manifestations may be the initial presentation of the disease. About 25% of patients with long-standing psoriasis develop psoriatic arthritis. It is **usually polyarticular.**

- Psoriatic arthritis **typically involves the small joints of the hands, especially the distal interphalangeal (DIP) joints.**

The imaging **hallmarks of psoriatic arthritis:**

- **Juxtaarticular erosions, especially of the DIP joints** of the hands
- **Bony proliferation at the sites of tendon insertions (enthesophytes)**
- **Periosteal reaction along the shafts of the bone** (not common)
- **Resorption of the terminal phalanges** or the distal interphalangeal joints with telescoping of one phalanx into another **(pencil-in-cup deformity)** (Fig. 25-18)
- **Absence of osteoporosis**
- In the **sacroiliac (SI) joints,** psoriasis can produce **bilateral, but asymmetric, sacroiliitis.**
- The **SI joints do not usually completely fuse as they do in ankylosing spondylitis.**

Ankylosing Spondylitis

- Ankylosing spondylitis is a chronic and progressive arthritis characterized by **inflammation** and **eventual fusion of the sacroiliac joints** and the spinal facet joints, as well as involvement of the paravertebral soft tissues.
- More **common in young males,** the disease **characteristically ascends the spine, starting in the SI joints** then moving to the lumbar, thoracic, and finally, cervical spine.
- Almost all patients with ankylosing spondylitis test positive for the **human leukocyte antigen B27 (HLA-B27),** in comparison with only about 5% to 10% of the general population.
- Conventional radiographs of the affected areas are the usual study performed for diagnosis and follow-up of patients with ankylosing spondylitis.

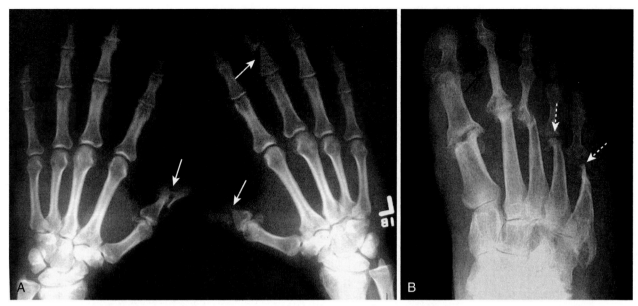

FIGURE 25-18 Psoriatic arthritis, hand (A) and foot (B). A, Psoriatic arthritis typically involves the small joints of the hands, especially the distal interphalangeal (DIP) joints *(white arrows)*, leading to telescoping of one phalanx into another **(*pencil-in-cup deformity*). B,** In the foot of another patient with psoriasis, there is ankylosis of the 2nd toe *(solid black arrow)* and more pencil-in-cup deformities *(dotted white arrows)*.

- Ankylosing spondylitis is an **enthesopathy,** a process that **produces inflammation with subsequent calcification** and ossification at and around the **entheses,** which are the insertion **sites of tendons, ligaments, and joint capsules.**

- Sacroiliitis is the hallmark of ankylosing spondylitis. It is usually **bilaterally symmetrical** and **eventually leads to bony fusion** or **ankylosis** of these joints, until they appear either as a thin white line (instead of a joint space) or they disappear altogether (Fig. 25-19).

- In the spine there is **ossification of the outer fibers of the annulus fibrosis,** producing **thin, bony bridges** from the corners of one vertebra to another called *syndesmophytes.* Progressive syndesmophyte production connecting adjacent vertebral bodies produces a **bamboo-spine** appearance (Fig. 25-20).

INFECTIOUS ARTHRITIS

- **Infectious arthritis** usually **occurs as a result of hematogenous seeding of the synovial membrane from an infected source elsewhere in the body,** such as a wound infection or **from direct, contiguous extension from osteomyelitis** adjacent to the joint.
- It is usually subdivided into **pyogenic (septic) arthritis,** due mostly to staphylococcal and gonococcal organisms, and **nonpyogenic arthritis,** caused mostly by infection with *Mycobacterium tuberculosis.*
- **Risk factors** include intravenous drug use, steroids either injected or orally administered, joint prostheses, and recent joint trauma, including joint surgery and diabetes.
- **In children** and **adults,** the **knee is frequently affected;** in **children,** the **hip is** another **common** site of infection. Hands may be infected from human bites, and the feet from diabetes.

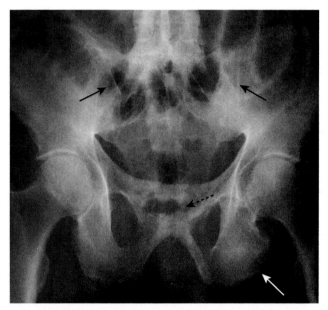

FIGURE 25-19 Ankylosing spondylitis. Ankylosing spondylitis is an **enthesopathy,** a process that produces inflammation with subsequent calcification and ossification at and around the **entheses,** which are the insertion sites of tendons, ligaments, and joint capsules *(solid white arrow points to bony overgrowth at the ischial tuberosity)*. Bilaterally symmetrical **sacroiliitis** is the hallmark of ankylosing spondylitis. This eventually leads to bony fusion or **ankylosis** of the sacroiliac joints until they disappear as joints altogether *(solid black arrows)*. The symphysis pubis is also ankylosed *(dotted black arrow)*.

- Although **conventional radiographs** are obtained as the initial study, they are **relatively insensitive** to the early findings of the disease except for soft tissue swelling and osteopenia.
- If septic arthritis is strongly suspected, aspiration of the joint will usually confirm the diagnosis.

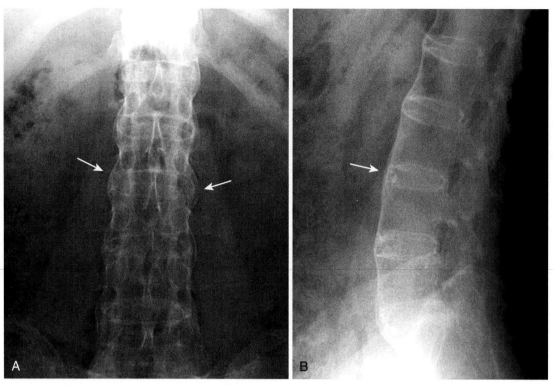

FIGURE 25-20 Ankylosing spondylitis, frontal (A) and lateral (B) spine. In the spine there is ossification of the outer fibers of the annulus fibrosus, producing thin, bony bridges joining the corners of adjacent vertebrae called *syndesmophytes (white arrows)*. Progressive ossification connecting adjacent vertebral bodies produces the **bamboo-spine** appearance seen in this case, which is characteristic of ankylosing spondylitis.

⊙ The **hallmark of infectious arthritis**, especially the pyogenic form, **is destruction of the articular cartilage** and **long, contiguous segments of the adjacent articular cortex** from proteolytic enzymes released by the inflamed synovium.

♦ This **destruction**, unlike other arthridities, **is usually quite rapid**.
♦ Infectious arthritis tends to be **monoarticular** and is associated with **soft tissue swelling** and **osteopenia** from the hyperemia of inflammation (Fig. 25-21).

■ Because of its sensitivity, MRI is now used extensively after conventional radiography in diagnosing septic joints. **Enhancement of the synovium** and the **presence of a joint effusion** have the best correlation with the clinical diagnosis of a septic joint.

■ **Nonpyogenic infectious arthritis** is most often caused by *Mycobacterium tuberculosis*, which **spreads via the bloodstream from lung**.
♦ Unlike pyogenic arthritis, **tuberculous arthritis** has an **indolent** and **protracted course** resulting in **gradual loss of the joint space** and **late destruction of the articular cortex**.
♦ Nonpyogenic infectious arthritis is **usually monarticular. Severe osteoporosis is a common finding. In children, the spine is most often affected; in adults, the knee.**

■ Healing with fibrous and bony ankylosis occurs in both pyogenic and nonpyogenic infectious arthritis.

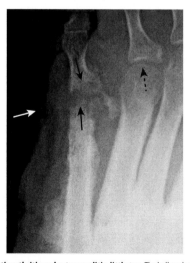

FIGURE 25-21 Septic arthritis and osteomyelitis, little toe. The hallmark of infectious arthritis, especially the pyogenic form, is destruction of the articular cartilage and long, contiguous segments of the adjacent articular cortex from proteolytic enzymes released by the inflamed synovium. This patient has septic arthritis (the metatarsal-phalangeal joint is destroyed), which has extended to the subjacent bone as osteomyelitis and also destroyed either side of the joint space *(solid black arrows)*. Note the normal white line representing the articular cortex of the head of the 4th metatarsal *(dotted black arrow)*. There is a large soft tissue ulcer overlying the lateral aspect of the left foot *(white arrow)*. This patient had diabetes.

TAKE-HOME POINTS

RECOGNIZING JOINT DISEASE: AN APPROACH TO ARTHRITIS

An arthritis is a disease of a joint that invariably leads to joint space narrowing and changes to the bones on both sides of the joint.

Arthritides can be roughly divided into **hypertrophic, infectious, and erosive** (inflammatory) categories.

Hypertrophic arthritis features subchondral sclerosis, marginal osteophyte production, and subchondral cyst formation.

Primary osteoarthritis, the most common form of arthritis, is a type of hypertrophic arthritis. It typically occurs on weight-bearing surfaces of the hip and knee and the distal interphalangeal joints of the fingers.

Other hypertrophic arthridities include calcium pyrophosphate deposition disease (CPPD), Charcot joints, and osteoarthritis either secondary to prior trauma or avascular necrosis or superimposed on another underlying arthritis.

CPPD occurs with the deposition of calcium pyrophosphate crystals (chondrocalcinosis). It can produce large and multiple subchondral cysts, narrowing of the patellofemoral joint space, metacarpal hooks, and proximal migration of the distal carpal row.

Charcot or neuropathic joints feature fragmentation, sclerosis, and soft tissue swelling. Diabetes is the most frequent cause of a Charcot joint today.

Erosive (or inflammatory) arthritis is associated with inflammation and synovial proliferation (pannus formation), which produces lytic lesions in or near the joint called *erosions*.

Rheumatoid arthritis, gout, and psoriasis are three examples of erosive arthritis; the site of involvement is helpful in differentiating among the causes of erosive arthritides.

Rheumatoid arthritis affects the carpals and proximal joints of the hand, can widen the predentate space in the cervical spine, and produces fusion of the posterior elements in the cervical spine.

Gout most often affects the metatarsal-phalangeal joint of the great toe with juxtaarticular erosions and little or no osteoporosis. Tophi are late manifestations of the disease and usually do not calcify.

Psoriatic arthritis usually occurs in patients with known skin changes and affects the distal joints primarily in the hands, producing characteristic erosions that resemble a pencil in a cup.

Ankylosing spondylitis is a chronic and progressive arthritis characterized by symmetric fusion of the sacroiliac joints and ascending involvement of the spine, eventually producing a bamboo-spine appearance.

Infectious arthritis features soft tissue swelling, osteopenia, and in the case of pyogenic arthritis, relatively early and marked destruction of most or all of the articular cortex. It is caused mostly by staphylococcal and gonococcal organisms.

 WEBLINK

Visit StudentConsult.Inkling.com for more information and a quiz on Manifestations of Arthritis.

For your convenience, the following QR Code may be used to link to **StudentConsult.com**. You must register this title using the PIN code on the inside front cover of the text to access online content.

CHAPTER 26

Recognizing Some Common Causes of Neck and Back Pain

CONVENTIONAL RADIOGRAPHY, MAGNETIC RESONANCE IMAGING, AND COMPUTED TOMOGRAPHY

- **Conventional radiographs** remain an important method of evaluating certain spinal lesions because they **are inexpensive** and **rapidly available,** even on a portable basis if needed, and they **demonstrate bony anatomy well.** They are utilized as a **screening method** for most spinal trauma.

- Nevertheless, they provide little, if any, information about the important soft tissues of the spine, and they use ionizing radiation to produce an image.

- ➡ **Magnetic resonance imaging (MRI),** with its superior soft tissue differentiation, **is the study of choice for most diseases of the spine** because of its **ability to visualize** and **detect abnormalities in soft tissues** such as bone marrow, the spinal cord, and the intervertebral disks, in addition to its **ability to display images in any plane** and the **lack of exposure to radiation**.

- But **MRI has its limitations.** The study **remains relatively expensive** and its **availability is not as widespread** as computed tomography or conventional radiographs. **Patients with pacemakers** and certain internal ferromagnetic materials (aneurysm clips) **are not able to be scanned,** the procedure **takes more time** to complete, and some patients cannot tolerate the **claustrophobia** they experience in some of the high-field–strength MRI scanners (see Chapter 22).

- **Computed tomography (CT),** with its ability to reformat images in different planes and provide additional information about the soft tissues, now supplements and, in some cases, **has replaced conventional radiography in the evaluation of spinal trauma.** CT is utilized to determine **the extent of the injury** in any patient in whom a traumatic spinal injury has already been demonstrated by conventional radiographs.

- CT is also utilized to **detect bony lesions not visible on conventional radiographs** and to **evaluate soft tissue abnormalities** in patients who are unable to undergo an MRI examination.

THE NORMAL SPINE

- There are normally seven cervical vertebrae, twelve thoracic vertebrae—all of which normally bear ribs—five lumbar vertebrae, and five fused sacral vertebrae.

Vertebral Body

- Each vertebra (well, almost every vertebra) has a **body** composed of inner cancellous bone and marrow and **posterior elements** made of compact, dense bone consisting of the pedicles, laminae, facets, transverse processes, and a spinous process (Fig. 26-1, *A*).

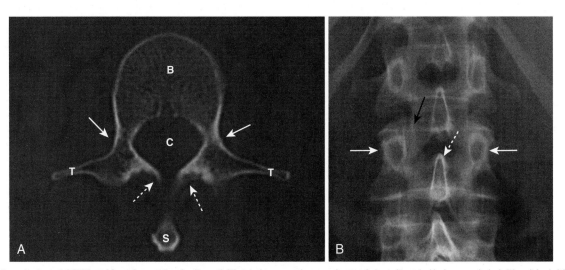

FIGURE 26-1   Normal spine, axial CT (A), and frontal conventional radiograph (B). A, In this computed tomography scan of a "typical" vertebral body, we see the body (B), pedicles *(solid white arrows)*, laminae *(dotted white arrows)*, transverse processes (T), spinal canal (C), and spinous process (S). **B,** Each vertebral body has two pedicles, which project as small ovals on either side of the vertebral body *(solid white arrows)*. The spinous process *(dotted white arrow)* may be visualized superimposed on the body to which it is attached. The facet joint is seen here *en face (solid black arrow)*.

- From the level of **C3 through the level of L5,** the vertebral bodies are more or less **rectangular in shape** and of about equal height posteriorly as anteriorly.
- The **end plates of contiguous vertebral bodies are roughly parallel to each other.**
- The articular facets of the **superior** and **inferior articular processes** are lined with cartilage, and these **facet joints** are **true synovial joints.**
- In the frontal projection, each vertebral body displays two ovoid **pedicles** visible on each side of the vertebral body. The pedicles of L5 are frequently difficult to visualize, even in normal individuals, because of the lordosis of the lumbar spine (see Fig. 26-1, *B*).

 In the cervical spine, three parallel arcuate lines should smoothly join: (1) all of the **spinolaminar white lines** (the junction between the lamina and the spinous process), (2) the **posterior aspects of the vertebral bodies,** and (3) all of the **anterior aspects of the vertebral bodies.** Alterations in the smooth parallel curvature of these three lines may indicate forward or backward displacement of all or part of vertebral body (Fig. 26-2).

- On conventional radiographs of the **lumbar spine** performed in the oblique projection, the anatomic structures normally superimpose to produce a shadow that resembles the front end of a Scottish terrier, the famous **Scottie dog** sign (Fig. 26-3).

Intervertebral Disks

- The intervertebral disks have a central gelatinous **nucleus pulposus** surrounded by an outer **annulus fibrosus,** which is in turn made up of inner fibrocartilaginous fibers and outer cartilaginous fibers (**Sharpey fibers**). The nucleus pulposus is located near the posterior aspect of the disk.

The **relative height of the disk space varies in each part of the spine.**

- In the **cervical spine,** the **disk spaces** are about **equal** to each other in height.
- In the **thoracic spine,** they are usually slightly **decreased** in size from the **cervical spine** but **equal** in height **to each other.**
- In the **lumbar spine,** the disk spaces **progressively increase in height** with each successive interspace, except for L5 to S1, which can be equal to or slightly less than the height of L4 to L5 on conventional radiographs.

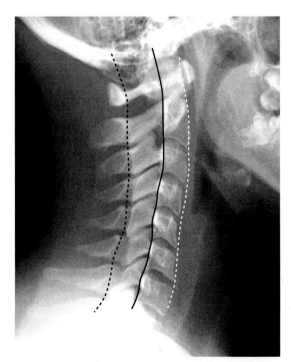

FIGURE 26-2 **The three cervical lines.** The lateral view of the cervical spine enables a quick assessment for spinal fracture/subluxation before further studies are performed, which might involve motion of the patient's neck. Three parallel arcuate lines should smoothly join all of the spinolaminar white lines, which are the junctions between the laminae and the spinous processes *(dotted black line)*; a second line should join all of the posterior aspects of the vertebral bodies *(solid black line)*; and a third should join all of the anterior aspects of the vertebral bodies *(dotted white line)*.

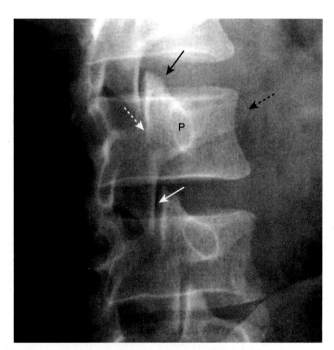

FIGURE 26-3 **Normal Scottie dog.** This is a left posterior oblique view of the lumbar spine (the patient is turned about halfway toward her own left). The "Scottie dog" is made up of the following: the "ear" *(solid black arrow)* is the superior articular facet, the "leg" *(solid white arrow)* is the inferior articular facet, the "nose" *(dotted black arrow)* is the transverse process, the "eye" (P) is the pedicle, and the "neck" *(dotted white arrow)* is the pars interarticularis. All of these structures are paired—an identical set should be visible on the patient's right side.

Spinal Cord and Spinal Nerves

- The spinal cord extends from the medulla oblongata to the level of L1 to L2, ending as the **conus medullaris.** The **cauda equina** extends inferiorly from that point as a collection of nerve roots, with each root exiting below its respectively numbered vertebral body.
- Each of the paired **neural foramina** of the spine contains a **spinal nerve, blood vessels,** and **fat.**
- **Spinal nerves** are named and numbered **according to the site they exit** from the spinal canal. **From C1 to C7** nerves exit **above** their respective vertebra. The **C8** nerve exits **between the seventh cervical** and **first thoracic** vertebrae. The **remaining nerves exit below** their respectively numbered vertebrae.

Spinal Ligaments

- There are several ligaments that traverse the spine (Table 26-1).

Normal MRI Appearance of the Spine

➡ On **T1-weighted** sagittal MRI images of the spine, the **vertebral bodies,** containing bone marrow, will normally be of **high signal intensity (bright),** the **disks** will be **lower in signal intensity,** and **cerebrospinal fluid (CSF)** in the thecal sac will have a **low signal intensity** (dark) (Fig. 26-4, *A*).

- On conventional **T2-weighted images,** the **vertebral body** will be slightly **lower in signal** intensity than the disks,

whereas the **CSF will appear bright** (the reverse of T1) (see Fig. 26-4, *B*).
- Cortical bone is dark (has a low signal) on all sequences.

BACK PAIN

- It has been estimated that **almost 80% of all Americans will have some episode of back pain** during their lifetime. The causes of back pain are numerous, and the anatomic and physiologic interrelationships that produce it in many cases are still not known.

TABLE 26-1 LIGAMENTS OF THE SPINE

Ligament	Connects
Anterior longitudinal ligament	Anterior surfaces of vertebral bodies
Posterior longitudinal ligament	Posterior surfaces of vertebral bodies
Ligamentum flavum	Laminae of adjacent vertebral bodies and lies in posterior portion of spinal canal
Interspinous ligament	Between spinous processes
Supraspinous ligament	Tips of spinous processes

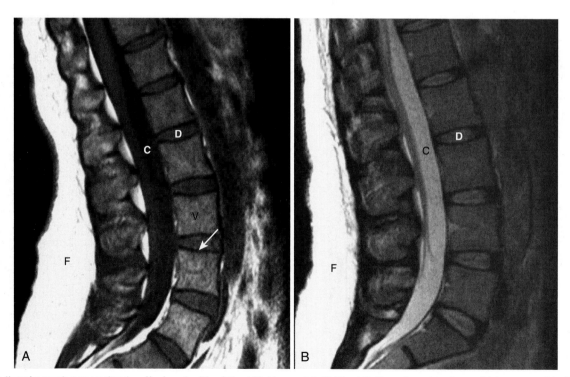

FIGURE 26-4 Normal magnetic resonance imaging of lumbar spine, T1 and T2 weighted images. A, Sagittal T1-weighted image demonstrates the normal dark appearance of the disks (D) relative to the vertebral body (V). The cerebrospinal fluid in the spinal canal is dark (C), and the subcutaneous fat of the back is bright (F). Cortical bone has a low signal *(white arrow).* **B,** Sagittal T2-weighted image demonstrates the normal appearance of the disks (D), which are slightly higher intensity (brighter) than the vertebral bodies. The cerebrospinal fluid (C) in the spinal canal is now bright, and the subcutaneous fat (F) of the back remains bright.

- Some causes of back pain:
  - ◆ **Muscle and ligament strain**
  - ◆ **Herniation of an intervertebral disk**
  - ◆ **Degeneration of an intervertebral disk**
  - ◆ **Arthritis involving the synovial joints of the spine**
  - ◆ **Compression fractures, usually from osteoporosis**
  - ◆ **Trauma to the spine**
  - ◆ **Malignancy involving the spine**
  - ◆ **Infection of the spine**
- We will discuss all except muscle and ligament strains.

Herniated Disks

- Only about 2% of patients with acute low back pain have a herniated disk. In the **lumbar region,** disk herniation may lead to **back pain** and **sciatica,** whereas herniation of a **cervical disk** may produce **radiculopathy** and **myelopathy. MRI is the study of choice for evaluating herniated disks.**
- In the cervical spine, disk herniations occur most frequently at **C4-C5, C5-C6, and C6-C7** (Fig. 26-5).
- Compared with the cervical and lumbar spine, **thoracic disks are very stable,** in part because they are protected by the rib cage that surrounds them.

> The majority of **disk herniations occur at** the lower three lumbar disk levels, **L3-L4, L4-L5 (most common), and L5 to S1. More than 60% of disk herniations occur posterolaterally,** the location of the herniation determining the clinical presentation depending on which nerve roots are compressed.

- **Degeneration** of the outer annular fibers of the disk or **trauma** can lead to an interruption in those fibers and allow

the disk material to bulge, which may or may not be associated with back pain.

- When the annular fibers rupture, the **nucleus pulposus may herniate** (usually posterolaterally) **through a weakened area of the posterior longitudinal ligament.**
- The herniated material may **protrude** but remain in contact with the parent disk from which it originated, or it may be completely **extruded** into the spinal canal.
- Symptoms are caused by acute compression of the nerve root.
- Disk herniations can be visualized on both **CT** and **MRI.** CT demonstrates disk material compressing nerve roots or the thecal sac. On MRI, the herniated disk material is usually a **focal, asymmetric protrusion of hypointense disk material** that extends beyond the confines of the annulus fibrosus (Fig. 26-6).
- **Postlaminectomy syndrome** (also called *failed back surgery syndrome*) is persistent pain in the back or legs following spine surgery. In some studies it has been estimated to occur in 40% of postoperative patients. Gadolinium-enhanced MRI studies of the spine are useful in differentiating persistent or recurrent disk herniation from scar formation as a cause of the pain.

Degenerative Disk Disease (DDD)

- With increasing age, the normally gelatinous nucleus pulposus becomes dehydrated and degenerates. This gradually leads to **progressive loss of the height of the intervertebral disk space.** At times, desiccation of the disk leads to **release of nitrogen** from tissues surrounding the disk resulting in the appearance of **air density in the disk space,**

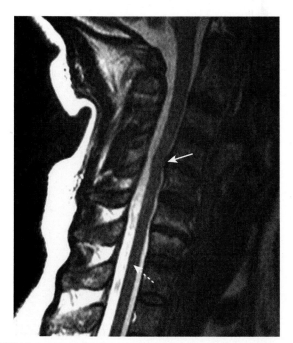

FIGURE 26-5 **Herniated disk, C4-C5, on magnetic resonance imaging.** The spinal cord is dark *(dotted white arrow)* relative to the high-intensity (whiter) signal surrounding it, which is the cerebrospinal fluid in the spinal canal. A herniated disk *(solid white arrow)* extends posteriorly from the C4-C5 disk space and compresses the cord.

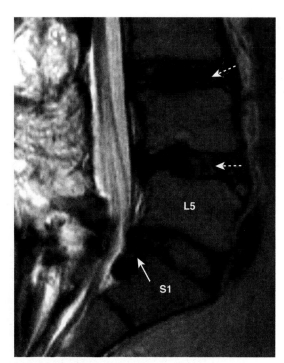

FIGURE 26-6 **Disk herniation, L5-S1.** Sagittal T2-weighted image of the lower lumbar spine demonstrates disk material *(solid white arrow)* beyond the confines of the L5-S1 intervertebral disk space, representing a disk herniation extending inferiorly. Notice that degeneration and desiccation of the other disks has lead them to become darker than normal on this T2-weighted image *(dotted white arrows).*

which is called a *vacuum disk phenomenon.* A vacuum disk represents a **late sign of a degenerated disk** (Fig. 26-7).

■ **Degenerative disk disease on MRI**
  ♦ The decrease in water content of the nucleus pulposus results in a lower signal intensity of the disk on T2-weighted images (see Fig. 26-6).

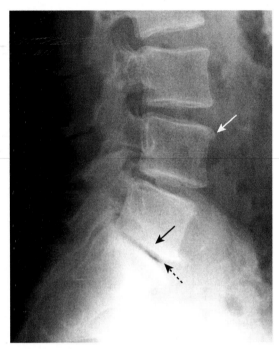

FIGURE 26-7 Degenerative disk disease. With increasing age, there is progressive loss of the height of the intervertebral disk space. The end plates of contiguous vertebral bodies become sclerotic *(solid black arrow),* small osteophytes are produced at the margins of the vertebral bodies *(solid white arrow),* and there is desiccation of the disk with a **vacuum disk phenomenon** (gas in the disk space) recognized by the air density in place of the disk seen at L5 to S1 *(dotted black arrow).*

■ **Degenerative disk disease on conventional radiographs**
  ♦ There is **disk space narrowing.** There are also changes in the vertebral bodies themselves.
  ♦ The end plates of contiguous vertebral bodies become **eburnated** or sclerotic. **Small osteophytes are produced** at the margins of the vertebral bodies at each disk space (see Fig. 26-7).
  ♦ At the same time, there is typically **degeneration of the outer annulus fibrosus.** This leads to the production of **larger marginal osteophytes** at the end plates than those seen with degeneration of the nuclear material.
■ It should be noted that osteophytes are an extremely common finding, that they increase in prevalence with increasing age, and that most patients with osteophytes of the spine are **asymptomatic.**

Osteoarthritis of the Facet Joints

■ The **facet joints** (also known as the *apophyseal joints*) are **true joints** in that they have cartilage, a synovial lining, and synovial fluid. As such, they are **subject to developing osteoarthritis,** similar to true joints in the appendicular skeleton.
■ Some consider the small jointlike structures at the lateral edges of C3 to T1, called the *uncovertebral joints* or the *joints of Luschka,* to be true joints, although others do not.
  ♦ In either case, they are frequent sites of osteophyte formation. **Osteophytes at the uncovertebral joints are frequently associated with both degenerative disk disease** and **osteophytes of the facet joints.**

➡ In the **cervical spine, osteophytes that develop at the uncovertebral joints** can produce **protrusions of bone into the** normally oval-shaped **neural foramina,** which are visualized on conventional radiographs taken in the oblique projection (Fig. 26-8, *A*).

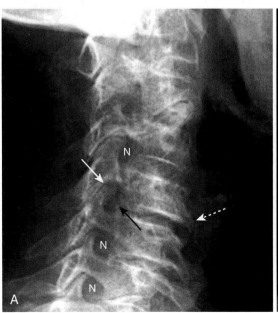
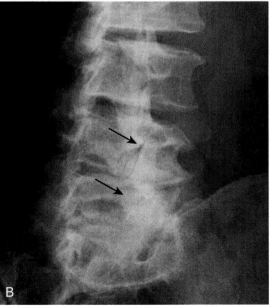

FIGURE 26-8 Uncovertebral osteophytes and facet arthritis. A, In the cervical spine, osteophytes may develop at the uncovertebral joints *(solid black arrow),* producing bony protrusions into the normally oval-shaped neural foramina (N) in this radiograph taken in the oblique projection. Osteophytes at the uncovertebral joints are frequently associated with both degenerative disk disease *(dotted white arrow)* and osteophytes of the facet joints *(solid white arrow).* **B,** There is sclerosis and osteophyte formation involving the lower facet joints of the lumbar spine *(solid black arrows).*

- There is usually a **complex interrelationship between degenerative disk disease** and **facet arthritis** such that the two frequently occur together. The osteophytes formed by osteoarthritis of the facet joints **may also encroach on the neural foramina** and **produce radicular pain.**
- In the **lumbar spine,** facet osteoarthritis may cause **narrowing** and **sclerosis of the facet joints,** best seen on oblique views of the spine. Facet arthritis is easier to visualize on CT scans of the spine than conventional radiographs, and actual nerve compression is easier to visualize on MRI of the spine (see Fig. 26-8, *B*).

Diffuse Idiopathic Skeletal Hyperostosis

- Diffuse idiopathic skeletal hyperostosis (DISH) is a common disorder characterized by bone or calcium formation at the sites of ligamentous insertions **(enthesopathy).**
- It usually affects **men** over the age of **50,** can occur **anywhere in the spine,** but **most often** affects the **lower thoracic** and/or **lower cervical spine.**
- Frequently, patients with DISH complain of back stiffness, but they may have **no pain** or only **mild back pain.**
- **Conventional radiographs of the spine are sufficient** to make the diagnosis of DISH.

➡ DISH is manifest by **thick, bridging**, or **flowing calcification/ossification** of the **anterior** or sometimes **posterior longitudinal ligaments** (Fig. 26-9).

- This ossification is visualized along the anterior or anterolateral aspects **of at least four contiguous vertebral bodies.** Unlike degenerative disk disease, **the disk spaces** and (usually) the **facet joints are preserved.** The flowing ossification seen in DISH is **separated slightly from the vertebral body** (see Fig. 26-9, *A*).
  - ◆ Unlike ankylosing spondylitis, which may radiographically resemble DISH in the spine, the **sacroiliac joints are normal.**
- **Ossification of the posterior longitudinal ligament (OPLL)** is often present with DISH and better visualized on CT and MRI than on conventional radiographs. It may cause compression of the spinal cord, especially in the cervical spine, as a result of narrowing of the spinal canal (see Fig. 26-9, *B*).

Compression Fractures of the Spine

- Vertebral compression fractures are **common,** affecting **women more than men,** and typically secondary to **osteoporosis.** They may be **asymptomatic,** or they may produce **pain** in the midthoracic or upper lumbar area that typically disappears in 4 to 6 weeks. Sometimes they are first noticed because of increasing **kyphosis,** or **loss of overall body height.**
- **Conventional spine radiographs are usually the study of first choice.** MRI can be utilized for differentiating osteoporotic compression fractures from malignancy. Both MRI and nuclear bone scans can help in establishing the age of a compression abnormality, which might be impossible on conventional radiographs alone.

➡ Osteoporotic compression fractures usually **involve the anterior** and **superior aspects** of the vertebral body **sparing the posterior body.** There will usually be a

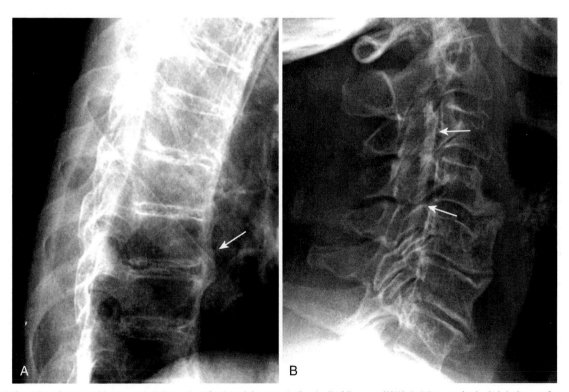

FIGURE 26-9 Diffuse idiopathic skeletal hyperostosis (DISH) and ossification of the posterior longitudinal ligament (OPLL). A, DISH is manifest by thick, bridging, or flowing calcification/ ossification of the anterior longitudinal ligaments *(white arrow).* This ossification is visualized along the anterior or anterolateral aspects of at least four contiguous vertebral bodies. **B,** OPPL *(white arrows)* can be associated with DISH and may contribute to the production of spinal stenosis.

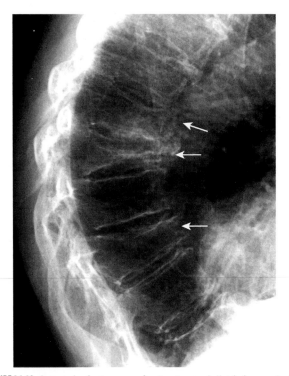

FIGURE 26-10 Compression fractures secondary to osteoporosis. Vertebral compression fractures are common, affecting women more than men, and typically secondary to osteoporosis. Osteoporotic compression fractures usually involve the anterior and superior aspects of the vertebral body sparing the posterior aspect *(white arrows)*. This produces a wedge-shaped deformity that leads to accentuation of the kyphosis in the thoracic spine. Progressive loss of overall body height is a common finding with compression fractures in older adults.

difference in the height between the **anterior** and **posterior** aspects of the same vertebral body **in excess of 3 mm.** Alternatively, the compressed body is typically >20% shorter than the body above or below it.

♦ This compression pattern produces a **wedge-shaped deformity** that leads to accentuation of the normal kyphosis in the thoracic spine (the so-called *dowager's hump*) and increases the lordosis in the lumbar spine (Fig. 26-10).

■ There is **usually no neurologic deficit** associated with an osteoporotic compression fracture because the fracture involves the anterior part of the vertebral body, away from the spinal cord.

Spinal Stenosis

■ Spinal stenosis refers to **narrowing of the spinal canal** or the neural foramina secondary to **soft tissue** or **bony abnormalities,** either on an **acquired** or **congenital** basis. **Acquired causes** such as those from **degenerative changes** are **more common** than congenital causes.

■ **Soft tissue abnormalities** that can lead to **spinal stenosis** include hypertrophy of the ligamentum flavum, bulging disk(s), and ossification of the posterior longitudinal ligament (OPLL).

■ **Bony abnormalities** that can lead to **spinal stenosis** include a congenitally narrow spinal canal, osteophytes, facet

osteoarthritis, or spondylolisthesis. A spinal canal that was borderline normal in size may become stenotic when any of these processes superimpose to further narrow the canal.

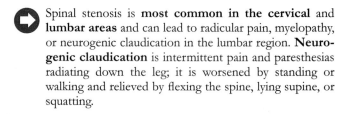 Spinal stenosis is **most common in the cervical** and **lumbar areas** and can lead to radicular pain, myelopathy, or neurogenic claudication in the lumbar region. **Neurogenic claudication** is intermittent pain and paresthesias radiating down the leg; it is worsened by standing or walking and relieved by flexing the spine, lying supine, or squatting.

■ **Conventional radiographs are usually obtained first** in evaluating for spinal stenosis, but **MRI is the study of choice** because the bony dimensions alone do not account for the soft tissues that can produce stenosis.

■ **Conventional radiographic** findings may include an **anteroposterior (AP) diameter of the spinal canal of 10 mm or less, facet joint arthritis,** and **spondylolisthesis** (Fig. 26-11).

■ **CT** provides an excellent method of demonstrating bony abnormalities, but **MRI is the imaging modality of choice for detecting lumbar spinal stenosis.**

MALIGNANCY INVOLVING THE SPINE

■ Metastases to bone are 25 times more common than primary bone tumors. Metastases usually occur where **red marrow** is found, with 80% of metastatic bone lesions occurring in the **axial skeleton** (i.e., spine, pelvis, skull, and ribs) since this is where, with normal aging, the red marrow exists.

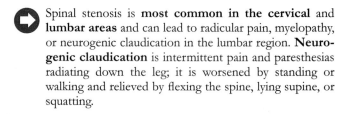 Because of the **rich blood supply** in the posterior portion of vertebral bodies, hematogenous **metastatic deposits to that part of the spine are common,** especially from **lung** and **breast** carcinoma.

■ In the spine, **metastases may lead to compression fractures. Metastases** tend to **destroy the vertebral body,** including the posterior aspect and the pedicles, which is different from **osteoporotic compression fractures** in which **the posterior vertebral body** and **pedicles remain intact** (see Fig. 23-19).

■ As discussed in Chapter 23, **metastatic disease** can either be mostly bone-producing, that is, *osteoblastic,* or mostly bone-destroying, that is, *osteolytic.* Metastases that contain both osteolytic and osteoblastic processes occurring simultaneously are called *mixed metastatic lesions.*

■ **Prostate cancer** is the prototype example of a primary malignancy that produces **osteoblastic** metastases; the prototype of such a primary malignancy in a female is **breast cancer.** The most common causes of primary malignancies that produce **osteolytic** metastatic lesions are **lung** and **breast cancer. Thyroid** and **renal carcinomas** may produce **osteolytic** lesions that are also **expansile** (see Fig. 23-18).

■ The spine is also a frequent site of **multiple myeloma,** the most common primary malignancy of bone. Multiple myeloma is known for its tendency to produce almost **totally lytic lesions.** One of the hallmarks of multiple myeloma is osteoporosis, so that myeloma may be associated with **diffuse spinal osteoporosis** and **multiple compression fractures** (Fig. 26-12).

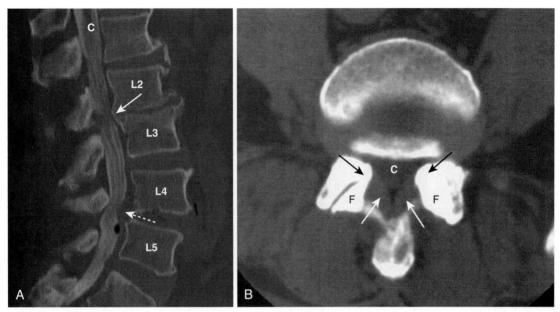

FIGURE 26-11 Spinal stenosis, computed tomography. A, At the level of L2-L3, there are osteophytes *(solid white arrow)* that narrow the canal (C). There is a protruding disk *(dotted white arrow)* at L4-L5 that reduces the size of the spinal canal at this level. **B,** The canal (C) is narrowed by thickening of the ligamenta flava *(white arrows)* and by bony overgrowth *(black arrows)* from osteoarthritis of the facet joints (F). Spinal stenosis refers to narrowing of the spinal canal or the neural foramina secondary to soft tissue or bony abnormalities.

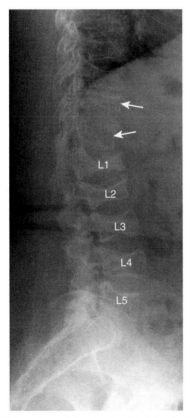

FIGURE 26-12 Multiple myeloma of the spine. One of the hallmarks of multiple myeloma is osteoporosis, so that myeloma may be associated with diffuse spinal osteoporosis *(white arrows)* and multiple compression fractures (L1 to L5). Notice how the posterior aspects of the vertebral bodies remain essentially normal in height in osteoporotic fractures and the central and anterior aspects collapse.

The current **screening study of choice** for the detection of spinal metastases is a **technetium-99m (^{99m}Tc) bone scintiscan** (radionuclide bone scan). The technetium is most commonly bound to **methylene diphosphonate (MDP)** that transports the ^{99m}Tc to bone. A radionuclide bone scan is relatively inexpensive, widely available, and screens the entire body. Although bone scans are **highly sensitive** to the presence of metastatic deposits, they are **not very specific.** In many cases a confirmatory study, usually a conventional radiograph, is needed to exclude other causes of abnormal radiotracer uptake such as fractures, infection, and arthritis (Fig. 26-13).

MRI IN METASTATIC SPINE DISEASE

■ MRI can detect changes of spinal metastases even **earlier than a radionuclide scan,** and techniques are being described that may allow for a rapid, whole-body screening similar to radionuclide bone scanning.

With neoplastic infiltration of the bone marrow, there is a **decrease** in the normally high signal of the vertebra on **T1-weighted images,** and there is usually a **high signal on T2-weighted** images (Fig. 26-14).

■ Unlike osteoporotic compression fractures, which tend to spare the posterior aspects of the bodies while the central and anterior portions collapse, **spinal metastases** will more often **affect the entire body,** including the posterior portion.

SPINAL TRAUMA

■ Fractures of the spine are infrequent compared with fractures of other parts of the skeleton. When they occur, they

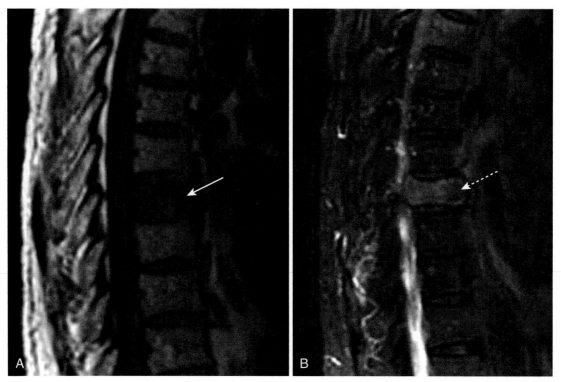

FIGURE 26-13 Metastases to the spine, magnetic resonance imaging (MRI). A, T1-weighted sagittal MRI of the thoracic spine demonstrates marrow replacement in the T8 vertebral body *(solid white arrow)*. The signal is decreased compared with the normal bodies above and below it. **B,** T2-weighted sagittal MRI shows abnormally high signal in the compressed vertebral body *(dotted white arrow)*. The entire vertebral body is involved. The patient had primary breast carcinoma.

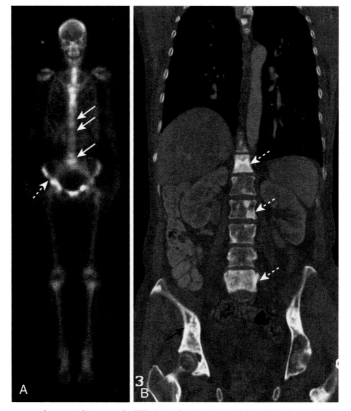

FIGURE 26-14 Metastases, radionuclide bone scan and computed tomography (CT). A, A technetium-99m methylene diphosphonate (MDP) bone scan demonstrates abnormal tracer uptake in numerous areas, including the spine *(solid white arrows)* and pelvis *(dashed white arrow)*. Because of the nonspecific nature of positive bone scans, a confirmatory study, usually conventional radiography, is also obtained. In this case the patient had undergone a CT scan, and the coronal reformatted image **(B)** shows numerous osteoblastic lesions in the spine *(dotted white arrows)* and the pelvis *(solid black arrow)* corresponding to the lesions seen on bone scan. This patient had known breast carcinoma.

have particular importance because of the implications for associated spinal cord injury. The **most commonly fractured vertebrae are L1, L2, and T12,** accounting for more than half of all thoracolumbar spine fractures. In the thoracic spine, compression fractures are the most common type of spinal fracture.

- **The three cervical lines** (see Fig. 26-2).
    - ◆ Many trauma protocols include a **cross-table lateral view of the cervical spine** (the patient's neck is immobilized, the patient remains on the stretcher or examining table, and the x-ray beam is directed **horizontally** so that the patient's **head is not moved**).
    - ◆ This view enables a **quick assessment for spinal fracture/subluxation** before further studies are performed, which might involve motion of the patient's neck.

➡️ At the beginning of this chapter, we talked about the three cervical lines: three parallel arcuate lines. **Deviation from the normal parallel curvature of these lines should suggest a subluxed and/or fractured vertebral body.**

- **Some of the more common spinal fractures:**
    - ◆ **Compression fractures,** the most common fractures of the spine, were previously discussed
    - ◆ **Jefferson fracture**
    - ◆ **Hangman's fracture**
    - ◆ **Burst fracture**
    - ◆ **Chance fracture**

Jefferson Fracture

- A Jefferson fracture is a **fracture of C1** usually involving both the anterior and posterior arches. In its classical presentation, there are **bilateral fractures of both the anterior** and **posterior arches of C1,** producing four fractures in all.
- It is **caused by an axial loading injury** (such as diving into a swimming pool and hitting one's head on the bottom).

➡️ On conventional radiographs, the hallmark of a Jefferson fracture is **bilateral, lateral offset of the lateral masses of C1 relative to C2 as seen on the open mouth view (atlantoaxial view)** of the cervical spine. The fracture is confirmed utilizing **CT** (Fig. 26-15).

- A Jefferson fracture is a "self-decompressing" fracture in that the spinal canal at the level of the fracture is wide enough to accommodate any swelling of the cord. There is usually **no neurologic deficit** associated with this type of fracture.

Hangman's Fracture

- A hangman's fracture is a **fracture of the posterior elements of C2.** Hangman's fractures **result from a hyperextension-compression injury** typically occurring in an unrestrained occupant in a motor vehicle accident who strikes his forehead on the windshield.
- Hangman's fractures are **best evaluated on the lateral view** of the cervical spine on conventional radiography and the **sagittal view** on CT.

➡️ The fracture effectively **separates the posterior aspect of the C2 vertebral body** from the **anterior aspect of C2,** allowing the **anterior aspect of C2 to sublux forward on the body of C3** (Fig. 26-16).

- Some hangman's fractures are less displaced, and CT may be needed for their detection.
- Because hangman's fractures lead to overall **widening of the bony canal,** they are usually **not associated with neurological deficits.** This is in contradistinction to the injury after which this fracture is named, the fracture incurred during a **judicial hanging,** in which there is hyperextension leading to a fracture of C2 and **marked distraction** of C2 from C3 with distraction of the spinal cord itself.

Burst Fractures

- Burst fractures can occur at any level but are most common in the **cervical spine, thoracic spine** and **upper lumbar spine.**
- They are high-energy **axial loading injuries,** typically secondary to motor vehicle accidents or falls in which the disk above is driven into the vertebral body below, and the vertebral body bursts. This in turn drives bony fragments **posteriorly into the spinal canal (retropulsed fragments),** and the anterior aspect of the vertebral body is displaced forward.
- Because these fractures involve incursion on the spinal canal, the **majority of burst fractures are associated with a neurologic deficit.**

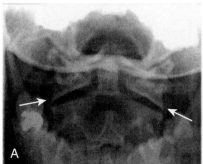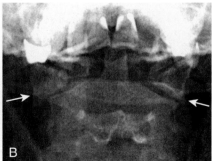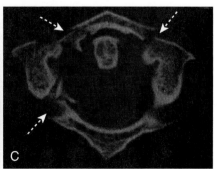

FIGURE 26-15 Normal open-mouth view, Jefferson fracture, open-mouth view, and computed tomography (CT). A Jefferson fracture is a fracture of C1 usually involving both the anterior and posterior arches. **A,** A normal "open mouth" view of C1 and C2 demonstrates that the lateral margins of C1 *(solid white arrows)* line up with the lateral margins of C2. **B,** The hallmark of a Jefferson fracture is bilateral, lateral offset of the lateral masses of C1 *(solid white arrows)* relative to C2. **C,** The fracture is confirmed utilizing CT, which shows fractures of both the right and left anterior arch of C1 and the right posterior arch *(dotted white arrows).*

➡ Findings of burst fractures include a **comminuted compression fracture of the vertebral body** in which the **posterior aspect of the body is bowed backward toward the spinal canal. CT** is the **best imaging modality** for identifying **bony fragments** in the spinal canal (Fig. 26-17).

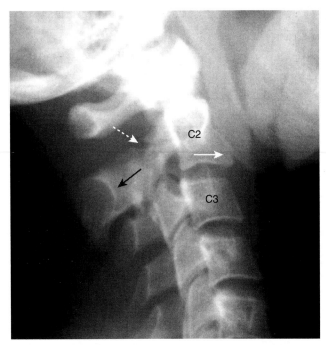

FIGURE 26-16 Hangman's fracture. A hangman's fracture results from a hyperextension-compression injury. It involves fractures through the posterior elements of C2, best evaluated on the lateral view. The fracture (*dotted white arrow* shows fracture line) effectively separates the posterior aspect of the C2 vertebral body from the anterior aspect of C2, allowing the anterior aspect to sublux forward on the body of C3. Notice that the spinolaminar line of C2 lies posterior to the other vertebral bodies (*solid black arrow*), and the anterior aspect of C2 lies forward to the body of C3 (*solid white arrow*).

Chance Fracture

♦ Chance fractures (named after George Q. Chance, the British radiologist who first described them) are **transverse fractures** through the **entire vertebral body, pedicles, and spinous process** frequently occurring as a result of a fall or motor vehicle accidents in which the occupant is using only a lap seatbelt. They typically occur in the **upper lumbar-lower thoracic spine** and are associated with a relatively high incidence of intraabdominal visceral injuries, especially to the duodenum, pancreas, and mesentery.

■ Radiographic findings include a horizontal fracture that shears through the vertebral body, pedicles, and spinous process. The vertebral body component may not be as visible as the fractures through the posterior elements (Video 26-1).

Locked Facets

■ Bilateral locking of the facets in the cervical spine can occur as a result of a **hyperflexion injury** in which the **inferior facets of one vertebral body slide over** and **in front of the superior facets of the body below.** In this position, the slipped facets cannot return to their normal position without medical intervention; thus, the term *locked.*

■ Locked facets occur with **forward slippage** of the affected vertebral body on the body below it **by at least 50% of its AP diameter.**

➡ On **lateral (sagittal) imaging** of the cervical spine, the **inferior articular facets will lie in front of the superior facets** of the body below. This is the reverse of the normal anatomic relationship between adjacent facets (Fig. 26-18).

♦ Since the superior articular facets are no longer "covered" by the inferior facets above them, this appearance is described on CT as the *naked facet sign.*

■ This injury virtually **always results in neurologic impairment.**

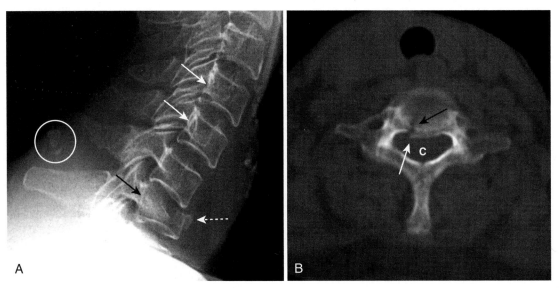

FIGURE 26-17 Burst fracture, radiograph and computed tomography (CT). A, Findings include a comminuted, compression fracture of the vertebral body in which the anterior vertebral body is displaced forward (*dotted white arrow*) and the posterior aspect of the body is propelled dorsally toward the spinal canal (*black arrow*). Notice how the posterior aspects of the normal vertebral bodies are concave inward (*solid white arrow*). There is calcification on the ligamentum nuchae, of no clinical consequence (*white circle*). **B,** Axial CT scan of the spine shows a fracture of the body (*black arrow*) and the retropulsed fragment (*white arrow*) protruding into the spinal canal (C).

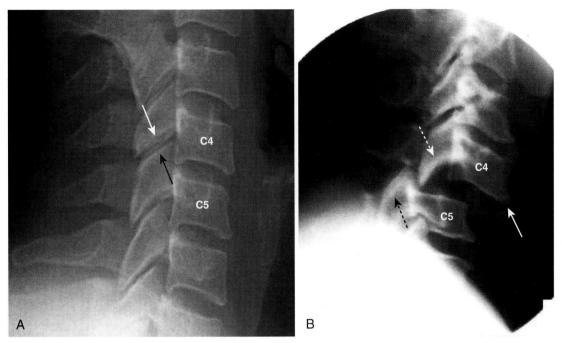

FIGURE 26-18 **Normal and bilateral locked facets. A,** Normally, the inferior articular facet of the body above (in this case, C4—*white arrow*) lies posterior to the superior articular facet of the body below (in this case, C5—*black arrow*). **B,** The inferior articulating facet of C4 *(dotted white arrow)* lies **anterior** to the superior articular facet of C5 *(dotted black arrow)*, the reverse of normal. Locking of the facets in the cervical spine can occur as a result of a hyperflexion injury, which results in forward slippage of the affected vertebral body on the body below it by at least 50% of its anterior-posterior diameter *(solid white arrow)*.

TAKE-HOME POINTS

RECOGNIZING SOME COMMON CAUSES OF NECK AND BACK PAIN

Conventional radiographs, computed tomography (CT), and magnetic resonance imaging (MRI) are all used to evaluate the spine, but MRI is the study of choice for most diseases of the spine because of its superior ability to display soft tissues.

Normal features of the vertebral bodies, intervertebral disks, the spinal cord, spinal nerves, and spinal ligaments are described.

Some of the more common causes of back pain are muscle and ligament strain, herniation of an intervertebral disk, degeneration of an intervertebral disk, arthritis involving the synovial joints of the spine, compression fractures from osteoporosis, trauma to the spine, and malignancy involving the spine.

Most **herniated disks** occur posterolaterally in the lower cervical or lower lumbar spine and are best evaluated using MRI.

Postlaminectomy syndrome is persistent pain in the back or legs following spine surgery. Gadolinium-enhanced MRI can be very helpful in detecting the cause.

With increasing age, the nucleus pulposus becomes dehydrated and degenerates, leading to changes of **degenerative disk disease** such as progressive loss of the height of the disk space, marginal osteophyte production, sclerosis of the end plates of the vertebral bodies, and occasionally, the appearance of a vacuum disk.

The facet joints are true joints and so are subject to changes of osteoarthritis; **facet osteoarthritis** is frequently associated with degenerative disk disease and can lead to radicular pain.

Diffuse idiopathic skeletal hyperostosis is manifest by thick, bridging, or flowing calcification/ossification of the anterior longitudinal ligaments usually occurring in men over the age of 50. The disk spaces and the facet joints are usually preserved.

Compression fractures of the spine are most often secondary to osteoporosis and seen more commonly in women. They can be seen on conventional radiographs and, because they usually disproportionately involve the anterior portion of the body, produce an exaggerated kyphosis in the thoracic spine.

Spinal stenosis is a narrowing of the spinal canal or the neural foramina secondary to soft tissue or bony abnormalities, either on an acquired (more common) or congenital basis; it is most common in the cervical and lumbar regions.

Metastatic lesions to the spine are common, especially to the blood-rich posterior aspect of the vertebral body, including the pedicles; lung (mixed), breast (mixed), and prostate (osteoblastic) metastases are the most common.

Multiple myeloma also frequently involves the spine either with severe osteoporosis, which can produce compression fractures, or lytic destruction of the vertebral body.

Three parallel arcuate lines should smoothly connect important landmarks on the lateral cervical spine view and allow for a quick assessment for fracture or subluxation before the patient is moved for further studies.

The findings in a Jefferson fracture, hangman's fracture, burst fracture, Chance fracture, and locked facets are discussed; the first two are self-decompressing injuries that are usually not associated with neurologic deficit.

WEBLINK

Visit StudentConsult.Inkling.com for videos and a quiz on the Causes of Back Pain.

For your convenience, the following QR Code may be used to link to **StudentConsult.com**. You must register this title using the PIN code on the inside front cover of the text to access online content.

CHAPTER 27
Recognizing Some Common Causes of Intracranial Pathology

- Advances in neuroimaging have had a remarkable impact on the diagnosis and treatment of neurologic diseases ranging from earlier detection and treatment of stroke to a more timely diagnosis of dementia and from the rapid detection and treatment of cerebral aneurysms to the ability to diagnose multiple sclerosis after a single attack.
- Both computed tomography (CT) and magnetic resonance imaging (MRI) are utilized for studying the brain and spinal cord, but MRI is the study of first choice for most clinical scenarios (Table 27-1). Conventional radiography has no significant role in imaging intracranial abnormalities.

NORMAL ANATOMY (Fig. 27-1)

- The normal anatomy of the brain is somewhat easier to understand with CT than MRI.
- In the posterior fossa the **4th ventricle** appears as an inverted U-shaped structure. Similar to all cerebrospinal fluid (CSF)–containing structures, it normally appears black on CT. Posterior to the 4th ventricle are the **cerebellar hemispheres,** and anteriorly lie the **pons** and **medulla oblongata**. The **tentorium cerebelli** separates the **infratentorial** components of the posterior fossa (medulla, pons, cerebellum, and 4th ventricle) from the **supratentorial** compartment.

- The **interpeduncular cistern** lies in the midbrain and separates the paired **cerebral peduncles,** which emerge from the superior surface of the pons. The **suprasellar cistern** is anterior to the interpeduncular cistern and usually has a five- or six-point starlike appearance.
- The **sylvian fissures** are bilaterally symmetrical and contain CSF. They separate the temporal from the frontal and parietal lobes.
- The **lentiform nucleus** is composed of the **putamen** (laterally) and **globus pallidus** (medially). The **third ventricle** is slitlike and midline. At the posterior aspect of the third ventricle is the **pineal gland**. Farther posterior is the **quadrigeminal plate cistern.**
- The **corpus callosum** connects the right and left cerebral hemispheres and forms the roof of the lateral ventricle. The anterior end is called the *genu,* and the posterior end is called the *splenium.*
- The **basal ganglia** are represented by the **subthalamic nucleus, substantia nigra, globus pallidus, putamen,** and **caudate nucleus**. The putamen and caudate nucleus are called the *striatum.*
- The **frontal horns** of the **lateral ventricles** hug the head of the **caudate nucleus**. The two frontal horns are separated by the midline **septum pellucidum**. The **temporal horns,**

TABLE 27-1   IMAGING STUDIES OF THE BRAIN FOR SELECTED ABNORMALITIES

Abnormality	Study of First Choice	Other Studies
Acute stroke	Diffusion-weighted imaging for acute or small strokes, if available	Noncontrast computed tomography (CT) can differentiate hemorrhagic from ischemic infarct
Headache, acute and severe	Noncontrast CT to detect subarachnoid hemorrhage	MR-angiography (MRA), or CT-angiography (CTA) to detect aneurysm if subarachnoid hemorrhage is found
Headache, chronic	Magnetic resonance imaging (MRI) without and with contrast	CT without and with contrast can be substituted
Seizures	MRI without and with contrast	CT without and with contrast can be substituted if MRI not available
Blood	Noncontrast CT	Ultrasound for infants
Head trauma	Nonenhanced CT is readily available and the study of first choice in head trauma	MRI is better at detecting diffuse axonal injury, but requires more time and is not always available
Extracranial carotid disease	Doppler ultrasonography	MRA excellent study; CTA best for preoperative stenosis evaluation
Hydrocephalus	MRI as initial study	CT for follow-up
Vertigo and dizziness	Contrast-enhanced MRI	MRA if needed
Masses	Contrast-enhanced MRI	Contrast-enhanced CT if MRI not available
Change in mental status	MRI without or with contrast	CT without contrast is equivalent

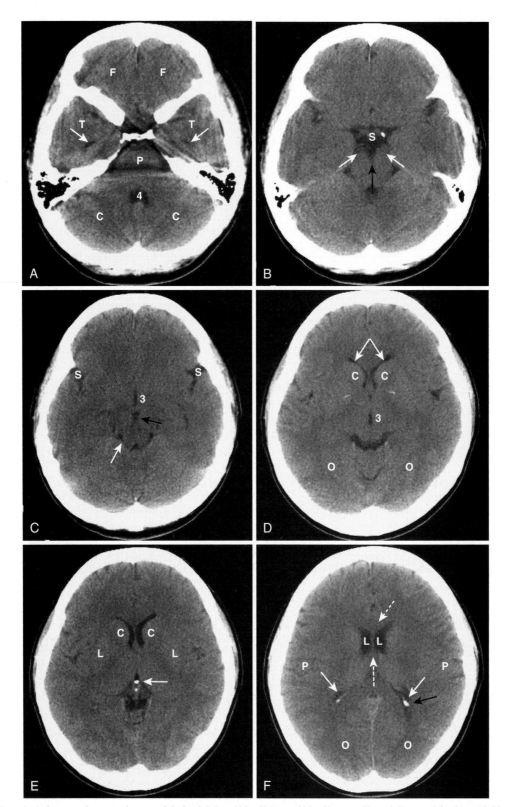

FIGURE 27-1 **Normal unenhanced computed tomography scans of the head. A,** Frontal lobes (F); temporal lobes (T); temporal horns *(white arrows);* fourth ventricle (4); cerebellum (C); pons (P). **B,** Suprasellar cistern (S); cerebral peduncles *(white arrows);* interpeduncular cistern *(black arrow).* **C,** Sylvian fissures (S); third ventricle (3); interpeduncular cistern *(solid black arrow);* quadrigeminal plate cistern *(white arrow).* **D,** Frontal horns of the lateral ventricles *(white arrow);* caudate nuclei (c); third ventricle (3); occipital lobes (0). **E,** Caudate nuclei (C); lentiform nuclei (L); calcified pineal gland *(white arrow).* **F,** Genu of corpus callosum *(dotted white arrow);* lateral ventricles (L); septum pellucidum *(dashed white arrow);* parietal lobes (P); occipital horn *(black arrow);* calcified choroid plexus *(solid white arrows);* occipital lobes (0).

which are normally very small, are more inferior and contained in the **temporal lobes.** The posterior horns of the lateral ventricle **(occipital horns)** lie in the occipital lobes. The most superior portion of the ventricular system are the **bodies** of the lateral ventricles.

■ The **falx cerebri** lies in the **interhemispheric fissure,** which separates the two **cerebral hemispheres** and is frequently calcified in adults.

■ The surface or **cortex** of the brain is made up of **gray matter** convolutions composed of **sulci** (grooves) and **gyri** (elevations). The medullary **white matter** lies below the cortex.

➡ On an unenhanced CT scan of the brain, anything that appears "white" will generally either be **bone (calcium)** density or **blood,** in the absence of a metallic foreign body (Table 27-2).

■ **Calcifications that may be seen on CT of the brain** and are nonpathologic (Fig. 27-2):
   ♦ Pineal gland (see Fig. 27-2, *A*)
   ♦ Basal ganglia (see Fig. 27-2, *A*)

   ♦ Choroid plexus (see Fig. 27-1, *F*)
   ♦ Falx and tentorium (see Fig. 27-2, *B*)
■ **Normal structures that can enhance** after administration of iodinated intravenous contrast:
   ♦ **Venous sinuses**
   ♦ **Choroid plexus**
   ♦ **Pituitary gland and stalk**
■ **Metallic densities** in the head can cause artifacts on CT scans. Dental fillings, aneurysm clips, and bullets can all cause **streak artifacts.**

MRI AND THE BRAIN

➡ In general, **MRI is the study of choice** for detecting and staging intracranial and spinal cord abnormalities. It is usually more sensitive than CT because of its **superior contrast** and **soft tissue resolution.** It is, however, **less sensitive** to CT in detecting **calcification** in lesions and **cortical bone,** which appear as signal voids with MRI. It cannot be used in most patients with pacemakers.

TABLE 27-2   COMPUTED TOMOGRAPHY DENSITIES

Hypodense (Dark) (Also Known as Hypointense)	Isodense	Hyperdense (Bright) (Also Known as Hyperintense)
Fat (not usually present in the head)	Normal brain	Metal (e.g., aneurysm clips or bullets)
Air (e.g., sinuses)	Some forms of protein (e.g., subacute subdural hematomas)	Iodine (after contrast administration)
Water (e.g., cerebrospinal fluid)		Calcium
Chronic subdural hematomas/hygromas		Hemorrhage (high protein)

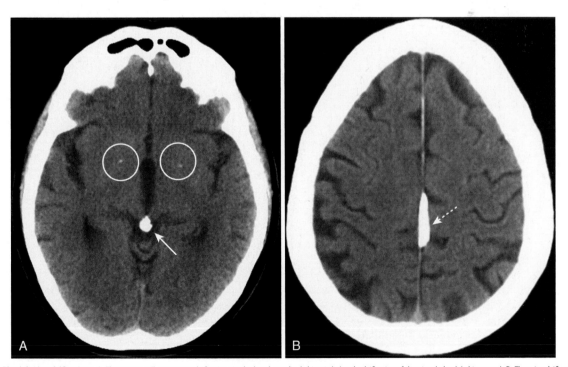

FIGURE 27-2 Physiologic calcifications. A, There are small, punctate calcifications in the basal ganglia *(white circles)* and calcification of the pineal gland *(white arrow).* **B,** There is calcification of the falx cerebri *(dotted white arrow).* Anything that appears white on a noncontrast enhanced computed tomography scan of the brain is either calcium or blood. Physiologic calcifications tend to increase in incidence with advancing age.

- MRI is more difficult to interpret than is CT. The same structure or abnormality will appear differently depending on the pulse sequence and the scan parameters; also, MRI is more variable in its depiction of differences that occur over the time course of some abnormalities (e.g., hemorrhage).
- **Initial evaluation of an MRI of the brain** might start with a T1-weighted sagittal sequence of the brain. On this sequence the brain resembles the anatomic specimens or diagrams you are accustomed to looking at (Fig. 27-3). We are fortunate that many structures in the brain are paired, so do not forget to compare one side with the other on axial scans of the brain (Fig. 27-4).
- Table 27-3 summarizes the signal characteristics of various tissues seen on MRI.

HEAD TRAUMA

- **Traumatic brain injuri**es extract a huge cost to the patient and society, not only as a result of the acute injury, but also for the long-term disability they produce. In the United States, motor vehicle accidents account for nearly half of traumatic brain injuries.
- **Unenhanced CT is the study of choice in acute head trauma.** The primary goal in obtaining the scan is to determine if there is a life-threatening, but treatable, lesion.

➡ **Initial CT evaluation of the brain** in the emergency setting focuses on whether there is (a) **mass effect** and (b) **blood.**

♦ To determine whether there is **mass effect,** look for a **displacement** or **compression** of key structures from

their normal positions by analyzing the location and appearance of the **ventricles, basal cisterns,** and the **sulci.**
- **Blood** will usually be hyperattenuating (bright) and might collect in the **basal cisterns, sylvian** and

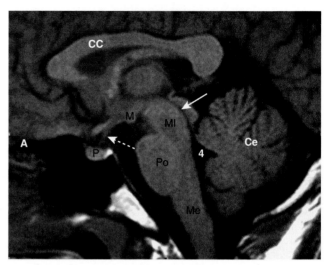

FIGURE 27-3 Normal midline magnetic resonance image. Sagittal, T1-weighted, close-up image demonstrates the midline structures of the brain. For orientation purposes, anterior (A) is to your left. The corpus callosum (CC) is located superiorly. The pituitary gland (P) sits in the sella turcica and connects to the hypothalamus via the pituitary stalk or infundibulum *(dotted white arrow)*. The mammillary bodies (M) are located anterior to the brainstem. The cerebral aqueduct *(solid white arrow)* is superior to the midbrain. The brainstem is composed of the midbrain (Mi), the pons (Po), and the medulla (Me). The 4th ventricle (4) communicates with the cerebral aqueduct and lies between the cerebellum (Ce) and the brainstem.

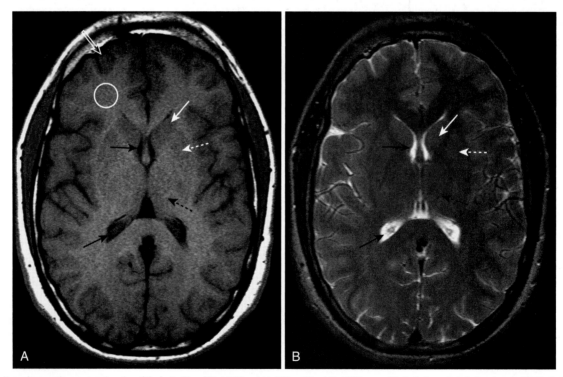

FIGURE 27-4 Normal magnetic resonance imaging of the brain, T1 and T2. Axial T1-weighted **(A)** and T2-weighted **(B)** images of the brain demonstrate that cerebrospinal fluid in the lateral ventricles is dark on T1 and bright on T2 *(solid black arrows)*. Gray matter, which contains the neuronal cell bodies, is actually gray on T1-weighted images *(open white arrow)*, and white matter, which contains myelinated axon tracts, is whiter *(white circle)*. Together, the caudate nucleus *(solid white arrows)* and lentiform nucleus *(dotted white arrows)* form the basal ganglia. The thalamus *(dotted black arrows)* is located posterior to the basal ganglia.

TABLE 27-3 SIGNAL CHARACTERISTICS OF VARIOUS TISSUES SEEN ON T1-WEIGHTED AND T2-WEIGHTED MAGNETIC RESONANCE IMAGING SCANS

Bright on T1	Dark on T1	Bright on T2	Dark on T2
Fat	Calcification	Water (edema, cerebrospinal fluid)	Fat
Gadolinium	Air		Calcification
High protein	Chronic hemorrhage	Hyperacute hemorrhage	Air
Subacute hemorrhage	Acute hemorrhage	Late subacute hemorrhage	Early subacute hemorrhage
Melanin	Water (edema, cerebrospinal fluid)		Chronic hemorrhage Acute hemorrhage High protein

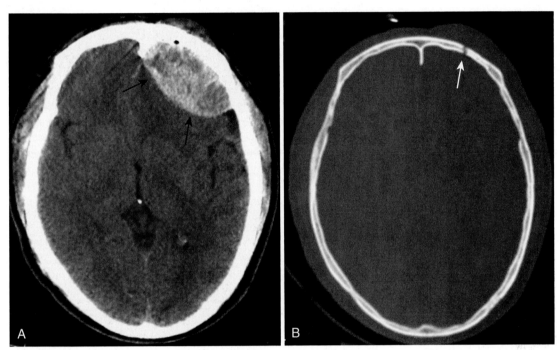

FIGURE 27-5 Computed tomography (CT) head, "brain," and "bone" windows. In order to visualize skull fractures, you must view the CT scan using the "bone" windows. **A,** Using the "brain" window, a lenticular-shaped, hyperattenuating lesion is seen in the left frontal region, displaying the typical appearance of an epidural hematoma *(black arrows)*. **B,** Viewing the same scan at the "bone" window setting shows a fracture *(white arrow)* of the left frontal bone at the site of the epidural hematoma.

interhemispheric fissures, ventricles, the subdural or **epidural spaces,** or in the brain parenchyma (intracerebral).

Skull Fractures

- Skull fractures are usually produced by **direct impact** to the skull and they most often occur at the point of impact. They are important primarily because their presence implies a **force substantial enough to cause intracranial injury.**
- In order to visualize skull fractures, you must view the CT scan using the "bone window" settings that optimize visualization of the osseous structures (Fig. 27-5).
- Skull fractures can be described as **linear, depressed, or basilar.**

Linear skull fractures

- **Linear skull fractures** are the most common and have little importance other than for the intracranial abnormalities that may have occurred at the time of the fracture, such as

an epidural hematoma. Fractures of the cranial vault are most likely to occur in the **temporal** and **parietal bones** (see Fig. 27-5, *B*).

Depressed skull fractures

- **Depressed skull fractures** (Fig. 27-6) are more likely to be associated with **underlying brain injury.** They result from a high-energy blow to a small area of the skull (e.g., from a baseball bat), most often in the **frontoparietal region,** and are **usually comminuted.** They may require **surgical elevation** of the depressed fragment when the fragment lies deeper than the inner table adjacent to the fracture (see Fig. 27-6, *A*).

Basilar skull fractures

- **Basilar fractures** are the **most serious** and consist of a linear fracture at the base of the skull. They can be associated with **tears in the dura mater** with subsequent **CSF leak,** which can lead to CSF rhinorrhea and otorrhea.

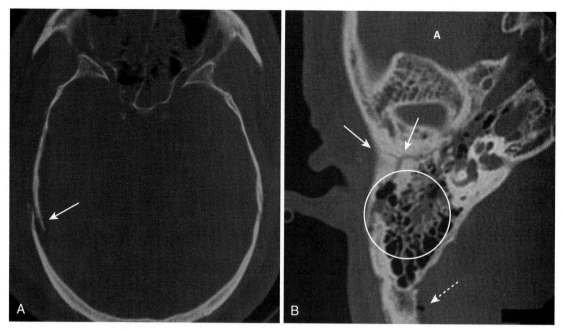

FIGURE 27-6 Skull fractures. A, There is a depressed fracture of the right parietal bone with the fragment lying deeper than the inner table of the adjacent bone *(white arrow)*. Depressed skull fractures occur more often in the frontoparietal region. **B,** There is a comminuted fracture of the right temporal bone *(solid white arrows)*, fluid in the mastoid air cells *(circle)*, and air in the brain **(pneumocephalus)** *(dotted white arrow)*. Basilar skull fractures can be associated with tears in the dura mater with subsequent cerebrospinal fluid leak. *A*, Anterior.

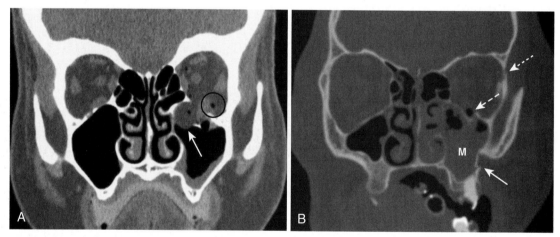

FIGURE 27-7 Facial bone fractures. A, Blow-out fracture. There is air in the left orbit representing **orbital emphysema** *(circle)*. There is a fracture of the floor of the orbit, and soft tissue (in this case, fat) extends inferiorly into the top of the maxillary sinus *(white arrow)*. **B, Tripod fracture.** There is diastasis of the frontozygomatic suture on the left *(dotted white arrow)*, a fracture of the floor of the orbit with orbital emphysema *(dashed white arrow)*, and a fracture *(solid white arrow)* through the lateral wall of the maxillary sinus (M), which is filled with blood.

They can be suspected if there is air seen in the brain **(traumatic pneumocephalus), fluid in the mastoid air cells,** or an **air–fluid level** in the sphenoid sinus (see Fig. 27-6, *B*).

Facial Fractures

■ **CT is the imaging study of choice** for evaluating **facial fractures.** Multislice scanners allow for reconstruction in the sagittal and coronal planes so that the patient does not have to be repositioned in the scanner.

 Care must be taken in diagnosing facial fractures based on viewing what appears to be a fracture on only one

image, since CT scans, by their nature, produce sections so thin that they may not demonstrate the entire contour of the bone in question. Look for a fracture to appear on several contiguous images.

■ The most common **orbital fracture** is the **blow-out fracture** (Fig. 27-7), which is produced by a **direct impact on the orbit** (e.g., a baseball strikes the eye) and causes a sudden increase in intraorbital pressure leading to a fracture of the inferior orbital floor (into the maxillary sinus) or the medial wall of the orbit (into the ethmoid sinus). Sometimes the **inferior rectus muscle** can be trapped in the fracture, leading to **restriction of upward gaze** and **diplopia.**

- **Recognizing a blow-out fracture of the orbit** (see Fig. 27-7, *A*):
  - ◆ **Orbital emphysema.** Air in the orbit from communication with one of the adjacent air-containing sinuses, either the ethmoid or maxillary sinus
  - ◆ **Fracture** through either the medial wall or floor of the orbit
  - ◆ **Entrapment of fat** and/or **extraocular muscle,** which projects downward as a soft tissue mass into the top of the maxillary sinus
  - ◆ **Fluid (blood)** in the maxillary sinus
- A **tripod fracture,** usually a result of blunt force to the cheek, is another relatively common facial fracture. This fracture involves **separation of the zygoma** from the remainder of the face by **separation of the frontozygomatic suture, fracture of the floor of the orbit,** and fracture of the **lateral wall of the ipsilateral maxillary sinus** (see Fig. 27-7, *B*).

INTRACRANIAL HEMORRHAGE

- Skull fractures may be associated with **intracranial hemorrhage** and/or **diffuse axonal injury.**
- There are **four types of intracranial hemorrhages** that may be **associated with head trauma:**
  - ◆ **Epidural hematoma**
  - ◆ **Subdural hematoma**
  - ◆ **Intracerebral hemorrhage**
  - ◆ **Subarachnoid hemorrhage** (discussed with aneurysms)

Epidural Hematoma (Extradural Hematoma)

- Epidural hematomas represent **hemorrhage** into the potential space **between the dura mater** and **the inner table of the skull** (Table 27-4).
- Most cases are caused by **injury to the middle meningeal artery** or **vein** from blunt head trauma, typically from a motor vehicle accident.

 Almost all epidural hematomas (95%) have an associated skull fracture, frequently the temporal bone. Epidural hematomas may also be caused by disruption of the dural venous sinuses adjacent to a skull fracture.

- **Recognizing an epidural hematoma**
  - ◆ They appear as a **high density, extraaxial, biconvex lens-shaped mass** most often found in the **temporoparietal region** of the brain (Fig. 27-8).
  - ◆ Since the dura is normally fused to the calvarium at the margins of the sutures, it is **impossible for an epidural hematoma to cross suture lines** (subdural hematomas can cross sutures).
  - ◆ Epidural hematomas can cross the tentorium, but subdural hematomas do not.

Subdural Hematoma (SDH)

- **Subdural hematomas are more common** than epidural hematomas and are usually not associated with a skull fracture. They are most commonly a **result of deceleration injuries in motor vehicle** or **motorcycle accidents** (younger patients) or **secondary to falls** (older patients).
- Subdural hematomas are usually produced by **damage to the bridging veins** that cross from the cerebral cortex to the venous sinuses of the brain. They represent hemorrhage into the potential space **between the dura mater** and **the arachnoid.**

 Acute subdural hematomas frequently herald the presence of more severe parenchymal brain injury, with increased intracranial pressure, and are **associated with a higher mortality rate.**

- **Recognizing an acute subdural hematoma** (Fig. 27-9)
  - ◆ **On CT acute subdural hematomas are crescent-shaped, extracerebral bands of high attenuation that**

TABLE 27-4   THE MENINGES

Layer	Comments
Dura mater	Composed of two layers: an **outer periosteal layer,** which cannot be separated from the skull, and an **inner meningeal layer;** the inner meningeal layer enfolds to form the tentorium and falx.
Arachnoid	The avascular middle layer; it is separated from the dura by a potential space known as the **subdural space.**
Pia mater	Closely applied to the brain and spinal cord, the pia mater carries blood vessels that supply both; separating the arachnoid from the pia is the **subarachnoid space;** together the **pia** and **arachnoid** are called the ***leptomeninges.***

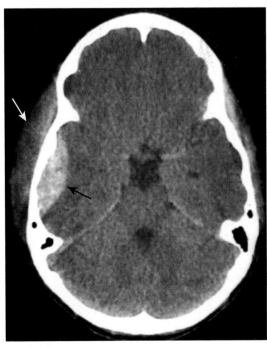

FIGURE 27-8 Epidural hematoma. Acute epidural hematomas usually occur as a result of a trauma-induced skull fracture. Findings include a high density, extraaxial, biconvex, lens-shaped mass lesion often found in the temporal-parietal region of the brain *(black arrow)*. There is a scalp hematoma *(white arrow)* also present. The patient had a temporal bone skull fracture seen on bone windows.

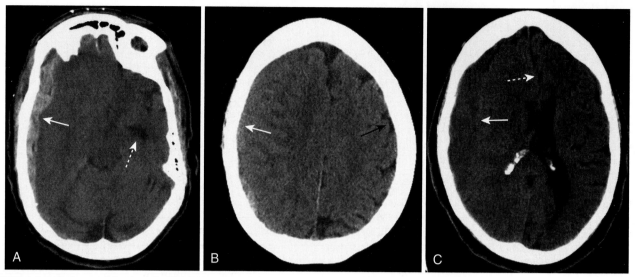

FIGURE 27-9 Acute, isodense, and chronic subdural hematoma. A, There is a crescent-shaped band of high-density blood concave inward toward the brain *(solid white arrow).* There is mass effect with herniation of the brain as indicated by the dilated contralateral temporal horn *(dotted white arrow).* **B,** As they become subacute, subdural hematomas become less dense and may be the same density (isodense) as the normal brain tissue *(white arrow).* You can recognize an isodense subdural by the unilateral absence or displacement of the sulci away from the inner table of the skull compared with the normal opposite side *(black arrow).* **C,** Chronic subdural hematomas (more than 3 weeks old) are usually of low density *(solid white arrow)* compared with the remainder of the brain. There is still mass effect demonstrated by displacement of the interhemispheric fissure *(dotted arrow)* and compression of the lateral ventricle.

may cross suture lines and enter the interhemispheric fissure. They **do not cross the midline.**

♦ Typically, an SDH is **concave** inward toward the brain (epidural hematomas are **convex** inward) (see Fig. 27-9, *A*).

♦ As time passes and they become subacute, or if the subdural blood is mixed with lower-attenuating CSF, they may appear **isointense (isodense)** to the remainder of brain, in which case you should look for **compressed** or **absent sulci** or **sulci displaced away from the inner table as signs of SDH** (see Fig. 27-9, *B*).

♦ Subdural collections may demonstrate a **fluid-fluid level after 1 week,** as the cells settle under serum.

■ **Chronic subdural hematoma**

♦ Chronic subdural hematomas are those present more than 3 weeks after injury.

♦ Chronic subdural hematomas are usually **low density** compared with the remainder of the brain (see Fig. 27-9, *C*).

Intracerebral Hematoma (Intracerebral Hemorrhage)

■ Trauma is only one of the mechanisms that can lead to intracerebral hemorrhage. Intracerebral hematomas can also occur from ruptures of **aneurysms, atheromatous disease in small vessels,** or **vasculitis.**

Injuries occurring at the **point of impact** (called *coup injuries*) and injuries occurring **opposite the point of impact** (called *contrecoup injuries*) are most common following trauma. **Coup** injuries are most often caused by **shearing** of small intracerebral vessels. **Contrecoup** injuries are **acceleration/deceleration injuries** that occur when the brain is propelled in the opposite direction and strikes the inner surface of the skull.

■ Either of these mechanisms can produce a **cerebral contusion. Hemorrhagic contusions** are **hemorrhages,** with **associated edema,** usually found in the **inferior frontal lobes** and **anterior temporal lobes on** or **near the surface of the brain** (Fig. 27-10).

■ **CT findings of intracerebral hemorrhage change over time** and **may not be immediately evident on the initial scan.** MRI typically demonstrates the lesions from the time of injury but may not be available on an emergency basis.

■ **Recognizing traumatic intracerebral hemorrhage on CT:**

♦ Cerebral hemorrhagic contusions may appear as multiple, small, well-demarcated areas of **high attenuation** within the brain parenchyma (see Fig. 27-10, *A*).

♦ They may be surrounded by a **hypodense rim** from **edema** (see Fig. 27-10, *B*).

♦ **Intraventricular blood** may be present (Fig. 27-11).

♦ **Mass effect is common.** The mass effect may produce **compression** of the **ventricles** and **shift of the 3rd ventricle** and **septum pellucidum** to the opposite side. Such displacement can produce severe brain or vascular damage.

♦ These displacements are called *herniations.* Patients with sufficient mass effect are at risk for **transtentorial** and **subfalcine brain herniation** and death (see Fig. 27-10, *B*).

♦ **The types of brain herniation are described in** Table 27-5.

DIFFUSE AXONAL INJURY

■ Diffuse axonal injury is **responsible for the prolonged coma following head trauma** and is the head injury with the poorest prognosis.

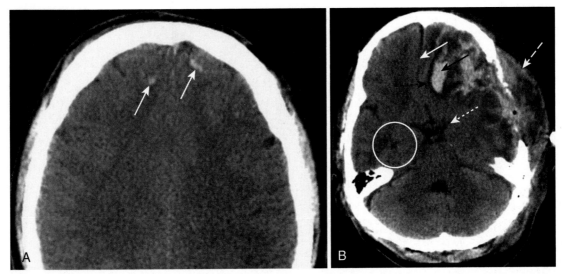

FIGURE 27-10 Cerebral contusions. A, Cerebral contusions are usually the result of trauma and can manifest by multiple areas of high-attenuation hemorrhage *(white arrows)* within the brain parenchyma on computed tomography. **B,** Contusions *(solid black arrow)* are frequently surrounded by a rim of hypoattenuation from edema *(dotted black arrow)*, and mass effect is common, as is demonstrated here by amputation of the ipsilateral basil cisterns *(dotted white arrow)*, midline displacement *(solid white arrow)* representing subfalcine herniation, and dilatation of the contralateral temporal horn *(circle)*. A portion of the left side of the skull has been surgically removed, and there is a large scalp hematoma present *(dashed white arrow)*.

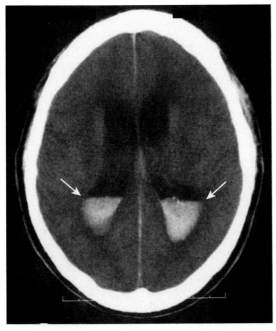

FIGURE 27-11 Intraventricular hemorrhage. Intraventricular hemorrhage *(white arrows)* is common in premature infants but less common in adults. It usually results from breakthrough bleeding from a hypertensive basal ganglia hemorrhage, brain contusion, or subarachnoid hemorrhage and requires a considerable amount of force to produce. Therefore it is typically associated with severe brain damage, results in communicating hydrocephalus, and has a poor prognosis.

■ Acceleration/deceleration forces **diffusely injure axons deep to the cortex,** producing unconsciousness from the moment of injury. This occurs most often as a result of a motor vehicle accident.

➡ The **corpus callosum is most commonly affected,** and the initial CT scan may be normal or may underestimate

TABLE 27-5   TYPES OF BRAIN HERNIATION

Type	Remarks
Subfalcine herniation	The supratentorial brain, along with the lateral ventricle and septum pellucidum, herniates beneath the falx and shifts across the midline toward the opposite side (see Fig. 27-12, *A*).
Transtentorial herniation	Usually, the cerebral hemispheres are displaced downward through the incisura beneath the tentorium, compressing the ipsilateral temporal horn and causing **dilatation of the contralateral temporal horn** (see Fig. 27-12, *B*).
Foramen magnum/ tonsillar herniation	Infratentorial brain is displaced downward through the foramen magnum.
Sphenoid herniation	Supratentorial brain slides over the sphenoid bone either anteriorly (in the case of the temporal lobe) or posteriorly (for the frontal lobe).
Extracranial herniation	Displacement of brain through a defect in the cranium.

the degree of injury. CT findings may be similar to those described for intracerebral hemorrhage following head trauma.

■ **MRI is the study of choice in identifying diffuse axonal injury.**
 ◆ The **small petechial hemorrhages** may be **bright on T1-weighted images.** (On some newer sequences, these petechial hemorrhages may appear dark.)
 ◆ The most common findings are multiple **bright areas on T2-weighted images** at the **temporal** or **parietal cervicomedullary junction** or in the corpus callosum (Fig. 27-13).

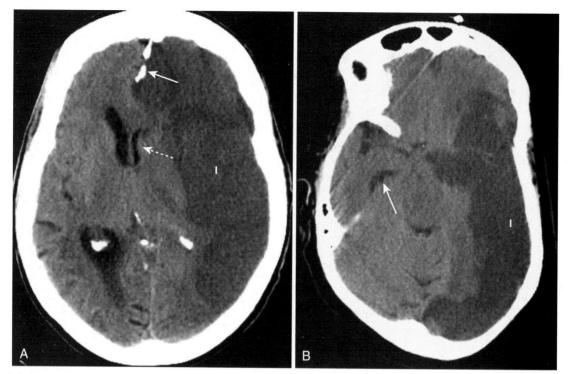

FIGURE 27-12 Brain herniations. A, Subfalcine herniation occurs when the supratentorial brain, along with the lateral ventricle and septum pellucidum, herniates beneath the falx *(solid white arrow)* and shifts across the midline toward the opposite side *(dotted white arrow)*. **B, Transtentorial herniation** usually occurs when the cerebral hemispheres are displaced downward through the incisura beneath the tentorium, compressing the ipsilateral temporal horn and causing dilatation of the contralateral temporal horn *(white arrow)*. Both patients had large cerebral infarcts (I) with cytotoxic edema.

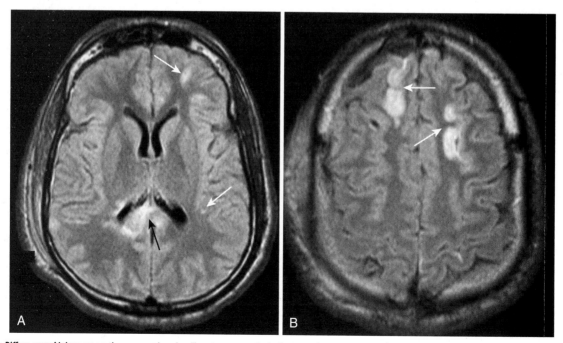

FIGURE 27-13 Diffuse axonal injury, magnetic resonance imaging. These images were obtained using a pulse sequence similar to T2 but with suppression of the bright signal from cerebrospinal fluid, which enhances areas of edema that appear bright. **A** and **B,** Axial images demonstrating multiple foci of abnormal increased signal at the gray-white matter junction *(white arrows)* and within the splenium of the corpus callosum *(black arrow)* in a patient with diffuse axonal injury.

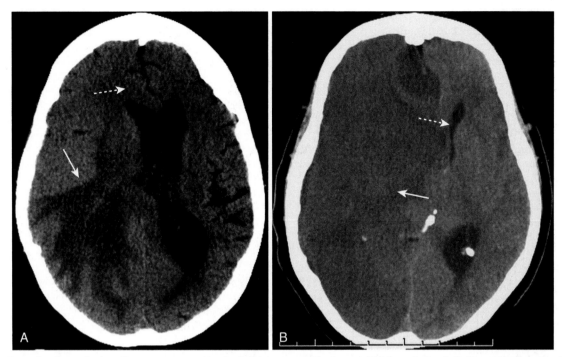

FIGURE 27-14 **Vasogenic and cytotoxic edema.** There are two major categories of cerebral edema: vasogenic **(A)** and cytotoxic **(B)**. **A,** Vasogenic edema *(solid white arrow)* represents extracellular accumulation of fluid and is the type that occurs with infection and malignancy, as in this unenhanced scan of a patient with a glioma. It predominantly affects the white matter. **B,** Cytotoxic edema *(solid white arrow)* represents cellular edema and affects both the gray and white matter. Cytotoxic edema is associated with cerebral ischemia, as in this patient with a very large ischemic infarct on the right. In both of these patients, there is increased intracranial pressure as manifest by the herniation of brain to the contralateral side *(dotted white arrows)*.

INCREASED INTRACRANIAL PRESSURE

- Some of the clinical signs of **increased intracranial pressure** are papilledema, headache, and diplopia.

➡ In general, increased intracranial pressure is caused by either **cerebral edema,** which leads to increased **volume of the brain,** or hydrocephalus, which is **increased size of the ventricles.**

Cerebral Edema

- In adults, **trauma, hypertension** (associated as it is with **intracerebral bleeds** and **stroke**), and **masses** are the most common causes of brain edema.
- Cerebral edema is divided into **two major types: vasogenic** and **cytotoxic** (Fig. 27-14).
  - **Vasogenic edema** represents extracellular accumulation of fluid and is the type that is associated with **malignancy** and **infection.** It is caused by **abnormal permeability of the blood-brain barrier.** It predominantly affects the **white matter** (see Fig. 27-14, *A*).
  - **Cytotoxic edema** represents cellular edema and is associated with **cerebral ischemia.** It is attributed to **cell death. It affects both the gray** and **white matter** (see Fig. 27-14, *B*).
- **Recognizing cerebral edema:**
  - There is a **loss of the normal differentiation between gray** and **white matter** in cytotoxic edema.
  - **There may be effacement** (compression or obliteration) of the normal **sulci.**
  - **The ventricles may be compressed** (Fig. 27-15).

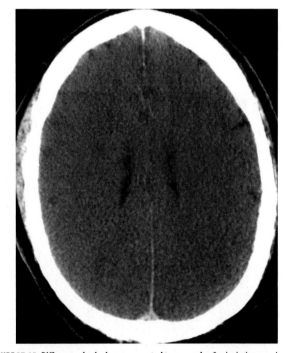

FIGURE 27-15 **Diffuse cerebral edema, computed tomography.** Cerebral edema produces loss of normal differentiation between gray and white matter, effacement (narrowing or obliteration) of the normal sulci, and ventricular compression, all of which are visible in this patient with anoxic encephalopathy.

♦ **Herniation** of the brain may manifest, in part, by **effacement of the basal cisterns** (see Fig. 27-10, *B*).

■ Increased intracranial pressure secondary to increase in the size of the ventricles is discussed in the **"Hydrocephalus"** section.

STROKE

General Considerations

■ Stroke is a nonspecific term that usually denotes an **acute loss of neurologic function** that occurs when the blood supply to an area of the brain is lost or compromised.

➡ The diagnosis of stroke is usually made clinically. Patients with suspected stroke are imaged to determine if there is **another cause of the neurologic impairment** besides a stroke (e.g., a brain tumor) and to **identify the presence of blood** so as to distinguish **ischemic** from **hemorrhagic** stroke. Whether or not hemorrhage is present may determine whether or not thrombolytic therapy will be instituted, and it may **identify** the **infarct** and **characterize it.**

■ **Most strokes are embolic in origin,** the emboli arising from the internal **carotid artery** or the common carotid bifurcation. Emboli can also arise in the **heart** and **aortic arch.**

■ The other common cause of stroke is **thrombosis,** representing in situ **occlusion** of the carotid, vertebrobasilar, or intracerebral circulations from **atheromatous lesions.** Thrombosis of the **middle cerebral artery** is particularly common.

■ Strokes are divided into two large groups: **ischemic** or **hemorrhagic.** Ischemic strokes are much more common. The classification is important because rapid treatment of

ischemic stroke with **tissue plasminogen activator (t-PA)** or another form of intraarterial recanalization technique can substantially improve the prognosis.

■ Most **acute strokes are initially imaged by obtaining a noncontrast-enhanced CT scan of the brain** (within 24 hours of the onset of symptoms), mostly because of its availability. CT findings may be present immediately after a hemorrhagic stroke and within hours after the onset of symptoms for ischemic stroke.

■ MR imaging has become more widely used for early diagnosis. **Diffusion-weighted MR** imaging is more sensitive and relatively specific for detecting early infarction, with the capacity to detect changes within 20 to 30 minutes of the onset of the event. On MRI, the temporal staging of hemorrhage can be identified based on chemical changes that occur in the hemoglobin molecule as the hemorrhage evolves (Fig. 27-16).

Ischemic Stroke

■ **Thromboembolic disease** as a consequence of **atherosclerosis** is the **most common cause of an ischemic stroke.** The source of the emboli can be from atheromatous debris, arterial stenosis, and occlusion or from emboli arising from the left side of the heart (e.g., atrial fibrillation).

■ **Vascular watershed** areas are the distal arterial territories that represent the **junctions between areas served by the major intracerebral vessels,** such as the region between the anterior cerebral artery distribution and the middle cerebral artery distribution. **Reduction in blood flow,** for whatever reason, **affects these sensitive** and **susceptible watershed areas the most.**

➡ The most common finding of an acute, nonhemorrhagic stroke is a normal CT scan (less than 24 hours old). If multiple vascular distributions are involved,

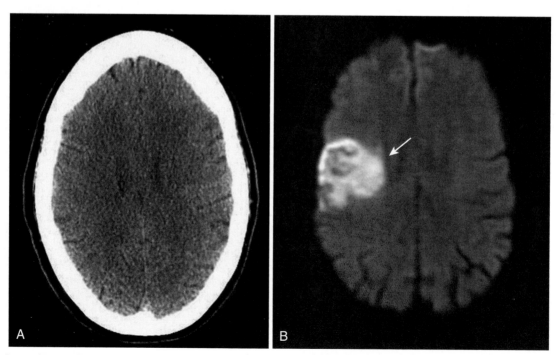

FIGURE 27-16 Computed tomography (CT) and diffusion-weighted magnetic resonance imaging (MRI) in acute stroke. A, The CT scan in this patient with symptoms for 2 hours prior to the study is normal. **B,** A diffusion-weighted MRI scan on the same patient a few minutes later shows an area of abnormally bright signal intensity in the right frontoparietal region *(white arrow)*. Diffusion-weighted imaging (DWI) is an MRI sequence that can be rapidly acquired and which is extremely sensitive to detecting abnormalities in normal water movement in the brain so that it can identify a stroke within 20 to 30 minutes after the event.

emboli or **vasculitis** should be thought of as the cause. If the stroke **crosses** or **falls between vascular territories**, then hypoperfusion caused by hypotension (**watershed infarcts**) should be considered.

■ Table 27-6 briefly summarizes the four major vascular distribution patterns of strokes and some of the symptoms associated with each.

■ **Recognizing ischemic stroke** (Fig. 27-17):
   ◆ On CT, the findings will depend on the amount of time that has elapsed since the original event.
   ◆ **12 to 24 hours:** Indistinct area of low attenuation in a vascular distribution.
   ◆ **>24 hours:** Better circumscribed lesion with mass effect that peaks at 3 to 5 days and usually disappears by 2 to 4 weeks (see Fig. 27-17, *A*).

TABLE 27-6   VASCULAR DISTRIBUTION TERRITORIES OF STROKE

Circulation	Anatomy Affected	Signs and Symptoms
Anterior cerebral artery (uncommon)	Supplies all of the frontal and parietal lobes (medial surfaces), the anterior four fifths of the corpus callosum, the frontobasal cerebral cortex, and the anterior diencephalon.	Can result in disinhibition with perseveration of speech; produces primitive reflexes (e.g., grasping or sucking), altered mental status, impaired judgment, contralateral weakness (greater in legs than arms).
Middle cerebral artery (common)	Supplies almost the entire convex surface of the brain, including the frontal, parietal, and temporal lobes (laterally), as well as the insula, claustrum, and extreme capsule. Lenticulostriate branches supply the basal ganglia, including the head of the caudate nucleus, and the putamen, including the lateral parts of the internal and external capsules.	Produces contralateral hemiparesis or hypesthesia, ipsilateral hemianopsia, and gaze preference toward the side of the lesion; agnosia is common; receptive or expressive aphasia may result if the lesion occurs in the dominant hemisphere; weakness of the arm and face is usually worse than that of the lower limb.
Posterior cerebral artery	Supplies portions of the midbrain, subthalamic nucleus, basal nucleus, thalamus, mesial inferior temporal lobe, and occipital and occipitoparietal cortices.	Occlusions affect vision, producing contralateral homonymous hemianopsia, cortical blindness, visual agnosia, altered mental status, and impaired memory.
Vertebrobasilar system	Perfuses the medulla, cerebellum, pons, midbrain, thalamus, and occipital cortex.	Occlusion of large vessels in this system usually leads to major disability or death; small lesions usually have a benign prognosis; may cause a wide variety of cranial nerve, cerebellar, and brainstem deficits; a hallmark of posterior circulation stroke are crossed findings: ipsilateral cranial nerve deficits and contralateral motor deficits (in contrast to anterior circulation stroke).

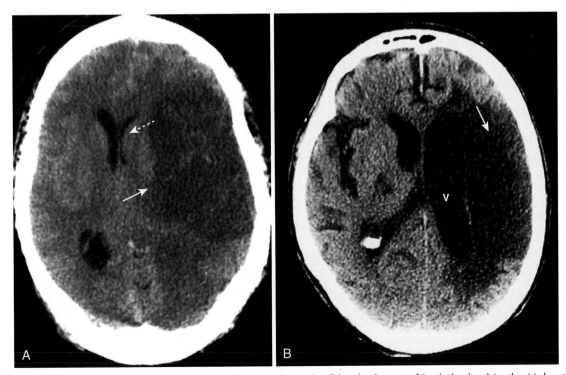

FIGURE 27-17 Computed tomography, ischemic stroke, newer and older. The findings in ischemic stroke will depend on the amount of time that has elapsed since the original event. **A,** At about 24 hours, the lesion becomes relatively well circumscribed *(solid white arrow)* with mass effect evidenced by a shift of the ventricles *(dotted white arrow)* that peaks at 3 to 5 days and disappears by about 2 to 4 weeks. **B,** As the stroke matures, it loses its mass effect, tends to become an even more sharply margined low-attenuation lesion *(solid white arrow)*, and may be associated with enlargement of the adjacent ventricle (V) due to loss of brain substance in the infarcted area.

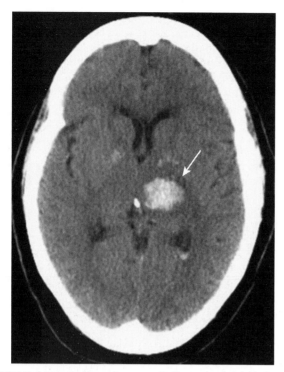

FIGURE 27-18 **Intracerebral hemorrhage, acute.** Freshly extravasated whole blood, as in this bleed into the thalamus *(white arrow),* will be visible as increased density on nonenhanced computed tomography scans of the brain due primarily to the protein in the blood (mostly hemoglobin). As the clot begins to form, the blood becomes denser for about 3 days because of dehydration of the clot. After day 3, the clot gradually decreases in density from the outside in and becomes invisible over the next several weeks.

♦ **72 hours:** Though contrast is rarely used in the setting of acute stroke, contrast enhancement typically occurs when the mass effect is waning or has disappeared.
♦ **>4 weeks:** Mass effect disappears; there is now a well-circumscribed, low-attenuation lesion with no contrast enhancement (see Fig. 27-17, *B*).

Hemorrhagic Stroke

■ Hemorrhage **occurs in about 15%** of strokes. Hemorrhage is **associated with a higher morbidity** and **mortality** than ischemic stroke. Hemorrhage from stroke can occur into the brain parenchyma or the subarachnoid space.
■ In the majority of cases, there is **associated hypertension. About 60% of hypertensive hemorrhages occur in the basal ganglia.** Other areas commonly involved are the thalamus, pons, and cerebellum (Fig. 27-18).

➡ The decision to utilize **thrombolytic** or another **intraarterial recanalization therapy** is based on algorithms formulated by the initial nonenhanced CT scan findings. The **more rapidly treatment is initiated** (usually less than 4 to 5 hours after the onset of symptoms), the **greater its potential benefit.**

■ **Recognizing intracerebral hemorrhage (in general):**
  ♦ Freshly extravasated whole blood with a normal hematocrit will be visible as **increased density** on nonenhanced CT scans of the brain immediately after

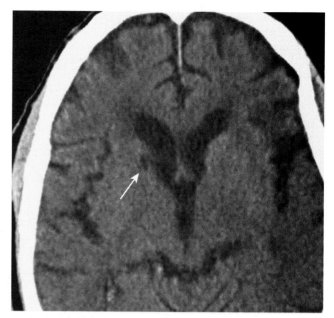

FIGURE 27-19 **Lacunar infarct.** A lacunar infarct, or **lacune,** is a small cerebral infarct produced by occlusion of an end artery. Lacunar infarcts have a predilection for the basal ganglia, internal capsule, and pons, primarily related to hypertension and atherosclerosis. The term *chronic lacunar infarct* is reserved for low-density, cystic lesions, 5 to 15 mm in size *(white arrow).*

BOX 27-1   LACUNAR INFARCTS

Small cerebral infarcts produced by occlusion of small end arteries accounting for up to 20% of all cerebral infarctions.

They have a predilection for the basal ganglia, internal capsule, and pons and occur in association with hypertension, atherosclerosis, and diabetes.

The term *chronic lacunar infarct* is reserved for low-density, cystic lesions, about 5 to 15 mm in size.

the event (see Fig. 27-18). This is attributed to the protein in the blood (mostly hemoglobin).
♦ **Dissection** of blood into the **ventricular system** can occur in hypertensive intracerebral bleeds (see Fig. 27-11).
♦ As the clot begins to form, the **blood becomes denser for about 3 days** because of dehydration of the clot.
♦ **After day 3,** the **clot decreases in density** and becomes invisible over the next several weeks. The clot loses density from the **outside in** so that it **appears to shrink.**
♦ **After about 2 months,** only a small hypodensity may remain (Fig. 27-19; Box 27-1).
■ On MRI, the changes in the appearance of hemorrhage over time are more dramatic. MRI is sensitive to changing effects in both the **iron** and **protein** portions of the hemoglobin molecule in the days and weeks following an acute bleed. Table 27-7 summarizes those changes.

RUPTURED ANEURYSMS

■ The most frequent central nervous system aneurysm is the **berry aneurysm,** which develops from a **congenital weakening in the arterial wall,** usually at the sites of vessel

TABLE 27-7 CHANGES IN THE APPEARANCE OF BLOOD OVER TIME ON MAGNETIC RESONANCE IMAGING

Phase	Time	T1	T2
Hyperacute	<24 hours	Isointense	Bright
Acute	1–3 days	Isointense	Dark
Early subacute	3–7 days	Bright	Dark
Late subacute	7–14 days	Bright	Bright
Chronic	>14 days	Dark	Dark

BOX 27-2 AMYLOID ANGIOPATHY

Amyloid is a proteinaceous material that can be deposited with increasing age in the media and adventitia of small and medium-sized intracranial vessels, mostly involving the frontal and parietal lobes.

This deposit produces a loss of elasticity of the vessels and increases their fragility.

Hemorrhages from amyloid angiopathy are usually large, involving an entire lobe; they may be multiple and occur in several areas simultaneously.

There is no association with hypertension, and it is not associated with amyloid elsewhere.

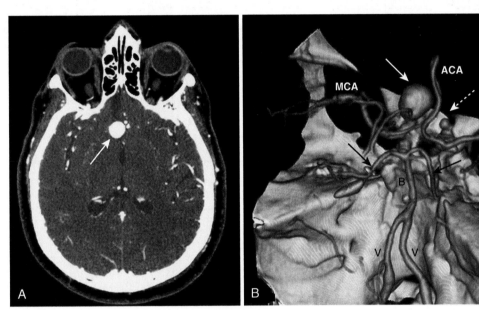

FIGURE 27-20 Berry aneurysm, axial computed tomography (CT), and 3D reconstruction. A, There is a 2-cm focal outpouching of contrast in the region of the right internal carotid artery (ICA) on this contrast-enhanced CT of the brain *(white arrow)*. This is consistent with an aneurysm. **B,** A 3D reconstruction of the circle of Willis from a CT-angiogram demonstrates the aneurysm arising from the supraclinoid segment of the right ICA *(solid white arrow)* and another smaller aneurysm arising from the supraclinoid segment of the left ICA *(dotted white arrow)*. *Black arrows* point to posterior cerebral arteries. *ACA,* Anterior cerebral arteries; *B,* basilar artery; *MCA,* middle cerebral arteries; *V,* vertebral artery.

branching in the **circle of Willis** at the base of the brain. It can be familial (about 10% of cases) or associated with connective tissue diseases.

- **Hypertension** and **aging** play a role in the growth of aneurysms. **Larger aneurysms bleed more frequently** than do smaller ones.
- The aim is to discover and treat the aneurysm **before** it has undergone a major bleed. In some studies, **10 mm was found to be the critical size for rupture.**
- The classical history of a patient who has had a ruptured aneurysm describes it as "the worst headache of my life."

⏩ When aneurysms rupture, the **blood usually enters the subarachnoid space.**

Rupture of an aneurysm is the **most common nontraumatic cause of a subarachnoid hemorrhage** (80%) but not the only cause as trauma, arteriovenous malformations, or breakthrough of an intraparenchymal bleed can also produce subarachnoid hemorrhage.

- Today, most aneurysms are detected by **either CT angiography or MR angiography. CT angiography** is performed using a power injector to deliver a rapid bolus intravenous injection of iodinated contrast, a CT scanner capable of rapid acquisition of data, and special computer algorithms and postprocessing techniques that can highlight the vessels and, if desired, display them three-dimensionally (Fig. 27-20).

- **MR angiography** is routinely done without contrast using a **time-of-flight** technique. This technique depicts flowing blood in the arteries as white. Computerized postprocessing of the images generates three-dimensional reconstructions.

- **Recognizing a subarachnoid hemorrhage (from a ruptured aneurysm)** (Fig. 27-21)
  - ◆ **On CT, acute blood is hyperdense** and **may be visualized within the sulci** and **basal cisterns** (see Fig. 27-21, *A* and 27-21, *B*).
  - ◆ The region of the **falx may become hyperdense,** widened, and irregularly margined (see Fig. 27-21, *C*).
  - ◆ Generally, the **greatest concentration of blood indicates the most likely site of the ruptured aneurysm.**

- Other causes of intracerebral hemorrhage besides aneurysms include arteriovenous malformations, tumors, mycotic aneurysms, and **amyloid angiopathy** (Box 27-2).

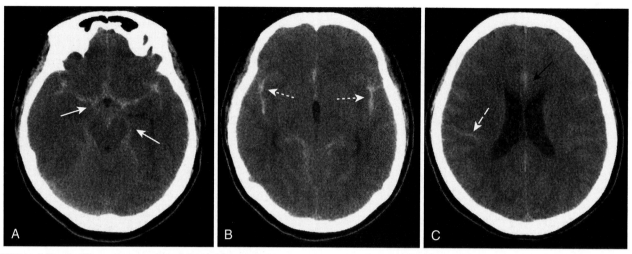

FIGURE 27-21 **Subarachnoid hemorrhage, unenhanced computed tomography (CT) scans.** Subarachnoid hemorrhage is frequently the result of a ruptured aneurysm. Blood may be most easily visualized within the basal cisterns *(white arrows)* **(A),** in the fissures *(dotted white arrows)* **(B),** and interdigitated in the subarachnoid spaces of the sulci *(dashed white arrow)* **(C).** The region of the falx may become hyperdense, widened, and irregularly marginated *(solid black arrow).*

HYDROCEPHALUS

- **Hydrocephalus is defined as an expansion of the ventricular system** on the basis of an increase in the volume of cerebrospinal fluid contained within it (Box 27-3).

➡ Hydrocephalus may be due to several factors:

- ◆ **Underabsorption of cerebrospinal fluid (communicating hydrocephalus)**
- ◆ **Restriction of the outflow of cerebrospinal fluid from the ventricles (noncommunicating hydrocephalus)**
- ◆ **Overproduction of cerebrospinal fluid** (rare)

- In hydrocephalus the **ventricles** are usually **disproportionately dilated** compared with the **sulci,** whereas **both the ventricles** and **sulci** are **proportionately enlarged** in **cerebral atrophy.**
- The **temporal horns are particularly sensitive** to increases in CSF pressure. In the absence of hydrocephalus, the **temporal horns are barely visible.** With hydrocephalus the temporal horns may be greater than 2 mm in size (Fig. 27-22).

Obstructive Hydrocephalus

- Obstructive hydrocephalus is divided into two major categories: **communicating (extraventricular obstruction)** and **noncommunicating (intraventricular obstruction).**
- **Communicating hydrocephalus** is caused by abnormalities that **inhibit the resorption of cerebrospinal fluid,** most often at the level of the arachnoid villi (Fig. 27-23).
  - ◆ **CSF flow** through the **ventricles** and **over the convexities** normally occurs **unimpeded.** Reabsorption through the arachnoid villi can become restricted by such things as **subarachnoid hemorrhage** or **meningitis.**

➡ Classically, the 4th ventricle is **dilated** in **communicating hydrocephalus** and **normal** in size in **noncommunicating** hydrocephalus.

BOX 27-3 NORMAL FLOW OF CEREBROSPINAL FLUID

Most cerebrospinal fluid is produced by the choroid plexuses in the ventricles, primarily the lateral and 4th ventricles.

The direction of flow is from the lateral ventricles through the foramina of Monro to the 3rd ventricle, then through the aqueduct of Sylvius to the 4th ventricle, and then into the basal subarachnoid cisterns through the two lateral foramina of Luschka and the medial foramen of Magendie.

Cerebrospinal fluid (CSF) can then take one of two paths. It can pass upward over the convexities to be reabsorbed into the bloodstream at the arachnoid villi.

Or CSF can also pass inferiorly down the spinal subarachnoid space, where it is either reabsorbed directly or ascends back to the brain to the arachnoid villi.

- ◆ Communicating hydrocephalus is **usually treated with a ventricular shunt.**
- **Noncommunicating hydrocephalus** occurs as a result of **tumors, cysts,** or other physically obstructing lesions that **do not allow cerebrospinal fluid to exit from the ventricles** (Fig. 27-24).
  - ◆ Congenital hydrocephalus is often produced by a blockage between the 3rd and 4th ventricles at the level of the **aqueduct of Sylvius (aqueductal obstruction).**
  - ◆ When the obstruction is caused by a tumor or a cyst, noncommunicating hydrocephalus is **usually treated by surgically** removing the obstructing lesion.
- **Nonobstructive hydrocephalus** from overproduction of CSF is rare and can occur with a **choroid plexus papilloma.**

Normal-Pressure Hydrocephalus

- Normal-pressure hydrocephalus (NPH) is a form of **communicating hydrocephalus** characterized by a classical triad of clinical symptoms: **abnormalities of gait, dementia,**

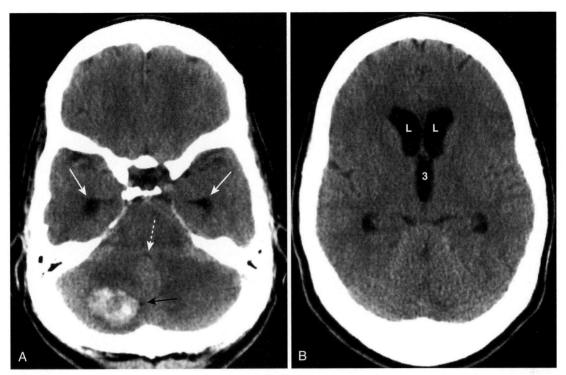

FIGURE 27-22 Noncommunicating hydrocephalus. A, There is dilatation of the temporal horns *(solid white arrows)*, and the 4th ventricle is compressed and nearly invisible *(dotted white arrow)*. There is a hemorrhagic metastatic lesion *(black arrow)* that is obstructing the 4th ventricle. **B,** The frontal horns of the lateral ventricles (L) and 3rd ventricle (3) are dilated, but note that the sulci are not dilated. This form of hydrocephalus is the result of obstruction to the outflow of cerebrospinal fluid from the ventricles.

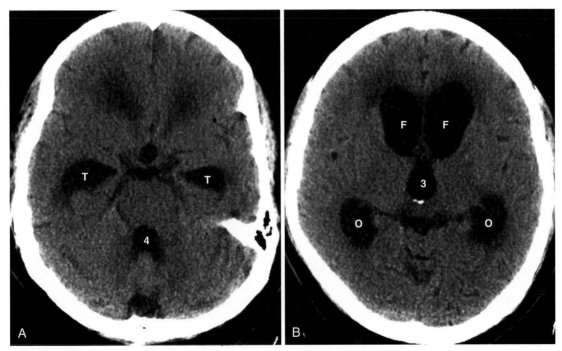

FIGURE 27-23 Communicating hydrocephalus. Communicating hydrocephalus is caused by abnormalities that inhibit the resorption of cerebrospinal fluid, most often at the level of the arachnoid villi. **A,** Classically, the 4th ventricle is dilated in communicating hydrocephalus (4) but normal in size in noncommunicating hydrocephalus. The temporal horns (T) are particularly sensitive to increases in intraventricular volume or pressure and are dilated here. **B,** The frontal horns (F), occipital horns (O), and 3rd ventricle (3) are markedly dilated. There is a disproportionate dilatation of the ventricles compared with the sulci (which are normal to small here). Communicating hydrocephalus is usually treated with a ventricular shunt.

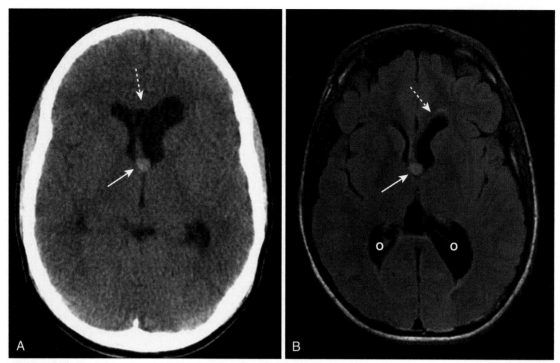

FIGURE 27-24 **Colloid cyst of the 3rd ventricle, computed tomography and magnetic resonance imaging.** A colloid cyst is an uncommon, benign lesion of the 3rd ventricle, which can cause obstructive hydrocephalus. **A,** There is a hyperdense mass in the anterior aspect of the third ventricle *(solid white arrow)*, causing asymmetric obstruction of the left foramen of Monro compared with the right *(dotted white arrow)*. **B,** On this T2-FLAIR sequence, the lesion has increased signal *(solid white arrow)* and is causing dilatation of the frontal *(dotted white arrow)* and occipital (O) horns of the lateral ventricle. *FLAIR,* Fluid-attenuated inversion recovery pulse sequence on MRI that suppresses signal from cerebrospinal fluid.

and **urinary incontinence.** The age of onset is typically between 60 to 70 years old.

➡ Its recognition is important because it is usually amenable to treatment using a one-way **ventriculoperitoneal shunt,** which allows the CSF to exit the ventricles and drain into the peritoneal cavity, where the CSF is reabsorbed.

■ Imaging findings are similar to other forms of communicating hydrocephalus and include **enlarged ventricles,** particularly the temporal horns, with **normal** or **flattened sulci** (Fig. 27-25).

CEREBRAL ATROPHY

■ Disorders associated with **gross cerebral atrophy** are also associated with **dementia, Alzheimer disease being one of the most common.** Atrophy implies a loss of both gray and white matter.
  ♦ The major finding in patients with Alzheimer disease (though not specific) is **diffuse cortical atrophy,** especially in the **temporal lobes.**

➡ As in hydrocephalus, the **ventricles dilate** in cerebral atrophy, but do so because a loss of normal cerebral tissue produces a vacant space that is filled passively with the CSF-filled ventricles. Unlike hydrocephalus, the **dynamics of CSF** production and absorption are **normal in atrophy.**

■ In general, cerebral atrophy leads to **proportionate enlargement of both the ventricles** and the sulci (Fig. 27-26).

BRAIN TUMORS
Gliomas of the Brain

■ **Gliomas are the most common primary, supratentorial, intraaxial mass in an adult. They account for 30% of brain tumors of all types and 80% of all primary malignant brain tumors. Glioblastoma multiforme** accounts for more than half of all gliomas, astrocytomas about 20%, with the remainder split among **ependymoma, oligodendroglioma,** and **mixed glioma (e.g., oligoastrocytoma).**
■ Glioblastoma multiforme occurs **more commonly in males** between 65 to 75 years of age, especially in the **frontal** and **temporal lobes.** It has the **worst prognosis** of all gliomas.

➡ It **infiltrates** adjacent areas of the brain along **white matter tracts,** making it difficult to resect, but like most brain tumors, it **does not produce extracerebral metastases.**

■ **Recognizing glioblastoma multiforme:**
  ♦ As the most aggressive of tumors, glioblastoma multiforme frequently demonstrates **necrosis** within the tumor.
  ♦ The tumor **infiltrates** the surrounding brain tissue, frequently **crossing** the white matter tracts of the **corpus callosum** to the opposite cerebral hemisphere, producing a pattern called a *butterfly glioma.*

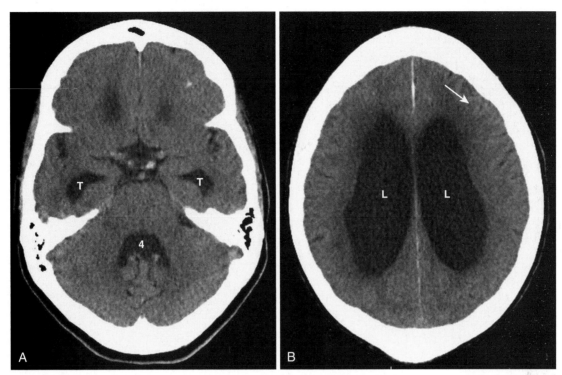

FIGURE 27-25 **Normal-pressure hydrocephalus (NPH).** NPH is a form of communicating hydrocephalus characterized by a classical triad of clinical symptoms: abnormalities of gait, dementia, and urinary incontinence. **A,** The ventricles are enlarged, particularly the temporal horns (T) and the 4th ventricle. **B,** The bodies of the lateral ventricles (L) are also markedly enlarged, but the sulci are normal or flattened *(white arrow).*

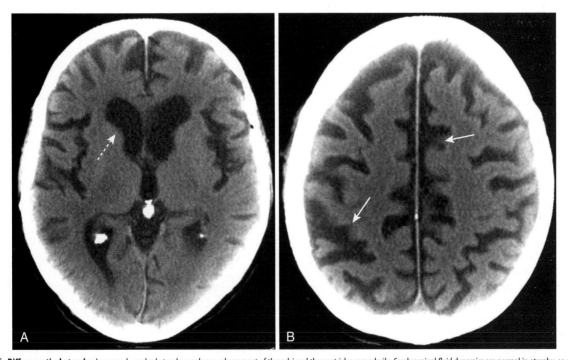

FIGURE 27-26 **Diffuse cortical atrophy.** In general, cerebral atrophy produces enlargement of the sulci and the ventricles secondarily. Cerebrospinal fluid dynamics are normal in atrophy, as compared with hydrocephalus. Disorders that cause gross cerebral atrophy are also associated with dementia, Alzheimer disease being one of the most common. **A,** The lateral ventricles *(dotted white arrow)* are enlarged. **B,** Unlike hydrocephalus, the sulci are also enlarged *(white arrows).*

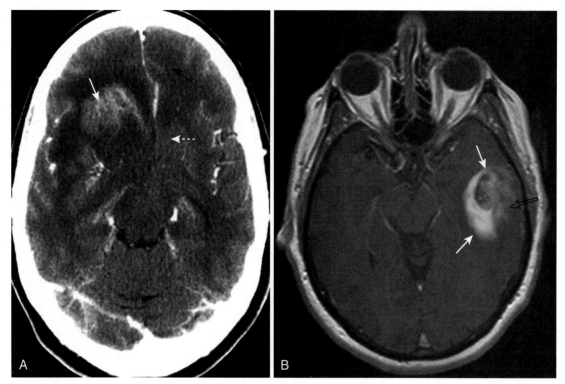

FIGURE 27-27 Glioblastoma multiforme, computed tomography (CT) and magnetic resonance imaging, two different patients. A, In this contrast-enhanced CT, the tumor enhances *(solid white arrow)*, produces considerable vasogenic edema *(dotted white arrow)*, and infiltrates the surrounding brain tissue. There is either edema or tumor that has crossed over to the left frontal lobe *(black arrow)*. **B,** Axial T1-weighted postgadolinium image in another patient demonstrates an enhancing mass in the left temporal lobe *(white arrows)*. The internal enhancement of the mass is somewhat heterogeneous *(open black arrow)*, which implies intratumoral necrosis or cystic change.

♦ It tends to produce considerable **vasogenic edema** and **mass effect** and **enhances with contrast,** at least in part (Fig. 27-27).

Metastases

■ Solitary intraaxial masses are about evenly split between **solitary metastases** and **primary brain tumors.** About **40% of all intracranial neoplasms** are metastases. **Lung, breast, and melanoma** are the most common primary malignancies to produce brain metastases.

 Recognizing metastases to the brain:

♦ Metastases to the brain are frequently **well defined, round masses near the gray-white junction.**
♦ They are **usually multiple but can be solitary.**
♦ They are **typically hypodense** or **isodense on nonenhanced CT.**
♦ With intravenous contrast they can **enhance,** sometimes with a pattern of **ring enhancement.**
♦ Most evoke some **vasogenic edema,** frequently out of proportion to the size of the mass (Fig. 27-28).

Meningioma

■ Meningiomas are the **most common extraaxial mass,** usually occurring in middle-aged women. Their most common locations are **parasaggital, over the convexities,** the sphenoid wing, and the cerebellopontine angle cistern, in decreasing frequency.

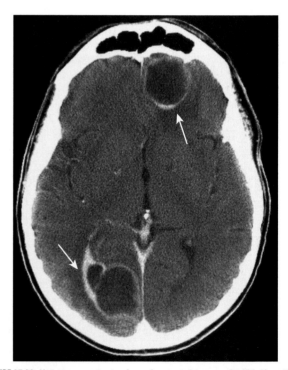

FIGURE 27-28 Metastases, contrast-enhanced computed tomography (CT). About 40% of all intracranial neoplasms are metastases. They typically produce well-defined, round masses near the gray-white junction and are usually multiple. With intravenous contrast, they can enhance and show ring enhancement *(white arrows)*. Lung, breast, and melanoma are the most common primary malignancies to produce brain metastases. This patient had a lung cancer.

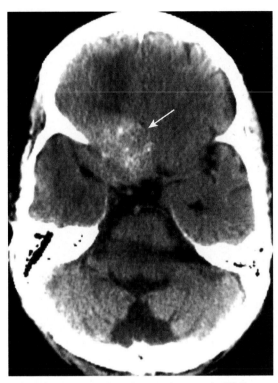

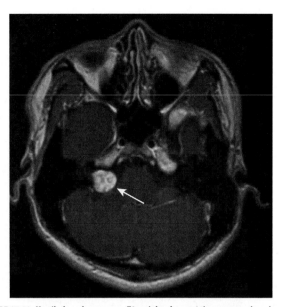

FIGURE 27-30 **Vestibular schwannoma, T1-weighted magnetic resonance imaging with contrast.** There is a homogeneously enhancing soft tissue mass at the right cerebellopontine angle *(white arrow),* classical for a vestibular schwannoma. These tumors occur most commonly along the course of the 8th nerve. Hearing loss is the most common presenting symptom.

FIGURE 27-29 **Meningioma, nonenhanced computed tomography (CT).** The most common extraaxial mass, meningiomas usually occur in middle-aged women. This meningioma is arising from the right sphenoid wing, a relatively common site of origin. On unenhanced CT, over half are hyperdense to normal brain, and about 20% contain calcification, as does this lesion that appears as a dense mass *(white arrow).* On contrast-enhanced studies, meningiomas markedly enhance.

■ They tend to be **slow-growing** with an **excellent prognosis** if surgically excised.

■ When **multiple,** they may have an association with **neurofibromatosis type 2.**

➤ **Recognizing a meningioma on CT scan of the brain:**

♦ **On unenhanced CT,** over **half are hyperdense** to normal brain and about **20% contain calcification** (Fig. 27-29).

♦ On contrast-enhanced studies, **meningiomas enhance markedly.**

♦ They **may induce vasogenic edema in the adjacent brain parenchyma.**

Vestibular Schwannoma (Acoustic Neuroma)

■ **Vestibular schwannomas** are the most common schwannomas of all of the cranial nerves. Their most common symptom is **hearing loss,** but they also produce tinnitus and disturbances in equilibrium.

■ They occur most commonly along the **course of the 8th nerve** within the **internal auditory canal** at the **cerebellopontine angle** (Fig. 27-30).

■ Similar to meningiomas, when they are multiple (i.e., bilateral), they are usually associated with **neurofibromatosis type 2.**

➤ **Contrast-enhanced MRI** is the **most sensitive** imaging study for detecting vestibular schwannomas, which

virtually **always enhance,** usually **homogeneously.** On thin-section T2-weighted images, they can be seen inside the internal auditory canal.

OTHER DISEASES
Multiple Sclerosis

■ Multiple sclerosis (MS) is considered to be **autoimmune** in origin and is the **most common demyelinating disease.** Any neurologic function can be affected by the disease, with some patients having mostly cognitive changes, whereas others present with ataxia, paresis, or visual symptoms.

■ Characterized by a **relapsing** and **remitting course,** a diagnosis of MS requires confirmation of specific clinical criteria, but imaging by MRI along with ancillary tests now permit the diagnosis to be made after a single episode.

■ It characteristically affects **myelinated (white matter) tracks** with lesions known as **plaques.** Lesions of multiple sclerosis have a **predilection for the periventricular area, corpus callosum, and optic nerves.**

➤ **MRI is the study of choice** in imaging multiple sclerosis because of its greater sensitivity than CT in demonstrating plaques both in the brain and spinal cord.

♦ The lesions produce **discrete, globular foci of high signal intensity (white) on T2-weighted images.**

♦ On T1-weighted, nonenhanced images, they are isointense to hypointense, but **in acute MS the lesions enhance with gadolinium on T1-weighted images.**

♦ The lesions tend to be oriented with their long axes perpendicular to the ventricular walls (Fig. 27-31).

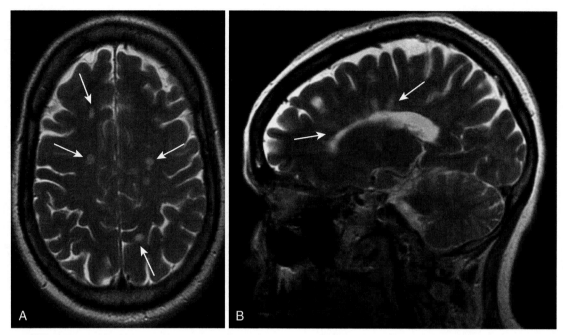

FIGURE 27-31 Multiple sclerosis, axial and sagittal magnetic resonance imaging (MRI). Lesions of multiple sclerosis have a predilection for the periventricular area, corpus callosum, and optic nerves *(white arrows).* **A,** The lesions produce discrete, globular foci of high signal intensity (white) on T2-weighted images. **B,** Ovoid lesions with their long axis perpendicular to the ventricular surface seen in multiple sclerosis are called **Dawson fingers** *(white arrows).* MRI is the study of choice in imaging multiple sclerosis because of its greater sensitivity than computed tomography in demonstrating plaques, both in the brain and spinal cord.

NEUROIMAGING TERMINOLOGY

Term	Remarks
Intraaxial/extraaxial	Intraaxial lesions originate in the brain parenchyma, and extraaxial lesions originate outside of the brain substance (i.e., meninges, intraventricular).
Infratentorial	Beneath the tentorium cerebelli, which includes the cerebellum, brain stem, fourth ventricle, and cerebellopontine angles.
Supratentorial	Above the tentorium cerebelli, which includes the cerebral hemispheres (frontal, parietal, occipital, and temporal lobes) and the sella.
Transient ischemic attack (TIA)	Sudden neurologic loss that persists for a short time and resolves within 24 hours.
Completed stroke	Neurologic deficit lasts for >21 days.
Open versus closed head injuries	**Open:** communication of intracranial material outside of the skull. **Closed:** no external communication.
Increased attenuation/hyperattenuation/hyperdense/hyperintense	On computed tomography (CT), tissue that is of increased attenuation hyperattenuates; hyperdense is whiter than surrounding tissues. *Hyperintense* refers to increased signal on magnetic resonance imaging (MRI).
Decreased attenuation/hypoattenuation/hypodense/hypointense	On CT, tissue that is of decreased attenuation hypoattenuates; hypodense is darker than surrounding tissues. *Hypointense* refers to decreased signal on MRI.
Diffusion-weighted imaging (DWI)	An MRI sequence that can be rapidly acquired and which is extremely sensitive to detecting abnormalities in normal water movement in the brain so that it can identify a stroke within 20 to 30 minutes after the event. DWI also helps differentiate acute infarction from more chronic infarction.
Time-of-flight magnetic resonance angiography	A method used to differentiate vessels containing moving blood from the adjacent stationary tissues, making the vessels much brighter than their surrounding structures.

TAKE-HOME POINTS

RECOGNIZING SOME COMMON CAUSES OF INTRACRANIAL PATHOLOGY

The normal computed tomography (CT) anatomy of the brain is discussed. Magnetic resonance imaging (MRI) is generally the study of choice for detecting and staging intracranial and spinal cord abnormalities, because of its superior contrast and soft tissue resolution.

Unenhanced CT is usually the study of first choice in acute head trauma. The search for findings should initially focus on the presence of mass effect or blood.

Linear skull fractures are important mainly for the intracranial abnormalities that may have occurred at the time of the fracture; **depressed skull fractures** can be associated with underlying brain injury and may require elevation of the fragment; **basilar skull fractures** are more serious and can be associated with cerebrospinal fluid (CSF) leaks.

Blow-out fractures of the orbit result from a direct blow and may present with orbital emphysema, fracture through either the floor or medial wall of the orbit, and entrapment of fat and/or extraocular muscles in the fracture.

There are four types of intracranial hemorrhages that may be associated with trauma: epidural hematoma, subdural hematoma, intracerebral hemorrhage, and subarachnoid hemorrhage.

Epidural hematomas represent hemorrhage into the potential space between the dura mater and the inner table of the skull and are usually caused by injuries to the middle meningeal artery or vein from blunt head trauma; almost all (95%) have an associated skull fracture. When acute, epidural hematomas appear as hyperintense collections of blood that typically have a lenticular shape; as they age, they become hypodense to normal brain.

Subdural hematomas most commonly result from deceleration injuries or falls; they represent hemorrhage into the potential space between the dura mater and the arachnoid; acute subdural hematomas portend the presence of more severe brain injury. Subdural hematomas are crescent-shaped bands of blood that may cross suture lines and enter the interhemispheric fissure, although they do not cross the midline; they are typically concave inward to the brain and may appear isointense (isodense) to remainder of brain as they become subacute, and hypodense when chronic.

Traumatic **intracerebral hematomas** are frequently caused by shearing injuries and present as petechial or larger hemorrhages in the frontal or temporal lobes; they may be associated with increased intracranial pressure and brain herniation.

Brain herniations include subfalcine, transtentorial, foramen magnum/tonsillar, sphenoid, and extracranial herniations.

Diffuse axonal injury is a serious consequence of trauma in which the corpus callosum is most commonly affected; CT findings are similar to those for intracerebral hemorrhage following head trauma; MRI is the study of choice in identifying diffuse axonal injury.

In general, **increased intracranial pressure** is caused by either increased volume of the brain (cerebral edema) or increased size of the ventricles (hydrocephalus).

There are two major categories of **cerebral edema:** vasogenic and cytotoxic.

Vasogenic edema represents extracellular accumulation of fluid; this type occurs with malignancy and infection, and it affects the white matter more.

Cytotoxic edema represents cellular edema, which is caused by cell death and affects both the gray and white matter; cytotoxic edema is associated with cerebral ischemia.

Stroke denotes an acute loss of neurologic function that occurs when the blood supply to an area of the brain is lost or compromised. MRI is more sensitive to the early diagnosis of stroke than is CT. There are certain patterns of ischemia that develop depending on which vascular territory is involved.

Strokes are usually **embolic** (more common) or **thrombotic** events and are typically divided into **ischemic** (more common) and **hemorrhagic** varieties (poorer prognosis); hypertension is frequently associated.

Intracerebral hemorrhage will display increased density on nonenhanced CT scans of the brain; as the clot begins to form, the blood becomes denser for about 3 days because of dehydration of the clot and then decreases in density, becoming invisible over the next several weeks; after about 2 months, only a small hypodensity may remain.

Berry aneurysms are usually formed from congenital weakening in the arterial wall; when they rupture, the blood typically enters the subarachnoid space, presenting in the basilar cisterns and in the sulci. The aneurysm itself can be detected on either CT angiography or MR angiography.

Hydrocephalus represents an increased volume of CSF in the ventricular system and may be caused by overproduction of cerebrospinal fluid (rare), underabsorption of cerebrospinal fluid at the level of the arachnoid villi **(communicating),** or obstruction of the outflow of cerebrospinal fluid from the ventricles **(noncommunicating).**

Normal-pressure hydrocephalus is a form of communicating hydrocephalus characterized by a classical triad of clinical symptoms: abnormalities of gait, dementia, and urinary incontinence; it may be improved by insertion of a ventricular shunt.

Cerebral atrophy is a loss of both gray and white matter that may resemble hydrocephalus, except that the CSF fluid dynamics are normal in atrophy and, in general, cerebral atrophy produces proportionate enlargement of both the ventricles and the sulci; there is diffuse cerebral atrophy associated with Alzheimer disease.

Glioblastoma multiforme is a highly malignant glioma that occurs most commonly in the frontal and temporal lobes, producing a very aggressive, infiltrating, partially enhancing, sometimes necrotic mass that may cross the corpus callosum to the opposite cerebral hemisphere.

Metastases to the brain are frequently well defined, round masses near the gray-white junction, usually multiple, typically hypodense, or isodense on nonenhanced CT that enhance with contrast; they can provoke vasogenic edema out of proportion to the size of the mass; lung, breast, and melanoma are the most frequent sources of brain metastases.

Meningiomas usually occur in middle-aged women in a parasagittal location; they tend to be slow growing with an excellent prognosis if surgically excised; on CT they characteristically can be dense without contrast because of calcification within the tumor and may enhance dramatically.

Vestibular schwannomas occur most commonly along the course of the 8th cranial nerve within the internal auditory canal at the cerebellopontine angle; they are best identified on MRI, where they homogeneously enhance.

Multiple sclerosis is the most common demyelinating disease, characterized by a relapsing and remitting course and a predilection for the periventricular area, corpus callosum, and optic nerves; it is best visualized on MRI and produces discrete, globular foci of high signal intensity (white) on T2-weighted images.

🌐 WEBLINK

Visit StudentConsult.Inkling.com for more information and a quiz on Neuroimaging Abnormalities.

For your convenience, the following QR Code may be used to link to **StudentConsult.com**. You must register this title using the PIN code on the inside front cover of the text to access online content.

CHAPTER 28
Recognizing Pediatric Diseases

Children differ physiologically from adults and are susceptible to abnormalities in development and maturation not seen in adults. They differ in anatomy (e.g., the thymus) and are more susceptible to the harmful effects of ionizing radiation. This chapter will highlight some of the more common pediatric diseases associated with imaging findings.

CONDITIONS DISCUSSED IN THIS CHAPTER

- **Newborn respiratory distress**
  - Transient tachypnea of the newborn
  - Neonatal respiratory distress syndrome (hyaline membrane disease)
  - Meconium aspiration syndrome
  - Bronchopulmonary dysplasia
- **Childhood lung disease**
  - Reactive airways disease/bronchiolitis
  - Asthma
  - Pneumonia
- **Soft tissues of the neck**
  - Enlarged tonsils and adenoids
  - Epiglottitis
  - Croup (laryngotracheobronchitis)
  - Ingested foreign bodies
- **Other diseases**
  - Cardiomegaly in infants
  - Salter-Harris epiphyseal plate fractures
  - Child abuse
  - Necrotizing enterocolitis
  - Esophageal atresia with/without tracheoesophageal fistula

NEWBORN RESPIRATORY DISTRESS

- Respiratory distress is one of the most common presenting problems of newborns. Clues to its cause include the infant's gestational age, the severity and the progression of symptoms, and the appearance of the chest radiograph. There are many causes of respiratory distress, including those produced by cardiac, metabolic, hematologic, and anatomic factors. Four pulmonary causes will be discussed.

Transient Tachypnea of the Newborn

- Occurring in full-term or in larger, slightly preterm infants, **transient tachypnea of the newborn (TTN)** is the **most common cause of respiratory distress in the newborn.** It is thought to be the result of delay in the resorption of

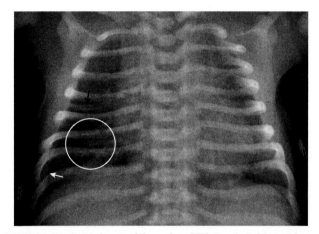

FIGURE 28-1 Transient tachypnea of the newborn (TTN). Occurring in full-term infants or larger, slightly preterm infants, transient tachypnea of the newborn is the most common cause of respiratory distress in the newborn. The lungs are usually hyperinflated. There may be streaky, perihilar, linear densities *(circle);* fluid in the fissures *(black arrow);* and/or laminar pleural effusions *(white arrow).*

fetal lung fluid. TTN is more common in shortened labor, when delivery is by cesarean section, and in mothers with diabetes or asthma.
- Clinically TTN is marked by the **immediate onset** of tachypnea and mild respiratory distress. Infants typically **improve over several hours** with oxygen and supportive therapy and completely recover by 48 hours.
- **Imaging findings of TTN**
  - The lungs are usually **hyperinflated.** There may be **streaky, perihilar, linear densities.** There may be **fluid in the fissures** and/or **laminar pleural effusions** (Fig. 28-1).

Respiratory Distress Syndrome of the Newborn (Hyaline Membrane Disease)

- **Respiratory distress syndrome (RDS)** is a disease of **premature infants,** usually less than 34 weeks of gestation. The incidence and severity of the disease worsens with increasing prematurity. There are numerous risk factors, including perinatal asphyxia and hypoxia and maternal diabetes.
- The major cause of this disorder is **surfactant deficiency.** Without surfactant the alveolar sacs have an increased tendency to collapse, leading to widespread atelectasis.

■ Typically, these premature infants have **severe respiratory distress after birth,** which **progressively worsens.** Clinical findings include cyanosis, grunting, nasal flaring, intercostal and subcostal retractions, and tachypnea.

 Imaging findings of RDS (Fig. 28-2)

◆ There is typically a diffuse "ground-glass" or **finely granular appearance** to the lungs in a **bilateral** and **symmetric** distribution. The granularity seen in the lungs is the interplay of air-distended bronchioles and ducts against a background of atelectasis of alveoli.

◆ **Air bronchograms are common,** especially extending peripherally. **Hypoaeration** is seen in nonventilated lungs. **Hyperinflation** (in nonventilated lungs) **excludes RDS.**

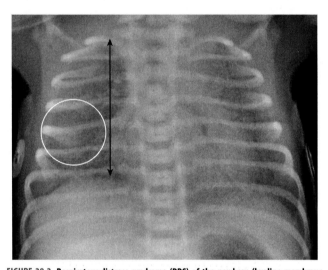

FIGURE 28-2 **Respiratory distress syndrome (RDS) of the newborn (hyaline membrane disease, HMD).** RDS is a disease of premature infants. There is usually a diffuse "ground-glass" or finely granular appearance *(white circle)* in a bilateral and symmetric distribution. Hypoaeration is seen in nonventilated lungs *(double black arrow)*. Hyperinflation essentially excludes RDS.

■ **Oxygen requirements progressively increase** over the first few hours after birth. The treatment of RDS generally includes **positive pressure ventilation** and may include **intratracheal surfactant replacement.**

■ Infants who are very premature (23 to 28 weeks gestation) may require positive pressure respiratory support for several weeks. They are at increased risk for the development of **bronchopulmonary dysplasia** (also known as **chronic respiratory insufficiency of the premature**) (see below). The ductus may fail to close, leading to **patent ductus arteriosus** and **pulmonary hemorrhage** can result.

■ Sudden worsening in symptoms suggests the possibility of an **air leak**—a complication of positive pressure ventilation in relatively noncompliant lungs. There are four manifestations of an air leak (Table 28-1).

■ The **mortality rate** from RDS in the newborn varies by country and is dependent in part on the rate of premature births and the availability of appropriate treatment interventions in those who develop the disease. In the United States the mortality rate has declined dramatically in the last several decades.

Meconium Aspiration Syndrome

■ Meconium aspiration syndrome is the most common cause of neonatal respiratory distress in **postmature infants.**

■ Meconium is found in the amniotic fluid in approximately 15% to 20% of pregnancies. As a consequence, **meconium aspiration** is considered to be a **relatively common event.** Meconium products produce bronchial obstruction, air trapping, and a chemical pneumonitis.

■ **Meconium aspiration syndrome,** on the other hand, usually occurs in **postmature infants.** There is **severe respiratory distress almost immediately,** differentiating it from RDS. However, while many infants have the onset of symptoms at birth, some infants have an asymptomatic period of several hours and then worsen over time.

■ Clinically, there may be tachypnea, hypoxia, and hypercapnia. Small airway obstruction may produce a ball-valve effect, leading to **air trapping, overdistension,** and **air leaks.**

■ Treatment is **supportive,** consisting of **antibiotics** and **oxygen,** inhaled nitric oxide, or extracorporeal membrane oxygenation (ECMO) if needed.

TABLE 28-1 COMPLICATIONS OF TREATMENT—BAROTRAUMA (AIR LEAKS)

Assisted ventilation with high peak inspiratory pressures and positive end-expiratory pressures may cause overdistension and rupture of alveoli

Complication	Remarks
Pulmonary interstitial emphysema	Rupture of an alveolus with dissection of the air back along the bronchovascular and perilymphatic interstitium may produce multiple small pockets of air in the lung called *pulmonary interstitial emphysema* (Fig. 28-3).
Pneumothorax	Rupture of an alveolus adjacent to the visceral pleural surface of the lung may dissect outward into the pleural space, producing a pneumothorax (Fig. 28-4).
Pneumomediastinum	Air leaks along the bronchovascular bundles of the lung will eventually reach the mediastinum and may produce a pneumomediastinum (Fig. 28-5).
Pneumopericardium	In infants there may be connections between the mediastinum and the pericardial sac. Air that surrounds the heart but does not extend above the level of the great vessels may be a sign of pneumopericardium (Fig. 28-6).

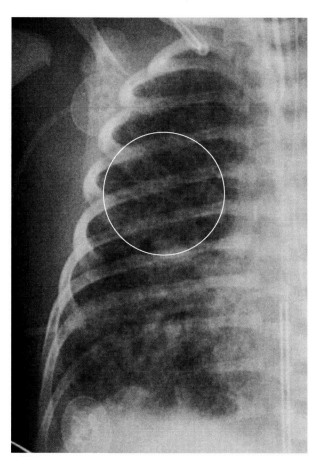

FIGURE 28-3 Pulmonary interstitial emphysema. Assisted ventilation with high peak inspiratory pressures and positive end-expiratory pressures may cause overdistention and rupture of alveoli. The air can rupture outward into the pleural space or dissect backward along the bronchovascular and perilymphatic interstitium toward the hilum. When the multiple small pockets of air *(white circle)* are seen in the interstitial tissues of the lung, it is called *pulmonary interstitial emphysema*. This baby has respiratory distress syndrome.

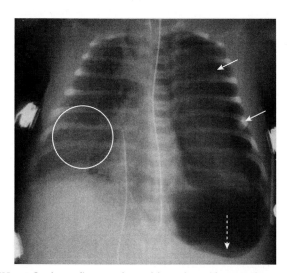

FIGURE 28-4 Respiratory distress syndrome of the newborn with pneumothorax. Rupture of an alveolus adjacent to the visceral pleural surface of the lung may dissect outward into the pleural space, producing a pneumothorax. There is a ground-glass appearance to the lungs *(circle)*. A left-sided pneumothorax is present *(solid white arrows)*. There is a deep sulcus sign of a pneumothorax on the left *(dotted white arrow)*.

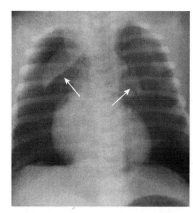

FIGURE 28-5 Thymic sail sign from pneumomediastinum. Air leaks along the bronchovascular bundles of the lung will eventually reach the mediastinum and may produce a pneumomediastinum. The right and left lobes of the thymus normally rest on the base of the heart. When pneumomediastinum forces the thymic lobes upward and outward, they can form a characteristic appearance shown here that has been likened to the spinnaker sail on a boat *(white arrows)*.

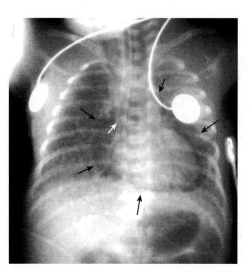

FIGURE 28-6 Pneumopericardium. This is a premature infant, with underlying respiratory distress syndrome, who is on a ventilator. There is a lucency surrounding the heart *(black arrows)*, representing air in the pericardial space. Notice how the air does not extend above the reflection of the aorta and main pulmonary artery. In infants pneumopericardium can occur from air that dissects along the bronchovascular bundles of the lungs *(pulmonary interstitial emphysema)*. The tip of the endotracheal tube extends too far and is in the right main bronchus *(white arrow)*.

- ■ **Imaging findings of meconium aspiration** (Fig. 28-7)
  - ♦ The lungs are hyperinflated with diffuse "ropey" densities (similar in appearance, but not in timing, to bronchopulmonary dysplasia). There may be patchy areas of **atelectasis** and **emphysema** from air trapping. Spontaneous pneumothorax and pneumomediastinum occur in 25% (Fig. 28-8). There may be an associated pneumonia, usually without air bronchograms. Small pleural effusions may be present in 20%.

Bronchopulmonary Dysplasia (Chronic Respiratory Insufficiency of the Premature)

- ■ Bronchopulmonary dysplasia (BPD) is a **consequence of early, acute lung disease,** frequently **respiratory distress syndrome.** It is a clinical diagnosis and has been defined

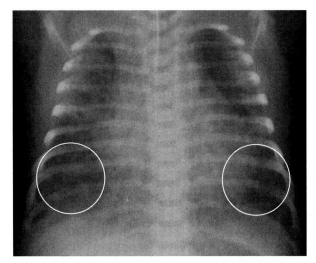

FIGURE 28-7 Meconium aspiration syndrome. Meconium aspiration syndrome usually occurs in postmature infants. There is severe respiratory distress almost immediately. Here the lungs are hyperinflated with diffuse "ropey" densities *(white circles)*. The patchy pattern is a mosaic of areas of atelectasis, along with emphysema from air trapping. This is a baby of 40+ weeks gestation with meconium staining.

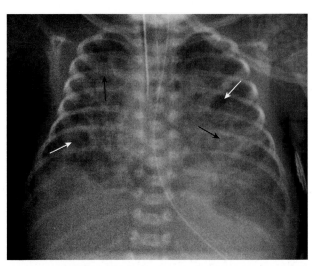

FIGURE 28-9 Bronchopulmonary dysplasia (BPD). BPD is a clinical diagnosis and has been defined as oxygen dependence at 28 days of life to maintain arterial oxygen tensions >50 mm Hg accompanied by abnormal chest radiographs. The lungs in this radiograph have a typical "spongelike" appearance. They are usually hyperaerated and contain coarse linear densities, representing atelectasis *(black arrows)* intermixed with lucent, cystlike foci from hyperexpanded areas of air trapping *(white arrows)*. BPD is manifesting in milder forms than previously and is also called *chronic respiratory insufficiency of the premature*.

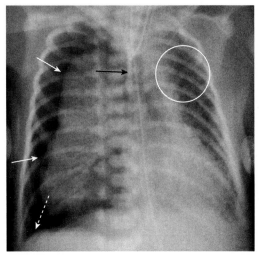

FIGURE 28-8 Meconium aspiration syndrome with pneumothorax. Spontaneous pneumothorax and pneumomediastinum occur in 25% of babies with meconium aspiration syndrome. The lungs show a coarse interstitial pattern *(white circle)*. There is a large right pneumothorax *(white arrows)*. A deep sulcus sign (see Chapter 10) is also present *(dotted white arrow)*. This is a tension pneumothorax, since the heart and trachea *(black arrow)* are pushed to the left.

as **oxygen dependence at 28 days of life to maintain arterial oxygen tensions >50 mm Hg** accompanied by **abnormal chest radiographs.** Associated with most cases of BPD is the prior use of oxygen, which was administered under positive pressure.

■ BPD may also complicate meconium aspiration syndrome and neonatal pneumonia.

■ Clinically infants with BPD have **oxygen dependence,** hypercapnia, and a compensatory metabolic alkalosis. They may develop pulmonary arterial hypertension and right-sided heart failure. Milder forms of BPD are now more common than the severe manifestations of the disease when it was first described.

■ **Imaging findings of BPD** (Fig. 28-9)
  ♦ It may be impossible to distinguish early stages of bronchopulmonary dysplasia from later stages of RDS. The lungs are usually **hyperaerated** overall. They may contain **coarse, irregular, ropelike, linear densities,** representing **atelectasis** or, later, **fibrosis.** These areas of atelectasis may be intermixed with **lucent, cystlike** foci, representing hyperexpanded areas of **air trapping.** Together, these processes give the lung a **spongelike** appearance.

■ Infants with BPD may require mechanical ventilation for months. Changes of BPD will revert to normal on the chest radiograph in most patients after the age of 2 years, although abnormalities may still be visible on computed tomography (CT) of the chest.

CHILDHOOD LUNG DISEASE

Reactive Airways Disease/Bronchiolitis

■ This is a general term for a group of diseases in the pediatric population featuring **wheezing, shortness of breath,** and **coughing.** Initial episodes are frequently referred to as *bronchiolitis.* Unlike asthma, which is chronic, reactive airways disease is usually transient, although it can progress over time to asthma.
  ♦ The **clinical findings** of reactive airways disease/ bronchiolitis are tachypnea, retractions, cough, fever, and rhinorrhea.

■ **Imaging findings of reactive airways disease** (Fig. 28-10)
  ♦ **Peribronchial thickening,** which involves primarily the lobar or segmental bronchi and is manifest by **visualization of the walls of the bronchi on conventional radiographs.**

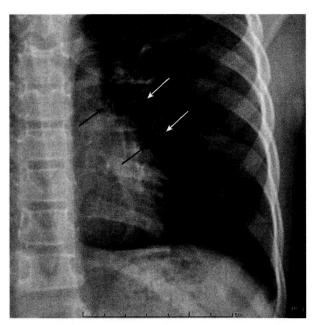

FIGURE 28-10 **Reactive airways disease.** There is peribronchial thickening, which involves primarily the lobar and segmental bronchi and is manifest by visualization of the walls of the bronchi on conventional radiographs. Here they are seen as small circles or doughnuts near the hilum of a child *(white and black arrows)*. Adults may normally demonstrate these *en face* bronchi in the hilar regions, but children usually do not.

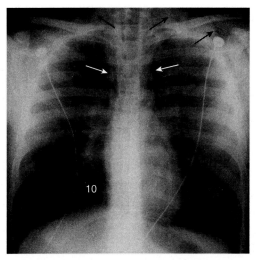

FIGURE 28-11 **Asthmatic with pneumomediastinum.** Chest radiographs can help in determining either the cause or the complications of an asthmatic episode. The edges of the mediastinum are visible as white lines *(white arrows)* because there is now air in the mediastinum. There is also subcutaneous emphysema *(black arrows)* in this patient, who was having an asthmatic attack. This patient has more than 10 posterior ribs visible, in this case a pathologic inspiratory effort.

♦ While adults may have *en face* bronchi visible in the hilar regions, children usually do not. Bronchi viewed *en face* appear as **small, doughnut-like** densities.

♦ Peribronchial thickening may also produce **tramtrack-like linear densities** in the lungs from the thickened bronchial walls visualized in profile.

♦ There may be **hyperinflation** of the lungs. **Atelectasis** from mucus plugging may be present.

■ **Treatment** includes bronchodilators, steroids, and oxygen.

Asthma

■ Asthma is a clinical, not a radiologic, diagnosis. Chest radiographs can help in determining either the **cause** or the **complications** of an asthmatic episode.

■ Inciting events include **pneumonia,** usually accompanied by a fever, which provides a clue to its presence. Complications include **atelectasis** secondary to plugging by mucus (see Fig. 7-2), **pneumothorax,** and **pneumomediastinum** (Fig. 28-11).

■ During or after an acute attack, the lungs may be **overaerated** with **flattening of the diaphragm.** There may be **peribronchial thickening,** as is seen with reactive airways disease.

Pneumonia

■ Age is a determining factor in both the causes and clinical presentation of childhood pneumonia.

■ In **neonates** group B beta hemolytic streptococcus is the most common cause of pneumonia. Its imaging appearance can mimic respiratory distress syndrome of the newborn.

■ In **older infants** the most common causes are **respiratory syncytial virus (RSV), respiratory viruses** (parainfluenza, influenza, and adenoviruses), and *Mycoplasma pneumoniae* **in children older than 5 years.**

■ Clinically, neonates may have only a fever. In older infants and children, **bacterial** pneumonia tends to produce **fever, chills, tachypnea, cough, pleuritic chest pain,** and **shortness of breath.** Viral pneumonia is more often associated with cough, wheezing, and stridor than fever.

■ **Imaging findings of pneumonia**
♦ Bacterial pneumonia (see Fig. 9-6)
• It characteristically produces lobar consolidation, or a round pneumonia, with pleural effusion in 10% to 30% of cases.
♦ Viral pneumonia
• It characteristically shows interstitial infiltrates or patchy areas of consolidation suggestive of bronchopneumonia.
♦ **Treatment** is supportive with antibiotics, as needed.

SOFT TISSUES OF THE NECK

Enlarged Tonsils and Adenoids

■ **Newborns do not have visible adenoids,** even though they are present at birth. They are usually not visible radiographically **until 3 to 6 months.** The adenoids can grow until about age 6 years. They then involute through adulthood. **Adults normally do not have visible adenoids** (Fig. 28-12). The **palatine tonsils** and **adenoids** frequently enlarge at the same time. As with enlarged adenoids, the significance of enlarged tonsils is best made in light of clinical symptoms and compression of the airway.

■ **Clinical findings of enlarged tonsils and adenoids** include nasal congestion, mouth-breathing, chronic or recurrent otitis media as a result of their proximity to the eustachian tubes, painful swallowing, and sleep apnea.

■ **Imaging findings of enlarged adenoids** (Fig. 28-13)
♦ The lateral neck radiograph is the main imaging study; the head should always be extended at the time of exposure.

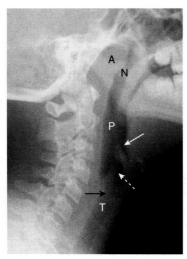

FIGURE 28-12 Normal soft tissue lateral neck radiograph, 4-year-old. The adenoids (A) are seen at the base of the skull and are adjacent to the nasopharyngeal airway (N). More distally is the pharynx (P). The epiglottis *(solid white arrow)* is bounded superiorly by air in the vallecula. The aryepiglottic folds are thin, paired structures *(dotted white arrow)*. The normal-sized laryngeal ventricle *(black arrow)* separates the false vocal cords above from the true cords below. The trachea (T) starts below the true cords.

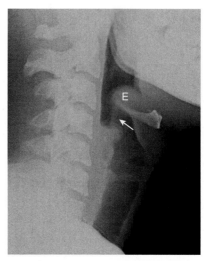

FIGURE 28-14 Epiglottitis. The classical triad of epiglottitis is drooling, severe dysphagia, and respiratory distress with inspiratory stridor. The epiglottis (E) should not be thumblike in appearance. It is very enlarged in this case. There is thickening of the aryepiglottic folds *(white arrow)*. Care must be taken to protect the airway from complete closure in any patient with epiglottitis.

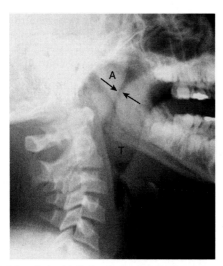

FIGURE 28-13 Enlarged adenoids. The lateral neck radiograph is the primary imaging study; the head should always be extended. Measurements of the adenoids (A) are not reliable. The main consideration is the degree to which they impinge on the nasopharyngeal airway. In this case there is marked narrowing of the nasopharynx *(black arrows)*. The palatine tonsils (T) are also enlarged. The tonsils and adenoids frequently enlarge at the same time. As with enlarged adenoids, the significance of enlarged tonsils is best made in light of clinical symptoms and compression of the airway.

 Measurements of the adenoids are not reliable. The size of the adenoids is less of a consideration than the degree to which they do or do not impinge on the nasopharyngeal airway. **Look for marked narrowing or obliteration of the nasopharyngeal airway.**

Epiglottitis

- Acute bacterial epiglottitis can be a **life-threatening medical emergency,** leading to airway obstruction caused by infection, with edema of the epiglottis and aryepiglottic folds.
- The most frequent causative organism had previously been *Haemophilus influenzae* type B, but introduction of the vaccine in 1985 has led to a marked decrease in the number of cases of epiglottitis. **H. influenzae still remains the most common cause,** although it may also be caused by *Pneumococcus, Streptococcus group A,* viral infection such as herpes simplex 1 and parainfluenza, thermal injury, or angioneurotic edema.
- Epiglottitis typically has a **peak incidence from about 3 to 6 years of age**.
- **Clinically** epiglottitis resembles **croup**, but the clinician should think of epiglottitis if the child cannot breathe unless sitting up, what appears to be croup seems to be worsening, or the child cannot swallow saliva and drools. Cough is rare. The classical triad of epiglottitis is **drooling, severe dysphagia, and respiratory distress with inspiratory stridor.**
- The patient should be accompanied everywhere by someone experienced in endotracheal intubation. Imaging studies are not always necessary for the diagnosis and may be falsely negative in early stages.

 The imaging study of choice is the lateral neck radiograph, which should be exposed in the **upright position only**, because the supine position may close off the airway.

- **Imaging findings of epiglottitis**
  - **Enlargement of the epiglottis.** It should not be larger than your thumb, no matter what the size of your thumb. There is **thickening of the aryepiglottic folds** (which is an important component of airway obstruction and the true cause of stridor) and sometimes circumferential **narrowing of the subglottic portion of the trachea** during inspiration (Fig. 28-14).

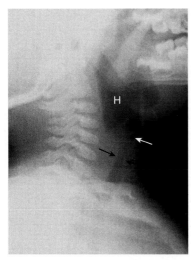

FIGURE 28-15 Laryngotracheobronchitis (croup). The three key findings of croup are all seen on this lateral soft tissue neck radiograph. There is distension of the hypopharynx (H), distension of the laryngeal ventricle *(white arrow)*, and haziness with narrowing of the subglottic space, that is, proximal trachea *(black arrows)*. This is an 11-month-old with a characteristic barking cough.

FIGURE 28-16 Ingested disk battery in colon. This is an extreme close-up of the right lower quadrant where the outline of a round metallic foreign body can be seen. It is in the approximate position of the right colon. The dark band *(white arrow)* inside the whiter rim and the object's size identify it as a disk or button battery. Disk batteries in the esophagus should be removed emergently because of the danger of perforation.

■ **Treatment** includes securing the airway, which may require intubation or emergency tracheostomy; some treatment protocols call for administering intravenous steroids and starting empiric antibiotic therapy.

Croup (Laryngotracheobronchitis)

■ Croup is usually viral in etiology, the most common causes being parainfluenza viruses (types 1, 2, and 3). Respiratory syncytial virus, influenza, and *Mycoplasma* are other common causes. It typically occurs from age **6 months to 3 years,** which is a younger age range than for epiglottitis. **It frequently follows a common cold.**

■ The diagnosis of croup is usually made on the basis of **clinical findings.** It may be difficult to distinguish from early retropharyngeal abscess; the differential can be aided by imaging. There is characteristically a **harsh cough** described as *barking* or *brassy,* associated with hoarseness, inspiratory stridor, low-grade fever, and respiratory distress.

■ **Imaging findings of croup** (Fig. 28-15)

➡ The **three key findings** seen on the lateral soft-tissue neck radiograph are **distension of the hypopharynx, distension of the laryngeal ventricle**, and haziness and/or **narrowing of the subglottic trachea.** The "steeple sign," which may be seen on the frontal radiograph of the neck, is by itself an unreliable sign of croup.

■ **Treatment** can consist of steroids, aerosolized epinephrine, and humidification.

INGESTED FOREIGN BODIES

■ The majority of foreign body ingestions occur between **6 months** and **6 years.** Over **80% pass spontaneously.** Food or true foreign body ingestions include coins, toys, chicken bones (opaque), and fish bones (nonopaque).

■ Most often, they impact **just below the cricopharyngeus** at the level of C5-C6 (70%), at the level of the thoracic inlet,

at the **aortic arch** (20%), or at the level of the **esophago-gastric junction** (10%). Once past the esophagus, most foreign bodies will pass through the gastrointestinal tract.

■ The **major complications** of ingested foreign bodies are perforation, obstruction, or stricture formation.

■ **Disk (button) batteries** and **magnets** pose **particular hazards.** Button batteries should be removed from the esophagus emergently because of their ability to produce perforation (Fig. 28-16). Multiple small magnets also pose the risk of perforation by drawing apposing loops of bowel together and should be removed, if possible.

■ **Clinical findings** of an impacted esophageal foreign body most commonly include dysphagia and odynophagia. Even after the foreign body passes, many complain of pain occurring in the cervical esophagus regardless of the site of impaction.

■ **Imaging findings of an impacted foreign body**
  ◆ The imaging findings will depend on whether the foreign body is opaque or not. Conventional radiographs of the neck and chest are usually obtained first. If no foreign body is visualized, the abdomen may also be radiographed.
  ◆ A coin may appear differently on conventional radiographs depending on whether it is impacted in the **esophagus** or in the **trachea** (Fig. 28-17).
  ◆ If conventional radiographs are negative and there is still a very high suspicion for a nonopaque ingested foreign body, then either contrast esophagram or CT may be considered.

■ Since children who ingest foreign bodies may also be prone to eating other foreign material (pica) such as paint, always look for dense lines in the metaphyses of visible bones on any foreign body study, a sign of **lead poisoning.**

■ **Treatment**
  ◆ Under most circumstances, watchful waiting will result in elimination of the foreign body. Endoscopy is performed if removal is considered clinically necessary (e.g., sharp objects, disk batteries, magnets).

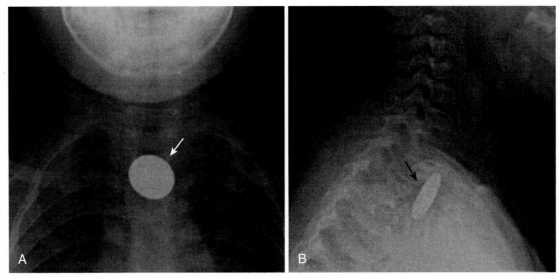

FIGURE 28-17 **Ingested coin in the esophagus.** A coin may appear differently on conventional radiographs, depending on whether it is impacted in the esophagus or in the trachea. In the esophagus, it will usually appear round on the frontal view *(white arrow)* and flat on the lateral view *(black arrow)* because of its orientation in the esophagus. A coin in the trachea will more likely appear on-end on the frontal view, and round on the lateral view because of the shape of the trachea. This coin was a U.S. nickel and it passed uneventfully a day later.

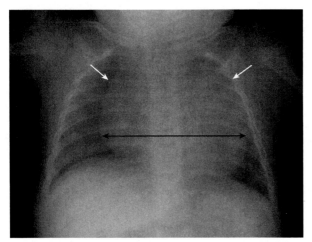

FIGURE 28-18 **Normal infant chest.** In the normal infant, the cardiothoracic ratio may be as large as 65% (compared with 50% in adults). Any assessment of cardiac enlargement in an infant should also take into account other factors such as the appearance of the pulmonary vasculature and any associated clinical signs or symptoms (such as a murmur, tachycardia, or cyanosis). Here the heart appears enlarged *(black arrows)* but is normal for a newborn. The thymus gland *(white arrows)* accounts for some of the apparent cardiomegaly.

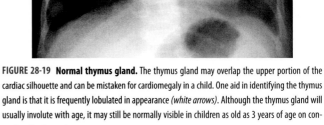

FIGURE 28-19 **Normal thymus gland.** The thymus gland may overlap the upper portion of the cardiac silhouette and can be mistaken for cardiomegaly in a child. One aid in identifying the thymus gland is that it is frequently lobulated in appearance *(white arrows)*. Although the thymus gland will usually involute with age, it may still be normally visible in children as old as 3 years of age on conventional radiographs.

OTHER DISEASES

Recognizing Cardiomegaly in Infants

■ It is important to remember that in **newborns** and **infants,** the **heart will normally appear larger** relative to the size of the thorax than it does in adults. Whereas a cardiothoracic ratio greater than 50% is considered abnormal in adults, the **cardiothoracic ratio may reach up to 65% in infants** and **still be normal,** because newborns cannot take as deep an inspiration as adults and the relative proportions in the size of their abdomen to chest are not the same as for adults (Fig. 28-18).

■ **Any assessment of cardiac enlargement in an infant** should **take into account other factors** such as the appearance of the **pulmonary vasculature,** and any **associated** **clinical signs** or **symptoms** (e.g., a murmur, tachycardia, or cyanosis).

 Also, in a child the thymus gland may overlap portions of the heart and sometimes mimic cardiomegaly. The normal thymus may be seen on conventional chest radiographs up to 3 years of age, and sometimes later. The **normal thymus** gland has a somewhat **lobulated** appearance, especially where it is indented by the ribs (Fig. 28-19).

Salter-Harris Classification of Epiphyseal Plate Fractures in Children

■ In growing bone, the hypertrophic zone in the **growth plate** (**epiphyseal plate** or **physis**) is most vulnerable to shearing

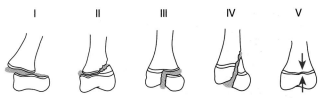

FIGURE 28-20 **Salter-Harris classification of epiphyseal plate fractures.** All of these fractures involve the epiphyseal plate (growth plate). **Type I** are fractures of the epiphyseal plate alone. **Type II** fractures, the most common of the epiphyseal plate fractures, involve the epiphyseal plate and metaphysis. These first two types have a favorable prognosis. **Type III** are fractures of the epiphyseal plate and the epiphysis and have a less favorable prognosis. **Type IV** is a fracture of the epiphyseal plate, epiphysis, and metaphysis. It has an even less favorable prognosis. **Type V** is a crush injury of the epiphyseal plate. It has the worst prognosis.

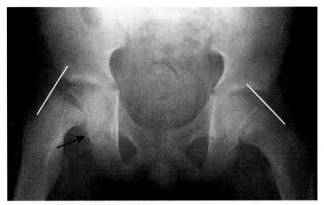

FIGURE 28-21 **Slipped capital femoral epiphysis (SCFE).** SCFE is a manifestation of a Salter-Harris type I injury. It occurs more often in taller and/or heavier teenage boys and produces inferior, medial, and posterior slippage of the proximal femoral epiphysis relative to the neck of the femur *(black arrow)*. A line drawn parallel to the neck of the femur called *Klein's line (white lines)* should intersect the lateral portion of the head. It does so on the normal left side but does not on the right side because the epiphysis has slipped.

TABLE 28-2 SALTER-HARRIS CLASSIFICATION OF EPIPHYSEAL PLATE FRACTURES

Type	What Is Fractured	Remarks
I	Epiphyseal plate	Seen in phalanges, distal radius, SCFE; good prognosis
II	Epiphyseal plate and metaphysis	Most common of all Salter-Harris fractures; frequently distal radius; displays **corner sign**; good prognosis
III	Epiphyseal plate and epiphysis	Intraarticular fracture, especially distal tibia; prognosis: less favorable
IV	Epiphyseal plate, epiphysis and metaphysis	Seen in distal humerus and distal tibia; poor prognosis
V	Crush injury of epiphyseal plate	Worst prognosis; difficult to diagnose until healing begins

injuries. Epiphyseal plate fractures are common and **account for as many as 30% of childhood fractures.** By definition, since these are all fractures through an open epiphyseal plate, they **can only occur in skeletally immature** individuals.

- The **Salter-Harris classification of epiphyseal plate injuries** is a commonly used method of describing such injuries that helps identify the type of treatment required and predicts the likelihood of complications based on the type of fracture (Fig. 28-20; Table 28-2).

 Types I and II heal well.

- ◆ Type III can develop arthritic changes or asymmetric growth plate fusion.
- ◆ Types IV and V are more likely to develop early fusion of the growth plate, with angular deformities, and/or shortening of that bone.
- **Type I—fractures of the epiphyseal plate alone**
  - ◆ Salter-Harris type I fractures are often **difficult to detect without the opposite side for comparison.** Fortunately, these fractures have a **favorable prognosis.**
  - ◆ **Slipped capital femoral epiphysis (SCFE)** is a manifestation of a Salter-Harris type I injury.
    - Slipped capital femoral epiphysis occurs most often in **taller** and/or **heavier teenage boys.** It can occur during periods of rapid growth from trauma, renal

osteodystrophy, and endocrine disorders such as hypothyroidism.
  - The proximal (capital) femoral epiphysis slips inferior, medial, and posterior relative to the neck of the femur (Fig. 28-21).
  - It is **bilateral in about 25% of cases,** and can result in **avascular necrosis** of the slipped femoral head because of interruption of the blood supply **in up to 15% of cases.**
- **Type II—fracture of the epiphyseal plate** and **fracture of the metaphysis**
  - ◆ This is the **most common type of Salter-Harris fracture (75%),** seen especially in the distal radius. Healing is typically rapid, and growth is rarely disturbed, except in the distal femur and proximal tibia where a residual deformity may occur.
  - ◆ The small metaphyseal fracture fragment of a Salter-Harris type II fracture produces the so-called *corner sign* (Fig. 28-22).
- **Type III—fracture of the epiphyseal plate** and **the epiphysis**
  - ◆ There is a **longitudinal fracture through the epiphysis itself,** which means the fracture invariably enters the joint space and fractures the articular cartilage.
  - ◆ This type of injury can have long-term implications for the development of **secondary osteoarthritis,** and can result in **asymmetric** and **premature fusion of the growth plate,** with subsequent deformity of the bone (Fig. 28-23). Its treatment requires early recognition for accurate reduction.
- **Type IV—fracture of the epiphyseal plate, metaphysis, and epiphysis**
  - ◆ Type IV fractures have a **poorer prognosis** than other Salter-Harris fractures (i.e., premature and possibly asymmetric closure of the epiphyseal plate, especially in bones of the lower extremity that may lead to differences in leg length, angular deformities, and secondary osteoarthritis) (Fig. 28-24).

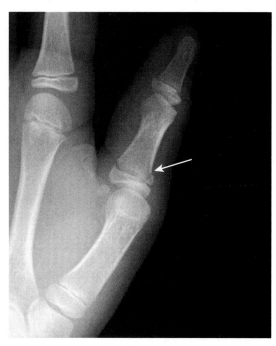

FIGURE 28-22 **Salter-Harris II fracture.** In Salter-Harris type II fractures, there is a fracture of the epiphyseal plate and a fracture of the metaphysis. This is the most common type of Salter-Harris fracture. The small metaphyseal fracture fragment *(white arrow)* produces the so-called ***corner sign.***

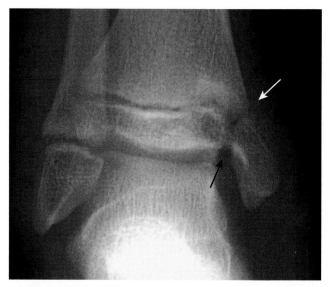

FIGURE 28-24 **Salter-Harris IV fracture.** In type IV fractures there is a fracture of the epiphyseal plate, metaphysis *(white arrow)* and the epiphysis *(black arrow).* These have a poorer prognosis than other Salter-Harris fractures because of increased likelihood of premature and possibly asymmetric closure of the epiphyseal plate. Salter-Harris IV fractures are most often seen in the distal humerus and distal tibia.

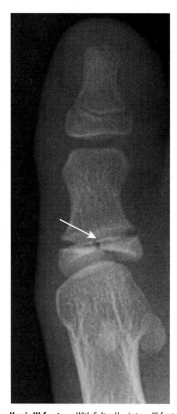

FIGURE 28-23 **Salter-Harris III fracture.** With Salter-Harris type III fractures, there is a fracture of the epiphyseal plate, as well as a longitudinal fracture through the epiphysis itself *(white arrow)*, which means the fracture invariably enters the joint space and fractures the articular cartilage. This can have long-term implications for the development of secondary osteoarthritis and can result in asymmetric and premature fusion of the growth plate with subsequent deformity of the bone.

♦ If the fracture is displaced, open reduction and internal fixation (ORIF) is usually performed, although growth deformities can occur even with perfect reduction.

■ **Type V—crush fracture of epiphyseal plate**
  ♦ Type V Salter-Harris fractures are **rare,** crush-type injuries of the epiphyseal plate, which are **associated with vascular injury** and almost always **result in growth impairment** through early focal fusion of the growth plate.
  ♦ They are **most common in the distal femur,** proximal tibia, and distal tibia. They are **difficult to diagnose on conventional radiographs** until later in their course when complications ensue (Fig. 28-25; see Table 28-2).

Child Abuse

■ Salter-Harris fractures are examples of accidental injuries in children. Certain fractures at other sites, or fractures of other types can be highly suggestive for nonaccidental injuries produced by abuse. **Radiologic evaluations are key in diagnosing child abuse.**

➡ There are several fracture sites and characteristics that should raise the suspicion for child abuse (Table 28-3; Fig. 28-26).

  ♦ **Metaphyseal corner fractures.** Small, avulsion-type fractures of the metaphysis, caused by rapid rotation of ligamentous insertions; corner fractures are considered **diagnostic of physical abuse.** They parallel the metaphysis and can have a **bucket-handle appearance** (see Fig. 28-26, *A*).
  ♦ **Rib fractures,** especially multiple fractures, and/or fractures of the posterior ribs (which rarely fracture, even by accidental trauma) (see Fig. 28-26, *B*).
  ♦ **Head injuries** are the most common cause of death in child abuse under the age of two years. Findings include

subdural and **subarachnoid hemorrhage** and **cerebral contusions. Skull fractures** tend to be comminuted, bilateral, and may cross suture lines.

Necrotizing Enterocolitis

- Necrotizing enterocolitis (NEC) is the **most common gastrointestinal** medical and/or surgical emergency occurring in **neonates.** Its cause remains unknown, although ischemia and/or reperfusion injury have been implicated, along with poorly developed intestinal motility with resultant stasis.

- It is **more common in premature** infants but can also be seen in term babies. Usually, its **onset** occurs within the **first week** of life. Infants who have had enteral feedings are at increased risk. Term infants have an increased risk for the disease if they have congenital heart disease or if there is a history of maternal cocaine use.

- **Clinical findings** may be subtle and can include feeding intolerance, delayed gastric emptying, abdominal distention, and/or tenderness, and decreased bowel sounds.

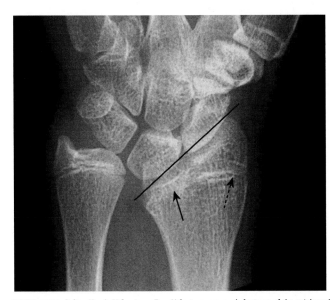

FIGURE 28-25 Salter-Harris V fracture. Type V fractures are crush fractures of the epiphyseal plate. They are frequently not diagnosed until after they produce their growth impairment through early focal fusion of the growth plate, leading to angular deformity. In this child the medial portion of the distal radial epiphyseal plate has fused *(solid black arrow)*, whereas the lateral portion remains open *(dotted black arrow)*. This premature fusion of the medial growth plate has resulted in an angular deformity of the distal radius *(black line)*.

TABLE 28-3 SKELETAL TRAUMA SUSPICIOUS FOR CHILD ABUSE

Site(s)	Remarks
Distal femur, distal humerus, wrist, ankle	Metaphyseal corner fractures
Multiple	Fractures in different stages of healing
Femur, humerus, tibia	Spiral fractures in children <1 year of age
Posterior ribs, avulsed spinous processes	Unusual "naturally occurring" fractures <5 years of age
Multiple skull fractures	Multiple fractures of occipital bone should suggest child abuse
Fractures with abundant callous formation	Implies repeated trauma and no immobilization
Metacarpal and metatarsal fractures Sternal and scapular fractures Vertebral body fractures and subluxations	Unusual "naturally occurring" fractures in children <5 years of age

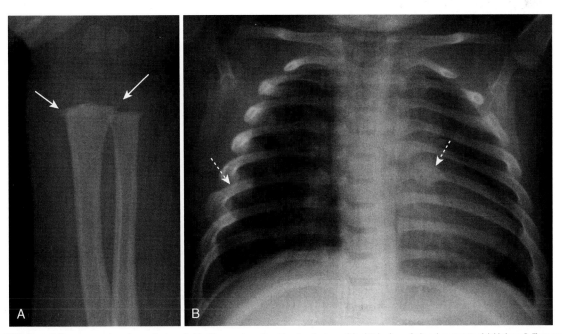

FIGURE 28-26 Child abuse. A, There are metaphyseal corner fractures *(solid white arrows)*, small avulsion-type fractures of the distal radius, a finding characteristic of child abuse. **B,** There are several healing rib fractures *(dotted white arrows)*, including one involving the left 6th posterior rib. Fractures of the posterior ribs are unusual, even in accidental trauma, and should raise suspicion for child abuse.

- Box 28-1 describes the imaging findings of a **normal** infant abdomen.
- **Imaging findings of necrotizing enterocolitis** (Fig. 28-28)
  - ♦ Conventional radiographs of the abdomen remain the modality by which the disease is diagnosed most often.

BOX 28-1 NORMAL INFANT ABDOMEN—CONVENTIONAL RADIOGRAPHS

Almost all infants will demonstrate gas in the stomach by 15 minutes after birth and air in the rectum by 24 hours of age.

It is virtually impossible to differentiate small from large bowel in infants because the haustra do not develop in the colon until about 6 months of age.

Most infants swallow a great deal of air, so expect to see many air-filled, polygonal-shaped loops of bowel with walls that closely appose each other (Fig. 28-27).

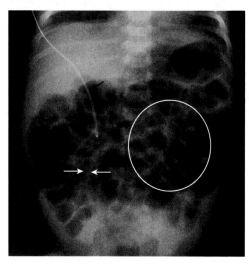

FIGURE 28-27 Normal infant abdomen. Almost all infants will demonstrate gas in the stomach by 15 minutes after birth and gas in the rectum by 24 hours of age. It is impossible to differentiate small from large bowel in newborns. Since most infants swallow a great deal of air, there are many air-filled, polygonal-shaped loops of bowel *(circle)* with walls that closely appose each other *(white arrows)*. There is an external body temperature sensor in place *(black arrow)*.

- The acute disease **most commonly affects the terminal ileum.**
- Early findings may be a few **distended loops of bowel.** A persistently dilated loop that remains unchanged in appearance is a marker of advanced disease, although it can be seen earlier in some.
- Thickened bowel walls are manifest by **separation of the bowel loops.**
- **Pneumatosis intestinalis** is **pathognomonic of NEC in the newborn.** A **linear** radiolucency within the bowel wall parallels the bowel lumen and represents **subserosal** air that has entered from the lumen (see Fig. 28-28, *A*). **Cystic** collections of air are **submucosal** in location but also a sign of pneumatosis.
- Abdominal **free air** is an ominous finding that usually requires emergency surgical intervention.
- **Portal venous gas,** originally thought to be ominous, is now considered less so. It appears as linear branching areas of decreased density over the periphery of the liver and represents air in the portal venous system (see Fig. 28-28, *B*).
- **Complications**
  - ♦ Of those who survive, 50% develop a long-term complication. The two most common complications are **intestinal stricture** and **short-gut syndrome.** Mortality rates range from 10% to 44% in infants weighing less than 1,500 g.

Esophageal Atresia with/without Tracheoesophageal Fistula (TEF)

- The etiology is not completely understood. By far the most common form is a **blind-ending esophagus (esophageal atresia),** with a **fistulous connection between the trachea** and **the distal esophageal remnant.**
- Esophageal atresia is commonly associated with **other abnormalities of the GI tract,** including imperforate anus, duodenal atresia, or stenosis. In 25% of cases, there are 13 or more pairs of ribs or 6 or more lumbar vertebral bodies. Infants with isolated esophageal atresia have an increased incidence of trisomy 21 and duodenal atresia.

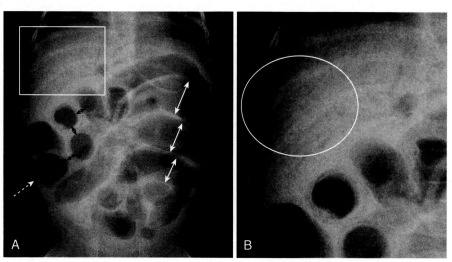

FIGURE 28-28 Necrotizing enterocolitis. This is a 1-week-old infant. **A,** The bowel is dilated *(double white arrows)*. There is a linear radiolucency paralleling the bowel wall, representing air in the bowel wall *(dotted white arrow)* and abnormal separation of the loops of bowel *(double black arrows)*. Numerous, small, peripherally located, air-containing vessels are seen in the liver (see **(B)** for inset), representing portal venous gas. **B,** Close-up of the liver shows multiple small lucencies of air in the portal venous system *(white oval)*.

In up to 50% of cases of esophageal atresia with/without TEF, there may be **associated congenital abnormalities.** An acronym to help remember the constellation of congenital abnormalities that may occur together is **VACTERL:** **V**ertebral anomalies, **A**nal atresia, **C**ardiac abnormalities, **TE**F and/or esophageal atresia, **R**enal agenesis or dysplasia, and **L**imb defects.

- **Clinical findings usually occur early in life** and include choking, drooling, difficulty handling secretions, regurgitation, aspiration, and respiratory distress.
- **Imaging findings** will depend on the type:
  - With esophageal atresia and **no fistula**, no air enters the GI tract, so the **abdomen is airless**. Normally, there should be air in the stomach within 15 minutes after birth.
  - With a **distal fistula** between the esophagus and trachea (the most common type), there is **gas in the bowel** that has entered via the trachea, and a radiolucent, **blind-ending, dilated pouch of upper esophagus** may be seen on chest x-ray representing the atretic esophagus. Further imaging studies are usually not necessary, but introduction of a soft catheter into the blind-ending pouch will prove the diagnosis (Fig. 28-29).
  - There may be **aspiration pneumonia**. Aspiration pneumonia often involves the right upper lobe when there is esophageal atresia.
  - Prenatal ultrasound can suggest the diagnosis as early as 24 weeks because of polyhydramnios.
- **Treatment**
  - Primary anastomosis of the proximal and distal esophagus is usually done when an infant is a few months old. Colonic interposition may be used if primary anastomosis is impossible.

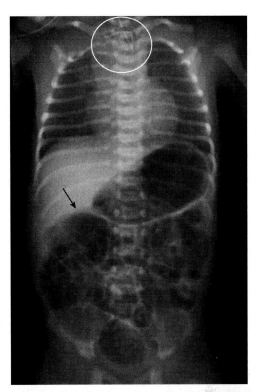

FIGURE 28-29 Esophageal atresia with a tracheoesophageal fistula (TEF). The tip of an orogastric tube is kinked in a blind-ending pouch in the upper esophagus and can pass no farther in the atretic esophagus *(circle)*. The air throughout the bowel *(black arrow)* arrived via a fistula that connects the trachea with the distal end of the esophagus. This is the most common type of esophageal atresia with a tracheoesophageal fistula.

TAKE-HOME POINTS

RECOGNIZING PEDIATRIC DISEASES

Transient tachypnea of the newborn is the most common cause of respiratory distress in the newborn, occurring in full-term or larger, slightly preterm infants. The lungs are usually hyperinflated with streaky, perihilar linear densities, fluid in the fissures, and/or laminar pleural effusions.

Respiratory distress syndrome (RDS) is a disease of premature infants. There is typically a diffuse "ground-glass" or finely granular appearance in a bilateral and symmetric distribution with air bronchograms. The lungs are frequently hypoaerated.

Meconium aspiration syndrome is the most common cause of neonatal respiratory distress in full-term/postmature infants. The lungs are hyperinflated with diffuse "ropey" densities. There may be patchy areas of atelectasis, along with emphysema from air trapping.

Bronchopulmonary dysplasia (BPD) is a consequence of early, acute lung disease, frequently respiratory distress syndrome. BPD may also complicate meconium aspiration syndrome and pneumonia. Infants with BPD have supplemental oxygen dependence. The lungs are usually hyperaerated and "spongelike" in appearance, containing both coarse linear densities from atelectasis intermixed with lucent foci from hyperexpanded areas of air trapping.

Complications of treatment of respiratory distress in the newborn are usually from **barotrauma** (air leaks) and include pulmonary interstitial emphysema, pneumomediastinum, pneumothorax, and pneumopericardium.

Reactive airways disease/bronchiolitis is a general term for a group of diseases in the pediatric population featuring wheezing, shortness of breath, and coughing. There may be peribronchial thickening and hyperaeration.

Asthma is a clinical diagnosis. Chest radiographs can help in determining either the cause or the complications of an asthmatic episode. Pneumonia is one cause, and pneumothorax and pneumomediastinum are amongst the complications of an asthmatic attack.

Clinically, neonates with **pneumonia** may have only a fever. Bacterial pneumonia characteristically produces lobar consolidation, or a round pneumonia, with pleural effusion in some cases. Viral pneumonia characteristically shows interstitial infiltrates or patchy areas of consolidation suggestive of bronchopneumonia.

The adenoids can grow until about age 6 years and then involute through adulthood. The key imaging finding for enlarged adenoids is

Continued

TAKE-HOME POINTS—cont'd

marked narrowing or obliteration of the nasopharyngeal airway on a soft-tissue lateral neck radiograph. The tonsils and adenoids frequently enlarge together.

Acute bacterial **epiglottitis** can be a life-threatening medical emergency. The epiglottis should not normally be thumblike in size or appearance. There is associated thickening of the aryepiglottic folds. It occurs at an older age (3 to 6 years) than croup (6 months to 3 years).

Croup usually has a viral cause, and the diagnosis is usually made on the basis of clinical findings. There are three key imaging findings seen on the lateral soft-tissue neck radiograph: distension of the hypopharynx, distension of the laryngeal ventricle and haziness, and/or narrowing of the subglottic trachea.

Most **foreign body ingestions** occur between 6 months and 6 years, with the vast majority passing spontaneously. Most often, they impact just below cricopharyngeus in the neck or at the level of the thoracic inlet, the aortic arch, or the esophagogastric junction. Disk (button) batteries and magnets pose particular hazards and should generally be removed.

The **cardiothoracic ratio** may reach up to 65% in infants and still be normal. In a child the thymus gland may overlap portions of the heart and sometimes mimic cardiomegaly.

Epiphyseal plate fractures are common in childhood. The Salter-Harris classification of epiphyseal plate injuries is a commonly used method of describing these injuries, identifying the type of treatment required, and predicting the likelihood of complications.

Certain kinds of fractures can be highly suggestive for nonaccidental injuries produced by **child abuse**. Radiologic evaluations are key in diagnosing child abuse. Injuries pointing to child abuse include metaphyseal corner fractures, rib fractures, and head injuries.

Necrotizing enterocolitis is the most common gastrointestinal medical and/or surgical emergency occurring in neonates, especially premature infants. There may be dilated loops of bowel, thickened bowel walls, pneumatosis intestinalis, and portal venous gas.

Esophageal atresia may occur with or without a tracheoesophageal fistula (TEF). The most common form is a blind-ending esophagus with a fistulous connection between the trachea and the distal esophageal remnant.

TEF may be associated with other congenital anomalies signified by the acronym VACTERL which stands for **V**ertebral anomalies, **A**nal atresia, **C**ardiac abnormalities, **TE**F and/or esophageal atresia, **R**enal agenesis or dysplasia, and **L**imb defects.

WEBLINK

Visit StudentConsult.Inkling.com for more information.

For your convenience, the following QR Code may be used to link to **StudentConsult.com**. You must register this title using the PIN code on the inside front cover of the text to access online content.

Appendix

What to order when

The links to the American College of Radiology's Appropriateness Criteria provided below explain which imaging study to order under certain clinical circumstances. These guidelines were developed by a series of expert panels consisting of diagnostic radiologists, interventional radiologists, and radiation oncologists, as well as leaders in other specialties. These are evidence-based guidelines designed to assist health-care providers in making the most appropriate imaging or treatment decision for a patient with a specific clinical condition.

	AMERICAN COLLEGE OF RADIOLOGY APPROPRIATENESS CRITERIA
Cardiac	Acute chest pain—suspected pulmonary embolism
	Chest pain suggestive of acute coronary syndrome
	Chronic chest pain—high probability of coronary artery disease
	Dyspnea—suspected cardiac origin
Gastrointestinal	Acute (nonlocalized) abdominal pain and fever or suspected abdominal abscess
	Acute pancreatitis
	Blunt abdominal trauma
	Dysphagia
	Jaundice
	Left lower quadrant pain—suspected diverticulitis
	Right lower quadrant pain—suspected appendicitis
	Right upper quadrant pain
	Palpable abdominal mass
	Suspected small-bowel obstruction
Musculoskeletal	Chronic ankle pain
	Chronic elbow pain
	Chronic foot pain
	Chronic hip pain
	Chronic neck pain
	Chronic wrist pain
	Low back pain
	Metastatic bone disease
	Nontraumatic knee pain
	Osteoporosis and bone mineral density
	Suspected spine trauma
Neurologic	Cerebrovascular disease
	Dementia and movement disorders
	Focal neurologic deficit
	Head trauma
	Headache
	Seizures and epilepsy
Pediatric	Fever without source—child
	Headache—child
	Limping child—ages 0-5 years
	Seizures—child
	Suspected physical abuse—child
	Urinary tract infection—child
	Vomiting in infants up to 3 months of age
Thoracic	Chronic dyspnea—suspected pulmonary origin
	Hemoptysis
	Blunt chest trauma
	Noninvasive clinical staging of bronchogenic carcinoma
	Radiographically detected solitary pulmonary nodule
	Routine chest radiographs in ICU patients
	Routine admission and preoperative chest radiography
	Screening for pulmonary metastases
Urologic	Acute onset flank pain—suspicion of stone disease
	Acute onset of scrotal pain—without trauma, without antecedent mass
	Acute pyelonephritis
	Hematuria
	Renal failure
	Renal trauma
	Renovascular hypertension

Bibliography

Texts

Arnold W, DeLegge MH, Schwaitzberg SD: *Enteral access: The foundation of feeding*, Dubuque, IA, 2002, Kendall/Hunt.

Blickman H: *Pediatric radiology: The requisites*, ed 2, St Louis, 1998, Mosby.

Fraser RS, Paré JAP, Fraser RG, et al: *Synopsis of diseases of the chest*, ed 2, Philadelphia, 1994, Saunders.

Goodman LR: *Felson's principles of chest roentgenology*, ed 3, Philadelphia, 2007, Saunders.

Greenspan A: *Orthopedic radiology: A practical approach*, ed 3, Philadelphia, 2000, Lippincott Williams & Wilkins.

Grossman RI, Yousem DM: *Neuroradiology: The requisites*, St Louis, 1994, Mosby.

Guyton AC, Hall J: *Textbook of medical physiology*, ed 11, Philadelphia, 2005, Saunders.

Harris JH, Harris WH: *The radiology of emergency medicine*, ed 4, Philadelphia, 2000, Lippincott Williams & Wilkins.

Juhl JH, Crummy AB: *Paul and Juhl's essentials of radiologic imaging*, ed 6, Philadelphia, 1993, Lippincott.

Love MB: *An introduction to diagnostic ultrasound*, Springfield, Ill, 1981, Charles C. Thomas.

Manaster BJ, Disler DG, May DA: *Musculoskeletal imaging: The requisites*, ed 2, St Louis, MO, 2002, Mosby.

McLoud TC: *Thoracic radiology: The requisites*, St Louis, 1998, Mosby.

Mettler FA, Guiberteau MJ: *Essentials of nuclear medicine imaging e-book*, ed 6, Philadelphia, 2012, Saunders.

Meyers MA: *Dynamic radiology of the abdomen: Normal and pathological anatomy*, ed 2, New York, 1982, Springer-Verlag.

Resnick D: *Diagnosis of bone and joint disorders*, ed 2, Philadelphia, 1981, Saunders.

Resnick D: *Diagnosis of bone and joint disorders*, ed 4, Philadelphia, 2002, Saunders.

Rumack CM, Wilson SR, Charboneau JW, editors: *Diagnostic ultrasound*, ed 2, St Louis, 1998, Mosby.

Schultz RJ: *The language of fractures*, Baltimore, 1972, Williams & Wilkins.

Shikora S, Martindale RG, Schwaitzberg SD, et al, editors: *Nutritional considerations in the intensive care unit: Science, rationale, and practice*, Dubuque, IA, 2002, Kendall/Hunt.

Swischuk LE: *Imaging of the newborn, infant, and young child*, ed 5, Philadelphia, 2003, Lippincott Williams & Wilkins.

Webb WR, Brant WE, Helms CA: *Fundamentals of body CT*, Philadelphia, 1991, Saunders.

Weissleder R, Wittenberg J, Harisinghani MG: *Primer of diagnostic imaging*, ed 3, St Louis, 2003, Mosby.

Journal Articles

Aberle DR, Wiener-Kronish JP, Webb WR, et al: Hydrostatic versus increased permeability pulmonary edema: diagnosis based on radiographic data in critically ill patients, *Radiology* 168:73–79, 1988.

Almeida A, Roberts I: Bone involvement in sickle cell disease, *Br J Haematol* 129:482–490, 2005.

ASGE Standards of Practice Committee, Ikenberry SO, Jue TL, Anderson MA, et al: Management of ingested foreign bodies and food impactions, *Gastrointest Endosc* 73:1085–1091, 2011.

Amjadi K, Alvarez GG, Vanderhelst E, et al: The prevalence of blebs or bullae among young healthy adults: a thoracoscopic investigation, *Chest* 132:1140–1145, 2007.

Bates D, Ruggieri P: Imaging modalities for evaluation of the spine, *Radiol Clin North Am* 29:675–690, 1991.

Boudiaf M, Soyer P, Terem C, et al: CT evaluation of small bowel obstruction, *Radiographics* 21:613–624, 2001.

Burney K, Burchard F, Papouchado M, et al: Cardiac pacing systems and implantable cardiac defibrillators (ICDs): a radiological perspective of equipment, anatomy, and complications, *Clin Radiol* 59:699–708, 2004.

Cohen SM, Kurtz AB: Biliary sonography, *Radiol Clin North Am* 29:1171–1198, 1991.

Cury RC, Budoff M, Taylor AJ: Coronary CT angiography versus standard of care for assessment of chest pain in the emergency department, *J Cardiovasc Comput Tomogr* 7:79–82, 2013.

Dalinka MK, Reginato AJ, Golden DA: Calcium deposition diseases, *Semin Roentgenol* 17:39–48, 1982.

de Jong EM, Felix JF, de Klein A, et al: Etiology of esophageal atresia and tracheoesophageal fistula: "mind the gap.", *Curr Gastroenterol Rep* 12:215–222, 2010.

Doubilet PM, Benson CB, Bourne T, et al: Diagnostic criteria for nonviable pregnancy early in the first trimester, *N Engl J Med* 369:1443–1451, 2013.

Dyer DS, Moore EE, Mestek MF, et al: Can chest CT be used to exclude aortic injury?, *Radiology* 213:195–202, 1999.

Edeiken J: Radiologic approach to arthritis, *Semin Roentgenol* 17:8–15, 1982.

Edward MO, Kotecha SJ, Kotecha S: Respiratory distress of the term newborn infant, *Paediatr Respir Rev* 14:29–37, 2013.

Garcia MJ: Could cardiac CT revolutionize the practice of cardiology?, *Cleve Clin J Med* 72:88–89, 2005.

Ellis K, Austin JH, Jaretzki A 3rd: Radiologic detection of thymoma in patients with myasthenia gravis, *AJR Am J Roentgenol* 151:873–881, 1988.

Gaskill MF, Lukin R, Wiot JG: Lumbar disc disease and stenosis, *Radiol Clin North Am* 29:753–764, 1991.

Haus BM, Stark P, Shofer SL, et al: Massive pulmonary pseudotumor, *Chest* 124:758–760, 2003.

Hendrix RW, Rogers LF: Diagnostic imaging of fracture complications, *Radiol Clin North Am* 27:1023–1033, 1989.

Henschke CI, Yankelevitz DF, Wand A, et al: Accuracy and efficacy of chest radiography in the intensive care unit, *Radiol Clin North Am* 34:21–31, 1996.

Henry M, Arnold T, Harvey J, et al: BTS guidelines for the management of spontaneous pneumothorax, *Thorax* 58(Suppl 2):ii39–ii52, 2003.

Horrow MM: Ultrasound of the extrahepatic bile duct: issues of size, *Ultrasound Q* 26:67–74, 2010.

Indrajit IK, Shreeram MN, d'Souza JD: Multislice CT: a quantum leap in whole body imaging, *Indian J Radiol Imaging* 14:209–216, 2004.

Ingram MD, Watson SG, Skippage PL, et al: Urethral injuries after pelvic trauma: evaluation with urethrography, *Radiographics* 28:1631–1643, 2008.

Johnson JL: Pleural effusions in cardiovascular disease: pearls for correlating the evidence with the cause, *Postgrad Med* 107:95–101, 2000.

Karlo CA, Leschka S, Stolzmann P, et al: A systematic approach for analysis, interpretation, and reporting of coronary CTA studies, *Insights Imaging* 3:215–228, 2012.

Kundel HL, Wright DJ: The influence of prior knowledge on visual search strategies during the viewing of chest radiographs, *Radiology* 93:315–320, 1969.

Lingawi SS: The naked facet sign, *Radiology* 219:366–367, 2001.

Neu J, Walker WA: Necrotizing enterocolitis, *N Engl J Med* 364:255–264, 2011.

Old JL, Calvert M: Vertebral compression fractures in the elderly, *Am Fam Physician* 69:111–116, 2004.

Pathria MN, Petersilge CA: Spinal trauma, *Radiol Clin North Am* 29:847–865, 1991.

Riddervold HO: Easily missed fractures, *Radiol Clin North Am* 30:475–494, 1992.

Shifrin RY, Choplin RH: Aspiration in patients in critical care units, *Radiol Clin North Am* 34:83–95, 1996.

Smith-Bindman R, Miglioretti DL, Larson EB: Rising use of diagnostic medical imaging in a large integrated health system, *Health Aff* 27:1491–1502, 2008.

Steenburg SD, Ravenel JG: Acute traumatic thoracic aortic injuries: experience with 64-MDCT, *AJR Am J Roentgenol* 191:1564–1569, 2008.

Thomas EL, Lansdown EL: Visual search patterns of radiologists in training, *Radiology* 81:288–291, 1963.

Tie MLH: Basic head CT for intensivists, *Crit Care Resusc* 3:35–44, 2001.

Tocino I, Westcott JL: Barotrauma, *Radiol Clin North Am* 34:59–81, 1996.

Veltman CE, de Graaf FR, Schuijf JD, et al: Prognostic value of coronary vessel dominance in relation to significant coronary artery disease determined with non-invasive computed tomography coronary angiography, *Eur Heart J* 33:1367–1377, 2012.

Yu S, Haughton VM, Rosenbaum AE: Magnetic resonance imaging and anatomy of the spine, *Radiol Clin North Am* 29:691–710, 1991.

Yuh WT, Quets JP, Lee HJ, et al: Anatomic distribution of metastases in the vertebral body and modes of hematogenous spread, *Spine* 21:2243–2250, 1996.

Chapter 1 Quiz Answers

Here you are at the end of the book. Finished the text already? That was speedy reading on your part. Here are the answers to the quiz that appears in Chapter 1.

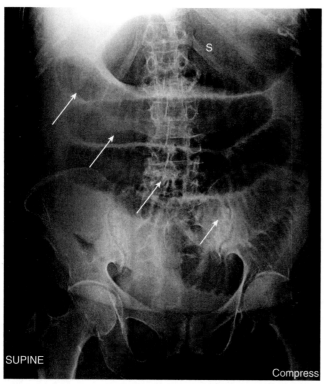

FIGURE 1-1 Small-bowel obstruction. There are multiple air-filled and dilated loops of small bowel *(white arrows)* with virtually no gas in the large bowel. The stomach (S) is also dilated. The disproportionate dilatation of small bowel is indicative of a mechanical small bowel obstruction caused, in this case, by adhesions from previous surgery.

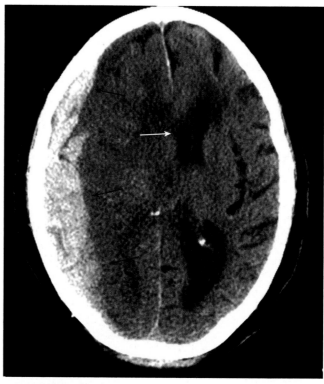

FIGURE 1-2 Subdural hematoma. A curvilinear band of increased attenuation in the right parietal region *(black arrows)* is causing a subfalcine shift of the midline structures to the left *(white arrow).* The crescentric increased density, paralleling the inner table, is classic for a subdural collection. The patient fell from a height and struck his head.

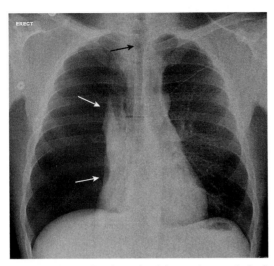

FIGURE 1-3 Pneumothorax, right. A large, right-sided pneumothorax completely collapsed the right lung toward the hilum *(white arrows)*. A slight shift of the trachea to the left *(black arrow)* raises suspicion that the pneumothorax is under slight tension. The patient had a spontaneous pneumothorax.

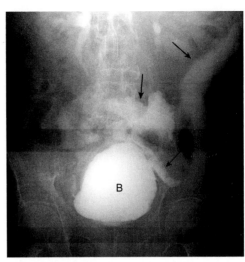

FIGURE 1-5 Extraperitoneal bladder rupture. This is a cystogram in which contrast is instilled into the urinary bladder (B) through a Foley catheter. Such images are obtained to determine if the contrast remains in the bladder as it should. In this case, contrast flows freely out of the bladder into the peritoneal cavity *(black arrows)* from a hole in the dome of the bladder. The patient had been in an automobile collision.

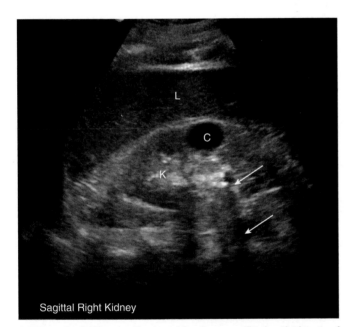

FIGURE 1-4 Simple kidney cyst. There is a round sonolucent mass (C) in the mid-right section of the kidney (K) with a strong back wall of echoes *(white arrows)*, indicating that the mass is a fluid-filled renal cyst. This was found incidentally on a scan of the kidneys performed for flank pain. *L,* Liver.

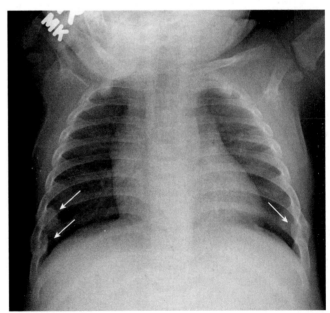

FIGURE 1-6 Child abuse. Child abuse may sometimes be suspected only on the basis of injuries seen on imaging studies that would be unusual for accidental trauma. This 4-month-old was brought to the Emergency Department for irritability, but a chest x-ray revealed bilateral healing rib fractures *(white arrows)*, an injury that is very unlikely to be accidental at this age. A thorough history confirmed the suspicion of child abuse.

Index

Note: Page numbers followed by *f* refer to figures, by *b* to boxes, and by *t* to tables. Page numbers preceded or followed by *e* refer to online-only content.